Natural Antioxidants and Food Quality in Atherosclerosis and Cancer Prevention

Organizers

Laboratory of Food Chemistry, Food Research Institute,
Agricultural Research Centre of Finland;
Research Institute of Public Health, University of Kuopio, Department of Nutrition,
National Public Health Institute Departments of Biomedicine and Public Health,
University of Tampere

Sponsors
Finnish Ministry of Commerce and Industry
The World Health Organization

Natural Antioxidants and Food Quality in Atherosclerosis and Cancer Prevention

Edited by

Jorma T. Kumpulainen
Agricultural Research Centre of Finland, Jokioinen, Finland

Jukka T. Salonen
University of Kuopio, Kuopio, Finland

THE ROYAL SOCIETY OF CHEMISTRY
Information Services

The Proceedings of two simultaneous conferences, the International Conference on Natural Antioxidants and Lipid Peroxidation in Atherosclerosis and Cancer (ALPAC) and the International Conference on Quality and Safety Aspects of Food and Nutrition in Europe '95 (QSFNE), held 22–25 August 1995 in Helsinki, Finland.

Special Publication No. 181

ISBN 0-85404-721-2

A catalogue record for this book in available from the British Library

© The Royal Society of Chemistry 1996

All rights reserved.

Apart from any fair dealing for the purpose of research or private study, or criticism or review as permitted under the terms of the UK Copyright, Designs and Patents Act, 1988, this publication may not be reproduced, stored or transmitted, in any form or by any means, without the prior permission in writing of The Royal Society of Chemistry, or in the case of reprographic reproduction only in accordance with the terms of the licences issued by the Copyright Licensing Agency in the UK, or in accordance with the terms of the licences issued by the appropriate Reproduction Rights Organization outside the UK. Enquiries concerning reproduction outside the terms stated here should be sent to The Royal Society of Chemistry at the address printed on this page.

Published by The Royal Society of Chemistry,
Thomas Graham House, Science Park, Milton Road,
Cambridge CB4 4WF, UK

Printed in Great Britain by Bookcraft (Bath) Ltd

Preface

International Conferences on "Quality and Safety Aspects of Food & Nutrition in Europe '95 (QSFNE): Natural Antioxidants, Anticarcinogens, Micronutrients and Chemical Contaminants" and, "Natural Antioxidants and Lipid Peroxidation in Atherosclerosis and Cancer (ALPAC)" were held in Helsinki, Finland, August 22-25, 1995. Scientists working in the medical, biomedical, nutritional and food sciences all share an increasing need to exchange ideas and discuss the many new findings on the role of nutrition in the ethiology of cardiovascular diseases and cancers, which affect European and US populations in particular. An attempt to address that need was reflected in the structural organization of these two conferences, which were held concurrently at the same venue. Emerging information on experimental and human research indicates that balanced nutrition has a more pronounced effect on the health and well-being of aging populations than has been previously understood. Recent findings have begun to reveal the underlying mechanisms of degenerative diseases and the role of nutrition in their prevention or modulation. The ALPAC and QSFNE Conferences brought together many of the world's leading scientists in the field to present their most recent findings. The conferences were attended by 280 participants from 31 countries and 201 papers were presented.

Fundamental new information on the relationship between dietary composition and human health and well-being is rapidly emerging. The QSFNE conference examined the current state of the art and confronted the challenges for future research in this dynamic field. To this end, the QSFNE conference brought into focus the role of natural, non-nutritive antioxidants, anticarcinogens, micronutrients and chemical contaminants in foods and nutrition.

The main topics of the ALPAC Conference included the health effects of cholesterol oxides, natural antioxidants and risk of coronary heart disease and cancer, the measurement of oxidative stress in humans and desirable vs. potentially harmful intakes of antioxidant vitamins, carotenoids and ubiquinone.

Contributions from both conferences have been merged into one edition in order to better elucidate multidisciplinary character of the natural antioxidant research field, as well as provide the reader with a balanced overview of the principal issues. The following topics from a total of 74 contributions have been selected:

1. Role of Oxidative Stress in Atherosclerosis and Cancers
2. Measurement of Oxidative Stress in Humans
3. Role of Antioxidative Vitamins and Carotenoids in Cardiovascular Diseases and Cancers
4. Selenium Intake and Status of Various Populations
5. Vitamins and Sesame Seed Lignans in Foods and Nutrition
6. Flavonoids in Foods and Role as Natural Antioxidants
7. Dietary Fiber and Inositol Phosphates in Foods and Nutrition
8. Natural Anticarcinogenic Compounds in Diets and Cancer Prevention
9. Contaminants in Foods and Diets

I sincerely hope that the present Proceedings will stimulate new multidisciplinary research into role of natural antioxidants and anticarcinogens in the prevention of cardiovascular diseases and cancers. I express my deepest gratitude to all the scientists who have contributed to the success of both conferences and this volume, and particularly to the members of the Scientific and Organizing Committees and invited speakers.

February 1996　　　　　　　　　　　　　　　　　　　　　　　　　　J.T. Kumpulainen

IN MEMORIAM Professor Timo Metsä-Ketelä 1947 -1995

SCIENTIFIC COMMITTEE

Adlercreutz H. (Finland)
Alho H. (Finland)
Appelqvist L-Å. (Sweden)
Aro A. (Finland)
Björkhem I.(Sweden)
Favier A. (France)
Hervonen A. (Finland)
Kromhout D. (The Netherlands)
Korpela H. (Finland)
Kumpulainen J.T. (Finland) (Chairman)
Levander O. (USA)
Metsä-Ketelä T.(Finland)
Nordberg G. (Sweden)
Parr R. (IAEA)
Pfannhauser W. (Austria)
Poulsen H. E. (Denmark)
Sies H. (Germany)
Slorach S. (Sweden)
Wagstaffe P. (EU)
Sandström B. (Denmark)
Salonen J.(Finland)
Vuori E. (Finland)
Weigert P.(WHO)

ORGANIZING COMMITTEE

Alfthan G. (Helsinki)
Alho H. (Tampere)
Aro A. (Helsinki)
Hervonen A. (Tampere)
Hietaniemi V. (Jokioinen), Secretary
Hägg M.(Jokioinen)
Hyvärinen H. (Jokioinen)
Korpela H. (Kuopio)
Kumpulainen J.T.(Jokioinen), Chairman
Metsä-Ketelä T.(Tampere)
Niemelä M. (Helsinki)
Plaami S.(Jokioinen)
Salonen J.(Kuopio)
Tahvonen R. (Jokioinen)
Vuori E. (Helsinki)

Contents

Role of Oxidative Stress in Atherosclerosis and Cancers

The Role of Lipid Peroxidation and Natural Antioxidants in Cardiovascular Diseases *Salonen J.T.*	3
LDL Oxidation and Atherogenesis *Ylä-Herttuala S.*	7
Oxysterols and Atherosclerosis - A Minireview *Björkhem I., Breuer O., Diczfalusy U. and Lund E.*	11
Characterization of the MM-LDL Fraction in Human Plasma *Nyyssönen K. and Salonen J.T.*	18
Exercise and Oxidative Stress in Diabetes Mellitus *Larson D., Atalay M., Khanna S. and Sen C.K.*	22
Lipid Peroxidation and Sod Activity in Diabetes *Freitas J.P., Silva J.N., Filipe P. and Guerra Rodrigo F.*	27
Increased Lipid Peroxidation in an Experimental Model of the Insulin Resistance Syndrome: The Effect of Antioxidant Therapy *Kazdova L., Vrana A., Matejckova M., Novakova V.*	31
Oxidative Stress Induced by Physical Exercise *Atalay M., Larson D., Khanna S., Hänninen O. and Sen C.K.*	36
Expression Kinetics of AcLDL Receptor(s) During the Differentiation Process of Peripheral Blood Monocytes in Vitro *Mangoni di S.Stefano C., Petillo O., Mangoni di S.Stefano G.S.R.C., Ruggiero A.M., Grippo P. and Peluso G.*	41
Blood Oxidative Stress Parameters and their Correlation with Indicators of Bone Marrow Recovery After Total Body Irradiation *Dubner D., Perez M. and Gisone P.*	46
Cohort Study on Serum Levels of Lipid Peroxide and Superoxide Dismutase and Cancer Mortality in Japan *Sasaki R., Ito Y., Suzuki S., Shinohara R., Yagyu K. and Endo N.*	52
Peroxidation of Proteins by Peroxynitrite *Lacsamana M. and Gebicki J.M.*	55
Dietary n-3 Polyunsaturated Fatty Acids Suppress Reparative Regeneration of the Rat Liver Connective Tissue *Arend A., Zilmer M. and Zilmer K.*	60

Measurement of Oxidative Stress in Humans

The Ferrylmyoglobin/ABTS Assay for Measuring Total Antioxidant Activity 69
Miller N.J.

Urinary Excretion of 5-Hydroxymethyluracil as Indicator of Oxidative DNA Damage and Repair 73
Bianchini F. and Cadet J.

Measurement of Oxidative DNA Injury in Humans: Evaluation of a Commercially Available ELISA Assay 78
Priemé H., Loft S., Cutler R.G. and Poulsen H.E.

Indicators of Oxidative Stress in Blood After the Administration of Farmorubicin and Alpha-Tocopherol 83
Birk.R., Paas L. and Muzyka V.

Failure of an Aerobic Iodometric Peroxide Assay 87
Gebicki J.M., Collins J., Baoutina A. and Phair P.

Role of Antioxidative Vitamins and Carotenoids in Cardiovascular Diseases and Cancers

Biological Activity of Carotenoids and their Bioavailability in the Human Organism 95
Stahl W. and Sies H.

Desirable Versus Potentially Harmful Intake Levels of Various Forms of Carotenoids 102
Stahl W.

Desirable Versus Potentially Harmful Levels of Vitamin E Intake 110
Korpela H.

The Suvimax Study: Scientific Justification and Design of a Large Randomized Trial for Testing the Health Effects of a Supplementation with Low Doses of Antioxidant Vitamins and Trace Elements 113
Favier A., Preziosi P., Roussel A.M., Malvy D., Paul-Dauphin A., Galan P., Briancon S. and Hercberg S.

Antioxidants, Diet and Mortality from Ischaemic Heart Disease in Rural Communities in Northern Finland 123
Luoma P.V., Näyhä S., Sikkilä K., Hassi J.

Calcium Pantothenate and Antioxidant Vitamins in Prevention of Surgical Stress 130
Dubrovshchik O.I., Tsilindz, I.T., Makshanov I.Y., Moiseenok A.G.

Serum ß-carotene Response to Chronic Supplementation with Raw Carrots, Carrot Juice or Purified ß-Carotene 135
Törrönen A.R., Lehmusaho M.H., Häkkinen S.H., Hänninen O. and Mykkänen H.M.

Effects of Dietary N-3 Fatty Acids and Vitamin E on Lipid Transport and Glucoregulation in Hypertriglycaridaemia 141
Vrana A., Kazdova L., Novakova V. and Matejckova M.

Enhanced Antioxidant Status in Long-term Adherents of a Strict Uncooked Vegan Diet (Living Food Diet) 145
Rauma A-L., Törrönen R., Hänninen O., Verhagen H., Mykkänen H.

Fruit and Vegetable Supplementation - Effect on ex Vivo LDL Oxidation in Humans *Chopra M., McLoone U., O'Neill M., Williams N., Thurnham D.J.*	150
Vitamin-C Reduces Uptake of Mercury Vapor in Rats *Rambeck W.A., Lohr B. and Halbach S.*	156

Selenium Intake and Status of Various Populations

Selenium Intakes and Plasma Selenium Levels in Various Populations *Alfthan G. and Neve J.*	161
Various Forms and Methods of Selenium Supplementation *Aro A.*	168
Low Dietary Selenium Intakes in Scotland and the Effectiveness of Supplementation *MacPherson A., Barclay M.N.I. and Molnár J.*	172
Selenium Deficient Status of Inhabitants of South Moravia *Kvícala J., Zamrazil V. and Jiránek V.*	177
Selenium Status and Cardiovascular Risk Factors in Populations from Different Portuguese Regions *Viegas-Grespo A.M., Pavao M.L., Santos V., Cruz M.L., Paulo O., Leal J., Sarmento N., Monteiro M.L., Amorim M.F., Halpern M.J. and Neve J.*	188
Alteration in Plasma Selenoprotein P Levels After Supplementation with Different Forms of Oral Selenium in Healthy Men *Persson-Moschos M., Alfthan G. and Åkesson B.*	195
Selenium in Food and Population of Serbia *Dujic I.S.*	199
Selenium Levels in Lead Exposed Workers *Giray B., Gürbay A., Basaran N., Hincal F.*	208

Vitamins and Sesame Seed Lignans in Foods and Nutrition

Effects of Processing, Packaging and Storage on the Levels of Vitamins in Foods and Diets *Hägg M.*	213
Analytical Methodology for Antioxidant Vitamins in Infant Formulas *Romera J.M., Angulo A.J., Ramirez M., Gil A.*	221
Vitamin E Contents of Breakfast Cereals and Fish *Hägg M. and Kumpulainen J.*	225
The in Vivo Antioxidant Properties of Sesame Seed Lignans *Kamal-Eldin A., Pettersson D. and Appelqvist L.Å.*	230
Sesame Seed and its Lignans Produce Marked Enhancement of Vitamin E Activity in Rats Fed a Low Alpha-tocopherol Diet *Yamashita K., Iizuka Y., Imai T. and Namiki M.*	236
Antioxidant Activity of Sesamin on NADPH-Dependent Lipid Peroxidation in Liver Microsomes *Akimoto K., Asami S., Tanaka T., Shimizu S., Sugano M., Yamada H.*	241

Dietary Flavonoids and Role as Natural Antioxidants in Humans

Dietary Flavonoids and Oxidative Stress 249
Pietta P.G., Simonetti P., Roggi C, Brusamolino A., Pellegrini N., Maccarini L. and Testolin G.

The Relative Antioxidant Activities of Plant-Derived Polyphenolic Flavonoids 256
Miller N.J.

Reduced Mortality from Cardiovascular and Cerebrovascular Disease Associated with
Moderade Intake of Wine 260
Grønbæk M., Deis A., Sørensen T.I.A., Becker U., Schnohr P. and Jensen G.

A New Sensitive and Specific HPLC Method for Measuring Polyphenol Plasma and Urinary Levels
After Green Tea Ingestion 264
Maiani G., Azzini E., Salucci M., Ghiselli A., Serafini M. and Ferro-Luzzi A.

Antioxidant Activity of Natural Food Supplementations: Flavonoid Rutin and Bio-Normalizer 273
Osato J.A., Afanas'ev I.B. and Korkina L.G.

Antioxidant Activity of Silibinin Against Haem-Protein Dependent Lipid Peroxidation and DNA
Damage Induced by Bleomycin-Fe(III) 281
Mira L., Silva M. and Manso C.F.

Influence of a Garlic Extract on Oxidative Stress Parameters in Humans 286
Scherat T.G., Siems W.G., Brenke R., Behrends H., Jakstadt M., Conradi E. and Grune T.

Prooxidant Activities of Flavonols: A Structure Activity Study 290
Gaspar J., Duarte Silva I., Laires A., Rodrigues A., Costa S. and Rueff J.

HPLC Method for Screening of Flavonoids and Phenolic Acids in Berries: Phenolic Profiles of
Strawberry and Black Currant 298
Häkkinen S.H., Mykkänen H.M., Kärenlampi S.O., Heinonen I.M. and Törrönen A.R.

Dietary Fibers and Inositol Phosphates in Foods and Nutrition

Dietary Fiber and its Applications: Definitions, Analytical Methods and their Applications 303
Prosky L., Lee S.C.

Dietary Fibre Intakes and Trends in European Countries 311
Mälkki Y.

Dietary Fibre and Phytic Acid: Potential Health Implications of their Increased Intake 317
Sandström B.

Analysis and Consumption of Dietary Fiber in Germany 322
Rabe E.

Dietary Fiber, ß-Glucan and Inositol Phosphate Contents of Some Breakfast Cereals 328
Plaami S.P. and Kumpulainen J.

Bioavailability of Essential Trace Elements with Special Reference to Dietary Fibre and Phytic Acid 334
Sandström B.

Dietary Fiber and Inositol Phosphate Contents of Finnish Breads 338
Plaami S.P., Kumpulainen J.T. and Tahvonen R.L.

Natural Anticarcinogenic Compounds in Diets and Cancer Prevention

Lignans and Isoflavonoids: Epidemiology and a Possible Role in Prevention of Cancer 349
 Adlercreutz H.

Anticancer Effect of Flaxseed Lignans 356
 Thompson L.U., Orcheson L., Rickard S., Jenab M., Serraino M., Seidl M. and Cheung F.

Compounds in Plants Inducing Detoxifying and Antioxidative Enzymes 365
 Dragsted L.O.

A Cross-sectional Study on Relationship Between Serum Levels of Carotenoids and Peripheral
Distribution of Lymphocyte Subsets in the Japanese Adults 373
 Ito Y., Sasaki R., Niiya Y.

Anticarcinogenic Properties of Lycopene 378
 Sharoni Y. , Levy J.

Flavonoids and Extracts of Strawberry and Black Currant are Inhibitors of the Carcinogen-Activating
Enzyme CYP1A1 in Vitro 386
 Kansanen L.A., Mykkänen H.M. and Törrönen A.R.

Coumarin 7-hydroxylation is not Altered by Long-Term Adherence to a Strict Uncooked Vegan Diet 389
 Rauma A-L., Rautio A., Pelkonen O., Pasanen M., Törrönen R. and Mykkänen H.

Effect of Chlorophyllin and Other Natural Antioxidants Upon the Respiratory Burst of Human
Phagocytes from Smokers and Non-Smokers 393
 Benitez L. and Wens M.A.

Contaminants in Foods and Diets

GEMS/Food-EURO: A WHO/FAO/UNEP Approach Towards Better Quality Foods in Europe 401
 Weigert P.

Risks of Detrimental Health Effects by Dietary Intakes of Heavy Metals in Various Populations 405
 Nordberg G.F.

Toxicity and Intakes of Various Forms of Heavy Metals 412
 Nordberg M., Nordberg G.F.

Cadmium Binding Compounds in Wheat: Occurrence and Partial Characterization 417
 Brüggemann J., Tümmers N., Thier H.P., Betsche T.

Lead, Cadmium, Mercury, Copper and Zinc Content in Polish Powdered Milk and Products for
Infants and Children 423
 *Wojciechowska-Mazurek M., Karlowski K., Kumpulainen J.T., Starska K.,
 Brulinska-Ostrowska E., Zwiek-Ludwicka K.*

Levels and Trends of PCBs, Organochlorine Pesticide Residues and Carcinogenic or Mutagenic PAH
Compounds in Finnish and Imported Foods and Diets 432
 Hietaniemi V.

Levels of Organocholoride and Organophosphorus Pesticide Residues in Greek Honey 437
 Tsipi D., Hiskia A. and Triantafyllou M.

Index of Authors	445
Subject Index	446

Role of Oxidative Stress in Atherosclerosis and Cancers

THE ROLE OF LIPID PEROXIDATION AND NATURAL ANTIOXIDANTS IN CARDIOVASCULAR DISEASES

J.T. Salonen, University of Kuopio, Kuopio, Finland

Research Institute of Public Health, University of Kuopio, P.O. Box 1627, SF 70211 Kuopio, Finland

1 LIPID PEROXIDATION AND ATHEROSCLEROSIS

Lipid peroxidation is a process that renders polyunsaturated fats rancid and toxic to human tissues. Lipid peroxidation is promoted by oxidative stress, which may be caused by either exogenous exposures such as cigarette smoke, UV radiation, excessive alcohol use or excessive intake or accumulation of catalytic transition metals, or by enhanced endogenous production of reactive oxygen species as a consequence of eg. inflammation or accelerated energy metabolism. Evidence supporting the role of lipid peroxidation and antioxidant deficiency in atherogenesis has accumulated both from basic research and epidemiology.

Basic research has shown that when low density lipoprotein (LDL) particles are oxidatively modified, it becomes more cytotoxic to arterial endothelial cells, may cause endothelial damage and initiate atherogenesis.

There is some epidemiologic evidence supporting the role of lipid peroxidation in atherogenesis. We observed in the Kuopio Ischaemic Heart Disease Risk Factor Study an association between titre of autoantibodies to oxidized LDL, reflecting lipid peroxidation in vivo, and the progression of carotid atherosclerosis in a 2-year follow-up study in 60 men[1]. This finding has been subsequently confirmed with different antigens and immunologic methods and in different kinds of subjects. In a nested case-control study in the Helsinki Heart Study subjects, the mean titre of antibodies to oxidized LDL was higher among subjects who experienced a cardiac death or myocardial infarction than among the controls[2].

Table 1 *Studies showing an association between antibodies against oxidised LDL and atherosclerosis or coronary heart disease*

Study	Reference	Study design and sample size	Ox-LDL titre (ratio)
KIHD 2-year follow-up	Salonen et al. Lancet 1992;339:883	Nested case-control, n= 30 men with fast atherogenesis and 30 controls	2.67 vs. 2.06, p=0.003
KAPS 1-year follow-up	Salonen et al. Eur Heart J 1992;13:391	Cohort, n=212 men	IMT incr. 0.25 vs. 0.13 mm/y, p<0.001
Italian	Maggi et al. Arterioscler Thromb 1994;14:1892	Cross-sectional, n=94 patients and 42 controls	2.39 vs. 2.04, p<0.01
Helsinki Heart Study	Puurunen et al. Arch Int Med 1994;154:2605	Nested case-control, n=135 men with AMI and 135 controls	0.412 vs. 0.356 p=0.002

2 TRANSITION METALS AND CORONARY HEART DISEASE

Redox-active forms of transition metals (e.g. iron and mercury) can catalyse the formation of free radicals and promote lipid peroxidation[3,4,5]. In our prospective study, the KIHD, both elevated body iron (serum ferritin $\geq$ 200 µg/l)[3] and accumulation of methylmercury (hair mercury $\geq$ 2.0 µg/g)[5] were associated with increased risk of myocardial infarction (MI) in almost 2000 men. The former finding was confirmed in the Nutrition Canada Survey cohort[6] and US Health Professionals' study[7]. Several additional epidemiologic studies to retest our finding concerning the role of mercury in coronary heart disease are under way.

3 NATURAL ANTIOXIDANTS AND CORONARY HEART DISEASE

There are prospective epidemiologic studies suggesting that high intakes of vitamin E[8,9], selenium[10], and possibly vitamin C[11] and flavonoids[12] are associated with reduced coronary heart disease (CHD) risk. A selection bias (vitamin users more health conscious, multicollinearity of nutrients) can not, however, be ruled out. In the US Health Professionals' study[9], high intake (>100 IU/d) of vitamin E (but not that of vitamin C) was associated with reduced risk of CHD. In the Finnish "Mobile Clinic" cohort study, CHD mortality was 32% lower in men and 65% lower in women in the highest third of baseline dietary vitamin E intake, compared to the lowest third[13]. In the "Euramic" case-control study, adipose tissue β-carotene but not α-tocopherol was associated with reduced CHD[14]. In the Basel Prospective Study there was increased CHD and stroke mortality in men in the lowest quarter of both plasma vitamin C and

lipid-standardized carotene[15]. Plasma vitamin E had no significant relation to CHD.
We reported 13 years ago a relationship between selenium deficiency and an excess risk of MI as well as death from CHD in Eastern Finland[10]. The finding was subsequently confirmed in another prospective population study[16]

4 NATURAL ANTIOXIDANTS IN THE PREVENTION OF CARDIOVASCULAR DISEASES

No conclusive clinical trials concerning the preventive effect of antioxidants in CHD have so far been reported. In the Chinese Linxian trial[17], there was reduced total mortality in a group that received a combination of α-tocopherol, β-carotene and selenium, as compared with three parallel randomized groups receiving other vitamins and minerals. In the Finnish α-tocopherol - β-carotene lung cancer trial[18], neither antioxidant had any significant effect on cardiovascular mortality. This study has been criticized because of the very small vitamin E dose (50 mg/d of racemic dl-α-tocopheryl acetate) and the selection of subjects.

"Antioxidant Supplementation in Atherosclerosis Prevention" ("ASAP") study is a 2x2 factorial double-masked randomized 3-year trial testing the effect of 200 mg/d of dα-tocopherol acetate and 500 mg/d of vitamin C on the progression of carotid atherosclerosis, blood pressure and cataract formation in 500 men and postmenopausal women aged 45-69 years. This trial and other on-going trials will eventually provide further necessary information about the role of antioxidants in arteriosclerosis and in the prevention of cardiovascular diseases.

References

1. J.T. Salonen, S. Ylä-Herttuala, R. Yamamoto, S. Butler, H. Korpela, R. Salonen, K. Nyyssönen, W. Palinski, J.L. Witztum. *Lancet*, 1992, **339**, 883.

2. M. Puurunen, M. Mänttäri, V. Manninen, L. Tenkanen, G. Alfthan, C. Ehnholm, O. Vaarala, K. Aho, T. Palosuo. *Arch. Intern. Med.*, 1994, **154**, 2605.

3. J. T. Salonen, K. Nyyssönen, H. Korpela, J. Tuomilehto, R. Seppänen, R. Salonen. *Circulation*, 1992, 86, 803.

4. J.T. Salonen JT. *Curr. Opinion Lipidol.* 1993, 4, 277.

5. J. T. Salonen, K. Seppänen, K. Nyyssönen, H. Korpela, J. Kauhanen, M. Kantola, J. Tuomilehto, H. Esterbauer, F. Tatzber, R. Salonen. *Circulation*, 1995, 91, 645.

6. H.I. Morrison, R. M. Semenciw, Y. Mao, D. T. Wigle. *Epidemiology*, 1994, 5, 243.

7. A. Ascherio, W. C. Willet, E. B. Rimm, E. Giovannucci, M. J. Stampfer. *Circulation*, 1994, 89, 969.

8. M. J. Stampfer, C. H. Hennekens, J. E. Manson, G. A. Coditz, B. Rosner, W. C. Willett. *N. Engl. J. Med.*, 1993, 328, 1444.

9. E. B. Rimm, M. J. Stampfer, A. Ascherio, E. Giovannucci, G. A. Colditz, W. C. Willet. *N. Engl. J. Med.*, 1993, 328, 1450.

10. J. T. Salonen, G. Alfthan, J. K. Huttunen, J. Pikkarainen, P. Puska. *Lancet*, 1982, II, 175.

11. J. E. Enstrom, L. E. Kanim, M. A. Klein. *Epidemiology*, 1992, 3, 194.

12 M. G. L. Hertog, E. J. M. Feskens, P. C. H. Hollman, M. B. Katan, D. Kromhaut. *Lancet*, 1993, 342, 1007.

13 P. Knekt, A. Reunanen, R. Järvinen, R. Seppänen, M. Heliövaara, A. Aromaa. *Am. J. Epidemiol.* 1994, 139, 1180.

14 A. F. Kardinaal, F. J. Kok, J. Ringstad, J. Gomez-Aracena, V. P. Mazaev, L. Kohlmeier, B. C. Martin, A. Aro, J. D. Kark, M. Delgado-Rodrigues, P. Riemersma, P. van't Veer, J. K. Huttunen, J. M. Martin-Moreno. *Lancet*, 1993, 342, 1379.

15 K.F. Gey, U. K. Moser, P. Jordan, H. B. Stähelin, M. Eichholzer, E. Lüdin. *Am. J. Clin. Nutr.*, 1993, 57 (suppl.), 787S.

16 P. Suadicani, O. Hein, F. Gyntelberg. *Atherosclerosis*, 1992, 96, 33.

17 W. J. Blot, J.-y. Li, P. R. Taylor, et al. *J. Natl. Cancer Int.*, 1993, 85, 1483.

18 The Alpha-Tocopherol, Beta Carotene Cancer Prevention Group. *N. Eng. J. Med.*, 1994, 330, 1029.

LDL OXIDATION AND ATHEROGENESIS

Seppo Ylä-Herttuala

A.I.Virtanen Institute and Department of Medicine
University of Kuopio
P.O.Box 1627
FIN-70211 Kuopio, Finland

1 INTRODUCTION

Atherosclerosis and cardiovascular diseases remain the major cause of mortality in industrialized countries [1]. High plasma total cholesterol and low-density lipoprotein (LDL) values show significant positive association with the development of atherosclerosis and cardiovascular diseases [2]. Recent evidence suggests that oxidation (Ox) of LDL plays an important role in the pathogenesis of atherosclerosis [2,3]. The purpose of this article is to review recent findings regarding the role of Ox-LDL in atherogenesis.

2 LDL OXIDATION

Lipid peroxidation is a chain reaction which can damage LDL particles. Several reactive radical species can initiate lipid peroxidation. These include reactions involving lipoxygenases, superoxide anion, hydroxyl radical, peroxinitrate, haem proteins, ceruloplasmin and myeloperoxidase. In vitro studies have demonstrated that at least 15lipoxygenase, superoxide anion, peroxinitrate and myeloperoxidase can oxidize LDL [4]. Peroxinitrate may be formed in the arterial wall from the reaction of superoxide anion with nitric oxide [5]. It is also possible that oxidized membrane lipids are transferred to LDL.

In addition to lipid peroxidation, LDL oxidation also involves the modification of apoprotein B. ApoB in Ox-LDL is fragmented and contains covalently bound malondialdehyde and 4-hydroxynovenal conjugates [2]. These reactions change LDL properties in a way that it is metabolized through macrophage scavenger receptors [6].

LDL antioxidant levels do not fully explain individual differences in the susceptibility of LDL to oxidative stress [7]. It is likely that other factors, such as fatty acid composition and LDL particle size may also affect LDL oxidation. It has been shown that small, dense LDL particles are more susceptable to oxidation than larger LDL subfractions [7]. On the other hand, LDL particles enriched with monounsaturated fatty acids are less prone to oxidation than particles enriched with polyunsaturated fatty acids [7].

3 ATHEROGENIC PROPERTIES OF OXIDIZED LDL

Oxidized LDL causes lipid accumulation in macrophages and form cell formation [2]. OxLDL is also cytotoxic to many cell types and chemotactic for monocyte macrophages. In addition, oxidized LDL can inactivate endothelial cell derived relaxing factor [2].

It is becoming clear that the biological properties of minimally oxidized LDL (MM-LDL) are different than those of fully Ox-LDL [8]. Table 1 summarizes some of the properties of MM-LDL and Ox-LDL. MM-LDL seems to have several specific effects on gene expression [8]. It can stimulate monocyte chemotactic factor-1 and macrophage colony stimulating factor-1 expression and activate prothrombotic properties in the vascular wall [8]. However, MM-LDL can not cause lipid accumulation in arterial cells since it is not metabolized through scavenger receptors [2]. Ox-LDL appears to be immunogenic causing autoantibody formation in humans and in experimental animals. According to preliminary results, presence of autoantibodies may predict the progression of atherosclerosis in human populations [9].

Table 1 *Properties of Minimally Oxidized LDL and Fully Oxidized LDL*

	Minimally Oxidized LDL	Oxidized LDL
Cellular lipid accumulation	No	Yes
Reactive aldehyde conjugates in ApoB	No	Yes
Metabolism through scavenger receptor	No	Yes
Metabolism through LDL receptor	Yes	No
Stimulates proinformatory cytokines	Yes	No
Inhibits endothelial-derived relaxing factor	No	Yes
Immunogenic	No	Yes
Cytotoxic	No	Yes
Chemotactic for plasma monocytes	No	Yes

4 EVIDENCE FOR THE PRESENCE OF OXIDIZED LDL IN ATHEROSCLEROTIC LESIONS

Substancial evidence now indicates that Ox-LDL is present in atherosclerotic lesions:

1) LDL isolated from human atherosclerotic lesions, but not from normal arteries, resembles Ox-LDL [3];
2) epitopes chracteristic of Ox-LDL can be demonstrated in atherosclerotic lesions by immunocytochemistry [10];
3) atherosclerotic lesions contain immunoglobulin which recognize Ox-LDL [11];
4) serum contains autoantibodies against oxidized LDL [9]; and
5) antioxidant treatment reduces the rate of development of atherosclerotic lesions in experimental animals (Table 2). Antioxidants shown to be effective in animal models include probucol [12], α-tocopherol [13], butylated hydroxytoluene [14], and diphenylphenylenediamine [15]. It remains to be determined whether small quantities of OxLDL are present in plasma.

Table 2 *Evidence for the Presence of Oxidized LDL in Atherosclerotic Lesions*

	Reference
LDL isolated from atherosclerotic lesions resembles Ox-LDL	3
Epitopes characteristic of Ox-LDL are present in atherosclerotic lesions	10
Atherosclerotic lesions contain immunoglobulins which recognize Ox-LDL	11
Serum contains autoantibodies against Ox-LDL	9,10
Antioxidant treatment reduces atherogenesis in experimental animals	12

5 CONCLUSIONS

Basic research has provided strong evidence that LDL oxidation plays an important role in the pathogenesis of atherosclerosis and cardiovascular diseases. There seems to be no doubt that lipid peroxidation and oxidative damage to LDL resembles chronic inflammation which causes various alterations in arterial gene expression and promotes lesion development. Based on the current knowledge about the role of Ox-LDL in atherogenesis, randomized placebo-controlled intervention trials using antioxidants are warranted to test the hypothesis whether increased antioxidant protection could be useful in the prevention of cardiovascular diseases.

Acknowledgements

Author wants to thank Mrs Marja Poikolainen for typing the manuscript. This work was supported by grants from Finnish Academy, Sigrid Juselius Foundation, Finnish Heart Foundation and research grant from the Finnish Insurance Companies.

References

1. WHO Monica, *Int J Epidemiol*, 1989, **18**, 38.
2. J.L. Witztum and D. Steinberg, *J Clin Invest*, 1991, **88**, 1785.
3. S. Ylä-Herttuala, W. Palinski et al, *J Clin Invest*, 1989, **84**, 1086.
4. S. Ylä-Herttuala, *Drugs of Today*, 1994, **30**, 507.
5. J.S. Beckman, T.W. Beckman et al, *Med Sci*, 1990, **87**, 1620.
6. T. Kodama, M. Freeman et al, *Nature*, 1990, **343**, 531.
7. H. Esterbauer, J. Gebicki et al, *Free Rad Biol Med*, 1992,**13**, 341.
8. J.A. Berliner and M.E. Haberland, *Curr Opin Lipidology*, 1993, **4**, 373.
9. J.T. Salonen, S. Ylä-Herttuala et al, *Lancet*, 1992, **339**, 883.
10. W. Palinski, M.E. Rosenfeld et al, *Proc Natl Acad Sci USA*, 1989, **86**, 1372.
11. S. Ylä-Herttuala, W. Palinski et al, *Arteriosclerosis Thromb*, 1994,**14**, 32.
12. T.E. Carew, D.C. Schwenke et al, *Proc Natl Acad Sci USA*, 1987, **84**, 7725.
13. A.J. Verlangieri and M.J. Bush, *J Amer Coll Nutr*, 1992, **11**, 131.
14. I. Björkheim, P. Henriksson et al, *Arteriosclerosis Thromb*, 1991,**11**, 15.
15. C.P. Sparrow, T.W. Doebber et al, *J Clin Invest*, 1992, **89**, 1885.

OXYSTEROLS AND ATHEROSCLEROSIS - A MINIREVIEW.

Ingemar Björkhem, Olof Breuer, Ulf Diczfalusy and Erik Lund

Division of Clinical Chemistry, Karolinska Institutet, Huddinge Hospital, Huddinge, Sweden.

1 INTRODUCTION

Oxysterols are defined as sterols containing one or more oxygen functionalities in addition to the oxygen function at C-3. These compounds are formed as a result of oxidation of cholesterol or some of its immediate precursors. Such reactions occur either during enzymatic cholesterol metabolism or as a result of autoxidation. For general reviews, the reader is referred to an extensive volume by L.L. Smith [1] and to two recent theses.[2,3]

The structures of cholesterol and the most common oxysterols are shown in Figure 1. Here a short review will be given on formation of oxysterols, analysis of such compounds, biological importance of oxysterols and the relevance of oxysterols in connection with atherosclerosis.

2 FORMATION OF OXYSTEROLS

Cholesterol autoxidation or oxidation of cholesterol secondary to peroxidation of unsaturated fatty acids occurs by the radical mechanism shown in Figure 2. This mechanism leads to formation of 7α-hydroperoxycholesterol [2] and 7α-hydroperoxycholesterol [3], which compounds are reduced to give 7β-hydroxycholesterol [4] and 7β-hydroxycholesterol [5], respectively, or dehydrated to give 7-oxocholesterol [6]. Cholesterol autoxidation also gives 5,6-oxygenated products: cholesterol-5α,6α-epoxide [7], cholesterol-5β,6β-epoxide [8] and cholestane-3β,5α,6β-triol [9]. Most probably the epoxides are formed by an attack of hydroperoxides on the Δ^5 double bond of cholesterol.

Polyunsaturated fatty acids constitute an integral part of many membranes. These compounds contain double-allylic hydrogen atoms which are easily abstracted because resonance stablilized reaction products are formed. The radical nature of these products promotes transformation to hydroperoxides and subsequent participation in free radicals chain reactions. Cholesterol also becomes a target of such reactions as it is located in close proximity to membrane fatty acids, thereby yielding the common autoxidation products 4-6 and 7-9. Furthermore, formation of these compounds has been demonstrated to be secondary to NADPH-dependent enzymatic lipid peroxidations in liver microsomes. These enzymatic reactions reduce Fe^{3+} to Fe^{2+} which in turn initiates autoxidation as described above. Lipid hydroperoxides seem to be involved in the formation of the cholesterol epoxides 7 and 8 by microsomal lipid peroxidation.

Enzymatic oxidation of cholesterol occurs by a number of cytochrome P-450 enzymes. These enzymes utilize molecular oxygen for the stereospecific addition of one oxygen atom to the substrate. One such important enzyme is the cholesterol 7α-hydroxylase in the liver, catalyzing the rate-limiting step in the degradation of cholesterol to bile acids. Another important enzyme is the mitochondrial sterol 27-hydroxylase present in most tissues and organs, catalyzing omega-hydroxylation of the steroid side-chain. This enzyme has also a potential to introduce a hydroxyl group at C-24 and C-25. In addition, specific enzymes are able to catalyze hydroxylations at C-22/C-20, and at C-24.

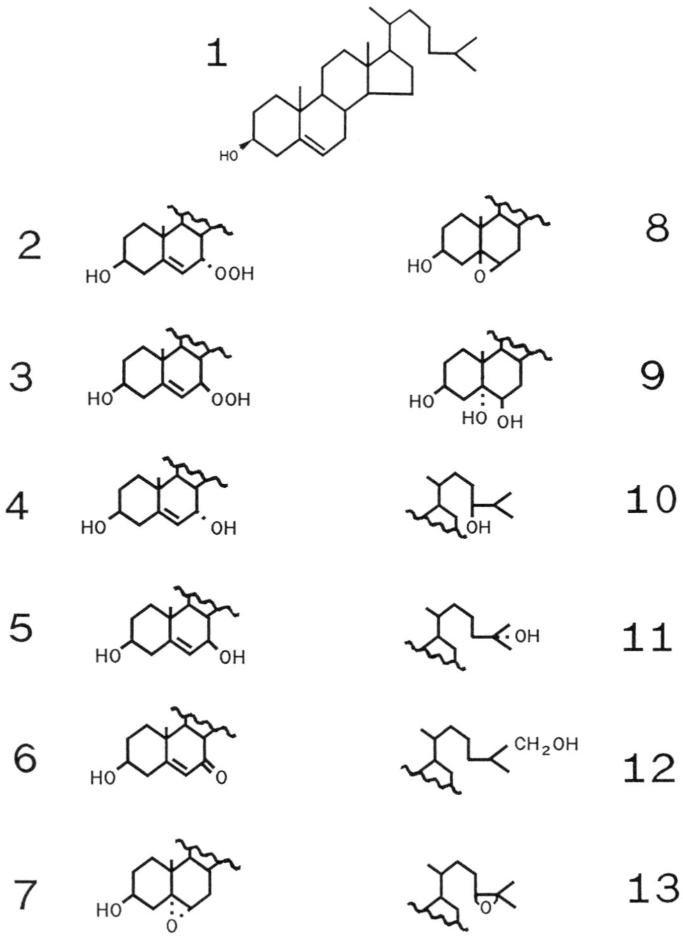

Figure 1. *Structures of cholesterol and oxysterols*

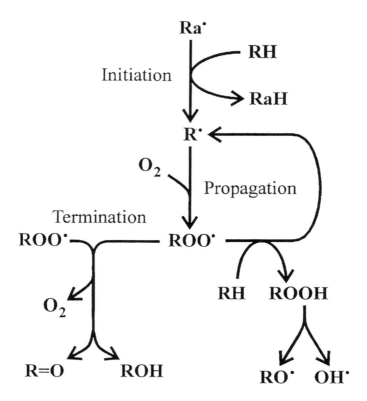

Figure 2. *Autoxidation reactions*

3 ANALYSIS OF OXYSTEROLS AND IN VITRO ARTEFACTS

Oxysterols are always present in the proximity of their parent compound, cholesterol. In most tissues, the concentration of cholesterol is 10^3-10^4 times higher than that of most oxysterols. This constitutes a large risk of quantitation artefacts, as even minor cholesterol autoxidation during sample processing would yield substantial elevations of oxysterol concentrations. Although a large number of quantitative studies of oxysterols have been made, it is difficult to interpret the results of many of these studies as antioxidative precautions often have been lacking. We [4,5] and others [6,7] have developed highly accurate and sensitive methods for analysis of oxysterols based on isotope dilution - mass spectrometry. In all quantitation methods, the applicability is limited by the risk of uncontrolled autoxidation occurring during sample collection, extractions and chromatographies. Based on a careful study in which all the critical analytical steps were performed in argone atmosphere, Kudo *et al.* claimed that a number of oxysterols reported to be present in human circulation are *in vitro* artefacts only.[8] Very recently, we have been able to demonstrate convincingly that all three 7-oxygenated cholesterol species, 4β-hydroxycholesterol, 24-,25- and 27-hydroxycholesterol are formed *in vivo*[9] in rats. The results do not exclude, however, that most part of the 5,6-oxygenated oxysterols are *in vitro* artefacts only. The experimental approach was to expose living rats to an $^{18}O_2$-containing atmosphere. Any formation of oxysterols *in vivo* during the $^{18}O_2$-exposure results in incorporation of an ^{18}O atom in the cholesterol molecule. Such incorporation cannot occur during sample processing.

With the very sensitive methods now available[4-7] and with the antioxidative precautions taken, we believe that it is now possible to accurately quantify the oxysterols in the human circulation and various tissues, and that the oxysterols measured are present also *in vivo*.

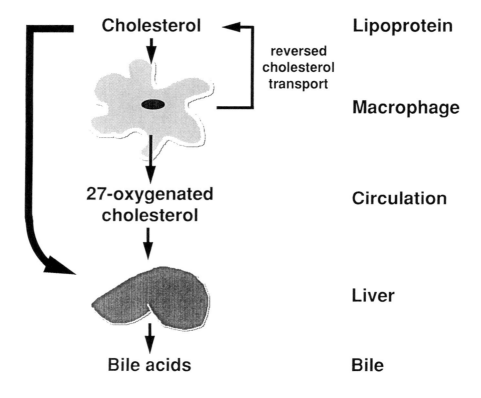

Figure 3. *Elimination of macrophage cholesterol by sterol 27-hydroxylase*

4 BIOLOGICAL IMPORTANCE OF OXYSTEROLS AND THE RELEVANCE OF OXYSTEROLS IN CONNECTION WITH DEVELOPMENT OF ATHEROSCLEROSIS

Many oxysterols are very potent inhibitors of HMG CoA reductase and also affect other important key enzymes in sterol metabolism including cholesterol 7α-hydroxylase, cholesterol 5,6-epoxide hydroxylase, HMG CoA synthase, acetoacetyl-CoA thiolase, mevalonate kinase and acyl-CoA:cholesterol-O-acyltransferase (ACAT). In spite of the very dramatic effects of some of the oxysterols *in vitro*, in particular in experiments with isolated cells, their importance *in vivo* has not been established with certainty (for a general review, see ref. 10).

With respect to development of atherosclerosis, it was early suggested that the atheromas developed by rabbits on a cholesterol-containing diet may be due to the oxysterols in the preparation rather than due to cholesterol itself. The high content of oxysterols (12% of the sterols) in the butter product ghee has been suggested to account for an otherwise unexplained high risk of atherosclerosis in certain Indian immigrant populations in Britain.[11]. Carefully controlled dietary studies have however not been able to confirm the hypothesis that oxysterols are atherogenic and according to one study, autoxidation products of cholesterol seem to have a protective effect on development of atherosclerosis in rabbits.[12] It is well established that some of the oxysterols have cytotoxic effects on endothelial cells [13,14] and arterial smooth muscle cells,[15,16] but the importance of this *in vivo* is difficult to evaluate. Due to the suspected atherogenicity of oxysterols, cholesterol oxidation products are continously being assayed in food.[17] It must be pointed out, however, that it is still uncertain whether dietary oxysterols actually contribute to atherosclerosis outside the laboratory.

There is a substantial accumulation of oxysterols in atheromas, in particular of 27-hydroxy-cholesterol. Recently an accumulation of 7β-hydroperoxycholesterol was established in human atheromas and oxidized LDL.[18] This compound is known to be highly cytotoxic, and it has been speculated that it may play a role in the development of atheromas.

According to the "oxidation hypothesis" of Steinberg, oxidation of the LDL-particle may be a critical step in the development of atherosclerosis.[19] The oxidation leads to formation of hydroperoxides of both unsaturated fatty acids and cholesterol and results in an oxidative change in the apolipoprotein that allows uptake of the particle by the scavenger receptor. The possible role of the oxidized cholesterol species in this process is not known, but it is evident that the levels of these compounds in the circulation may reflect the severity of the lipid peroxidation. In preliminary studies together with a Finnish group we have seen a correlation between the levels of some of the oxygenated sterols and the progress of atherosclerosis.

Recently, it was reported that 25-hydroxycholesterol is able to stimulate calcification of osteoblast-like cells from bovine arteries.[20] This may be of importance for the calcification of advanced atheromas in human atherosclerosis.

5 ELIMINATION OF CHOLESTEROL IN MACROPHAGES BY A NOVEL OXIDATIVE MECHANISM AND ITS POSSIBLE RELEVANCE FOR DEVELOPMENT OF ATHEROSCLEROSIS

Recently, we described a novel oxysterol-dependent mechanism for removal of cholesterol from human macrophages, that may be of some importance for atherosclerosis.[21] Sterol 27-hydroxylase was found to be very active in lung alveolar macrophages isolated from bronchio-alveloar lavage fluid. We have shown that this hydroxylase has the unique property to introduce three hydroxyl groups at the terminal methyl group of cholesterol yielding 3β-hydroxy-5-cholestenoic acid as product.[22] This acid product is considerably more polar than cholesterol and is easily eliminated from the macrophages. Cultured human macrophages were shown to take up cholesterol from the medium, convert it into 27 hydroxycholesterol and 3β-hydroxy-5-cholestenoic acid, and secrete these compounds into the medium.[21] The daily flux of 27-oxygenated cholesterol species from the macrophages corresponds to a conversion of about 40% of the intracellular cholesterol. In accordance with this, a specific inhibition of the sterol 27-hydroxylase in cultured macrophages by cyclosporin A led to a substantial intracellular accumulation of cholesterol (unpublished observation). The magnitude of this accumulation corresponded to the flux of 27-oxygenated cholesterol species in the control macrophages.

The products of the peripheral conversion of cholesterol are present in the circulation and are rapidly taken up by the liver. In the liver 27-hydroxycholesterol and 3β-hydroxy-5-cholestenoic acid are converted into bile acids. Up to 4% of the overall formation of bile acids may be due to this mechanism as calculated from the arteriovenous difference in concentration of the 27-oxygenated compounds over the liver (unpublished observation).

The novel oxidative mechanism may be regarded as a defense mechanism, similar to reverse cholesterol transport (Figure 3). The importance of this defence mechanism is difficult to evaluate at the present state of knowledge. About 15 years ago our laboratory demonstrated in collaboration with a Norwegian group that patients with the rare disease Cerebrotendinous Xantho-matosis (CTX) have a genetic lack of the sterol 27-hydroxylase (for a review - see ref. 23). It is well documented that patients with CTX suffer from premature atherosclerosis in spite of normal or even low levels of LDL.[23] It is thus tempting to suggest that the development of atherosclerosis in these patients is related to the lack of the above defense mechanism.

References

1. L. L. Smith. Cholesterol autoxidation. Plenum Press, New York, 1981.
2. E. Lund. Oxysterols: Studies on biosynthesis and regulatory importance. Thesis, Karolinska Institutet, Stockholm, 1993.
3. O. Breuer. Oxysterols: Analysis, Occurrence and Biological Effect. Thesis, Karolinska Instuitutet, Stockholm, 1995.
4. O. Breuer and I. Björkhem. *Steroids*. 1990. **55**, 185.
5. S. Dzeletovic, O. Breuer, E. Lund and U. Diczfalusy. *Anal. Biochem.*. 1995. **225**, 73.
6. L. D. Gruneke, D. C. Craig, N. L. Petrakis and M. B. Lyon. *Biomed. Mass Spectrom.*. 1987.**14**, 335.

7. N. B. Javitt, E. Kok, S. Burstein, B. Cohen and J. Kutscher. *J. Biol. Chem..*, 1981. **256**, 12644.
8. K. Kudo, G. T. Emmons, E. W. Casserly, D. P. Via, L. C. Smith, J. Pyrek and G. J. Schroepfer. *J. Lipid Res..* 1989. **30**, 1097.
9. O. Breuer and I. Björkhem. 1995. *J. Biol. Chem.* - in press.
10. E. Lund and I. Björkhem. 1995. *Acc. Chem. Res.* **28**, 241.
11. M. S. Jacobson.. *Lancet* 1987. 656.
12. N. A. Highley, J. T. Beery, S. L. Taylor, J. W. Porter, J. Dzubia and J. J. Lalich., *Atherosclerosis.* 1986. **62**, 91.
13. S. K. Peng, C. B. Taylor, J. C. Hill and R. J. Morin. *Atherosclerosis.*, 1985. **54**, 121.
14. G. A. Boissonneault, B. Hennig and C. M. Quyang. *Proc. Soc. Exp. Biol. Med.*, 1991 **196**, 338.
15. Q. Zhou, T. L. Smith and F. A. Kummerow.. *Proc. Soc. Exp. Biol. Med.*, 1993. **202**, 75.
16. H. Hughes, B. Mathews, M. L. Lenz and J. R. Guyton.. *Arteriosc. Thromb.*, 1994. **14**, 1177.
17. J. Nourooz-Zadeh and Å. Appelqvist.. *J. Food Sci..* 1987. **52**, 57.
18. G. M. Chisolm, G. Ma, K. C. Irwin, L. L. Martin, K. G. Gunderson, L. F. Linberg, D. W. Morel and P. E. DiCorleto. *Proc. Natl. Acad. Sci. USA.* 1994. **91**, 11452.
19. D. Steinberg, S. Parthasarathy, T. E. Carew, J. C. Khoo and J. L. Witztum., *N. Engl. J. Med.* 1989. **320**: 915.
20. K. E. Watson, K. Bergström, R. Ravindranath, T. Lam, B. Norton and L. Demer., *J. Clin. Invest.* 1994. **93**, 2106.
21. I. Björkhem, O. Andersson, U. Diczfalusy, B. Sevastik, R. J. Xiu, C. Duan and E. Lund.. *Proc. Natl. Acad. Sci. USA.* 1994. **91**, 8592.
22. I. Holmberg-Betsholtz, E. Lund, I. Björkhem and K. Wikvall. *J. Biol. Chem..*, 1993. **268**, 11079.
23. I. Björkhem and K. M. Boberg. In: The Metabolic Basis of Inherited Diseases (CR Scriver, A. L. Beaudet, W. S. Sly, D. Valle, eds) McGraw Hill, New York, 1994, pp 2073-2099.

CHARACTERIZATION OF THE MM-LDL FRACTION IN HUMAN PLASMA

K. Nyyssönen and J. T. Salonen

Research Institute of Public Health
University of Kuopio
P.O. Box 1627
70211 Kuopio
Finland

1. INTRODUCTION

Oxidative stress and lipid peroxidation have been suggested to have a role in the etiology of a number of major diseases including cancer and atherosclerosis (1). However, the direct measurement of lipid peroxidation is problematic, as the half-life of the end-products of lipid peroxidation is very short due to their high toxicity and because of the potential artefactual oxidation of lipids ex vivo (2). The oxidative modification of the low-density lipoprotein (LDL) has been assessed eg. by measuring autoantibodies against aldehyde-modified LDL (3). We and others have observed an association between the ratio of titre of antibodies against MDA (malondialdehyde) -modified LDL to the titre of antibodies to native LDL (4).

Electronegatively charged, minimally modified LDL ("MM-LDL") is a form of LDL that is supposed to be exposed to mild oxidation. It has been separated and chemically characterized in native human plasma (5) and after mild oxidation of LDL with copper in vitro (6).

We applied a quick measurement of MM-LDL to a liquid chromatography system (FPLC) equipped with an autosampler, and chemically characterized the MM-LDL and native LDL fractions.

2. METHODS

2.1 Subjects

Ten normolipidemic subjects, 7 male and 3 female, aged 54-65 years, were included in the MM-LDL characterization study.

2.2 MM-LDL isolation from plasma

LDL was separated from fresh EDTA plasma by ultracentrifugation. The plasma samples were adjusted to a density of 1.24 g/ml by potassium bromide and layered underneath a solution with density of 1.006 g/ml. The tubes were centrifuged for 2.5 hours at 10°C and 100 000 rpm with a Beckman Optima TLX ultracentrifuge and TLA-100.4 rotor (Beckman, Palo Alto, CA). The LDL fraction was seen in the middle of the tube and was removed with a needle in a total volume of 800 μl. For desalting, the LDL fraction was dialyzed overnight against 0.01 M Tris buffer, pH 7.4, containing 1 mM EDTA. The LDL fraction was shown to be free of other lipoproteins by gel permeation chromatography (7).

The desalted LDL fraction was injected into a liquid chromatographic system with an autosampler, a 500 μl loop injector and a Frac-200 fraction collector (FPLC, Pharmacia), and native LDL and MM-LDL were separated in an anion-exchange column according to a method of Vedie et al (6). Quantities were measured using an UV detector set at 280 nm for protein detection. The areas of native and MM-LDL peaks were analyzed by Beckman System Gold integration software (San Ramon, CA). MM-LDL was expressed as percentage from the total LDL area.

2.3 Characterization of the MM-LDL and native LDL fractions

Total cholesterol from native LDL and MM-LDL fraction was determined enzymatically (Kone Specific, Espoo, Finland). Fraction phospholipids (BioMérieux, Marcy l'Etoile, France), triglycerides and free cholesterol (Boehringer Mannheim) were determined with enzymatic tests photometrically (Kone Specific).

3. RESULTS

3.1 Method reproducibility

The between day coefficient of variation (CV) of the MM-LDL determination was tested with a plasma pool frozen in aliquots at -80° and analyzed daily within three months. The CV was 12.8% at the MM-LDL level of 7.1% (n=22). No drift in the analytical level was found during the three month period. The anion exhange column was washed weekly with Triton X 100 in 3M acetic acid to remove fats and heavy metals possibly accumulating on the column matrix.

3.2 Characterization of the MM-LDL fraction

The MM-LDL fraction was 10.5 $\pm$ 1.0% (mean $\pm$ SD) of the total plasma LDL in nine normolipidemic men as integrated from the UV curve. Triglyceride and free cholesterol contents were significantly higher and phospholipid and total cholesterol were significantly lower in MM-LDL than in native LDL fraction (p=0.005 for comparisons between native LDL and MM-LDL in Wilcoxon matched-pairs signed ranks test, Table 1).

Table 1. *Components of native LDL and mm-LDL fraction in 10 normolipidemic men (mean ± SEM).*

Component	Native LDL	MM-LDL
Triglyceride (mmol/g protein)	5.6 ± 3.7	50.9 ± 38.4
Free cholesterol (mmol/g protein)	1.07 ± 0.04	1.24 ± 0.02
Phospholipid (mmol/g protein)	1.00 ± 0.03	0.73 ± 0.06
Total cholesterol (mmol/g protein)	3.33 ± 0.10	2.66 ± 0.13

p=0.005 for all comparisons between native LDL and MM-LDL in Wilcoxon signed rank tests

4. DISCUSSION

Our method for determination of MM-LDL percentage in plasma has proved to be fast, simple and reproducible for analyzing up to sixteen samples in a day. The washing of the analytical column once a week (after 20 samples) was found necessary, because the MM-LDL levels tended to increase if the column was not cleaned regularly. For integration of the data from UV monitor we used a separate integration software instead the FPLC inbuilt integrator, which was found unaccurate for our purposes.

The MM-LDL fractioned and collected by liquid chromatographic methods has previously been characterized by lower phospholipid content, higher free cholesterol and triglyceride content, lower lecithin/lysolecithin ratio, lower content of vitamin E and higher content of thiobarbituric acid reactive substances (TBARS) and conjucated dienes, compared with the native LDL (5,6,8). The electronegatively charged form of plasma LDL, separated by our procedure, has parallel properties such as high triglyceride and free cholesterol, and low phospholipid and total cholesterol contents. The composition of plasma MM-LDL fraction resembles the characteristic of LDL isolated from atherosclerotic lesions (9). The previous findings have also shown increased electrophoretic mobility of human or rabbit lesion LDL compared with normal intimal LDL or with plasma LDL (9). The MM-LDL fractions separated with anion-exchange chromatography from human plasma have proved to have increased eletrophoretic mobility compared with native LDL fraction (6,8). Thus plasma MM-LDL fraction may represent oxidatively modified LDL released into circulation from atherosclerotic process in the arterial wall.

References

1. D. Steinberg and Workshop Participants, *Circulation*, 1992, **85**, 2337.
2. B. Halliwell, *Lancet*, 1994, **344**, 721.
3. J. L. Witztum, *Lancet* 1994, **344**, 793.

4. J. T. Salonen, S. Ylä-Herttuala, R. Yamamoto, S. Butler, H. Korpela, R. Salonen, K. Nyyssönen, W. Palinski, J. L. Witztum, *Lancet* 1992, **339**, 883.
5. G. Cazzolato, P. Avogaro, G. Bittolo-Bon, *Free Rad. Biol. Med.* 1991, **11**, 247.
6. B. Vedie, I. Myara, M. A. Pech, J. C. Maziere, C. Maziere, A. Caprani, N. Moatti, *J. Lipid Res.*, 1991, **32**, 1359.
7. K. Nyyssönen, J. T. Salonen, *J. Chromatogr.*, 1991, **570**, 382.
8. P. Avogaro, G. Bittolo Bon, G. Cazzolato, *Arteriosclerosis*, 1988, **8**, 79.
9. S. Ylä-Herttuala, W. Palinski, M. E. Rosenfeld, S. Parathasathy, T. E. Carew S. Butler, J. L. Witztum, D. Steinberg, *J. Clin. Invest.*, 1989, **84**, 1086.

EXERCISE AND OXIDATIVE STRESS IN DIABETES MELLITIS

D. E. Larson, M. Atalay, S. Khanna, C. K. Sen

Department of Physiology
University of Kuopio
70211 Kuopio
Finland

1. INTRODUCTION

Oxidative stress has been increasingly implicated in numerous diseases and conditions, including atherosclerosis,[1] the accelerated atherosclerosis and microvascular complications of diabetes[2,3,4] and exercise[5,6,7]. Increased oxidative stress can result in widespread lipid, protein and DNA damage[8] including oxidation of low density lipoprotein (LDL) cholesterol, believed to be central in the pathogenesis of atherosclerosis and endothelial dysfunction.[1,3,4]

During moderate exercise oxygen consumption increases by around 8-10 folds, and oxygen flux through the muscle may increase by 90-100 folds. Even moderate exercise increases free radical production and overwhelms antioxidant defences, resulting in oxidative damage[5,7,9,10]. Training or regular exercise, on the other hand, appears to protect against oxidative stress.[5,6,7,9,11] Large prospective studies[12,13] suggest a protective effect of regular exercise and physical fitness as measured by maximal oxygen consumption on cardiovascular disease and mortality independent of effects on known cardiovascular risk factors. Diabetic patients were not studied, however, and the mechanisms by which exercise lowers cardiovascular mortality remained unclear. Exercise as a form of preventive health has been widely recommended, including for diabetic patients.[14] The relative benefits or risks of acute and chronic exercise in relation to oxidative stress in groups with increased susceptibility to oxidative stress such as diabetic patients is unknown.

Clearly these issues must be addressed in order to maximize the the benefits of exercise in diabetic patients. We will briefly highlight major aspects of oxidative stress and its mechanisms in diabetes and exercise induced oxidative stress and the effect of training, touching on our own results in the field of exercise and more recently in diabetic patients.

2. ELEVATED OXIDATIVE STRESS IN DIABETES

Oxidative stress as measured by indices of lipid peroxidation has been shown to be elevated in both non-insulin dependent (NIDDM)[15,16] and insulin dependent diabetes (IDDM),[17,18] even in patients without complications.[15,18] Increased oxidized LDL or susceptibility to oxidation has also been shown in diabetes.[19,20,21]

The mechanisms behind the apparent increased oxidative stress in diabetes are not entirely clear. Accumulating evidence points to many, often interrelated mechanisms,[3,4] increasing production of free radicals such as O_2-[22] or decreasing antioxidant status.[23] These mechanisms include glycoxidation[24] and formation of advanced glycation products (AGE),[3,20] activation of the polyol pathway[25] and altered cell26 and glutathione redox status[26] and ascorbate metabolism[27] and perturbations in nitric oxide and prostaglandin metabolism.[4]

Insulin resistance may be an additional cause of elevated oxidative stress. Lipid peroxidation as measured by plasma thiobarbituric acid reactive substances (TBARS) was shown to be similarly increased in both NIDDM and glucose intolerant patients (many of whom later develop diabetes) compared to control subjects.[16] Multivariate regression analysis of clinical and biochemical characteristics relevant to diabetes and cardiovascular risk status of the combined groups showed only fasting glucose and insulin levels to be determinants of plasma TBARS. This may be especially important given that NIDDM patients comprise about 90 % of all diabetic patients and that glucose intolerance is even more common.[28] A correlation between insulin resistance as measured by the euglycemic clamp and blood superoxide production has been shown.[22] Recent in vitro evidence also supports a role of hyperinsulinemia in increasing susceptibility of LDL cholesterol to oxidation.[21] Studies examining the role of hyperinsulinemia in oxidative stress in either diabetic or glucose intolerant subjects are otherwise quite limited.

The apparent role of hyperinsulinemia as a mechanism of increased oxidative stress in NIDDM patients emphasizes the possibility that different mechanisms or variation in the relative importance of mechanism of oxidative stress exist between IDDM and NIDDM. The high incidence of co-existent hyperlipidemia, obesity and hypertension in NIDDM but not IDDM and the marked improvements in insulin sensitivity with exercise in NIDDM[29] make this is even more likely.

3. EXERCISE INDUCED OXIDATIVE STRESS AND THE EFFECT OF TRAINING

Recent studies have established that even moderate exercise increases free radical production and overwhelms antioxidant defences, resulting in oxidative damage.[5,7,9] We have shown increased lipid,[30,31] protein (submitted) and DNA[31] markers of oxidative stress in response to exercise in plasma, blood or skeletal muscle, in agreement with others.[32,33,34] Increase in blood glutathione disulphide or disulphide-total glutathione ratio has also been shwn to be a sensitive index to exercise induced oxidative stress.[30,31,35,36] Further supporting the role of oxidative stress in exercise are antioxidant supplementation studies showing a marked decrease in indices of oxidative stress in response to exercise.[31,33]

Training or regular exercise conversely appears to increase antioxidant defenses and reduce oxidative stress. $VO_{2\,max}$ has also been found in healthy young men to correlate with vastus lateralis muscle activities of superoxide dismutase and catalase, major endogenous antioxidant enzymes.[11] Exercise training in non-diabetic animals has been shown to reduce oxidative stress as measured by lipid peroxidation indices in heart[37] and skeletal muscle.[32] Training has also been shown to have favorable effects on oxidative stress and antioxidant status as measured by TGSH and GSSG[36,37] and glutathione peroxidase, catalase and superoxide dismutase activities[36] Increased activity of antioxidant enzymes has been reported in muscles with high oxidative capacity.[32,36] Training has been most beneficial in these same muscles.[32,36] Regular physical activity and higher $VO_{2\,max}$ may thus lower oxidative stress by strengthening antioxidant defenses and permitting more efficient use of oxygen.

The potential benefits and risks of exercise may be particularly important for those who at rest already show higher levels of oxidative stress, such as diabetic patients. We recently showed markedly increased resting oxidative stress as measured by plasma TBARS in otherwise healthy young insulin-dependent (IDDM) diabetic men compared to healthy men of similar age, body mass index and fitness (submitted). As a new finding, plasma TBARS was also elevated in diabetic men after moderate exercise, with a similar relative post-exercise increase in both groups. Potentially the most important new findings were the strong inverse correlation in the diabetic group between resting oxidative stress as measured by plasma TBARS and maximal oxygen uptake, and the strong correlation betwen relative increase in plasma TBARS and oxygen consumption. Thus in otherwise healthy IDDM patients fitness and regular exercise may protect against oxidative stress at rest, and partially offset the oxygen consumption-related increase in oxidative stress with exercise. We are currently undertaking further studies to test this hypothesis.

4. CONCLUSIONS

Regular exercise has potentially enormous therapeutic value in the normal population, although its long term benefit is poorly defined in disease states such as diabetes. The relative benefits and risks of exercise may be markedly different in diabetic patients, who have increased oxidative stress. Preliminary data suggest a protective effect of fitness in young, otherwise healthy IDDM men. The relationships between resting and exercise induced oxidative stress and physical fitness and regular exercise must be further explored in order to maximize the benefits of exercise in IDDM and NIDDM patients.

5. REFERENCES

1. D. Steinberg, S. Parthasarathy, T. E. Carew, J. C. Khoo and J. L. Witztum, *N. Engl. J. Med.*, 1989, **320**, 915.

2. J. W. Baynes, *Diabetes.*, 1991, **40**, 405.

3. A. M. Schmidt, O. Hori, J. Brett, S. D. Yan, J. L. Wautier and D. Stern, *Arterioscler. Thromb.*, 1994, **14**, 1521.

4. B. Tesfamariam, *Free. Radical. Biol. Med.*, 1994, **16**, 383.

5. H. M. Alessio, *Med. Sci. Sports. Exerc.*, 1993, **25**, 218.

6. C. K. Sen and O. Hänninen, "Exercise and oxygen toxicity", C. K. Sen, L. Packer and O. Hänninen, Elsevier, Amsterdam, 1994, 89.

7. C. K. Sen, *J. Appl. Physiol.*, 1995, **79**, 675.

8. B. Halliwell, *Lancet.*, 1994, **344**, 721.

9. K. J. Davies, A. T. Quintanilha, G. A. Brooks and L. Packer, *Biochem. Biophys. Res. Commun.*, 1982, **107**, 1198.

10. N. Haramaki and L. Packer, "Exercise and oxygen toxicity", C. K. Sen, L. Packer and O. Hänninen, Elsevier, Amsterdam, 1994,, 77.

11. R. R. Jenkins, R. Friedland and H. Howald, *Int. J. Sports. Med.*, 1984, **5**, 11.

12. R. S. Paffenbarger Jr., R. T. Hyde, A. L. Wing and C. H. Steinmetz, *Jama.*, 1984, **252**, 491.

13. T. A. Lakka, J. M. Venalainen, R. Rauramaa, R. Salonen, J. Tuomilehto and J. T. Salonen, *N. Engl. J. Med.*, 1994, **330**, 1549.

14. American Diabetes Association, *Diabetes. Care.*, 1993, **16**, 54.

15. A. Collier, A. Rumley, A. G. Rumley, J. R. Paterson, J. P. Leach, G. D. Lowe and M. Small, *Diabetes.*, 1992, **41**, 909.

16. L. Niskanen, J. T. Salonen, K. Nyssönen and M. I. Uusitupa, *Diabet. Med.*, 1995, **in press**.

17. S. K. Jain, R. McVie, J. Duett and J. J. Herbst, *Diabetes.*, 1989, **38**, 1539.

18. M. Yaqoob, P. McClelland, A. W. Patrick, A. Stevenson, H. Mason, M. C. White and G. M. Bell, *Q. J. Med.*, 1994, **87**, 601.

19. D. W. Morel and G. M. Chisolm, *J. Lipid. Res.*, 1989, **30(12)**, 1827.

20. R. Bucala, Z. Makita, T. Koschinsky, A. Cerami and H. Vlassara, *Proc. Natl. Acad. Sci. U. S. A.*, 1993, **90**, 6434.

21. V. A. Rifici, S. H. Schneider and A. K. Khachadurian, *Atherosclerosis.*, 1994, **107**, 99.

22. G. Paolisso, A. D'Amore, C. Volpe, V. Balbi, F. Saccomanno, D. Galzerano, D. Giugliano, M. Varricchio and F. D'Onofrio, *Metabolism.*, 1994, **43**, 1426.

23. K. Asayama, N. Uchida, T. Nakane, H. Hayashibe, K. Dobashi, S. Amemiya, K. Kato and S. Nakazawa, *Free. Radical. Biol. Med.*, 1993, **15**, 597.

24. J. V. Hunt, R. T. Dean and S. P. Wolff, *Biochem. J.*, 1988, **256**, 205.

25. D. R. Tomlinson, *Diabet. Med.*, 1993, **10**, 214.

26. R. W. Grunewald, I. I. Weber, E. Kinne-Saffran and R. K. Kinne, *Biochim. Biophys. Acta.*, 1993, **1225**, 39.

27. A. J. Sinclair, A. J. Girling, L. Gray, C. Le-Guen, J. Lunec and A. H. Barnett, *Diabetologia.*, 1991, **34**, 171.

28. M. Laakso and K. Pyörälä, *Diabetes. Care.*, 1985, **8**, 114.

29. S. H. Schneider, L. F. Amorosa and A. K. Khachadurian, *Diabetologia.*, 1984, **26**, 355.

30. C. K. Sen, M. Atalay and O. Hänninen, *J. Appl. Physiol.*, 1994, **77**, 2177.

31. C. K. Sen, T. Rankinen, S. Vaisanen and R. Rauramaa, *J. Appl. Physiol.*, 1994, **76**, 2570.

32. H. Alessio and A. H. Goldfarb, *J. Appl. Physiol.*, 1988, **64**, 1333.

33. M. M. Kanter, L. A. Nolte and J. O. Holloszy, *J. Appl. Physiol.*, 1993, **74**, 965.

34. A. Hartmann, A. M. Neiss, M. Gruenert-Fuchs and G. Speit, *Mutation.Res.*, 1995, **346**, 195.

35. K. Gohil, C. Viguie, W. C. Stanley, G. A. Brooks and L. Packer, *J. Appl. Physiol.*, 1988, **64**, 115.

36. C. K. Sen, E. Marin, M. Kretzschmar and O. Hanninen, *J. Appl. Physiol.*, 1992, **73**, 1265.

37. M. Kihlström, *J. Appl. Physiol.*, 1990, **68**, 1672.

LIPID PEROXIDATION AND SOD ACTIVITY IN DIABETES

J.P. Freitas, J.N. Silva, P.M. Filipe and F. Guerra Rodrigo

Department of Dermatology
Faculdade de Medicina de Lisboa
Av. Prof. Egas Moniz
1699 Lisboa Codex

1 INTRODUCTION

Glycation and free radical reactions may play important roles in complications of diabetes[1-3]. Increased free radical activity may contribute to the higher prevalence of vascular diseases in diabetic patients[2-4]. Sources of oxidative stress in diabetes include free radicals generated by autoxidation of glycated proteins[3,5]. An increased lipid peroxidation, due to an altered ratio between free radicals and antioxidant systems, has been related to diabetes[2,4,6]. Superoxide dismutase (SOD) is an antioxidant enzyme that plays a great role in inhibiting the oxidative stress in the living systems[7]. In diabetic patients the effect of blood glucose control on plasma lipid peroxide concentrations, measured as thiobarbituric acid reacting substances (TBARS), is controversial[6,8].

To study the possible relationship between lipid peroxidation and metabolic control, we measured the plasma concentrations of malondialdehyde (MDA)[9], end product of the oxidation of polyunsaturated fatty acids, in 40 type 2 diabetic patients and 40 comparable non-diabetic control subjects. Cu,Zn-SOD activity[10] in red cells was determined in 20 patients with and without retinopathy. Diabetes control was assessed by serum fructosamine levels[11].

2 MATERIALS AND METHODS

Forty type 2 diabetic patients, 18 men and 22 women, aged between 47 and 85 years were admitted to the trial. The results were compared to those of 40 age-matched healthy normoglicaemic subjects. Patients were excluded if they had hypertension or a history of hyperlipidemia. After an overnight, fast venous blood samples were drawn. Plasma was immediately separated from whole blood, by centrifugation at 2000 g at 4°C, and used. Plasma TBARS levels were determined by fluorometry using the method of Yagi (515 nm excitation and 553 nm emission) with malondialdehyde as a standard[9].

Results are expressed in terms of nmol of malondialdehyde per ml of plasma. Plasma fructosamine was measured by the nitroblue tetrazolium (NBT) colorimetric test (Isolab Inc., Akron, Ohio, USA)[11]. Red cell SOD activity was evaluated by the method of Beyer[10] in 10 patients with retinopathy and in 10 patients without retinopathy. Results are expressed in SOD units/g Hb[10]. Statistical analysis was performed using Student's t-test. Correlation coefficients were calculated. Values are expressed as mean ± SD.

3 RESULTS

An elevated level of lipid peroxides, measured as TBARS by fluorometry, was observed in the plasma of diabetic patients (4.51±1.29 nmol/ml) as compared to healthy subjects (3.54±1.00 nmol/ml) ($p<0.001$) (Figure 1). Fructosamine level was 3.25±1.15 mM in diabetic patients and 1.92±0.37 mM in controls ($p<0.001$). Lipid peroxides did not correlate with the fructosamine levels. Red cell SOD activity was higher in the diabetics with retinopathy ($p<0.02$) (Figure 2).

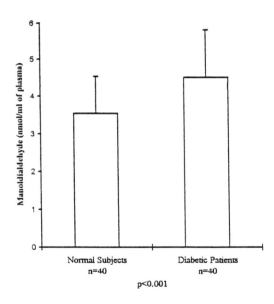

Figure 1 *Plasma lipid peroxide levels of 40 diabetic patients and 40 normal subjects. Plasma lipid peroxides are expressed in terms of malondialdehyde concentration (nmol/ml). Mean±SD ($p<0.001$)*

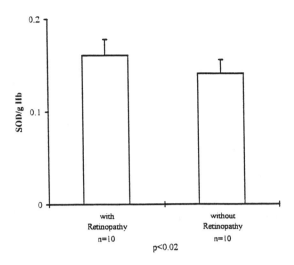

Figure 2 *Red cell SOD activity (SOD/g Hb) in 20 diabetics with and without retinopathy. Mean±SD (p<0.02)*

4 DISCUSSION

The data reported in this study show a significant increase in plasma concentrations of lipid peroxides in type 2 diabetics compared to non-diabetics. The results are consistent with those of other reports[2,6,8]. We didn't find any correlation between plasma MDA and fructosamine as previously reported[6,8]. Glycation may affect proteins and be partly responsible for free radical-mediated damage in diabetics[1,2,5,12]. Oxidative stress in diabetes may be induced not only by glycation, but also by oxidative metabolism resulting from changes in cellular energy balance, deficiency in the antioxidant systems activity[4,13], and tissue damage resulting from hypoxia and ischaemia/reperfusion injury[14]. In diabetes, red cell Cu,Zn-SOD can be glycated resulting in a decrease of its activity[15]. The cellular oxidative stress could be antagonised, partially, by the SOD increased activity in the red cells of diabetics with retinopathy compared those without retinopathy. The association between lipid peroxides and atherosclerosis is supported by several studies indicating raised levels of TBARS in patients with occlusive arterial disease and diabetes[2,5,16]. However, it is difficult to conclude whether increased lipid peroxidation is a cause or an effect of complications in diabetes[4,17]. Finally, glucose itself has been reported to undergo transition metal-catalysed autoxidation, and the resultant production of hydrogen peroxide and hydroxyl radicals could be implicated in the increasing of

TBARS in plasma[5,18]. Probably the mechanism for high lipid peroxides in plasma from diabetics is multifactorial.

References

1. M. Brownlee, *Diabetes*, 1994, **43**, 836.
2. D. Armstrong, N. Abdella, A. Salman, N. Miller, E. A. Rahman and M. Bojancyzk, *J. Diab. Comp.*,1992, **6**, 116.
3. J. P. Freitas, P. Filipe, J. Silva, E. Vendrell and F. Guerra Rodrigo, *Horm. Metab. Res.*, 1995, **suppl 1**, 63.
4. J. W. Baynes, *Diabetes*, 1991, **40**, 405.
5. C. J. Mullarkey, D. Edelstein and M. Brownlee, *Biochem. Biophys. Res. Commun.*, 1990, **173**, 932.
6. E. Altomare, G. Vendemiace, D. Chicco, Y. Procacci and F. Cirelli, *Diabetes Metab.*,1992, **18**, 264.
7. K. Brawn and I. Fridovich, *Acta Physiol. Scand.*, 1980, **492**, 9.
8. A. Collier, R. Wilson, H. Bradley, J. A. Thomson and M. Small, *Diabet. Med.*, 1990, **7**, 27.
9. K. Yagi, *Biochem. Med.*, 1976, **15**, 212.
10. W. F. Beyer, Y. Wang and I. Fridovich, *Biochemistry*, 1986, **25**, 6084.
11. R. N. Johnson, P. A. Metcalf and J. R. Baker, *Clin. Chim. Acta*, 1982, **127**, 87.
12. P. Gillery, J. C. Monboisse, F. X. Maquart and J. P. Borel, *Med-Hypothesis*, 1989, **29**, 47.
13. J. Jos, M. Rybak, P. H. Patin, J. J. Robert, C. Boitard and R. Thevenin, *Diabetes Metab.*,1990, **16**, 498.
14. J. Martins e Silva, F. Levy-Cruz, J. P. Freitas and P. Souza-Ramalho, *Acta Diabetol.*, 1984, **21**, 133.
15. K. Arai, S. Maguchi, S. Fugii, H. Ishibashi, K. Oikawa and N. Taniguchi, *J. Biol. Chem.*, 1987, **262**, 16969.
16. M. D. Stringer, P. G. Gorog, A. Freeman and V. V. Kakkar, *B. M. J.*, 1989, **298**, 281.
17. P. Faure, P. Corticelli, M. J. Richard et al., *Clin. Chem.*, 1993, **39**, 789.
18. C. Smith, M. J. Mitchinson, O. I. Aruoma and B. Halliwell, *Biochem. J.*,1992, **286**, 901.

INCREASED LIPID PEROXIDATION IN AN EXPERIMENTAL MODEL OF THE INSULIN RESISTANCE SYDROME: EFFECT OF ANTIOXIDANT THERAPY

L.Kazdová, A.Vrána, M.Matějčková, and V. Nováková

Institute for Clinical and
Experimental Medicine,
Prague, Czech Republic.

1 INTRODUCTION

Recent epidemiologic studies suggest that the insulin resistance syndrome, which may affect one fifth of the adult Western population, increases the risk of atherosclerosis and cardiovascular disease.[1] In this syndrome, insulin resistance (the need for increased insulin levels to maintain blood glucose homeostasis) is associated with hypertriglyceridaemia, hyperinsulinaemia and hypertension.[2] Although induvidually each of these metabolic disorders is often mild, they may synergically increase the risk of cardiovascular disease by mechanisms which still remain unknown.

There are growing evidence that the free radical process of lipid peroxidation is implicated in many of the key events in the pathogenesis of cardiovascular disease.[3] In addition to lipoprotein retention in the arterial wall, they are cytotoxic, may increase procoagulant activity and induce vasoconstriction N0 synthesis inhibition.[4] Plasma lipid peroxides have been reported to be elevated in patients with ischaemic heart disease, peripheral arterial disease[5] and in the children of parents suffering a myocardial infarction.[6] Moreover, in hypercholesterolaemic animals, vitamin E has been shown to have antiatherogenic effects.[7,8] Lipid peroxidation or metabolic effects of antioxidant therapy has not been tested in insulin resistance syndrome yet. One of the reasons for this was that no suitable animal model was available. Recently, a rat model of the insulin resistance syndrome has been developed in our laboratory. The line of hypertriglyceridaemic rats (HTg), originating from the Wistar strain, exhibits insulin resistance, hyperinsulinaemia and elevated blood pressure.[9] To get more information about the posssible synergic action of these metabolic disorders on the processes involved in the pathogenesis of cardiovascular diseases associated with the insulin resistance syndrome, we investigated the serum lipids parameters, lipid peroxidation, and the effects of antioxidant therapy in an animal model of this syndrome.

2 METHODS

Adult males of a hereditary hypertriglyceridaemic line originating from the Wistar strain, weighing 260 - 275 g, were used. Random bred males of the Wistar strain of a similar weight were used as controls. All rats were fed standard rat chow ad libitum without or with alpha-tocopherol acetate supplementation at a dose 500 mg/kg diet. In addition vitamin E-treated rats were injected 60 mg alpha-tocopherol/100g body weight, s.c., twice weekly. After 4 weeks, the rats were killed by decapitation and blood was centrifuged, aliquoted and immediately assayed for conjugated diene concentration. Liver samples was homogenized and supernatants were stored at -20° C.

Serum levels of triglycerides and glucose were measured by commercially available enzymatic kits (Lachema, Czech Republic). Insulinaemia was determined by radioimmunoassay (RIA-SAX-100 insulin, Germany). Serum free fatty acids (FFA) were estimated by titration.[10] Lipid peroxide concentration was measured fluorometrically (excitation at 515 nm and emission at 550 nm) as thiobarbituric acid reactive substances (TBARS) using tetramethoxypropan (Sigma) as a standard.[11] Conjugated dienes were measured by UV spectrophotometry at 235 nm.[12] Alpha-tocopherol in serum was estimated by HPLC analysis.[13]

Data are expressed as means ± SEM. Significant difference between mean values was determined by Student's t test for unpaired data.

3 RESULTS AND DISCUSSION

Data presented in Table 1 indicated that serum alpha-tocopherol levels were not different in controls and in the HTg line. Daily administration of vitamin E for 4 weeks doubled serum alpha-tocopherol levels.

As has been previously reported,[9] the HTg line exhibited serum triglycerides levels above those of control rats. Vitamin E treatment induced a mild decrease in serum triglycerides. This decrease was significant only in nonfasted animals (Table 1).

Table 1. *Serum concentrations of alpha-tocopherol, triglycerides (TG) and free fatty acids in controls and HTg rats (HTg) without or with vitamin E treatment (HTg+E)*

Groups	C	HTg	HTg+E
Alpha-tocopherol, µmol/L	18.70 ± 2.11	19.22 ± 1.93	$38.5 \pm 2.98^*$
Fasted TG, mmol/L	0.95 ± 0.06	$1.69 \pm 0.13^+$	$1,41 \pm 0.12^{**}$
Nonfasted TG, mmol/L	1.39 ± 0.15	$2.26 \pm 0.21^{++}$	$1.68 \pm 0.13^{**}$
Free fatty acids, µmol/L	548 ± 29	$747 \pm 64^{+++}$	$558 \pm 34^{**}$

Values are means ± SEM, 10 - 12 animals per group. HTg vs C $^+p < 0.05$, $^{++}p < 0.01$, $^{+++}p < 0.02$. HTg+E vs HTg $^*p < 0.001$, $^{**}p < 0.05$.

At present it is not known whether vitamin E has any effect on hypertriglyceridaemia. In some studies, vitamin E treatment lowered hypertriglyceridaemia in patients with non insulin-dependent diabetes,[14] in Watanabe heritable hyperlipidaemic rabbits[7] or in streptozotocin-induced diabetic rats.[15] Other studies failed to demonstrate a positive effect of vitamin E on diabetes-induced hypertriglyceridaemia in experimental animals.[16,17] The reason for these discrepancies remains to be determined.

Significant increases in serum FFA concentrations were observed in the HTg line. Vitamin E treatment lowered FFA levels to those found in control animals (Table 1). A similar beneficial effect of vitamin E supplementation on FFA values was observed in diabetic patients.[14] This effect of vitamin E seems to be very important as elevated FFA levels may not only contribute to glucose intolerance and insulin resistance[18] but, also, may be involved in the pathogenesis of vascular disease. It has been reported that elevated FFA may increase platelet aggregation[19] and cause dose-dependent inhibition of prostacyclin synthesis in the aorta.[20]

The results of serum and tissue analysis for lipid peroxidation are shown in Table 2. Estimation of lipid peroxidation is complicated by the large number of peroxidation products and their reactivity. We therefore measured as indicators of lipid peroxidation, both the precursors of cytotoxic hydroperoxides - conjugated dienes and the terminal product of lipoperoxidation - TBARS. The serum concentrations of conjugated dienes and TBARS were significantly higher in the HTg line compared with controls. Vitamin E treatment decreased elevated serum peroxide levels in the HTg line compared to those in the control group. This effect of vitamin E was pronounced only in the HTg line, in control rats, vitamin E has no effect on serum peroxide levels (data not shown). There is accumulating evidence that lipid peroxidation is a contributor to the pathogenesis of cardiovascular disease. Recently, increased levels of plasma lipid peroxides have been observed in association with hypercholesterolaemia,[21] atherosclerosis[5] and diabetes.[22]

Table 2. *Serum and liver concentrations of lipid peroxides in controls and HTg rats without or with vitamin E treatment (HTg+E)*

Groups	C	HTg	HTg + E
Serum conjugated dienes, abs. U	0.18 ± 0.01	$0.28 \pm 0.02^{++}$	$0.21 \pm 0.01^{*}$
Serum TBARS nmol/L	2.12 ± 0.02	$3,12 \pm 0.04^{+}$	$1.75 \pm 0.02^{**}$
Liver TBARS nmol/g	85.2 ± 5.6	$110.3 \pm 4.7^{+}$	$97.2 \pm 2.1^{*}$

Values are means $\pm$ SEM, 10 - 12 animals per group. HTg vs C $^{+}p < 0.05$, $^{++}p < 0.02$. HTg+E vs Htg $^{*}p < 0.05$, $^{**}p < 0.02$.

Both the liver and the arterial wall are the sites for the accumulation of lipids and lipid peroxides in hyperlipidaemia. As shown in Table 2, the concentration of liver lipoperoxidation products, as estimated by the levels of TBARS, was significantly higher in the HTg line than in controls. Vitamin E treatment resulted in a decrease in liver TBARS concentrations in the HTg line. It is noteworthy that the accumulation of liver lipid peroxides in a fatty streak-susceptible mouse strain fed an atherogenic diet, was associated with inflammatory gene activation and atherosclerotic lesion formation. In a control non-susceptible animal strain on atherogenic diet the accumulation lipid peroxides in liver was much lower and did not induce the above mentioned changes.[23] These observations suggest further mechanisms by which lipid peroxidation may contribute to cardiovascular disease development.

As far as the parameters of glucoregulation are concerned, in our experiments vitamin E treatment had no significant effect on blood glucose (HTg 4.52 ± 0.21 vs HTg+E 4.65 ± 0.24 mmol/l) or serum insulin (HTg 55.2 ± 7.4 vs HTg+E 59.5 ± 8.9 µU/ml). On the other hand, in some studies, vitamin E has recently been reported to reduce plasma glucose[14] and inhibit nonenzymatic glycosylation in diabetic patients.[24]

In conclusion, the results of the present study indicate that clustering of several risk factors for cardiovascular disease - hypertriglyceridaemia, insulin resistance, hyperinsulinaemia and hypertension, in an experimental model of insulin resistance were associated with increased lipid peroxidation which might be involved in the development of vascular disease associated with this syndrome. Vitamin E, by inhibiting lipid peroxidation and by its mild hypotriglyceridaemic effect might be potentially beneficial in preventing cardiovascular disorders associated with the insulin resistance syndrome.

REFERENCES

1. R. P. Donahue, *The Endocrinologist*, 1994, **4**, 112, .
2. G. M. Reaven, *Diabetes*, 1988, **37,** 1595.
3. J. A. Berliner, M. Navab, A.M. Fogelman, J. S. Frank, L. L. Demer, P. A. Edwards, A. D. Watson and A. J. Lusis, *Circulation,* 1994, **91**, 2488.
4. P. Holvoet and D. Collen, *FASEB J.,* 1994, **8,** 1279.
5. M. D. Stringer, P. G. Gorog, A. Freeman and V. V. Kakkar, *Br. Med. J.*,1989, **298**, 281.
6. T. Szamosi, I. Gara and I. Venekei, *Atherosclerosis*, 1987, **68**, 111.
7. R. J. Williams, J. M. Motteram, C. H. Sharp and P. J. Gallagher, *Atherosclerosis*, 1992,**94**, 153.
8. Y. Qiao, M. Yokoyama, K. Kameyama and G. Asano, *Arterioscler. Thromb.*, 1993, **13**,1885.
9. A. Vrána and L. Kazdová, *Transplant. Proc.*, 1990, **22**, 2579.
10. V. P. Dole, J. Clin. Invest. 1957, **35**, 150.
11. M. Yokode, T. Kita, Y.Kihawa, *J. Clin. Invest.*, 1988, **81**, 720.
12. P. J. Ward, G. O. Pill, J. R. Hatherill, H. Annensley and R. G. Kunkel, *J. Clin. Invest.*, 1985, **76**, 517.
13. G. L. Catignani, *Meth. Enzymol.,* 1986, **123**, 215.
14. G. Paolisso, A. D'Amore, D. Galzerano, V. Balbi, D. Giugliano, M.Varrocchio, F. D'Onofrio, *Diabetes Care*, 1993, **16**, 1433.

15. J. C. Ruff, M. Ciavatti, T. Gustavson, S. Renaud, *Diabetes*, 1991, **40**, 233,
16. D. W.,Morel, G. M.Chisolm, *J. Lipid Res.*, 1989, **30**, 1827.
17. C. Douillet, Y. Chancerelle, C. Cruz, C. Maroncles, J. F. Kergonou, S. Renaud, M. Ciavatti, *Biochem. Med. Metab. Biol.*, 1993, **50**, 265.
18. M. Laville, V. Rigalleauo, J. P. Rion, M. Beylot, *Metabolism*, 1995, **44**, 639.
19. J. C. Hoak, A. A. Spector, G. L. Fry, E. D. Warner, *Nature*, 1970, **228**, 1330.
20. J. Y. Jeremy, D. P. Mikhailidis, P. Dandona, *Diabetes*, 1983, **32**, 217.
21. A.Yalcin, N. Sabuneu, A. Kilinc, G. Gulcan, K. Emerk, *Atherosclerosis*, 1989, **80**, 169.
22. A. D. Moordian, *J. Am. Geriatr. Soc.*, 1991, **39**, 571.
23. F. Liao, A.Andalibi, F.C. deBeer, A. M. Fogelman, A. J. Lusis, *J. Clin. Invest.*, 1993, **91**, 2572.
24. A. Ceriello, D. Giugliano, A. Quatraro, C. Donzella, G. Dipalo, P. J. Lefebvre, *Diabetes Care*, 1991, **14**, 68.

Acknowledgment

This study was supported by grants 2016-3/94 and 3102-3/95 from the Internal Grant Agency of the Ministry of Health of the Czech Republic.

OXIDATIVE STRESS INDUCED BY PHYSICAL EXERCISE

M. Atalay, D. Larson, S. Khanna, O. Hänninen and C. K. Sen*

Department of Physiology, Faculty of Medicine
University of Kuopio
70211 Kuopio, Finland

1 INTRODUCTION

Physical exercise is associated with a 10-15 fold increase in oxygen consumption. Oxygen flux through the active skeletal muscle tissue may increase by an estimated 100 folds. A large number of recent studies have shown that physical exercise induces oxidative stress, a state where the pro-oxidant forces overwhelm the antioxidant defense of the body.[1-6]

A variety of endo- and exo-genous antioxidants act in synergism to protect against oxidative stress. Glutathione (L-γ-glutamyl-L-cysteinylglycine) is well established as being important in the circumvention of cellular oxidative stress, detoxification of electrophiles and maintenance of intracellular thiol redox status. Glutathione peroxidase is specific for its hydrogen donor glutathione, but may use a wide range of substrates extending from H_2O_2 to organic hydroperoxides. The cytosolic and membrane-bound monomer glutathione phospholipid hydroperoxide- glutathione peroxidase, and the distinct tetramer plasma glutathione peroxidase are able to reduce phospholipid hydroperoxides without the necessity of prior hydrolysis by phospholipase A_2. The protective action of phospholipid hydroperoxide-glutathione peroxidase against membrane damaging lipid peroxidation has been directly demonstrated. Glutathione is a major cellular electrophile conjugator as well. Glutathione S-transferases catalyze the reaction between the -SH group of glutathione and potential alkylating agents thereby neutralizing their electrophilic sites and rendering them more water soluble. Apart from participating in the glutathione peroxidase catalyzed detoxification of H_2O_2, glutathione can also spontaneously react with and scavenge a number of reactive oxygen species. As a major byproduct of such enzymatic and non-enzymatic antioxidative reactions glutathione is transformed to its oxidized form, glutathione disulfide. Intracellular glutathione disulfide thus formed is cytotoxic and may be reduced back to glutathione by glutathione reductase in the presence of NADPH. When the glutathione disulfide reductase dependent capacity to reduce glutathione disulfide fails to match the rate of glutathione disulfide formation, glutathione disulfide is exported to the extracellular compartment. Therefore, oxidative stress in tissues is often reflected as high glutathione disulfide level in the serum.[3-7]

*to whom correspondence should be addressed

Various endo- and exo-genous antioxidants are known to act in concert in the form of an antioxidant chain reaction.[1-4] Scavenging of lipid peroxyl radicals and breaking of the lipid peroxidation chain reaction by α-tocopherol is accompanied by the oxidation of the vitamin E compound to a radical configuration - the α-tocopheroxyl radical (vitamin E•). When accumulated sufficiently, the vitamin E• radical may be stressful for other cells. In this way, vitamin E in human low density lipoprotein may actually act as a pro-oxidant. To avoid such toxicity and utilize vitamin E in the most beneficial manner it is imperative that the radical form of vitamin E be recycled to its native antioxidant form. The water soluble antioxidants, ascorbic acid and reduced glutathione are suggested to be involved in regenerating α-tocopherol from its radical byproduct.[1-3,8] The pK_a of ascorbic acid being 4.25, vitamin C in physiological fluids is predominantly in the anionic form (AH⁻). While scavenging reactive oxygen species (*i.e.,* donating its reducing power to reactive oxygen species) AH⁻ gets oxidized to a radical structure, the ascorbyl radical (vitamin C•). Glutathione is also suggested to regenerate ascorbate from its oxidized byproduct. It is thus evident that apart from contributing to the enzymatic and non-enzymatic decomposition of reactive oxygen species, glutathione plays a central role in co-ordinating the activities of crucial exogenous antioxidants. To obtain best results, this synergism should be considered with particular care especially when designing antioxidant therapy protocols.

2 EXPERIMENTAL PROCEDURES AND RESULTS

Our studies have shown that glutathione dependent antioxidant protection is considerably affected by the state of physical activity; endurance training enhances and chronic inactivity diminishes such protection[9]. Beagle dogs were treadmill-trained 40 km/day at 5.5-6.8 km/h, 15% uphillgrade, 5 days/wk, 55 wks. On training, hepatic and red gastrocnemius glutathione increased, glutathione peroxidase and glutathione reductase increased in all the leg muscles, and hepatic glutathione S-transferases• activity increased. Physical inactivity (fibre cast immobilization of pelvic limb; 11 wks) did not affect glutathione peroxidase, glutathione reductase and glutathione S-transferases of red gastrocinemius but total glutathione content decreased. Wistar rats were treadmill-trained 2h/day at 2.1 km/h, 5 days/wk, 8 wks. On training, hepatic total glutathione and leg muscle glutathione peroxidase increased but glutathione reductase of red gastrocnemius decreased perhaps due to an increased muscle flavoprotein breakdown during exhaustive training. Glutathione as well as other γ-glutamyl containing compounds, including glutathione disulfide and γ-glutamylglutathione react with γ-glutamyltranspeptidase at the outer cell surface. The γ-glutamyl moiety is transferred to a suitable amino acid acceptor, and both the γ-glutamyl amino acid and cysteinylglycine are transported into the cell and reused for glutathione generation. Thus, γ-glutamyltranspeptidase supplies substrates to regenerate intracellular glutathione. γ-glutamyltranspeptidase was higher in the trained leg muscles. Exhaustive exercise, decreased muscle γ-glutamyltranspeptidase of only control leg muscle, depleted muscle (lesser extent in trained rats) and liver glutathione of both groups, decreased glutathione reductase only in untrained red gastrocnemius, and increased hepatic glutathione S-transferases[9].

Recently we also examined the influence of sprint type high intensity interval training on rat skeletal muscle and heart antioxidant enzymes. Activities of glutathione peroxidase, glutathione reductase, glutathione S-transferases and superoxide dismutase in different skeletal muscle types and heart were compared in sprint trained and control sedentary animals. Sprint training led to a favourable adaptation of the glutathione redox cycle in heart and in the agonist-antagonist gastrocnemius and extensor digitorum longus muscles. These skeletal muscles are rich in fast glycolytic fibers that are recruited during anaerobic treadmill running.

Experiments with cultured muscle cells (L6 myoblasts) suggested that exercise associated increase in plasma levels of oxidized glutathione is perhaps mainly contributed by glutathione disulfide expelled from the active skeletal muscle[10]. Muscle cells were observed to be very active in glutathione synthesis. Using 3-O-[^{14}C]-methyl-D-glucose we determined the intracellular water content of L6 cells and thus calculated that the intracellular concentration of glutathione in the cells was ~3 mM.[11] Free radicals have been suggested to be implicated in the development of oxidative skeletal muscle fatigue. Potassium efflux is crucial for a number of physiological control processes, however, exercise induced perturbation of K^+ homeostasis in skeletal muscle is suggested to be implicated in the generation of muscle fatigue. To reveal the dose dependent effect of oxidant exposure on inward and outward K^+ (^{86}RbCl) transporting systems, skeletal muscle derived L6 cells were treated with different concentrations of the oxidant tert-butylhydroperoxide.[12] We documented that even very low doses of oxidant had remarkable specific effects on the different components of K^+ influx in the skeletal muscle derived cells. However, K^+ efflux mechanisms appeared to be rather insensitive to the extracellular oxidant challenge.[11]

In a recent human experiment[4], we investigated the association between exercise intensity and related oxidative stress in nine healthy young men who exercised for 30 mins at their aerobic and anaerobic thresholds one week apart (2nd and 3rd weeks). We also tested the effect of oral N-acetylcysteine on exercise-associated rapid blood glutathione oxidation in the subjects who performed two identical maximal bicycle ergometer exercises three weeks apart (1st and 4th weeks). A single bout of exercise, as carried out during the maximal, aerobic and anaerobic thresholds tests, induced blood glutathione oxidation. Compared to that following aerobic threshold, exercising at the anaerobic threshold was associated with a marginally higher extent of oxidative stress. A single bout of exercise at maximal, aerobic threshold or anaerobic threshold did not influence the net peroxyl radical scavenging capacity of the plasma. An association between a single bout of exercise and leukocyte DNA damage was apparent. Plasma thiobarbituric acid-reactive substances did not change following either the control or post supplementation maximal test. N-acetylcysteine supplementation resulted in an increase in pre-exercise peroxyl radical scavenging capacity indicating a higher net antioxidant capacity of the plasma but did not affect blood total glutathione. Max associated rapid decline in blood thiol redox status was markedly attenuated by N-acetylcysteine supplementation indicating that the supplementation may have spared exercise-associated blood glutathione oxidation and the thiol redox status perturbation.[4]

Current reports claimed that exogenously administered glutathione may remarkably enhance endurance to exhaustive exercise; biochemical parameters were not presented however. We[5] examined *1)* how exogenous glutathione and N-acetylcysteine may affect exhaustive exercise induced changes in tissue glutathione status and lipid peroxide, and endurance, and *2)* the relative role of endogenous glutathione in the circumvention of exercise induced oxidative stress using glutathione deficient (L-buthionine-[S,R]-sulfoximine treated, glutathione synthesis blocked) rats. Glutathione injection remarkably increased plasma glutathione and was followed by a rapid clearance. Excess post-injection plasma glutathione was rapidly oxidized. Exogenous glutathione *per se* was an ineffective delivery agent of glutathione to tissues. Following repeated (1/d, 3 d) administration of glutathione, blood and kidney total glutathione were increased. N-acetylcysteine provided a partial protection against exercise associated glutathione oxidation in the lung and blood. Exogenous glutathione or N-acetylcysteine did not influence endurance to exhaustion on the treadmill.[5] In a previous study, we observed that glutathione supplementation (0.5 g/kg body weight; i.p.) to rats did not affect exercise induced increase of immunoreactive plasma Mn-superoxide dismutase concentration.[13] Glutathione deficiency (L-buthionine-[S,R]-sulfoximine treatment) decreased total glutathione pools in the *i)* liver, lung, blood and plasma by ~50%, and *ii)* skeletal muscle and heart by 80-90%. Endurance to exhaustion (treadmill run; 10% uphill grade; 1.2 km/h) in glutathione deficiency was cut short by ~50%, suggesting a critical role of endogenous glutathione in the circumvention of exercise induced oxidative stress and as a determinant of exercise performance.[4]

In muscle biopsies taken after eccentric exercise it has been shown that neutrophils rapidly infiltrate damaged muscle tissue, where the magnitude of infiltration appeared to be proportional to the extent of z-band damage. A number of studies shows that reactive oxygen species mediated chemotactic factor appears to play a major role in communication in neutrophil-mediated inflammatory events. In the model of glutathione -supplementation, -deficiency and N-acetylcysteine supplementation we tested the role of antioxidant capacity and increased free radical formation on exercise induced leukocyte margination and on neutrophil oxidative burst activity in order to evaluate the importance of antioxidant protection in these processes. We found that glutathione and also N-acetylcysteine can induce leukocyte and neutrophil locomotion and decrease leukocyte margination, suggesting that glutathione and N-acetylcysteine supplementation may be useful in the prevention of exercise induced muscle damage via decreasing excess leukocyte infiltration into damaged tissues.

3 CONCLUSION

The study of exercise induced oxidative stress is important to maximize the benefits of physical exercise as a therapeutic tool. Exhaustive exercise triggers lipid peroxidation, protein oxidation and DNA damage,[3,4] and in these ways can actually contribute to the progress of certain diseases *e.g.*, atherosclerosis. Thus, effective strategies to control exercise induced oxidative stress should have considerable clinical implications.

Work done by CKS and colleagues was supported by research grants from the Finnish Ministry of Education, University of Kuopio, and Juho Vainio Foundation, Helsinki.

References

1. A. Goldfarb, and C. K. Sen. In: *Exercise and Oxygen Toxicity* (eds. C. K. Sen, L. Packer and O. Hänninen), Elsevier Science Publishers B.V, Amsterdam, 1994, 163-190.
2. C. K. Sen, L. Packer and O. Hänninen, 'Exercise and Oxygen Toxicity'. Elsevier Science B.V, Amsterdam: 1994 .
3. C. K. Sen, *J. Appl. Physiol.*, 1995, **79(3)**, 675-686.
4. C. K.Sen, T. Rankinen, S. Väisänen and R. Rauramaa. *J. Appl. Physiol.*1994, **76**, 2570-2577.
5. C. K. Sen, M. Atalay and O. Hänninen. *J. Appl. Physiol.* 1994 **77(5)**, 2177-2187.

6   K. Gohil, C. Viguie, W. C. Stanley, G. A. Brooks, L. Packer. Blood glutathione oxidation during human exercise, *J. Appl. Physiol* 1988, **64**, 115-119.
7   C. A. Viguie, B. Frei , M. K. Shigenaga , B. N. Ames , L. Packer, G. A. Brooks. *J. Appl. Physiol*. 1993; **75**, 566-572.
8   A. Constantinescu , D. Han, L. Packer. *J Biol Chem* 1993; **268**, 10906-10913.
9. C. K. Sen, E. Marin, M. Kretzschmar and O. Hänninen. *J. Appl. Physiol.* 1994, **73**, 1265-1272.
10 C. K. Sen, P. Rahkila and O. Hänninen. *Acta Physiol. Scand.* 1993, **148**, 21-26.
11 C. K. Sen, I. Kolosova, O. Hänninen and S. N. Orlov. *Free.Rad. Biol. Med.*, 1995, **18 (4)**, 795-800.
12. C. K Sen., O. Hänninen and S. N. Orlov.. *J. Appl. Physiol.*, 1995, **78(1)**, 272-281.
13. C. K Sen, T. Ookawara, K. Suzuki, N. Taniguchi, O. Hänninen and H. Ohno. Pathophysiology, 1994, **1,** 165-168.

EXPRESSION KINETICS OF AcLDL RECEPTOR(S) DURING THE DIFFERENTIATION PROCESS OF PERIPHERAL BLOOD MONOCYTES IN VITRO.

C. Mangoni di S. Stefano*, O. Petillo°, G.S.R.C. Mangoni di S. Stefano*, A.M. Ruggiero*, P. Grippo' and G. Peluso°

*Dpt. of Human Physiology and Integrated Biological Functions, 'Inst.of Bioch. of Macromecules -Faculty of Medicine Second University of Naples; °Biochemistry of Proteins and Enzymology Institute-CNR-Naples

1 INTRODUCTION

Generation of foam cells in the arterial wall characterizes the early stage of atherosclerotic lesions[1]. A significant number of these cells are derived from blood monocytes, which differentiate to macrophages after migrating into the sub endothelial space[2]. Besides to well described LDL receptor, macrophages are able to bind modified lipoproteins such as oxidized LDL via acetyl-LDL receptor(s) (Macrophage Scavenger Receptor MSR)[3]. The unregulated uptake of modified LDL via these receptors leads to massive intracellular deposition of cholesterol esters this results in formation of foam cells[4].

The increase of negative surface charge determine the bound to the macrophage scavenger receptors (MSRs), internalization and degradation in a lysosomal compartment. The modified-LDL degradation results in an increased cellular cholesterol content and formation of cholesteryl ester droplets in the cytoplasm. The expression of the receptor is not affected by cholesterol increase, and lipoprotein uptake continues and finally the macrophage is transformed into a lipid-loden foam cell. This is in contrast with the very high homeostatic regulation of the receptor for the native LDL[5]; in fact, this is downregulated by an excess of intracellular cholesterol. The MSR expression is clearly related to the stage of differentiation and/or activation of macrophage.

MSRs are trimeric glycoproteins that bind acetylated LDL (Ac-LDL) and other ligands via collagen-like coiled-coil domains[6]. The MSRs gene products belong to an evolutionarily conserved family of proteins with cysteine-rich extracellular C-terminal domains[7].

We studied Ac-LDL receptor expression during the differentiation process of peripheral blood monocytes *in vitro* to elucidate whether these cells could uptake Ac-LDL at the same extent in the different phase of activation.

2 MATERIALS AND METHODS

2.1 Subjects

20 healthy subjects, randomly selected from the ambulance Dietetic Service of Medicine and Surgery Faculty of the Second University of Naples, were included in this study. A basal hematological outline was made. A 10 ml sample of peripheral blood was collected and used.

20 hypercholesterolemic patients, without familiar hypercholesterolemia history were randomly selected from the patients that come to the ambulance of the same Service, and treated as above.

10 patients with epidemiological and laboratory evidences of familiar hypercholesterolemia were selected from all the Clinical and Surgical Divisions and from the Dietetic Service of Medicine and Surgery Faculty of the Second University of Naples.

A basal hematological outline was requested to include into the study. Also these patients were requested to compile a dietological chart.

2.2 Cells

Buffy coats freshly prepared from the peripheral blood of healthy donors and hypercholesterolemic patients were provided from clinical division of the Medicine and Surgery Faculty of the Second University of Naples.

A clinical and dietological cart of both healthy and pathological subject is collected. Routinely analysis were made in all groups of the subjects, particularly pointed out to the serum lipid fractions.

2.3 Lipoprotein Binding and Internalization

Ac-LDL labeled with 1,1'-dioctadecyl-3,3,3',3'-tetramethyl-indocarbocyanine perchlorate (DiI) was used for analysis of MSR activity. Cells were incubated with DiI-Ac-LDL at 5 mg/ml in medium for 30 minutes at 4°C to detect binding and for 30 minutes at 37°C to visualize internalization. The specificity was verified by competition experiments with unlabeled Ac-LDL. Cells were analyzed either by fluorescence microscopy using rhodamine filter set or by FACS (see below).

2.4 FACS

After labeling with DiI-Ac-LDL or immunostaining, cells were immediately fixed in 1% paraformaldehyde in PBS and analyzed in an FACScan flow cytometer (Becton Dickinson).

2.5 Statistical Analysis

Student's t test was used for evaluation of differences between means. P values below 0.05 were considered significant (double-tailed test).

3 RESULTS

3.1 Serum Cholesterol, Lipoproteins and Apoproteins in Normal and Hypercholesterolemic Patients with or without Familiar Hypercholesterolemia

Laboratory lipid-related outline made from all three groups of patients were reported in table 1. It is clear that differences between hypercholesterolemic subjects with or without familiar hypercholesterolemia come from the familiar history, the resistance to pharmacological care and from dietary habit revealed by dietary cart.

Table 1. *Mean serum levels of lipids in subjects included in the study*

Subjects Value +/- SD	Normal (n. 35)	Hypercholester. (n. 35)	Hyperchol+arter. occlusions (n. 14)
Cholesterol (tot) mmol/L	5.24 +/- 0.78	6.17 +/- 0.83	6.52 +/- 0.91
LDL cholesterol mmol/L	3.36 +/- 0.55	3.85 +/- 0.54	4.48 +/- 0.55
HDL cholesterol mmol/L	1.29 +/- 0.22	1.18 +/- 0.24	1.13 +/- 0.27
Apolipoprot. A$_1$ g/L	1.57 +/- 0.32	1.54 +/- 0.22	1.62 +/- 0.39
Apolipoprot. B g/L	1.24 +/- 0.33	1.21 +/- 0.32	1.59 +/- 0.27
Triglicerides mmol/L	1.38 +/- 0.22	1.63 +/- 0.27	1.83 +/- 0.24

In table 2 we report determinations made in the same groups of patients on LDL monocyte receptor and on Ac-LDL MSRs. These results demonstrate that the mean fluorescence intensity (F.I.) for labelled LDL is close to normal in non-familiar hypercholesterolemic patients, being the ratio between LDL and Ac-LDL F.I. nearly 1. In the subjects with familiar hypercholesterolemia, there is an higher Ac-LDL receptor(s) expression, with a down regulated amount of LDL receptor. It is also clear an inverse correlation between this index and the eventual arterial obstruction(s).

Table 2. *Mean fluorescence intensity of monocytes from peripheral blood cells stained with LDL and Ac-LDL*

Subjects	Normals (n. 35)	Hypercholester. (n. 35)	Hyperchol+arter.oc clusions n. 14)
LDL-receptors %	35.21 +/- 4.2	34.14 +/- 3.9	11.2 +/- 3.8
Ac-LDL receptor %	27.16 +/- 3.4	55.86 +/- 4.3	88.8 +/- 4.5
LDL/Ac-LDL ratio	1.2	0.6	0.13

N.B.: it is possible that the arterial distric interested by obstruction were single or multiple; the value of the stryke was measured by arteriography and/or ecodoppler.

3.2 MSRs Are Present in Monocytes but Not in Other Leukocytes

The MSRs in peripheral circulating leukocytes were determined by assessing the binding of Ac-LDL, labeled with the fluorochrome DiI, on cell plasmamembranes. After incubation with DiI-Ac-LDL, leukocytes freshly isolated from peripheral blood were examined by fluorescence microscopy and flow cytometry. DiI-Ac-LDL binding cells were detected, indicating the presence of MSRs on these cells. To further characterize the cell type responsible for the DiI-Ac-LDL uptake, we performed an immunophenotyping assay using several monoclonal antibodies. Almost all DiI-Ac-LDL-labeled cells were located in the monocytic population, expressing the monomyeloid antigens CD13 and CD14. However, 25% of the freshly isolated CD14$^+$ monocytes exhibited low MSR activity. The same cell showed a high level of the myeloid carbohydrate antigen CD15, indicating the presence of an MSR^{low}CD15^{high} subpopulation of monocytes. In contrast to monocytes, there was no MSR activity present in neutrophils and lymphocytes.

Figure 1 a) shows the gate for monocytes, the only subset of leukocytes positively stained by fluorescente DiI-ac-LDL. In fact when the cells are plotted in the histograms against the fluorescence intensity, as shown in figure 1 b) lymphocytes (LYM)s and polymorphonuclear cells (PMN)s revealed the same F.I. of control. In contrast, the histogram for monocytes (MON)s was shifted toward higher fluorescence intensity if compared to the control.

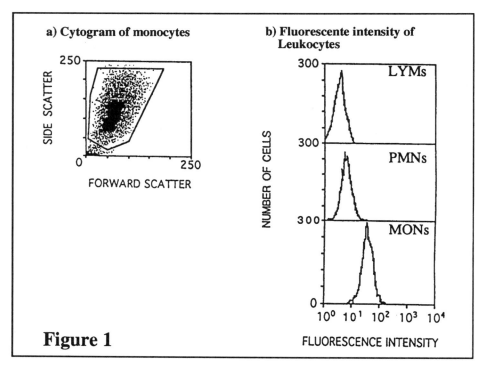

Figure 1

4 DISCUSSION

The formation of macrophage-derived, lipid-laden foam cells in the arteria intima, particularly that mediated by low density lipoprotein (LDL), is a key process contributing to the development of atheroscerosis and subsequent heart disease. Paradoxically, LDL in its native form lacks the capacity to generate foam cells from either macrophages or smooth muscle cells, implying that the receptor for native LDL is not required for foam-cell formation. This is most evident in patients with familial hypercholesterolemia who lack functional LDL receptors yet develop foam cells and atherosclerotic lesions similar to those in patients with normal LDL receptors.

MSR is a trimeric glycoprotein on the surface of mononuclear phagocytes that mediates the endocytosis of denatured and chemically modified macromolecules including modified LDL. MSR activity is increased severfold during the differentiation of a blood monocyte to macrophage. This is accompanied by an increased capacity to internalize modified LDL and by the tendency of macrophages to transform into foam cells when exposed to high concentrations of modified lipoproteins in the extracellular *milieu*.

The transformation of LDL to ox-LDL probably via Ac-LDL is a mechanism employed by the body to diminish the serum LDL level. Thus results in an increased cholesterol cell level that downregulates the LDL receptor expression and activity, and determines higher expression of modified lipoprotein receptor(s) that are independent from the cholesterol levels. The monocyte is transformed into macrophage and then in foam cell, that initiate the atherosclerotic process.

Furthermore, studies on atherosclerotic process, especially arterial plaque formation, that result in an blood flux retard in one or more arteries, demonstrated that there is a direct interconnection between LDL to Ac-LDL or Ox-LDL transformation, monocytes to macrophages evolution and MSR(s) expression and activity.

The diagnosis of familiar hypercholesterolemia comes from a sanitary familiar history, serum lipid habit and dietetic and/or pharmacological responsiveness.

Our results establish also that when the arterial obstruction strikes an artery as coronary or carotid, the only possible way of care is the by-pass operation; but, it is possible that some years after the by-pass, a new arterial obstruction can result.

In a patient reported in this study, after a by-pass operation to the left carotid artery, with an obstruction degree of 85 %, two years last a right femoral obstruction was made. Because the collateral circulation was sufficient and the LDL/Ac-LDL receptors ratio was very low, we treated the patient with an hypocholesterolemic diet, a hypocholesterolemizing pharmacological approach and an increased physical activity. At time, 5 years later, the patient conditions are still favorable.

Finally, we underline that the lipoprotein receptor activity is unimportant to the familiar hypercholesterolemia diagnosis, but assumes an high importance for prognostic and care choice (medical or surgical ?).

References

1. R. Ross, Am. J. Pathol., 1993, **143**, 987
2. J. Auwerx, Experientia, 1991, **47**, 22
3. M.S. Brown and J.L. Goldstein, Nature, 1990, **343**, 508
4. M.S. Brown and J.L. Goldstein, Ann. Rev. Biochem., 1983, **52**, 223
5. M.S. Brown and J.L. Goldstein, Science, 1986, **232**, 34
6. L. Roher, M. Freeman, T. Kodama, M. Penman, M. Krieger, Nature, 1990, **343**, 570
7. A. Aruffo, M.B. Melnick, P.S. Linsley, B. Seed, J. Exptl. Med., 1991, **174**, 94

BLOOD OXIDATIVE STRESS PARAMETERS AND THEIR CORRELATION WITH INDICATORS OF BONE MARROW RECOVERY AFTER TOTAL BODY IRRADIATION

D. Dubner, M.R. Pérez and P. Gisone.

Laboratorio de Radiopatología, Departamento de Apoyo Científico
Ente Nacional Regulador Nuclear (ENREN)
Centro Atómico Ezeiza C.C. 40 (1802) Aeropuerto Internacional de Ezeiza
Buenos Aires, ARGENTINA.

1 INTRODUCTION

In a situation of accidental irradiation with high dose total body overexposure, early diagnostic methods to assess damage in haematopoietic tissue and to predict whether recovery can occur are recquired.

Therapeutic total body irradiation (TBI), in the conditioning for bone marrow transplantation (BMT), provides a useful model of accidental irradiations. Such patients achieve a iatrogenic state of reversible bone marrow aplasia. The management of these patients is largely affected by the duration of the aplastic period. Hence, the availability of laboratory tests that reliably indicate succesful engraftment would contribute significantly to the clinical management of bone marrow transplant patients. [1]

It has been previously demonstrated that the conditioning therapy given to BMT recipients creates a high oxidant stress resulting in a reduction of endogenous antioxidants [2]

The presence of highly reactive free radicals might play a role in the peroxidation of membrane lipids. It may be particularly relevant in the setting of high dose chemotherapy or radiotherapy which depend to some degree upon the generation of free radicals for their biological effects. [3]

The purpose of this investigation was to examine the time course of lipid peroxides by the detection of thiobarbituric reactive substances (TBARS). We have prospectively measured plasma TBARS levels as a prognostic marker in patients undergoing BMT. Changes in TBARS levels were compared with antioxidant status of erythrocytes in the considered period. Results were analyzed by dividing patients in two subsets according to the clinical course of BMT. Engraftment was monitored by periodic blood cell counts and by evaluating a reticulocyte maturity index (RMI) that is considered to be the most sensitive indicator of engraftment. [1]

2 MATERIALS AND METHODS

2.1 Patients and treatments

Twenty patients undergoing BMT, 14 males and 6 females, mean age 30.4 years (range 6-54 years), were enrolled for this study. Sixteen patients underwent allogeneic BMT and three patients had autologous BMT. In one case the patient received a singeneic BMT (monozygotic twin donor). The diagnosis included 7 patients with acute myelogenous leukemia, 8 patients with chronic myelogenous leukemia, 2 patients with non-Hodgkin lymphoma, one patient with Wiscott-Aldrich syndrome, one patient with ß-talasemia and one patient with multiple mieloma.

Pretransplant conditioning regimes consisted in chemotherapy (CHT) in five patients and CHT plus radiotherapy (TBI) in fifteen patients. In these patients, TBI was administrated at 3 Gy/day for 4 days in twice daily fractionated doses, at a dose-rate of 0.04 Gy/min.

2.2 Samples Collection and Storage

Blood samples were obtained the day before starting CHT, daily during conditioning treatment and three times weekly until patient's discharge. For reticulocyte analysis, EDTA anticoagulated blood samples were stored at 4°C until 48 hs after collection. For TBARS determinations, plasma samples were frozen at -70°C until measurements. For the evaluation of antioxidant status of erythrocytes, hemolysates were stored at -18°C until Catalase (CAT) and Superoxide Dismutase (SOD) activities were measured.

2.3 Plasma TBARS Levels

Plasma levels of lipid peroxides were estimated in terms of TBARS. Briefly, the elimination of TBA-reacting substances other than lipid peroxides in plasma was performed by the treatment with phosphotungstic acid-sulfuric acid system according to the procedure of K Yagi. [4] The standard TBA assay was carried out according to the method of Ohkawa et al. [5] with the modifications of Kikugawa et al. [6] with fluorometric measurement at 553 nm with 515 nm excitation. Results were expressed as nmol/L.

2.4 Antioxidant Status of Erythrocytes

SOD activity was determined in hemolysates, using a modification of epinephrine-adenochrome detection system. [7] Briefly, the reaction mixture contained buffer glycine 50 mM (ph 10.0 adjusted with NaOH) and epinephrine bitartrate 60 mM (ph 2 adjusted with Hcl). The inhibition of adenochrome formation was following at 480 nm. The activity was expressed in units/ mg of protein.

CAT activity of erythrocytes was measured on aliquots obtained as hemolysates and expressed as the first-order kinetic constant of the rate of disappearance of hydrogen peroxyde, measured by absorbance at 240 nm, according to Aebi. [8] Results were expressed as nmol/mg of protein. Protein concentration was determined by Lowry method.

2.5 Engraftment Monitoring

2.5.1. Blood cells counts: blood counts changes were automated evaluated three times a week. A post-BMT increase in absolute neutrophil count to > 50/µL in two consecutive samples, was considered as an indicator of mielopoiesis recovery.

2.5.2. Flow cytometric reticulocyte analysis: reticulocyte percentage and absolute reticulocyte count were determined by flow cytometry according to Davis. [9]

The proportion of reticulocytes with high fluorescence intensity referenced to a cursor position prior established in samples from a normal control population, was used to derive an reticulocyte maturity index (RMI). A RMI increase by 20 % from the aplasic nadir was considered as indicator of erythropoietic recovery. [1]

2.6 Statistical Analysis

Statistical analysis of the results were performed using the Wilcoxon matched pairs signed rank sum test and one way ANOVA test.

3 RESULTS

Plasma TBARS levels were significantly increased immediately after TBI compare to before starting conditioning treatment in all patients (n=20, day -9 vs day 0 p< 0,05).

In patients who presented a succesful BMT (n=16), plasma TBARS levels returned towards baseline levels between the first and the second week (day 0 vs day 14 p < 0,05).(Figure 1A).

Patients who presented unfavorable post-BMT course (n=4), lipoperoxides levels increased in the post-transplant period. This increase was particularly evident during the third week (day 18 vs day 0 p < 0,05) (Figure 1B).

Changes in the antioxidant status in the erythrocytes were not significant in the two studied groups along the considered period,

Temporal behaviour of SOD and CAT activities are shown in figures 2 and 3.

The RMI indicated engraftment earlier than neutrophil count (17,5 ± 4,9 days vs 21,4 ± 5,3 days, p < 0,05).

There was no correlation with results and the conditioning regime (CHT vs. CHT + TBI).

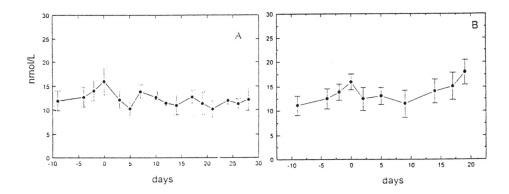

Figure 1. *TBARS values in BMT patients (• mean ± S.D.)*
A) Patients with succesful engraftment(n=16): values on day -9 (11.9 ±2.1 nmol/L) are significantly lower (p<0.05) from values on day 0 (16.0 ±2.7 nmol/L).
B) Patients with unfavorable evolution (n=4): TBARS levels increased significantly (p < 0.05) during the third week. Day -9 (11.1 ±2.0 nmol/L), day 19 (18.1 ±2.8 nmol/L.

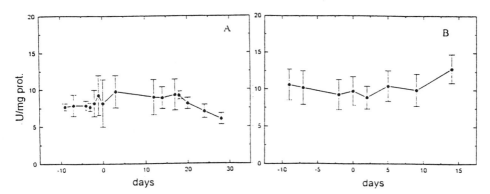

Figure 2. *Erythrocyte SOD activity (• mean ± sd) did not change significantly during conditioning therapy and after BMT.*
A) Patients with succesful engraftment (n=16).
B) Patients with unfavorable evolution (n=4).

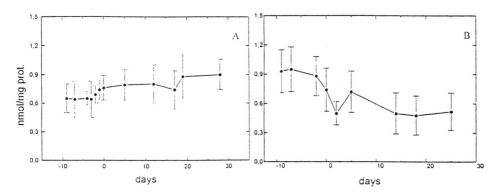

Figure 3. *Erythrocyte CAT activity (• mean ± sd)*
A) Patients with succesful engraftment(n=16). Values did not change significantly during conditioning therapy and after BMT.
B) Patients with unfavorable evolution (n=4).The changes were not significant but data showed a trend to decrease during the post transplant period

4 DISCUSSION

Since the first radiation accidents which resulted into severe health effects, great progress has been made in the medical management of disease such as aplasia. After whole body overexposure, and whichever being the therapeutic decision: bone marrow transplantation or other strategies, an early indicator or predictor of bone marrow recovery failure may allow the introduction of appropiate treatment. [10]

A role for free radicals has been proposed for the pathogenesis of radiation injury [11,12] as well as for the deletereous effects of chemotherapeutic agents.[13] Iron overload seen in these patients must be too considered as a possible mechanism for oxidative damage[3].

It is assumed that lipid peroxidation (LPO) can be set into motion whenever conditions of increased oxidative stress occur in the cell. At the level of cellular membranes, peroxidation of phospholipid fatty acid is a mechanism generally recognized of most importance[14].

The content of TBARS has been frequently used to assess lipid peroxide formation. It has been proposed that the increase in plasma TBARS may be related to endothelial and epithelial cells damage[3].

We have observed an early increase in LPO plasma levels suggesting a free radical-induced oxidative stress during the myeloablative therapy. Baseline TBARS levels were regained after 2 weeks post-transplantation in the subset of patients who presented a succesful engraftment. This fact suggest a temporal correlation between the return towards basal TBARS levels and the moment of bone marrow recovery determined by RMI.

On the other hand, some authors have shown changes in antioxidant status of erythrocytes in different pathological situations [15, 16]. Erythrocytes are exposed to oxygen radicals to a greater degree than most other cells.

Red cells CAT and SOD activities were not significantly modified by the oxidative challenge during the studied period. Since CAT handles about half of the generated H_2O_2 and the glutathione peroxidase (GPX)/reductase mechanism account for the other half [17], the interpretation of these results require caution and measuremens of GPX and glucose-6-phosphate dehydrogenase should be carried out.

Data obtained in this study suggest that LPO may work as a prognostic marker. Further studies involving more patients should be undertaken in order to confirm these results and to investigate the relationship between other parameters of oxidative stress and the clinical evolution in patients undergoing BMT.

Acknowledgements

This work was partially supported by a Research Contract from IAEA (8368/RB)

References

1. BH Davis,N Bigelow, E D Ball, L Mills and G G Cornwell, *Am J Hematol*, 1989, **32**, 81.
2. A Hunniset, S Davies,J Mc Laren-Howard, P Gravett, M Fin, D Gueretwardle, *Biol Tr Elem Res.*, 1995, **47**, 125.
3. LI Gordon, SG Brown, M S Tallman, A W Rademaker, S A Weitzman, H M Lazarus et al., *Free Rad Biol Med.*, 1995, **18**, 383.
4. K Yagi, Methods in Enzymology, *Academic Press, Inc.*, 1984, Vol 105, p. 328.
5. H Ohkawa, N Ohishi, K Yagi, *Anal. Biochem*, 1979, **95**, 351.
6. K Kikugawa, T Kojima, S Yamaki, H Kosugi, 1992, *Anal Biochem*, **202**,249.
7. A Boveris, CG Fraga, AI Varsavsky and O Koch, *Arch Bioch Bioph* ,1983, **227**,534.
8. H Aebi. Methods in Enzymology, *Academic Press, Inc.*, 1984, Vol 105, p.121.

9. BH Davis and NC Bigelow, *Clinical Flow Cytometry, NY Acad Scie*, 1993, Vol 677, p. 281.
10. S.V. Davies, I. Cavill, N. Bentley, C. D. Fegan and J.A. Whittaker, *Br. J. Haematol.*,1992, **81**,12.
11. P.A. Riley, *Int. J. Radiat. Biol.*, 1994, **65**, 27.
12. J.P. Kehrer, *Crit. Rev. Toxicol.*, 1993, **23**, 21.
13. M. Dürken, J. Agbenu, B. Finckh, C. Hubner and A. Kohlschutter, *Bone Marrow Transplant*, 1995,**15**, 757.
14. M. Comporti, *Lab Invest*, 1985, **53**, 599.
15. C. Reides, M. Repetto and S. Llesuy, Proc. Int. Symposium Oxygen Radicals in Biochem. Bioph. Med., 1994, p 32.
16. E. Peuchant, M.A. Carbonneau, L. Dubourg, C. Vallot and M. Clerc, *Free Radic. Biol. Med.*, 1994,**16**, 339.
17. G. Gaetani, H. Kirkman, R. Mangerini and A. Ferraris, *Blood*, 1994, **84**, 325.

COHORT STUDY ON SERUM LEVELS OF LIPID PEROXIDE AND SUPEROXIDE DISMUTASE AND CANCER MORTALITY IN JAPAN

R.Sasaki [1], Y.Ito [2], S.Suzuki [1], R.Shinohara [2], K.Yagyu [1], N.Endo [1]

1: Dept. of Public Health, Aichi Medical Univ., Nagakute-cho, Aichi, 480-11, Japan
2: School of Health Science, Fujita Health Univ., Toyoake-shi, Aichi, 470-11, Japan

1 INTRODUCTION

Many styles of epidemiologic and experimental studies to examine the protective effects for cancer of antioxidants such as carotenoids have shown favorable results except one report from Finland in 1994 [1]. This negative results [1] from chemoprevention study using beta-carotene to heavy smokers in Finland conducted by The Alpha-Tocopherol, Beta Carotene Cancer Prevention Study Group surprised many researchers in the fields of antioxidant studies in the world. This report showed harmful effects of high dose of beta-carotene to male smokers. On the other hand, this report shows, as another report showed, within control group low incidence rate of lung cancer in the group of higher levels of serum beta-carotene at the entry.

It is natural to consider from these results that effects of antioxidants to cancer prevention may be doubtful. So to confirm the occurrence relation between serum levels of lipid peroxide (LPO) and superoxide dismutase (SSOD) and cancer mortality, cohort study was conducted in northern part of Japan.

2 SUBJECTS AND METHODS

609 healthy Japanese (279 males and 330 females) aged 40 years and older who participated health check-up program in northern part of Japan were enrolled in this cohort study after getting informed consent for this study in 1982.

As shown in table 1, 58.1% among males and 4.9% among females were current smokers, and percentage of regular drinker were 63.4% for males and 13.9% for females, respectively.

Using fasting sera of participants, levels of LPO and SSOD were detected by XOD-NH_2OH method of R.Shinohara [2]. Mean levels of LPO at entry were 5.1 (S.D. 1.75) μmol/l for males and 5.0 (1.34) μmol/l for females, and those of SSOD were 0.26 (0.13) μg/ml and 0.26 (0.13) μg/ml, respectively.

Participants were followed-up untill the end of 1993 using special register system of Japan named "Koseki" after getting the permission from government. During follow-up periods, 39 males and 13 females were died, 23 from cancer (stomach 6, colon 3, liver 3, pancreas 3, lung 2 and other sites 6), 9 from circulatory diseases, 7 from accidents and 13 from other causes of diseases.

Role of Oxidative Stress in Atherosclerosis and Cancers

To examine occurrence relation between levels of LPO and SSOD and mortality, subjects whose levels of LPO and SSOD were highest quarter classified as high group and lowest quarter as low group and the others as middle group. Using Cox's proportional hazard model, age, smoking and alcohol drinking adjusted hazard ratio were calculated.

Table 1. *Characteristics of Subjects at entry to cohort study*

		Male N=279	Female N=330
Age at entry	-39 (yrs.)	55 (19.7%)	63 (19.1%)
	0-49	46 (16.5%)	55 (16.7%)
	50-59	81 (29.0%)	128 (38.8%)
	60-	97 (34.7%)	84 (25.4%)
Smoking status	never smoker	99 (35.5%)	311 (94.2%)
	ex-smoker	18 (6.5%)	3 (0.9%)
	current smoker	162 (58.1%)	16 (4.9%)
Alcohol drinking status	never drinker	90 (32.3%)	283 (85.8%)
	ex-drinker	12 (4.3%)	1 (0.3%)
	current drinker	177 (63.4%)	46 (13.9%)

Table 2. *Serum levels of LPO and SSOD and cancer death in Japan*

				all death except accidents		cancer death	
			No.of subj.	No.of death	H.R.(95% C.I.)*	No.of death	H.R.(95% C.I.)*
Male	LPO	low	81	6	1	3	1
		middle	121	18	1.93 (0.74-5.03)	10	2.11 (0.56-8.02)
		high	77	10	1.59 (0.55-4.59)	6	1.50 (0.35-6.36)
	SSOD	low	80	10	1	7	1
		middle	125	15	0.79 (0.35-1.77)	9	0.61 (0.22-1.73)
		high	74	9	0.67 (0.26-1.71)	3	0.26 (0.06-1.13)
Female	LPO	low	96	1	1		
		middle	151	6	2.87 (0.33-25.0)	1	0.47 (0.03-8.06)
		high	83	4	3.24 (0.35-24.8)	2	2.06 (0.18-24.1)
	SSOD	low	114	5	1	4	1
		middle	136	4	0.28 (0.05-1.45)	0	
		high	80	2	0.27 (0.05-1.54)	0	
Both	LPO	low	177	7	1	4	1
		middle	272	24	2.35 (0.99-5.58)	11	1.66 (0.52-5.34)
		high	160	14	1.89 (0.73-4.87)	8	1.53 (0.44-5.29)
	SSOD	low	194	15	1	11	1
		middle	261	19	0.69 (0.35-1.38)	9	0.41 (0.16-1.04)
		high	154	11	0.59 (0.27-1.34)	3	0.18 (0.05-0.71)

* Cox's proportional hazard ratio adjusting age, smoking and drinking status

3 RESULTS

As shown in table 2, Cox's proportional hazard model suggested following results.

1. High levels of serum LPO may cause harmful effects to mortality of all causes of death except death from accidents.
2. High levels of serum SOD might play some role in preventing cancer death as well as all causes of death except death from accidents.
3. Neither serum levels of LPO and SOD showed any causal relation between death from accidents.

4 DISCUSSION

As far as I found, this is the first cohort study to examine the occurrence relation between serum levels of LPO and SOD and cancer death. But unfortunately, before to confirm the hypothesis that we intended in this study, there will be some problem to solve in future. First problem is concerning too small size of this cohort study. At the first step of this study, because a lack of the data, we could not estimate suitable sample size. To solve this problem at least more than 5 years of follow-up will be need.

Second point to discuss is as to the meaning of "serum levels of LPO and SOD" which measured in this study. We could not find longitudinal study that showed stability or moveability of serum levels of LPO and/or SOD. To examine this point, serum levels of LPO and SOD of 96 subjects among survivors in this study were detected in 1993 and compared those in 1982. Results from this study will be published near future, but in brief, serum levels of LPO was stable as serum levels of GOT and SOD as serum level of total protein.

However because of small size this study could not confirm the hypothesis, it is suggested that lower serum levels of LPO and higher serum levels of SOD might play some role lengthening cancer death as well as death from aging.

Reference

1. The Alpha-Tocopherol, Beta Carotene Cancer Prevention Study Group, *N Engl J Med*, 1994, **330**, 1029.
2. R.Shinohara, *Kensa to Gijutsu*, 1990, **18**, 1608 (in Japanese).

PEROXIDATION OF PROTEINS BY PEROXYNITRITE

M. Lacsamana and J. M. Gebicki

School of Biological Sciences, Macquarie University, Sydney 2109, Australia

1 INTRODUCTION

There is much evidence in support of the notion that oxygen and its reactive derivatives play a significant role in initiation and potentiation of various forms of biological damage.[1] While there is little doubt that such intermediates are generated *in vivo* and that their reactions can lead to detectable biological changes, the relative importance of the many possible interactions of biomolecules with reactive oxygen species is not known.

Most of the attempts to discover sensitive biological targets for oxidants have involved studies of the oxidation of lipids, because of their potential ability to generate high yields of stable and reactive oxidised products. It is, however, far from certain that lipid oxidation constitutes the most significant damaging process for living organisms exposed to oxidants.

We have recently shown that oxidation of proteins can result in formation of protein peroxides.[2,3] The peroxides can oxidise ascorbate and glutathione and their decomposition leads to the release of a range of free radicals,[4] so that formation of protein peroxides *in vivo* could result in significant biological damage. Several oxidants were shown to be effective in peroxidation of a range of proteins: the hydroxyl free radical, products of the Fenton reaction, hypochlorite, and oxidants generated by activated neutrophils.[3,5] All of these are known to form in living organisms, especially in localised regions subjected to oxidant stress. In this paper we report that peroxynitrite, a known oxidant produced in living organisms in a reaction between the superoxide and nitric oxide free radicals,[6] can also induce the formation of peroxide groups in several proteins.

2 MATERIALS AND METHODS

Solutions were made up in acid-cleaned glassware in water purified by passage through a Millipore (Sydney) filtration system. Fatty acid-free bovine serum albumin (BSA, Cohn fraction V) and other proteins were from Boehringer-Mannheim (Sydney). All other chemicals and solvents were obtained from the Sydney agents of Merck, BDH or Malinckrodt and were of analytical or HPLC grade.

Peroxynitrite was synthesised essentially as recommended by Beckman et al.[6] Solutions of 0.6 M HCl/0.7 M H_2O_2 and 0.6 M $NaNO_2$ flowing at 18.3 ml/min were mixed in a 3mm diameter tube and the reaction quenched with 1.4 M NaOH 2 cm from the point of mixing. Excess peroxide was removed with MnO_2 and the mixture was fractionated by freezing at -20º. Concentration of peroxynitrite obtained in a yellow top layer of the frozen solution was determined from 302 nm absorbance, using a molar extinction coefficient of 1670 $M^{-1}cm^{-1}$ and acidified peroxynitrite solution as blank. Solutions containing 5 mg/ml of proteins were made up in 0.5 M phosphate buffer pH 7.3. Peroxynitrite was added from a microsyringe to rapidly stirred protein solutions, usually to a final concentration of 15 mM. The mixture was dialysed overnight at 4º against 10 mM phosphate. Protein peroxides were measured by the tri-iodide method.[7]

3 RESULTS

Solutions of BSA treated with peroxynitrite and then dialysed extensively, as outlined above, gave a positive peroxide test. In chromatography on Sephadex-G20, the peroxide-positive and the protein fractions eluted together (data not shown), indicating that the peroxynitrite-induced peroxide groups were attached to the protein. This agrees with earlier results of protein peroxidation by free radicals generated by γ radiation and by other oxidising systems.[3] Qualitatively similar results were obtained with human and dog albumins and with lysozyme, insulin, ovalbumin and lactoglobulin.

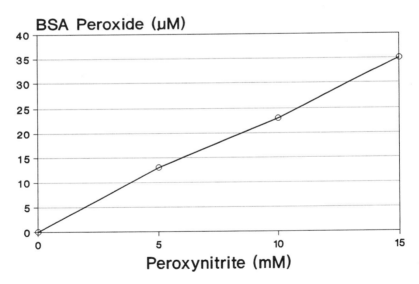

Figure 1 *Oxidation of BSA by peroxynitrite. 10 mg/ml protein in 0.5 M phosphate buffer was mixed with peroxynitrite to final concentration of 15 mM.*

The effect of increasing concentrations of peroxynitrite on the amounts of BSA peroxides formed are shown in Figure 1. The data points are averages of 4 experiments which gave standard deviations of up to 22%. This relatively low reproducibility was due to subtle differences in the mechanics of mixing of the protein and peroxynitrite solutions, which were difficult to control. Increasing the final peroxynitrite concentration beyond the 15 mM shown did not lead to corresponding linear increases in the amounts of BSA peroxide produced. This suggests that all of the available oxidant derived from peroxynitrite was scavenged by the protein. Similarly, 10 mg/ml concentrations of BSA were sufficient to react with all of the oxidant produced (resuts not shown).

As in our previous studies of protein peroxides,[3] we confirmed the identity of the protein peroxide groups by their response to reducing agents. Solutions containing 10 mg/ml of BSA oxidised by peroxynitrite were treated with the reductant (10 mM, except for 20 mM ascorbate) for 2 hrs at 20°. This was followed by dialysis and peroxide assay. The results (Figure 2) show the protein peroxide remaining after this treatment as % of the amount of peroxide in control solutions exposed to buffer instead of the reducing agent.

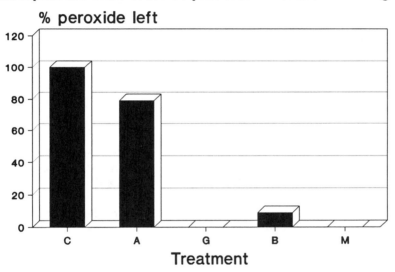

Figure 2 *Decomposition of BSA peroxide groups. 10 mg/ml of the oxidised protein was treated with buffer (C), 20mM ascorbate (A), or 10 mM GSH (G), borohydride (B) or mercaptoethanol (M). Data shows concentrations of peroxides remaining as % of control. G and M had no peroxides detectable.*

The proposition that the effective oxidant derived from peroxynitrite is produced only after protonation[6] was studied by examining the effect of pH on the amounts of protein peroxides formed. The average results of 3 experiments are shown in Figure 3. The highest standard deviation was 14%. In one series of tests, the protein was mixed with peroxynitrite at pH below 8, followed by dialysis at pH 7. In a second series, the dialysis pH was 11. In other tests, the pH at the time od mixing was 11, followed by dialysis at pH 7 or 11. The results show that the highest amounts of protein peroxides were obtained when the pH at the time of reaction and dialysis was near neutrality. The lowering of the yields of peroxides by dialysis at pH 11 suggests accelerated peroxide decomposition in alkaline solutions.

An experimental test of the possibility that the effective oxidising agent in these studies was the hydroxyl free radical derived from peroxynitrous acid[6] was carried out by performing the protein-peroxynitrite reaction in presence of scavengers of this radical.

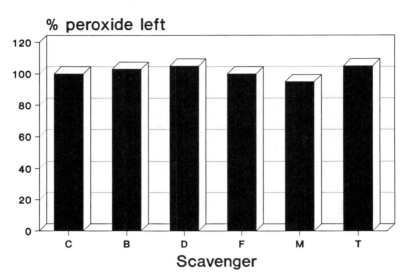

Figure 3 *The effect of scavengers of the HO· radicals on the peroxidation of BSA by peroxynitrite. Scavenger concentration was 10 mM. C - control; B - benzoate; D - dimethyl sulphoxide; F - formate; M - mannitol; T - tert-butanol.*

The concentration of the scavengers was 10 mM. The average results of their influence on the amounts of BSA peroxides produced by peroxynitrite in two experiments are shown in Figure 4.

4 DISCUSSION

The results obtained in this study show that all the proteins tested can be peroxidised by an oxidant derived from peroxynitrite. While not all the tests identifying the new chemical groups formed on the proteins treated with peroxynitrite have so far been carried out, all the results obtained are consistent with their peroxide function. They are decomposed by the reducing agents (Figure 2) and give a positive tri-iodide test under conditions applicable to protein peroxides (Figure 1). At present, it is not known on which of the constituent amino acid residues the peroxide groups are located. In our earlier work, only 6 amino acids gave high peroxide yields: glutamic acid, lysine, valine, proline, leucine and isoleucine.[3] It is likely that those are also the residues suscceptible to peroxidation by peroxynitrite derivatives, but this needs experimental testing because the oxidising agents used in the earlier and current studies were different. In previous work, the hydroxyl free radical was identified as the agent responsible for protein peroxidation.[3] Here, results shown in Figure 4 appear to exclude this oxidant. This agrees with recent calculations of the free energy of the homolytic formation of HO· from HONOO at pH 7, giving the value of 21 kcal/mol.[6] The small amounts of protein peroxides generated compared to the amounts of peroxynitrite used here suggest that the oxidant generated from HONOO is a minor product of its decay in aqueous solutions.

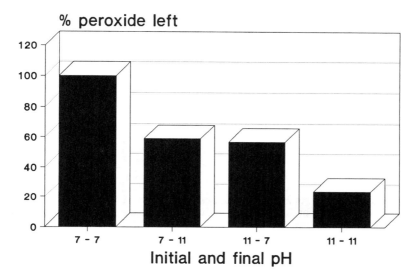

Figure 4 *The influence of pH on the amounts of BSA peroxides formed. Numbers below the bars show the pH at the time of addition of peroxynitrite to the protein solution (initial) and during dialysis (final).*

The possibility that protein peroxides formed from HONOO produced *in vivo* are an intermediate in the processes linking formation of ONOO⁻ and biological damage needs much further study. However, there is evidence that conditions favouring release of the precursors of ONOO⁻, the superoxide and nitric oxide free radicals, produce biochemical and biological effects.[8] The high probability of interaction of any reactive species formed in cells with proteins suggests that studies of the products of such reactions may disclose new pathways linking the generation of primary oxidants with their biological end effects.

References

1. B. Halliwell, J. M. C. Gutteridge and C.E. Cross, *J. Lab. Clin. Med.*, 1992, **119**, 598.
2. J. A. Simpson, S. Narita, S. Gieseg, S. Gebicki, J. M. Gebicki and R. T. Dean, *Biochem. J.*, 1992, **282**, 621.
3. S. Gebicki and J. M. Gebicki, *Biochem. J.*, 1993, **289**, 743.
4. M. J. Davies, S. Fu and R. T. Dean, *Biochem. J.*, 1995, **305**, 643.
5. A. V. Babiy, S. Gebicki and J. M. Gebicki, "Free Radicals: From Basic Science To Medicine", Birkhauser Verlag, Basel, 1993, p. 340.
6. Koppenol W. H., Moreno J.J., Pryor W.A., Ishiropoulos H. and Beckman J. S., *Chem. Res. Toxicol.* 1992, **5**, 834.
7. M. Hicks and J. M. Gebicki, *Anal. Biochem.* 1979, **99**, 249.
8. Darley-Usmar V., Wiseman H and Halliwell B., *FEBS Lett.*, **369**, 131.

DIETARY n-3 POLYUNSATURATED FATTY ACIDS SUPPRESS
REPARATIVE REGENERATION OF THE RAT LIVER CONNECTIVE TISSUE

A. Arend[1], M. Zilmer[2] and K. Zilmer[2]

Departments of Anatomy[1] and Biochemistry[2]
University of Tartu
Tartu EE2400, Estonia

1 INTRODUCTION

The n-3 and n-6 PUFAs are considered essential fatty acids (EFAs) because they cannot be synthesized in humans. It is well known that the linoleic acid (LA,n-6) is necessary for biosynthesis of arachidonic acid (AA,20:4 n-6), eicosatrienoic acid (dihomoγ-linolenic acid, DGLA,20:3 n-6) whereas the alpha-linolenic acid (ALA,n-3) is precursor for eicosapentaenoic acid (EPA,20:5 n-3). All abovementioned C_{20} PUFAs, in turn, are the immediate precursors for several short-lived regulatory lipid mediators collectively named as eicosanoids (prostaglandins, thromboxanes, leukotrienes, lipoxines). These eicosanoids exhibiting a broad spectrum of vaso- and immunoregulatory actions by their effects on blood vessel and inflammatory cells are formed mainly from AA via cyclo-oxygenase and lipoxygenase pathways. It is known that the metabolites of the n-3 and n-6 fatty acids can interfere with metabolism of an other. For example, EPA competitively inhibits AA metabolism and gives rise of the 3-series prostaglandins (PGs) and the 5-series leukotrienes (LTs). These eicosanoid analogues have diverse biological activities when compared to AA-derivates[2]. The n-3 PUFAs are much more effective in inhibiting n-6 PUFAs metabolism than vice versa[1]. In summary, all eicosanoids formed in human body are generated from PUFAs that must be derived from the diet[3] and different dietary intake of n-3 and n-6 PUFAs causes different content/spectrum of PGs in body.
  Despite of the considerable early enthusiasm about the beneficial effects on symptoms of atherosclerosis, heart diseases and inflammation several reports have appeared showing that fish oils may have undesirable effects[1]. It is possible that dietary intake of oils rich in n-3 PUFAs is followed by increased lipid peroxidation in tissues which, in turn, underlays this undesirability.
  Inflammatory response following any type of tissue injury is normal event of repair process. Wounding is immediately followed by coagulation, altered vascularity, and inflammation. Acute phase of inflammation is characterized by vasodilatation, increased vascular permeability, migration of neutrophils and then monocytes into the wound. Proliferation of fibroblasts in the wound starts regeneration or repair phase of inflammation. These two phases overlap and therefore are not sharply demarcated[4-6]. The connective tissue proliferation is also used to take as hall-mark for chronic inflammation[7]. Eicosanoids synergistically with other factors are involved in the regulation of inflammation and connective tissue proliferation and, thus, play an important role in the regulation of wound healing and may be involved in several chronic inflammatory diseases such as rheumatoid arthritis, systemic lupus erythematosus, systemic sclerosis or psoriasis[7-11].
  On the ground of the above mentioned aspects as: (i) possible undesirable effects

of rich n-3 PUFAs dietary intake, (ii) phenomenon that all eicosanoids formed are generated from PUFAs that must be derived from the diet whereas directed dietary intake of n-3 or n-6 PUFAs causes different spectrum of PGs in the body, (iii) our previous works[12, 13] which indicate the crucial role of PGs in the regulation of the connective tissue proliferation, the purpose of the present study was to investigate the effects of dietary PUFAs (n-3 and n-6) on the connective tissue proliferation (wound healing).

2 MATERIAL AND METHODS

2.1 Animals and Feeding Procedures

Experiments were performed on 17 male albino Wistar rats (weight 220-250 g). The first group of animals served as a control-group kept on the standard diet [RM1(C) 3/8" pellets, 801183W, Special Diet Services (SDS) and PMI Feeds]. The pellets contain low levels of fatty acids (0.13% arachidonic acid, 0.06% linolenic acid, 0.71% linoleic acid, and 0.76% oleic acid). The second group (n-6 group) was on the diet where to the standard RM1 pellets was added 10% of sunflower oil (100% pure refined sunflower oil, Unoli, AOH Algemene Oliehandel, Holland). The third group (n-3 group) was similarly fed with RM1 pellets with 10% addiction of cod liver oil (Peter Müller, Norway; containing over 25% of n-3 fatty acids). The animals were fed with above mentioned diets (fresh-made daily) for three weeks before and throughout experiment period. All animals had free access to food and tap water. The guidelines for the care and use of the animals were approved by the Ethical Committee of the University of Tartu.

2.2 Test Model

Test model used was a thermic liver wound of a standard size (10x1x1 mm) which has been used for decades in the University of Tartu[14]. Surgical procedure was performed under general ether anaesthesia. The wounds (two in each animal) were made after laparotomy by the application of the galvanic cauter's needle to the surface of the liver. This particular test model is suitable for investigating the connective tissue proliferation because in the liver of experimental animals (white rats, guinea pigs, and rabbits) there is relatively small amount of pre-existing connective tissue. Therefore starting from Day 3 and especially on Day 6 the proliferating connective tissue forms a zone distinctly delimited from the central necrotic focus from one side and the hepatic parenchyma from the other side. The above described circumstances allow easy and distinct histomorphological measurements of the width of the connective tissue zone.

2.3 Sampling

Animals were sacrificed on Day 6 by decapitation and as much blood as possible was collected. After the opening of abdomen the liver was rapidly excised. One wound was taken for the lipid peroxidation measurements and from the second wound one half was taken for measurements of tissue PGs and the other half for morphological investigations. Tissues, intended for PGs determination, were immediately chilled in ice-cold indomethacin-solution in 0.154 M NaCl. After blotting on filter paper tissues were weighed and then immersed in liquid nitrogen. Samples were kept at -70° until extraction of PGs. Samples from non-affected lobe of liver were treated in similar way and run in parallel.

2.4 Extraction of prostaglandins

Tissues were homogenized in glass-glass homogenizers after adding 2.0 ml phosphate-buffered saline (10 mM, pH 4.0). The homogenate obtained was applied to a C18 solid phase extraction cartridge (Bond-Elute, Harbor City, Analytichem, CA) together with four 2.0 ml washings of the homogenizer with the same buffer. The cartridges were rinsed from proteins and neutral lipids by eluting with 5.0 ml of filtered water and 5.0 ml HPLC-grade 10% (v/v) acetonitrile at a constant flow-rate of approx. 500 µl/min using a vacuum-station. PGs were then eluted with 4.0 ml HPLC-grade methanol, collected in siliconized glass test-tubes and taken to dryness under nitrogen.

Care was taken to ensure that any remaining PGs were rinsed from walls of the homogenizer by washing twice with 2.0 ml of methanol and subsequently run through the column. After evaporation, dried extract residues were redissolved in 1.0 ml Tris-HCl (10 mM, pH 7.4, with 0.1% w/v gelatin added), vortexed, and stored at -70° C until further analysis.

2.5 Radioimmunoassay of prostaglandins

PGs antibodies were purchased from Advanced Magnetics Inc., MA. Standards for the standard-curve were assayed in triplicate and samples in duplicates. 100 µl of thawed sample, 100 µl of antibody-solution in Tris-HCl (10 mM, pH 7.4, with 0.1% w/v gelatin) and 100 µl (approx. 5000 cpm) of respective tritiated tracer were incubated for four hours in room temperature in a 5 ml polyethylene minivial. At the end of incubation, 500 µl of a sepharose suspension with covalently linked anti-rabbit-IgG (Pharmacia Ltd, Uppsala, Sweden) were added and tubes incubated for 30 minutes. After centrifugation (10 min, 400 x g) tubes were left upside down for 15 min since the sepharose gel firmly adhered to the minivial bottom. 500 µl of 0.8 M NaOH was added and tubes were vortexed. 4.0 ml of scintillation-fluid was added and samples were counted for 5 min. Extraction blanks were assayed in each RIA giving values below the sensitivity of the standard curve. Also in each RIA before and after samples, controls in duplicate with low, medium and high concentration of PGs were assayed. As recovery with this extraction procedure exceeds 90%[15] the tissue content of PGs was not corrected for procedural losses.

2.6 Lipid peroxidation assay

Liver wounds were excised and washed by ice-cold 1.15 % KCl. Then 80-100 mg tissue was homogenized by glass-teflon homogenizer in 10 volumes ice-cold 1.15 % KCl, containing antioxidant butylated hydroxytoluene, BHT (v/v ratio 11 µl : 1000µl KCl) to suppress artifactual changes during handling and assay procedures. Lipid peroxidation was assessed via the level of thiobarbituric acid (TBA) reactive substances (TBARS) by method compiled on the ground of method Okhawa et al.[16] and our experiments[17]. Samples (250µl+ 250µl 0.9% NaCl) were incubated at 37° C for 30 min, 0.25% BHT (15µl) was added to interrupt the reaction and pH of mixture was reduced with acetate buffer to 3.5. Then 1% TBA (1000µl) solution was added, heated at 80° C for 70 min and after carefully cooling (5 min) the reaction mixture was acidified by 500 µl 5N HCl. After extraction with cold butanol (1700 µl), samples were centrifuged (for 10 min at 3000 rpm) and absorbance of butanol fraction was measured at 534 nm. The peroxidizability of lipids was assessed via iron induced TBARS (Fe-TBARS) as follows: 100µl pro-oxidant (475µM $FeSO_4$) was added to sample (250µl+ 150µl 0.9% NaCl), incubated (37° C, 30 min) and after adding 0.25 % BHT (15µl) assessment follows as for TBARS. All measurements of TBARS and Fe-TBARS were performed in duplicate, the mean calculated and used for statistical analysis.

2.7 Protein determination

100 μl of the homogenate was precipitated in 1.0 ml 10% (w/v) trichloroacetic acid overnight and centrifuged. The protein content was determined by the method of Lowry et al.[18], using bovine serum albumin as a standard.

2.8 Morphological investigations

Tissue samples for morphological investigations were fixed in 10% neutral formalin, embedded in paraffin, stained with hematoxylin-eosin and by the method of van Gieson. Ten measurements from 2-3 different sections of the width of the connective tissue zone in the liver wound were performed in each animal, the mean calculated and used for statistical analysis.

2.9 Statistics

Values are presented as mean±SEM. Significant differences between control and experimental group were evaluated by the Student t-test.

3 RESULTS

In the n-3 group (animals fed with n-3 PUFAs rich diet) compared with the control group significant retardation of the connective tissue proliferation as decided by the width of the connective tissue zone on Day 6 (Table 1) in the thermic liver wound was seen. At the same time the n-6 group (n-6 PUFAs rich diet) and the control group do not reveal significant differences in the width of the connective tissue zone. In addition, the suppressive effect of n-3 PUFAs rich diet on the connective tissue proliferation is obvious also when compared to the n-6 group (Table 1).

Table 1 *Effects of different diets on the connective tissue proliferation (width of the connective tissue zone) and on the level of lipid peroxidation (TBARS) and on the peroxidizability of lipids (Fe-TBARS) in the liver wounds on Day 6*

Treatment groups	No. of rats	Width of the connective tissue zone, μm	TBARS μmol/g protein	Fe-TBARS
Control	6	90.16±8.07	1.99±0.16	1.38±0.24
N-6	5	82.76±7.58	2.00±0.14	1.10±0.14
N-3	6	43.73±6.35[1,4]	3.14±0.32[2,5]	2.94±0.46[3,6]

Significant differences versus control group are indicated as [1] $P<0.001$; [2] $P<0.01$; [3] $P<0.02$; versus n-6 group as [4,5,6] $P<0.01$

The basal level of lipid peroxidation (TBARS) in the liver wound was practically similar in the case of the n-6 group and the control group. At the same time significant differences between the n-3 group and the control group, as well as between n-3 and n-6 groups were established (Table 1). Additionally the peroxidizability of lipids (Fe-TBARS) was also significantly increased in the n-3 group in comparison with the control group.

PGs (PGE_2 and $F_{2\alpha}$) were measured in parallel in the liver wound and in the intact liver tissue. The content of PGs in three experimental groups are shown on Figure 1. No changes in the levels of PGs were established between the control group and the n-6 group. In the case of the n-3 group significant decrease in the contents of PGs both in the liver wound and in the intact liver tissue was found.

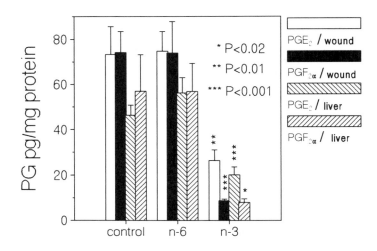

Figure 1 *Effects of different diets on contents of prostaglandins in the 6-day liver wound and in the intact liver tissue*

4 DISCUSSION

The present study demonstrates that the three week feeding with diet enriched with n-3 PUFAs suppress the connective tissue proliferation, increases the basal of lipid peroxidation, the peroxidizability of lipids and leads to decreased contents of 2-series prostaglandins (PGE_2 and $PGF_{2\alpha}$). The latter fact is in accordance with the study of Broughton and Morgan[19] showing significant correlation between frequency of n-3 PUFAs consumption and contents of n-3 PUFAs, PGs and LTs in hepatic tissue and peritoneal cells. Dietary intake of n-3 PUFAs at least more often than twice a week leads to alteration of eicosanoids synthesis (decrease in PGE, 6-keto-$PGF_{1\alpha}$ and 4-series LTs, increase in 5-series LTs). Decreased levels of PGE_2 and $F_{2\alpha}$ in our experiments after daily consumption for three weeks of n-3 but not n-6 PUFAs is evidently reason for suppression of the connective tissue proliferation by fish oil. Such conclusion refer to crucial role of endogenous PGs in the regulation of connective tissue proliferation. In addition, it is known that inflammatory response following any type of tissue injury is normal event in repair process. Wounding is immediately followed by coagulation, altered vascularity, and inflammation. Eicosanoids synergistically with other factors are involved in the regulation of both inflammation and connective tissue proliferation. In connection with the latter the low content of PGE_2 and $F_{2\alpha}$ in the case of the n-3 group is real. Firstly, the n-3

PUFAs are the precursors of EPA. Secondly, the anti-inflammatory effect of 2 weeks feeding with EPA on acute skin inflammatory reactions has been shown[20]. Thirdly, EPA inhibits competitively metabolism of AA. Fourthly, the fish oil (10%) diet reduces the activity of delta-6-desaturase followed by decrease of AA[21] which is precursor for PGE_2 and $F_{2\alpha}$.

By using the same test model it was established that cyclo-oxygenase inhibitor indomethacin causes similar retardation of connective tissue proliferation[12, 13] and decreases content of PGs (article submitted). In both cases - administration of indomethacin and consumption of fish oil - altered eicosanoids synthesis is apparent. It is shown that indomethacin inhibiting formation of PGs diminish their cytoprotective effect in the gastric mucosa[22]. Decreased formation of 2-series PGs by fish oil may also alter cytoprotective role of PGs as it is found that the n-3 PUFAs can enhance indomethacin-induced gastric ulceration[23]. Blocking PGs formation indomethacin directs the metabolism of AA on the lipoxygenase pathway giving relatively extensive formation of LTs and peroxidative derived oxyradical. The LTs, in turn, inhibit lymphocyte proliferation[24], cause vascular constriction (LTC_4), enhance vascular permeability and the adherence of leucocytes to the vascular wall (LTC_4, LTD_4) whereas the hydroperoxy products inactivate prostacyclin synthetase[25]. The latter phenomenon refer to the possibility that some products of lipid peroxidation can induce the spectrum of eicosanoids in tissues which is undesirable in respect of the connective tissue proliferation. Such situation is real since the basal level of lipid peroxidation in the n-3 group was significantly increased (Table 1). Additionally, it is known that the liver responds quite rapidly to the change of fish oil diet[21]. Hence, in the case of n-3 PUFAs rich dietary intake their incorporation into tissues (into several lipids etc.) is increased. Since the n-3 PUFAs are a good peroxidation substrate, the peroxidizability of lipids in the liver wound is significantly increased (Table 1.). It is known that proliferation of fibroblasts in the wound starts regeneration or repair phase of inflammation. The high level of lipid peroxidation and increased peroxidizability of lipids in the n-3 group indicate that abundant formation of lipid peroxidation products after rich fish oil consumption take place. Due to the facts that several products of lipid peroxidation are potent chemoatractants the normal migration of monocytes and proliferation of fibroblasts is disturbed. It is shown that dietary lipid unsaturation increases oxidative DNA damage[26]. Such phenomenon refer to possibility that a high dietary intake of fish oil can suppress the proliferation also via DNA damages. Here should be mentioned that several reports describing undesirable effects of fish oil consumption has been recently published. The n-3 fatty acids have shown to increase the lipid peroxidation on volunteers[27], to support persistence of tuberculosis infection[28], decrease resistance to *Salmonella thyphimurium* infection[29] and enhance bovine serum albumin induced immune complex nephritis[30].

Finally, it can be concluded, that feeding rats with diet significantly enriched with n-3 PUFAs suppress reparative regeneration of the connective tissue. This conclusion clearly supports reports by the other authors revealing undesirable effects of fish oil rich diets. However, our finding may be useful regarding the dietary intervention in treatment of chronic inflammatory diseases.

5 ACKNOWLEDGEMENTS

The authors wish to thank prof. Gunnar Selstam, University of Umeå, Sweden, for helpful support in prostaglandin assays. This research was in part supported by grants from the Royal Swedish Academy of Sciences and the Estonian Research Foundation (No. 1093).

References

1. D.F. Horrobin, *Prostaglandins Leukot. Essent. Fatty Acids*, 1991, **44**, 127.
2. C. Galli, F. Marangoni, G. Galella, *Prostaglandins Leukot. Essent. Fatty Acids*, 1993, **48**, 51.
3. S. Fischer, *Advances Lipid Res.*, 1989, **23**, 169.
4. T.J. Carrico, A.I. Mehrhof, I.K. Cohen, *Surg. Clin. North Am.*, 1984, **64**, 721.
5. O.M. Alvarez, "Connective tissue disease", edited by J. Uitto and A.J. Perejda. Marcel Dekker, Inc., New York and Base, 1987, p. 367.
6. J.I. Gallin, "Fundamental Immunology", edited by W.E. Paul, Raven Press Ltd., New York, 1989, p. 721.
7. I.L. Bonta, M.J. Parnham, *Biochem. Pharmac.*, 1978, **27**, 1611.
8. U.N. Das, *Prostaglandins Leukot. Essent. Fatty Acids*, 1991, **44**, 201.
9. P. Davies and D.E. MacIntyre, "Inflammation: Basic Principals and Clinical Correlates" edited by J.I. Gallin, I.M. Goldstein and R. Snyderman, Raven Press, New York, 1992, p. 123.
10. Williams K.J. and Higgs G.A., *J. Pathol.* 1988; **156**: 101-110.
11. F. Grimminger and P. Mayser, *Prostaglandins Leukot. Essent. Fatty Acids*, 1995, **52**, 1.
12. Ü. Arend and A. Arend, *Acta et Commentationes Universitatis Tartuensis*, 1987 **862**, 3.
13. Ü. Arend and A. Arend, "Tissue Biology", Reports of the 5th scientific meeting, June 5, 1990. Tartu University, Tartu, 1992, p. 25.
14. Ü. Arend, *Acta et Commentationes Universitatis Tartuensis*, 1966, **191**, 15.
15. J. Olofsson, E. Norjavaara and G. Selstam, *Biol. Reprod.*, 1990, **42**, 792.
16. H. Ohkawa, N. Ohishi and K. Yagi, *Anal. Biochem.*, 1979, **95**, 351.
17. M. Zilmer, K. Zilmer, V. Kask and A. Allmann, *Estonian Physician*, 1994, **1**, 15.
18. O.H. Lowry, N.J. Rosenbrough, A.L. Farr, R.J. Randall, *J. Biol. Chem.*, 1951, **193**, 265.
19. K.S. Broughton and L.J. Morgan, *J. Nutr.*, 1994, **124**, 1104.
20. K. Danno, K. Ikai and S. Imamura, *Arch. Dermatol. Res.*, 1993, **285**, 432.
21. L. Dinh, J.M. Bourre, O. Dumont and G. Durand, *Ann. Nutr. Metab.*, 1995, **39**, 117.
22. A. Robert, *Klin. Wochenschr.*, 1986, **64 (Suppl. VII)**, 40.
23. J.J. Turek and I.A. Schoenlein, *Prostaglandins Leukot. Essent. Fatty Acids*, 1993, **48**, 229.
24. A.C. Shapiro, D. Wu and S.N. Meydani, *Prostaglandins*, 1993, **45**, 229.
25. K. Gyires, *Agents Actions*, 1994, **41**, 73.
26. A.D. Haegele, S.P. Briggs and H.J. Thompson, *Free Rad. Biol. Med.*, 1994, **16**, 111.
27. A. Zang, *Clinica-Chimica Acta*, 1994, **227**, 159.
28. E. Mayatapek, K. Paul, M. Leichsering, M. Pfisterer, D. Wagner, M. Domann et al., *Infection*, 1994, **22**, 106.
29. H.R. Chang, A.G. Bullo, A.G. Gladoianu, D. Arsenijevic, L. Girardier and J.C. Pechere, *Metabolism*, 1991, **41**, 1.
30. S. Tateno and Y. Kobayashi, *Agents Actions*, 1994, **41**, Special Conference Issue: C212.

Measurement of Oxidative Stress in Humans

THE FERRYLMYOGLOBIN/ABTS$^{\bullet+}$ ASSAY FOR MEASURING TOTAL ANTIOXIDANT ACTIVITY

Nicholas J. Miller

Free Radical Research Group,
Division of Biochemistry and Molecular Biology,
UMDS, University of London,
Guy's Hospital, London SE1 9RT.

1. INTRODUCTION

The ferrylmyoglobin/ABTS$^{\bullet+}$ assay for measuring total antioxidant activity (i.e. the total potential of the constituent antioxidants) is a measure of the collective hydrogen-donating abilities of the antioxidants in the sample. It is based on the interaction between antioxidants in the sample with the ABTS$^{\bullet+}$ [2,2'-azinobis(3-ethylbenzothiazoline-6-sulphonic acid)] radical mono-cation, which is highly chromogenic and has absorbance maxima at 417, 635, 734 and 815 nm. Trolox (™ Hoffman-La Roche), the water-soluble vitamin E analogue, is used as an antioxidant standard. Quantitation of the reaction using the absorbance maximum at 734 nm in the near-I.R. region of the spectrum minimises interference from other light-absorbing components and from sample turbidity. The original method used[1], which is easy to automate with an enzyme analyser, is based on the inhibition by antioxidants of the ABTS$^{\bullet+}$ radical as it is generated in the reaction cuvette. However, several different analytical strategies are possible with the same reagents[2]. These include:

1. Inhibition assay, measuring at a fixed time point,
2. Inhibition assay, measuring reaction rate,
3. Decolorisation assay,
4. Lag phase measurement.

In systems 1. and 2. the reagents are mixed and the reaction started with hydrogen peroxide; in 1. the change in asorbance is measured after a pre-determined time interval (the lag time to absorbance change of the most concentrated standard is taken as the measuring tme), while in 2. the rate of the reaction is monitored. In both cases the presence of antioxidants in the reaction mixture reduces the absorbance change that is observed. In system 3. the ABTS radical cation is generated until a stable absorbance has been achieved, and then the addition of an antioxidant mixture decolorises the reaction mixture[3]. In system 4. the length of the lag time to absorbance change is measured: samples with higher antioxidant activity suppress the absorbance change for longer than those with less activity.

2. APPLICATIONS OF TAA MEASUREMENT

These factors make the assay adaptable to a wide range of specimens and it has been successfully applied to serum and plasma[1,4-7], saliva[8], low density lipoproteins[9], beverages and food extracts[10,11], as well as solutions of pure substances in non-polar solvents using a decolorisation system[3].

3. COMPARISON OF TAA METHODS

3.1 Spectrophotometric inhibition method for total antioxidant activity

The principle is that metmyoglobin (MetMb) and hydrogen peroxide form the ferrylmyoglobin radical which is then free to react (at a higher reaction rate) with ABTS (absorbance maximum 342 nm) to produce the ABTS radical cation (ABTS•+), a blue/green chromogen with characteristic absorption maxima in the wavelength regions of 645 nm, 734 nm, and 815 nm as well as at the commonly used maximum at 417 nm. The formation of this coloured radical can be suppressed by the presence of antioxidant reductants and the extent of the suppression directly related to the antioxidant capacity (activity) of the sample being investigated.

Reagents:
- Phosphate buffered saline 5mM pH 7.4 (PBS) which should not contain sodium azide or any other additional substances.
- Trolox (® Hoffman-La Roche) (6-hydroxy-2,5,7,8-tetramethylchroman-2-carboxylic acid. 2.5 mM Trolox is prepared in PBS. Frozen Trolox (-20°C) at this concentration is stable for more than 12 months. Fresh working standards are prepared daily by mixing 2.5 mM Trolox with PBS.
- ABTS (2,2'-azinobis(3-ethylbenzothiazoline-6-sulphonic acid) diammonium salt). 5 mM ABTS is prepared in PBS. This solution should be kept at 4°C in the dark and is stable for several days.
- Myoglobin (Metmyoglobin from horse heart, Sigma M-1882). Myoglobin is supplied as metmyoglobin, but it is, however, prudent to purify it before use by reaction with potassium hexacyanoferrate[III] and elution through a Sephadex G15-120 column. The concentration of myoglobin in the column eluate is then calculated using the extinction coefficients of met-, oxy- and ferrylmyoglobin at 490, 560 and 580 nm [12]. Purified metmyoglobin stored at 4°C is stable for at least 8 weeks and at -20°C for more than 6 months.
- Hydrogen peroxide. Working solutions of hydrogen peroxide are freshly prepared from stock H_2O_2.

In the manual assay system 8.4 µl of sample is added to a glass culture tube, followed by PBS to give a final volume of 1000 µl, metmyoglobin (2.5 µM) and ABTS (150 µM) (final concentrations). The tubes are mixed and brought to 30°C and then the reaction started with hydrogen peroxide (75 µM) after a time determined to be equivalent to the lag phase observed for a 21 µM Trolox standard. In the absence of antioxidants (i.e. in an analytical blank) an immediate and continuous absorbance change will be observed after the addition of hydrogen peroxide. No change in absorbance at 734 nm will be initially recorded using a aliquot of a 2.5 mM Trolox standard (final concentration 21 µM), until after 6 minutes, at which time the colour will start to develop. The exact timing of the end of the lag phase depends on temperature as well as reagent concentrations. The highest absorbance value will be recorded in the buffer blank. A quantitative relationship exists between the percentage inhibition of absorbance at 734 nm and the antioxidant activity of the added sample or standard.

The method described above is simple to automate with a spectrophotometric kinetic analyser system[1]. Precise timing and automated sample and reagent additions make the assay more practicable to perform. Minor fluctuations in assay conditions (for example caused by inter-instrumental variations) can be compensated for by appropriate adjustment of the measuring time. A non-linear curve fit is required to plot final absorbance at 734nm against standard concentration and a logit/log 4 plot has been found to be satisfactory. Analyticals can be added to the incubation mixture as an aliquot of an aqueous or an ethanolic solution (ethanol does not react with the ABTS radical cation).

3.2 Salivary TAA measurement

Salivary total antioxidant activity has been estimated by an adaptation of the assay [8]. TAA levels in saliva are relatively low compared to serum and, as a further complication, saliva also contains significant endogenous peroxidase activity. A standard curve using Trolox solutions up to 8.4 µM final concentration was used, together with a final hydrogen peroxide concentration of 375 µM. This ensures a fast analytical reaction before the interfering effect of the peroxidase activity of saliva becomes apparent. The measuring time for $\Delta 734nm$ was 60 seconds.

3.3 Direct measurement of LDL TAA

The TAA of plasma LDL fractions (prepared by density gradient ultracentrifugation) has been measured [9]. A larger sample fraction was used than in the original assay system (14.3%), with final reagent concentrations of 4.36 µM metmyoglobin, 436 µM ABTS, and 180 µM hydrogen peroxide. A Trolox standard curve up to 14.3 µM (final concentration) was used with a measuring time of 248 seconds. This assay is thus more sensitive than the original system, and is being used to explore the effect of dietary antioxidant supplementation on LDL oxidisability.

3.4 Decolorisation antioxidant assay

$ABTS^{\bullet +}$ is prepared and allowed to develop by mixing reagents to the following final concentrations: metmyoglobin - 2.5µM, ABTS - 250µM, hydrogen peroxide - 125µM, in 5mM PBS, incubating the solutions for 60 minutes at 25°C. Blank absorbances are read at 734 nm. After addition of the sample, analyticals are vortex mixed for 15 seconds and a further absorbance reading taken after 2 minutes (test). If the analytical is dissolved in a highly non-polar solvent, such as hexane, the phases will separate during this 2 minutes, leaving a clear light path for absorbance readings. The % absorbance remaining (blank - test) is plotted against final antioxidant concentration.

4. DISCUSSION

The concentration of ABTS is the limiting factor in determining the sensitivity of the assay: sensitivity increases as [ABTS] decreases. However the stabilisation of ABTS•+ (λ_{max} 417 nm, peaks at 645, 734 and 815 nm) depends on the presence of a mass of unreacted ABTS [13]. A resonance-stabilised mono-radical cation is formed if an excess of ABTS substrate is used. Under different conditions the reaction may also proceed further to form a radical dication (λ_{max} 513 nm). The appearance of this species will, of course, invalidate the assay, but spectral scanning has shown that it is not formed under the reaction conditions described.

The analytical strategies described, together with measurement in the near-infra red region of the spectrum, make this assay adaptable to the measurement of antioxidant activity in a variety of biological fluids and complex mixtures.

5. REFERENCES

1. Miller NJ, Rice-Evans CA, Davies MJ, Gopinathan V, Milner A, *Clin Sci* 1993, 84, 407-412.
2. Rice-Evans CA, Miller NJ, *Methods in Enzymology* 1994, 234, 279-293.
3. Miller NJ, Rice-Evans CA, *Redox Report*, 1995, in press.
4. Gopinathan V, Miller NJ, Milner AD, Rice-Evans CA, *FEBS Lett* 1994, 349,197-200.
5. Bird S, Miller NJ, Collins JE, Rice-Evans CA, *J Inher Metab Dis* 1995, 18, 123-126.
6. Goode HF, Richardson N, Myers DS, Howdle PD, Walker BE, Webster NR, *Ann Clin Biochem* 1995, 32, 413-416.
7. Morrison D, Rahman I, Wehbe L, Ramsay C, MacNee W, *Scotish Medical Journal*, 1995, in press.
8. Moore S, Kalder KAC, Miller NJ, Rice-Evans CA, *Free Rad Res* 1994,21,417-425.
9. Miller NJ, Paganga G, Wiseman S, Van Nielen W, Tijburg L, Chowienczyk P, Rice-Evans CA, *FEBS Letts* 1995, 365,164-166.
10. Miller NJ, Diplock AT, Rice-Evans CA,. *J Agric Food Chem* 1995, in press.
11. Salah N, Miller NJ, Paganga G, Tijburg L, Rice-Evans CA, *Arch Biochem Biophys* 1995, in press.
12. Whitburn KD, Shieh JJ, Sellers RM, Hoffman MZ, Taub IA, *J Biol Chem* 1982, 257,1860-1869.
13. Wolfenden BS, Willson RL, *J Chem Soc Perkin Trans*,1982,II,805-812.

URINARY EXCRETION OF 5-HYDROXYMETHYLURACIL AS INDICATOR OF OXIDATIVE DNA DAMAGE AND REPAIR

F. Bianchini and J. Cadet

Département de la Recherche Fondamentale sur la Matière Condensée, SESAM/LAN
CEA/Grenoble
17, Rue des Martyrs
38054 Grenoble Cedex 9, France

1 INTRODUCTION

Reactive oxygen species, formed either endogenously during normal oxidative processes or by external agents including solar light and ionizing radiations, can modify DNA bases leading to the formation of oxidized derivatives, which seem to be involved in ageing and carcinogenesis[1]. DNA repair processes, namely base excision repair and nucleotide excision repair, lead to the removal of the modified DNA base or nucleoside, which are then released into urine[2]. Urinary excretion of oxidatively modified DNA bases and nucleosides has therefore been suggested as a non invasive biomarker for assessment of oxidative stress in human studies[3]. Up to now most studies have focused on the determination of urinary excretion of 8-oxo-7,8-dihydro-2`-deoxyguanosine (8-oxodGuo[4,6]. However, other DNA modifications, out of the approximately 20 which have been identified so far, may have a biological relevance.

5-Hydroxymethyl-2`-deoxyuridine (5-HMdUrd), one of the major modifications of thymidine[7], is formed in animal and human DNA at levels comparable to 8-oxodGuo[8-11] and a repair enzyme excising the corresponding base 5-hydroxymethyluracil (5-HMUra) has been demonstrated in eukariotic[12,13]. This DNA modification could be therefore a possible biomarker for monitoring oxidative DNA damage and repair in human studies. We describe here a method developed for the determination of 5-HMUra and the corresponding nucleoside 5-HMdUrd; the application to human samples is also presented.

2 METHODS

The developed method is based on HPLC prepurification of urine and GC/MS analysis of the modified nucleobase or nucleoside (Figure 1).

2.1 HPLC prepurification

The prepurification of urine samples was achieved by a semipreparative octadecylsilyl silica gel column (250 x 10 mm, 5µm particle size) (Interchim, Montluçon, France), equipped with a RP-18 guard column. Urines (1 ml) to which 100 pmoles of the

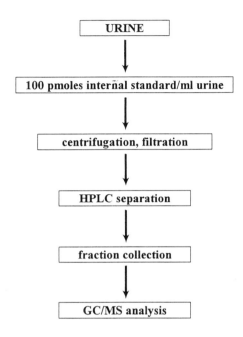

Figure 1 *Experimental procedure for measurement of 5-HMUra and 5-HMdUrd in urine*

isotopically enriched (M+4) internal standard was added were centrifuged, filtered and purified by HPLC. The following elution programme was used (flow rate 2 ml/min): 15 min 50 mM ammonium formate, 2 min gradient to 95:5 (v/v) 50 mM ammonium formate/MeOH, 20 min 95:5 (v/v) 50 mM ammonium formate/MeOH; the column was then washed with 100% MeOH before reconditioning to the initial conditions. 5-HMUra eluted at approximately 13 min and 5-HMdUrd at approximately 32 min; fractions corresponding to the compounds, namely fractions 12 to 15 min and fractions 31 to 34 min, respectively, were then collected.

2.2 GC/MS analysis

Separations were achieved on a 5890 Serie II gas-chromatograph (Hewlett-Packard, Les Ulis, France) equipped with a capillary column (30 m x 0.25 mm) coated with film (0.25µm thickness) of methylsiloxane substituted by 5% phenylsyloxane (HP5-MS, Hewlett-Packard, Les Ulis, France), using helium as the gas carrier (constant flow=1.6 ml/min). 100 µl of a mixture of acetonitrile / N-(*tert*-butyldimethilsilyl)-N-(methyl-trifluoroacetamide) (1:1, v/v), were added to the dried samples and incubated at 110°C for 20 min. Detection was obtained by MS analysis performed on a Hewlett Packard, Model 5972A detector using electron impact mass ionization (Hewlett-Packard, Les Ulis, France) in the selective ion monitoring (SIM) mode. As the nucleoside is converted into the corresponding base during the derivatization step, the final analytical procedure was the same for 5-HMUra and 5-HMdUrd. The injector (splitless mode) and detector temperatures were 250°C and 280°C respectively; the oven temperature, after 5 min at 70°C, was increased up to 275°C, at a rate of 15°C/min. The ions recorded were 427.4 and 431.4 for 5-HMUra and the (M+4) 5-HMUra internal standard, respectively. These ions result from the loss of a *tert*-butyl from the molecular ion, obtained by the attach of three *tert*-butyldimethylsilyl groups.

3 RESULTS AND DISCUSSION

Figures 2 and 3 report the calibration curves for the quantitation of 5-HMUra and 5-HMdUrd relative to the internal standard, respectively. The detection limit of the analytical procedure was 5 fmoles 5-HMUra and 50 fmoles 5-HMdUrd injected on the GC/MS, and the response was linear up to 5 pmoles. A typical GC/MS chromatogram of urinary 5-HMUra is reported in Figure 4. The intra-experimental variability, measured in four urine samples in triplicate, was low, the coefficient of variation (CV) being 1.1% +/- 0.7%.

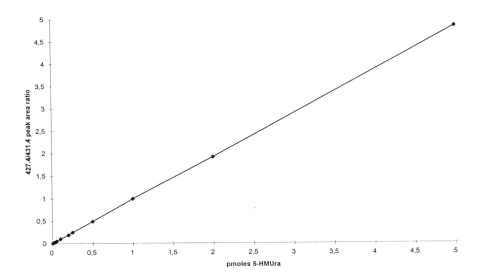

Figure 2 *Calibration curve for the determination of 5-HMUra relative to 5-HMUra (M+4)*

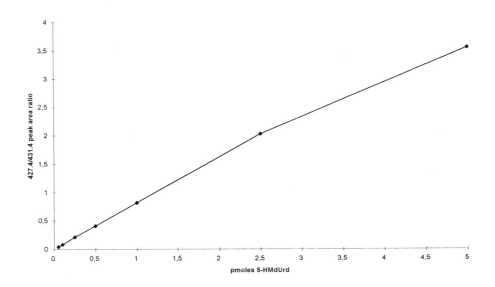

Figure 3 *Calibration curve for the determination of 5-HMdUrd relative to 5-HMdUrd (M+4)*

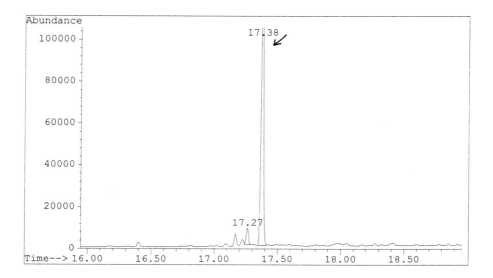

Figure 4 *GS/MS SIM chromatogram of urinary 5-HMUra*

The developed method was then applied to the measurement of 5-HMUra in 11 human urine samples, collected twice over a 24 hour period. The excretion varied either from 57 to 129 total nmoles 5-HMUra excreted or from 7.2 to 12.1 nmoles 5-HMUra/mmoles creatinine. Intra-individual variability was relatively low, the CV between different urine collections being 7.8% +/- 4.3 or 6.3% +/- 3.6 when results were or were not corrected by creatinine excretion, respectively. Preliminary data on the urinary excretion of 5-HMdUrd show that it is at the limit of detection of our assay. This indicates that the nucleoside is excreted at much lower levels, at least one tenth, than the corresponding base.

In conclusion, the results presented here show that the method developed for the analysis of 5-HMUra is suitable for application in human studies. Only 1 ml of urine is in fact required for the analysis and the experimental variability is extremely low. In addition, the level of urinary 5-HMUra seems to be different between individuals, while it is sufficiently stable with time for the same individual.

References

1. B. N. Ames, M. K. Shigenaga and T. M. Hagen, *Proc. Natl. Acad. Sci.*, 1993, **90**, 7915.
2. B. Demple and L. Harrison, *Ann. Rew. Biochem.*, 1994, **63**, 915.
3. D. S. Bergtold and M. G. Simic, 'Oxygen Radicals in Biology and Medicine', M. G. Simic, K. A. Taylor, J. F. Ward and C. von Sonntag (eds), Plenum Press, New York, 1988, p. 483.
4. S. Loft, K. Vistisen, M. Ewertz, A. Tjonneland, K. Overvad and H. E. Poulsen, *Carcinogenesis*, 1992, **13**, 2241.
5. G. van Poppel, H. E. Poulsen, S. Loft and H. Verhagen, *J. Natl. Cancer Inst.*, 1995, **87**, 310.
6. H. Verhagen, H. E. Poulsen, S. Loft, G. van Poppel, M. I. Willems and P. J. Bladeren, *Carcinogenesis*, 1995, **16**, 969.
7. R. Téoule and J. Cadet, 'Molecular Biology, Biochemistry and Biophysic', Vol 27, Chapter II, Springer-Verlag, Berlin, Heidelberg, New York, 1978, 171.
8. Z. Djuric, L. K. Heilbrun, B. A. Reading, A. Boomer, F. A. Valeriote and S. Martino, *J. Natl. Cancer Inst.*, 1991, **83**, 766.
9. Z. Djuric, M. H. Lu, S. M. Lewis, D. A. Luongo, X. W. Chen, L. K. Heilbrun, B. A. Reading, P. H. Duffy and R. W. Hart, *Toxicol. Appl. Pharmacol.*, 1992, **115**, 156.
10. R. Olinski, T. Zastawny, J. Budzbon, J. Skokowski, W. Zegarski and M. Dizdaroglou, *FEBS Lett.*, 1992, **309**, 193.
11. D.M. Malins, *J. Toxicol. Env. Health*, 1993, **40**, 247.
12. M. C. Hollstein, P. Brooks, S. Linn and B. N. Ames, *Proc. Natl. Acad. Sci.*, 1984, **81**, 4003.
13. R. J. Boorstein, D. D. Levy and G. W. Teebor, *Mutat. Res.*, 1987, **183**, 257.

MEASUREMENT OF OXIDATIVE DNA INJURY IN HUMANS: EVALUATION OF A COMMERCIALLY AVAILABLE ELISA ASSAY

H. Priemé[1], S. Loft[1], R. G. Cutler[2] and H. E. Poulsen[1,3]

[1] Department of Pharmacology, Health Science Faculty,
Panum Institute 18-5, 3 Blegdamsvej, DK-2200 Copenhagen N, Denmark.
Fax (+45)35327610. E-mail fihep@farmakol.ku.dk
[2] Genox Corporation, 1414 Key Highway, Baltimore MD 21230, USA.
Fax (410) 347-7617. E-mail genox@AOL.com. Internet home page http://genox.com.
Correspondence and reprint request to Professor Henrik E. Poulsen, M.D.,
Panum Institute 18-5-50, 3 Blegdamsvej, DK-2200 Copenhagen N, Denmark.
Fax (+45) 3532 7610. E-mail fihep@farmakol.ku.dk

1 INTRODUCTION

DNA modification can originate from reaction between DNA and a variety of chemicals, e.g. aromatic carcinogens. Most often the reaction is not between DNA and the chemicals per se but metabolic products which are more reactive. The DNA modifying agents of interest now expand to oxygen. Recently a variety of oxidative modifications of DNA have been demonstrated[1-6] and it appears that their occurrence is more frequent than modifications from other chemicals. Furthermore, it has now been demonstrated that modification from reactive oxygen species (ROS) can result in the same mutations as resulting from aromatic carcinogens, e.g. G-T transversion mutation in tumor suppressor gene p53 and codon 12 activation of c-Ha-ras or K-ras oncogenes[7-10].

The oxidative modifications have been estimated to approximately 10^4-10^5 base modifications per cell per day[11]. If unrepaired this would mean almost total oxidation of all DNA bases in the human maximum lifespan of about 100 years. It is therefore, and from a long list of other observations reasonable to assume a very efficient and almost complete repair of oxidative DNA modifications. Unrepaired oxidative modifications are probably few, but the rate at which the oxidation occurs is presumably predictive for the risk of mutation whereas the tissue levels of oxidized bases are determined by the balance between oxidation and repair.

Oxidation of the guanine base is the most prominent and mutagenic oxidative modification and its repair results in urinary excretion of the oxidatively modified base 8-oxoguanine (8-oxoGua) and the nucleoside 8-oxodeoxyguanosine (8-oxodG), the latter presently the most promising candidate as a biomarker of the rate of DNA oxidation also for estimation in specific tissues[12].

The assay possibilities for 8-oxodG in urine or tissue range from advanced HPLC-MS, GC-MS and multidimentional HPLC-EC to commercially available ELISA assay. In general the urine analysis presents more methodological problems than tissue analysis. We compared a commercial ELISA method to the HPLC-EC method on urine samples.

2 MATERIALS AND METHODS

Urine samples were collected as part of a large ongoing randomized smoking cessation study. From this study three healthy subjects were selected on the basis of high, intermediate and low urine concentrations of 8-oxodG estimated by HPLC. As part of the study the volunteers gave urine samples before entering the smoking cessation programme and after four and 26 weeks.

2.1 HPLC Assay to Measure 8-oxodG

As described in details elsewhere urine samples mixed with TRIS-HCl pH 2.9 are injected onto a 3 column HPLC system with computerized control and integration[13]. On the first column urine is chromatographed at pH 7.9 and from this "extraction column" a fast operating switching valve directs a small fraction containing the 8-oxodG to be trapped onto a small cation exchange column. From this column the fraction is eluted at pH 2.8 onto a high resolution C18 column for quantification by electrochemical detection from individual standards. Samples were analysed in duplicates with an intra and inter assay coefficient of variation of 10% and 13%, respectively. For analysis of the standards (see below) a single column HPLC assay was used[12].

2.2 ELISA Assay to Measure 8-oxodG

A commercial assay from GEN:OX was used (Genox Corporation, Baltimore, US). The kit includes 8-oxodG standards, reagents and a 96 well microtiter plate precoated with 8-oxodG. The procedure follows conventional ELISA methodology with application of 50 µl of sample or standard. After reaction the plate is read at 492 nm. The primary antibody is monoclonal, the secondary is horse radish peroxidase-conjugated antimouse polyclonal antibody. Lyophilized urine samples (5x concentration) were also prepared but the readings were optimal using untreated urine. The coefficient of variation for duplicate samples was 9.9% (n=12). Uric acid [1, 4, 10 mM], guanine, guanosine, deoxyguanosine, deoxyadenosine, and 8-oxoGua [2.3, 11.3, 546.5, 283 and 1413 nM] were tested in the ELISA assay but did not give any reaction.

2.3 Standards

The HPLC assay was calibrated by 8-oxodG samples with close to 100% purity kindly donated by Dr. Per Leanderson, Department of Occupational Medicine, University of Linköping, Sweden and by Dr. David W. Potter, Rohm and Haas Co., Spring House, P.A., US. The two standards deviated by a few per cent only. Standards from the ELISA kit and the former mentioned standards differed in the HPLC assay with at factor 2.13 (ELISA standard vs HPLC standard) evaluated by the slopes of standard curves up to 400 nM.

3 RESULT

The mean 8-oxodG concentration estimated by HPLC was 15.0 nM (sd=10.2) which was significantly different from the ELISA mean concentration of 48.6 nM (sd=39.0), p=0.008 by paired t-test. The ratios of ELISA to HPLC concentrations ranged 3 to 30, with a mean of 8.3 and sd=7.4, taking into account the difference in standards used for the two assays.

The actual concentrations of 8-oxodG estimated by ELISA and HPLC without correcting for difference in standards are depicted in Figure 1. The correlation between the ELISA and the HPLC estimate was 0.42 (r) and regression analysis gave a slope of 1.60 and an intercept of 24.39 nM.

The urinary 24 hour 8-oxodG excretion before and at 4 and 26 weeks after smoking cessation is depicted in Figure 2. By HPLC estimation there is a tendency to lowered 8-oxodG excretion after smoking cessation. By ELISA the values are considerably higher and appear more variable.

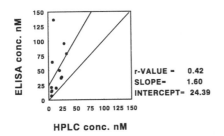

Figure 1
Correlation between concentration of 8-oxodG determined by HPLC and ELISA technique in 4 subjects at 3 occasions

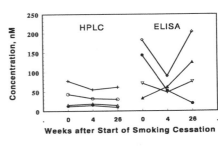

Figure 2
Urinary excretion of 8-oxodG in 4 smokers following smoking cessation measured by HPLC and ELISA technique

Table 1 Published Values Regarding Urinary Biomarkers of Oxidative DNA-damage in Humans

Experimental Protocol	Age	Lesion	Assay	Excretion (mean_SD or range) per 24 hours	Ref.
5 healthy subjects	unknown	8-oxodG	HPLC	323±23 pmol/kg	16
63 healthy subjects	unknown	8-oxodG	HPLC	172±79 pmol/kg	17
23 healthy subjects	unknown	8-oxodG	GC/MS	300±100 pmol/kg	18
53 healthy non-smokers vs.	40-64	8-oxodG	HPLC	213±84 pmol/kg	13
30 healthy smokers	40-64	8-oxodG	HPLC	320±99 pmol/kg	13
21 healthy non-smokers vs.	33_10	8-oxodG	HPLC	318±130 pmol/kg	15
12 healthy smokers	33_10	8-oxodG	HPLC	431±168 pmol/kg	15
300 g vegetable diet vs.	young	8-oxodG	HPLC	300±630 pmol/kg	19
300 g Brussels sprouts diet *	do.	do.	do.	210±490 pmol/kg	19
16 healthy non-smokers	35-50	8-oxodG	HPLC	666±202 pmol/kg	20
100% vs 60% energy in diet **	50	8-oxodG	GC/MS	345 vs. 110 pmol/kg	18
2 cancer patients before	57-59	8-oxodG	GC/MS	8-14 nmol	21
and after radiotherapy	do.	do.	do.	31-40 nmol	21

* n=5
** n=1

4 DISCUSSION

We tested an ELISA kit to determine 8-oxodG concentration in urine by comparing it to a multidimensional HPLC assay. The ELISA assay gave about eight fold higher results on average, however with high variation. The average level is compatible with the six fold higher levels found in another ELISA assay by Yin et al.[14] used on tissue, however. The higher degree of variation we observed in urine presumably relates to the larger variation in cross-reacting substances in urine due to differences in water intake, diet and environmental influence.

The HPLC assay is highly reproducible as can bee seen in Figure 2. In different studies both from our laboratory and those of others, urinary excretion rates in the same range have been found, Table 1, and we have been able to confirm increased excretion rates in different groups of smokers[13,15]. From a theoretical point of view HPLC is a highly selective and specific methodology. In the ELISA method specificity relies on the specificity of the reaction between the 8-oxodG and the antibody. DNA bases, RNA bases and oligonucleotides with or without oxidative modifications may cross-react. In larger series it could be of interest to determine differences between the ELISA detectable substances and the HPLC detectable substances to see if they represent other products of interest in relation to oxidative stress.

The commercially available assay is promising but presently it is too unspecific compared to the HPLC assay. Development of more specific antibodies may increase the specificity of the 8-oxodG ELISA assay, or combination with e.g. immunoaffinity-HPLC may make the ELISA assay applicable to molecular epidemiological studies.

Further advancement in the methodology - including verification that the 8-oxodG concentration in spot urine corrected by the urinary creatinine concentration is equivalent to the 8-oxodG level measured in 24 hours urine - is necessary for investigating the important question whether the oxidative DNA modification rate is predictive for development of degenerative and malignant diseases and ageing in man.

Acknowledgement

We than Genox for making the assay available. We thank BAT, UK and the Danish Environmental Programme for financial support. We also thank Bruce Ames for donating the 8-oxoGua.

References

1. Spector and H. G. William, *Exp. Eye. Res.*, 1981, **33**, 673.
2. B. N. Ames, *Free. Radic. Res. Commun.*, 1989, **7**, 121.
3. C. Fraga, M. K. Shigenaga, J.-W. Park, P. Degan, B. N. Ames, *Proc. Natl. Acad. Sci. USA*, 1990, **87**, 4533.
4. J. Lunec, *Ann. Clin. Biochem.*, 1990, **27**, 173.
5. P. M. Reilly, H. J. Schiller, G. B. Bulkley, *Am. J. Surg.*, 1991, **161**, 488.
6. R. A. Greenwald, *Semin. Arthritis. Rheum.*, 1991, **20**, 219.
7. S. Shibutani, M. Takeshita, A. P. Grollman, *Nature.*, 1991, **349**, 431.
8. K. C. Cheng, D. S. Cahill, H. Kasai, S. Nishimura, L. A. Loeb, *J. Biol. Chem.*, 1992, **267**, 166.

9. H. Kamiya, K. Miura, H. Ishikawa, S. Nishimura, E. Ohtsuka, *Cancer Res.*, 1992, **52**, 3483.
10. K. G. Higinbotham, J. M. Rice, B. A. Divan, K. S. Kasprzak, C. D. Reed, A. O. Perantoni, *Cancer Res.*, 1992, **52**, 4747.
11. R. Adelman, R. L. Saul, B. N. Ames, *Proc. Natl. Acad. Sci. USA*, 1988, **85**, 2706.
12. A. Fischer-Nielsen, G. B. Corcoran, H. E. Poulsen, L. M. Kamendulis, S. Loft, *Biochem. Pharmacol.*, 1995, **49**, 1469.
13. S. Loft, K. Vistisen, M. Ewertz, A. Tjønneland, K. Overvad, H. E. Poulsen, *Carcinogenesis*, 1992, **13**, 2241.
14. B. Yin, R. M. Whyatt, F. P. Perera, M. C. Randall, T. B. Cooper, R. M. Santella, *Free Radic. Biol. Med.*, 1995, **18**, 1023.
15. S. Loft, A. Astrup, B. Buemann, H. E. Poulsen, *Faseb J.*, 1994, **8**, 534.
16. M. K. Shigenaga, C. J. Gimeno, B. N. Ames, *Proc. Natl. Acad. Sci. USA*, 1989, **86**, 9697.
17. E.-M. Park, M. K. Shigenaga, P. Degan, T. Korn, J. W. Kitzler, C. M. Wehr, P. Kolachana, B. N. Ames, *Proc. Natl. Acad. Sci. USA*, 1992, **89**, 3375.
18. M. G. Simic and D. S. Bergtold, *Mutatation Res.*, 1991, **250**, 17.
19. H. Verhagen, H. E. Poulsen, S. Loft, G. van Poppel, M. I. Willems, P. J. van Bladeren, *Carcinogenesis*, 1995, **16**, 969.
20. S. Loft, E. J. M. V. Wierik, H. van der Berg, H. E. Poulsen, *Cancer Epidemiol. Biomarkers Prev.*, 1995, **4**, 515.
21. D. S. Bergtold, C. D. Berg, Simic;MG, *Adv. Exp. Med. Biol.*, 1990, **264**, 311.

INDICATORS OF OXIDATIVE STRESS IN BLOOD AFTER THE ADMINISTRATION OF FARMORUBICIN AND ALPHA-TOCOPHEROL

R. Birk, L. Paas and V. Muzyka

Department of Experimental Oncology and Laboratory of Environmental Carcinogenesis, Institute of Experimental and Clinical Medicine, Tallinn EE00016, Estonia

1 INTRODUCTION

Anthracyclines like some other anticancer antibiotics exert their toxic side effect (myelo-, hepato-, and cardiotoxicity) by the oxygen radical mediated mechanism[1,2]. The objective of the present study was to determine a complex of the biochemical markers that would render feasible the evaluation of the oxidative stress *in vivo* and through it the possible chemosensitivity of the organism to the toxic effect of anthracyclines during cancer chemotherapy, concurrently with the efficiency of protectors (alpha-tocopherol).

2 MATERIAL AND METHODS

In our experiments, farmorubicin (4'-epidoxorubicin, epirubicin; FR) was used as an anthracycline anticancer drug. FR was administered in 0.9% NaCl solution into the ear vein of rabbits at a dose of 2.5 mg per kg body weight once a week. As an antioxidant, the alpha-tocopherol acetate (vitamin E; vit. E) in oil solution was applied subcutaneously at a dose of 25 mg per kg body weight twice a week. FR was administered on the first day and vit. E on the first and third days of the week.

The objects of the study were blood plasma, erythrocytes and leukocytes of 10 male and 10 female rabbits. Blood samples for the determination of the oxidative stress and lipid peroxidation were obtained from the ear vein of rabbits before experiments and then on the second day after the administration of FR or on the third day of the week before the administration of vit. E. The administration of FR was suspended for three weeks after the third and for two weeks after the fourth administration of FR because of the steep decrease in the number of leukocytes in the blood of rabbits. The administrations of FR and vit. E are presented in Table 1.

Table 1 *Administrations of FR and vit. E*

	Weeks														
	0	1	2	3	4	5	6	7	8	9	10	11	13	16	19
FR	+	+	+	-	-	+	-	+	+	+	-	-	-	-	-
vit. E	++	++	++	++	++	++	++	++	++	++	-	-	-	-	-

The level of thiobarbituric acid reactive substances (TBARS) in blood plasma, the percentage of glucose-6-phosphate dehydrogenase (G6PD) dimer oxidized form I in erythrocytes, the induction of DNA alkali-labile sites, and the rate of unscheduled DNA synthesis in leukocytes were determined as indicators of the oxidative stress. The toxic side effect of FR was ascertained through the diminishing of the number of leukocytes and erythrocytes in the blood of rabbits. TBARS was determined in blood plasma by the method of Ohkawa et al.[3]; the percentage of G6PD dimer oxidized and reduced forms in erythrocytes by the method of Grigor[4]. Leukocytes were isolated for the determination of the DNA alkali-labile sites using Ficoll-Paque and suspended in the basic solution. Nucleoproteins were precipitated with trichloroacetic acid and washed with ethanol-CH_3COONa solution (pH 5.3) and organic solvents. The precipitate was dissolved in 0.3 N KOH solution and DNA was hydrolysed for 1 hour at $37°C$. A part of this hydrolysate was taken for the determination of protein. In the other part, the polymeric DNA was precipitated by 60% perchloric acid and the deoxyribose of DNA fragments was measured in supernatant by the diphenylamine method of Richards[5]. The precipitated DNA was dissolved in 0.5 N perchloric acid and incubated for 45 min. at $96°C$ and then quantified spectrophotometrically[6], determining the absorbancy at 260 and 280 nm. For the determination of the rate of the unscheduled DNA synthesis, a part of leukocytes was incubated with [^{3}H]thymidine and its incorporation into the polymeric DNA was measured by the liquid scintillation method.

3 RESULTS AND DISCUSSION

As an indication of the toxic side effect, FR diminished the number of leukocytes and erythrocytes to 58% and 82% from the basic level, respectively, after the third administration of the drug. Therefore the administration of FR was temporarily interrupted after the third and fourth injections. The co-administration of vit. E revealed a moderate diminishing effect on the myelosuppression and favoured the regeneration of leukocytes. Changes in the erythrocyte count were less expressed and the effect of vit. E was not significant.

FR caused an enhanced lipid peroxidation in blood plasma of the rabbits and produced an about 2-fold increase of TBARS after the first intravenous injection (Figure 1). The level of TBARS decreased in blood after the suspension of FR administrations and increased again after the continuation of the administration. Vit. E diminished the level of TBARS in blood and protected for FR-induced lipid peroxidation.

The percentage of the G6PD oxidized and reduced dimer forms in erythrocytes was determined by the method of polyacrylamide gel electrophoresis as another indicator of the oxidative stress in the organism. These forms represent posttranslational different oxidation states of G6PD, with the fastest migrating form I containing oxidized SH-groups (disulphide bonds), form II being partially oxidized, and the slowest form III containing only reduced SH-groups[4]. It has been suggested that the pattern of the G6PD dimer molecular forms is determined by the ratio of G6PD activity to the content of the tissue reduced glutathione (GSH), whereas the shift from the more reduced forms to the oxidized form I is an indicator of the NADPH and GSH deficit in cells[7]. The percentage of the G6PD dimer oxidized form I enhanced from 5% to 15% in erythrocytes after the first, and to 22% after the third administration of FR, decreasing again after the suspension of the drug administration (Figure 1).

The significant increase of DNA alkali-labile sites in the leukocytes of rabbits, as an indicator of the DNA damage by oxygen radicals, was detected after the third

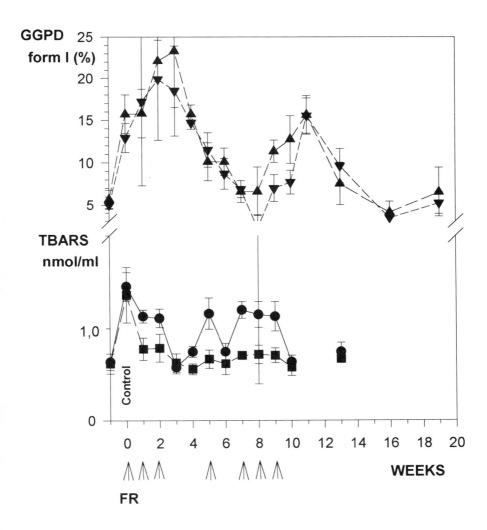

Figure 1. *The percentage of G6PD oxidized form I in erytrocytes of rabbits after the administration of FR (▲) and after co-administration of vit. E (▼), and the concentration of TBARS in blood(nmol/ml) after the administration of FR (●) and after co-administration of vit. E(■), mean ± SEM. Administrations of FR are presented by arrows.*

administration of FR (Table 2). Our results on the [^{3}H]thymidine incorporation into the polymeric DNA of leukocytes showed a significant increase in the unscheduled synthesis of DNA after the prolonged administration of FR, whereas the co-administration of vit. E promoted this process (Table 2, part II).

Table 2 *Content of alkali-labile DNA fragments and the rate of unscheduled DNA synthesis in the leukocytes of rabbits (mean ± SEM)*

I Content of alkali-labile fragments (mg DNA per 1 mg of nucleoprotein)

Weeks	FR	FR ± vit. E
Control	2.3 ± 0.6	-
3	14.0 ± 2.7	12.6 ± 3.1
6	8.1 ± 2.5	8.4 ± 2.2
8	7.5 ± 2.1	6.7 ± 1.5
10	3.4 ± 0.6	3.2 ± 0.7

II Rate of unscheduled DNA synthesis according to the incorporation of [^{3}H]thymidine into polymeric DNA (imp/mg DNA)

Weeks	FR	FR ± vit. E
Control	1210 ± 120	-
3	910 ± 235	925 ± 250
6	980 ± 260	1150 ± 285
8	1950 ± 180	2300 ± 195
10	2700 ± 190	3200 ± 210

The results of our experiments provide evidence that the changes in the oxidative stress and peroxidative status in blood may be markers of the ability of the organism to detoxify reactive oxygen species, generated by anthracyclines, and to serve as possible indicators of the chemosensitivity to anthracyclines during cancer chemotherapy.

References

1. G. Powis, 'Toxicity of Anticancer Drugs', Pergamon Press, New -York, 1991.
2. E. G. Mimnaugh, M. A. Trush, E. Ginsburg and T. E. Gram, *Cancer Res.*, 1982, **42**, 3574.
3. H. Ohkawa, N. Ohishi and K. Yagi, *Anal. Biochem.*, 1979, **95**, 351.
4. M. R. Grigor, *Arch. Biochem. Biophys.*, 1984, **229**, 612.
5. G. M. Richards, *Anal. Biochem.*, 1974, **57**, 369.
6. R. W. Wannemacher, W. L. Banks and W. H Wunner, *Anal.Biochem.*, 1965, **11**, 320.
7. A. Watanabe, K. Taketa and K. Kosaka, *J. Biochem.*, 1972, **72**, 695.

FAILURE OF AN AEROBIC IODOMETRIC PEROXIDE ASSAY

J.M. Gebicki, J. Collins, A. Baoutina and P. Phair*

School of Biological Sciences, Macquarie University, Sydney 2109, Australia
*Dept. of Surgery, Royal Children's Hospital, Melbourne 3052, Australia

1 INTRODUCTION

The abundant evidence implicating lipid peroxidation in the genesis of many forms of biological damage has resulted in the development of numerous peroxide assays. Of the methods in common use, the iodometric assay has the advantages of well-understood chemistry, exact stoichiometry, and reasonable specificity.[1] The assay uses the quantitative reduction of acidified iodide solutions in polar solvents by the peroxide, which leads to liberation of iodine:

$$2I^- + ROOH = I_2 + ROH + OH^-$$

Older protocols used titration with standard thiosulphate to quantitate the liberated iodine, and thus measure the amount of peroxide present. In more recent procedures, the reaction is performed in presence of excess iodide, which converts the iodine to the tri-iodide anion:[2]

$$I^- + I_2 = I_3^-$$

The concentration of the tri-iodide is then determined spectrophotometrically near 360 nm. The molar extinction coefficient of the anion is not very sensitive to the nature of the solvent,[1] with the best values lying near 29,000 M^{-1} cm^{-1}.

Although this last reaction is reversible, its K_{eq} is 714 M^{-1} at 25°,[3] so that at the usually recommended assay conditions of 4-8% final KI concentrations, the ratio I_3^-/I_2 lies between 200 and 400. It is important to note that, even though this results in virtually complete conversion of I_2 to I_3^-, some I_2 is always present. This can lead to errors in analysis of samples reactive towards iodine.

The major disadvantage of the tri-iodide peroxide assay is its sensitivity to oxygen. Various methods have been devised for its exclusion during the reaction. We found that the most reliable required totally anaerobic conditions.[2] For routine application, this can be very inconvenient. Much interest was therefore generated by a report by El-Saadani et al [4] that the oxygen problem can be overcome by the use of a commercially available reagent (called here the CHO reagent) developed for the assay of H_2O_2 produced in enzymatic oxidation of cholesterol. In this method, the oxygen sensitivity of the iodide was reduced by the use of neutral pH. The problem of low velocity of the iodide-peroxide reaction at neutral pH is overcome in the CHO reagent by inclusion of molybdate which catalyses the reaction with H_2O_2. The original publication[4] described the use of the reagent for assay of H_2O_2, t-butyl and cumene hydroperoxides and for oxidized low density lipoprotein. However, since then, other workers have recommended it for use with systems as chemically complex as plasma.[5]

Since our experience with iodometric peroxide assay suggested that the aerobic method may give false peroxide values, we tested the reliability of the new procedure[4] extensively with a variety of oxidised lipids. The results were compared with peroxide values measured with the standard anaerobic assay.[2] Tests carried out with t-butyl, cumene, linoleic acid and plasma hydroperoxides showed that the recommended procedure consistently underestimated the amounts of peroxides actually present. The method does give correct results for H_2O_2 and for low density lipoprotein hydroperoxides. However, its future use in the assay of other organic peroxides needs to be critically evaluated with well-tested procedures.

2 MATERIALS AND METHODS

Hydrogen peroxide was supplied by BDH (Sydney), cumene and t-butyl hydroperoxides by Koch-Light (Colnbrook, UK), and linoleic acid by Lipid Products (Redhill, UK). Fresh human blood plasma was isolated as the supernatant of gently centrifuged blood. Low density lipoprotein from human blood was the gift of Dr. Wendy Jessup of the Heart Research Institute, Sydney. Several batches of the cholesterol assay (CHO) reagent were obtained from Merck (Darmstadt). They were stored near 0° and used before expiry date. All other chemicals were of at least analytical grade. Solutions were made up in glass containers cleaned with hot nitric acid , thoroughly rinsed, and dried. Water was purified by passage through a reverse-osmosis system, followed by a 4 stage Millipore (Sydney) system equipped with a final 0.2 µm filter. Concentrations of all peroxides were measured by a slightly modified anaerobic tri-iodide method,[2] using either 1:3 methanol : acetic acid or water : acetic acid solutions as solvent and 10% KI as the reactant. In both cases peroxide concentrations were measured at 358 nm and calculated with molar extinction coefficient of 29,000 M^{-1} cm^{-1}.[1] In experiments using the procedure recommended by El-Saadani et al ,[4] 2 ml of the CHO reagent were incubated in 1 cm lightpath quartz cells with solutions of various peroxides at concentrations resulting in 365 nm absorbance readings of < 1.0.

3 RESULTS

Measurements of the kinetics of development of tri-iodide absorbance with H_2O_2 and the CHO reagent under conditions recommended by El-Saadani et al [4] showed that maximum readings at 365 nm developed within 30 min incubation at 20° and remained constant for at least 2 hrs. The accuracy of the values calculated for the concentration of the peroxide were confirmed by the anaerobic assay.[2] However, the CHO reagent failed to give even approximately correct concentrations for the organic peroxides used by these workers.[4] The course of absorbance development with t-butyl hydroperoxide at 20° (Figure 1, lower curve) shows that no plateau was reached even after 5 hrs and that the reading recorded after the recommended 30 min incubation clearly did not give the right peroxide values (Figure 4). In this and in other figures we show absorbance readings unconverted to peroxide concentrations, because the molar extinction coefficients listed by El-Saadani et al for the tri-iodide produced in the peroxide-CHO reaction are not correct.[1]

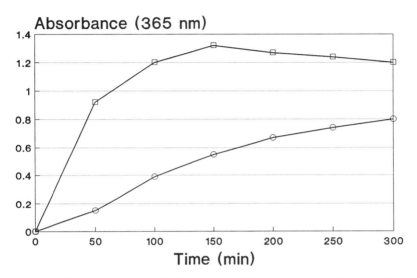

Figure 1 *Kinetics of colour development at 365 nm in reaction between the CHO reagent and t-butyl hydroperoxide. Upper curve, incubation at 50°. Lower curve, incubation at 20°.*

We also carried out the reaction between all the peroxides tested and the CHO reagent at 50°, because in earlier studies[1] we found that even in acid conditions the reaction of organic peroxides with iodide was slow. The upper curve in Figure 1 shows that the *t*-butyl hydroperoxide-CHO reaction was indeed faster at 50° than at 20°, but it still needed 150 mins to reach a maximum. The result also shows a decline of absorbance with time at 50°; this proved to be a general feature of the reactions of all other peroxides tested is this study. Similar results (not shown) were obtained with cumene hydroperoxide.

Qualitatively, a different profile of the kinetics of tri-iodide formation in the reaction of peroxides with the CHO reagent was obtained with oxidised blood plasma (Figure 2). The plasma was oxidised by exposure to γ radiation for 60 min in a cobalt-60 source at 30 Gy/min under an oxygen atmosphere. Iodine was then used to saturate iodine-binding sites in the plasma and the excess iodine and the H_2O_2 generated by radiolysis were removed by dialysis. Incubation with the CHO reagent at 20° produced an absorbance maximum after about 100 min (Figure 2, upper curve). However, the subsequent decline seen with other organic peroxides was also evident. At 50°, the absorbance increase was much slower and smaller (Figure 2, lower curve). Neither procedure gave the correct peroxide concentrations, as determined by the anaerobic iodide technique [3] (Figure 4).

In view of these results, we were surprised to find that reaction between oxidised low-density lipoprotein (LDL) and the CHO reagent at room temperature gave a reliable measure of the amount of peroxide present (Figure 3, upper curve, and Figure 4). At 50°

the absorbance increase was slower, did not reach a stable maximum, and failed to give the right peroxide concentration (Figure 3, lower curve, and Figure 4).

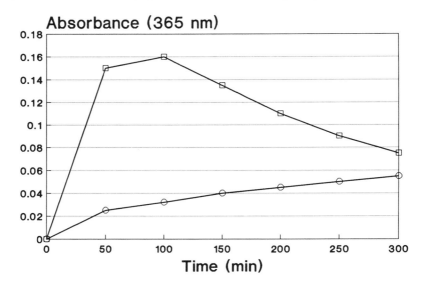

Figure 2 *Kinetics of absorbance development at 365 nm in reaction between the CHO reagent and oxidised human blood plasma. Upper curve, incubation at 20°. Lower curve, incubation at 50°.*

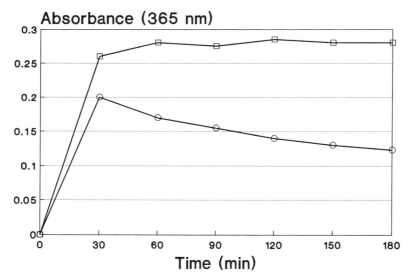

Figure 3 *Kinetics of absorbance development in reaction between the CHO reagent and oxidised low-density lipoprotein. Upper curve, incubation at 20°. Lower curve, incubation at 50°.*

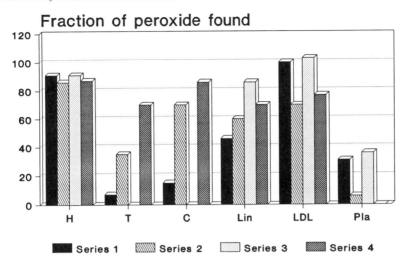

Figure 4 *Fraction of peroxide actually present (as determined by the anaerobic tri-iodide assay) found by the CHO reagent. The peroxides tested are indicated at the bottom of the graph: H = H_2O_2; T = t-Butyl hydroperoxide; C = Cumene hydroperoxide; Lin = linoleic acid hydroperoxide; LDL = low-density lipoprotein hydroperoxide; Pla = oxidised plasma. The bars show results obtained after different incubation times either at 20° or at 50°. Series 1 - 30 min at 20°. Series 2 - 30 min at 50°. Series 3 - maximum absorbance at 20°. Series 4 - maximum absorbance at 50°.*

4 DISCUSSION

The results of this study show that, contrary to many earlier claims, the CHO reagent can provide a reliable measure of peroxide concentrations only for H_2O_2, as originally designed, and for oxidised LDL. While the neutral pH of the reagent solution virtually eliminates its oxygen sensitivity, it greatly reduces the speed of reduction of the peroxides by the iodide. The result is a reaction requiring much longer incubation times than used by El-Saadani *et al.*[4] In the case of H_2O_2, this is overcome by the molybdate catalyst. However, molybdate is ineffective with the organic peroxides used in this study (results not shown). The use of 50° to accelerate the reaction, effective with anaerobic assays,[1] resulted in observable loss of absorbance throughout the incubation, giving unreliable absorbance maxima and lowered peroxide values. This effect was due to a reaction between the I_2 produced in the peroxide-CHO reagent reaction and the Triton detergent included in the reagent. In the case of plasma, there was additional absorption of I_2, apparent even at 20°, and greatly accelerated at 50°.

It may still be possible to obtain a qualitative indication of the presence of peroxides in solutions by the use of the CHO reagent. Our results with oxidised LDL also suggest that other organic peroxides may be capable of accurate determination with this method. However, in each case the kinetics of the peroxide-CHO reaction must be determined, the apparent peroxide concentration in the sample tested calculated with a molar extinction coefficient of 29,000 M^{-1} cm^{-1} and the result compared with parallel measurement obtained with a reliable peroxide method. For this, we recommend the anaerobic technique, in spite of its relative inconvenience.

References

1. W. Jessup, R. T. Dean and J. M. Gebicki, *Meth. Enzymol.*, 1994, **233**, 289.
2. M. Hicks and J. M. Gebicki, *Anal. Biochem.* 1979, **99**, 249.
3. A. D. Awtrey and R. E. Connick, *J. Amer. Chem. Soc.* 1951, **73**, 1842.
4. M. El-Saadani, H. Esterbauer, M. El-Sayed, M. Goher, A. Y. Nassar and G. Jurgens, *J. Lipid Res.*, 1989, **30**, 627.
5. P. Gorog, D. C. Kotak and I. B. Kovacs, *J. Clin. Pathol.*, 1991, **44**, 765.

Role of Antioxidative Vitamins and Carotenoids in Cardiovascular Diseases and Cancers

BIOLOGICAL ACTIVITY OF CAROTENOIDS AND THEIR BIOAVAILABILITY IN THE HUMAN ORGANISM

W. Stahl and H. Sies

Institut für Physiologische Chemie I
Heinrich-Heine-Universität Düsseldorf
P.O. Box 10 10 07
D-40001 Düsseldorf, Germany

1 INTRODUCTION

Carotenoids are structurally unique molecules consisting of a system of conjugated double bonds which is substituted at both ends. The biological properties of these compounds are closely related to their structure; e.g. the provitamin A activity is associated with the presence of at least one unsubstituted ß-ionone ring. The number of conjugated double bonds influences the antioxidant activity of carotenoids with respect to singlet oxygen quenching. Carotenoid distribution in body compartments as well as its biokinetics depend at least in part on their physicochemical properties. They are very lipophilic molecules absorbed and transported via lipid pathways.

2 BIOLOGICAL ACTIVITIES

With respect to biological effects of carotenoids it has been suggested that one differentiates between their functions, actions, and associations[1,2]. According to the following summary, the biological functions of carotenoids have been grouped into these categories.

A. Functions: accessory pigments in photosynthesis, protection against photo-sensitization, vitamin A activity mediated by enzymatic cleavage of the carotenoid.
B. Actions: antioxidant activity, enhancement of immune response, inhibition of mutagenesis and transformation, induction of gap junctional communication.
C. Associations: decreased risk of some cancers, decreased risk for some cardiovascular diseases, decreased risk of cataracts and macular degeneration.

In context with human health, actions of carotenoids as antioxidants and their influence on intercellular communication via gap junctions have been discussed as possible biochemical mechanisms underlying the protective effects observed in epidemiological studies. In vitro and in vivo studies added evidence to such suggestions.

2.1. Carotenoids as antioxidants

As an attribute to aerobic life the human organism is exposed to a variety of different prooxidants capable to damage biologically relevant molecules, such as DNA, proteins, carbohydrates, and lipids[3]. Among the various defense strategies, carotenoids are most likely involved in the scavenging of two of the reactive oxygen species, singlet molecular oxygen ($^1O_2$), and peroxyl radicals. Further they are effective deactivators of electronically excited sensitizer molecules which are involved in the generation of radicals and singlet oxygen[4,5].

The interaction of carotenoids with $^1O_2$ depends largely on physical quenching[6]. The energy of singlet molecular oxygen is transferred to the carotenoid molecule to yield ground state oxygen and a triplet excited carotene. Instead of further chemical reactions, the carotenoid returns to ground state dissipating its energy by interactions with the surrounding solvent. In contrast to physical quenching, chemical reactions between the excited oxygen and carotenoids is of minor importance, contributing less than 0.05% to the total quenching rate. Since the carotenoids remain intact during physical quenching of $^1O_2$ or excited sensitizers, they can be reused severalfold in such quenching cycles. Among the various carotenoids, xanthophylls as well as carotenes proved to be efficient quenchers of singlet oxygen interacting with reaction rates that approach diffusion control. The efficacy of carotenoids for physical quenching is related to the number of conjugated double bonds present in the molecule. Lutein, zeaxanthin, cryptoxanthin, α-, and ß-carotene, all of which are detected in human serum and tissues, belong to the group of highly active quenchers of $^1O_2$. The most efficient carotenoid is the open ring carotenoid lycopene, which contributes up to 30% to total carotenoids in the human[7]. From a clinical point of view, the most important use of ß-carotene is to treat the hereditary photosensitivity disease erythropoietic protoporphyria[8]. It is suggested that the carotenoid intercepts the protoporphyrin triplet formed via electronic excitation of protoporphyrin by light and thus prevents the formation of singlet oxygen which is thought to be the damaging agent responsible for the skin lesions observed in EPP[9].

Carotenoids are also able to deactivate radical-mediated reactions and thus inhibit lipid peroxidation. In a model system using the formation of radical-initiated hydroperoxides of methyl linoleate, canthaxanthin and astaxanthin were shown to be better and longer lasting antioxidants than ß-carotene and zeaxanthin[10]. The antioxidant activity of carotenoids regarding the deactivation of peroxyl radicals likely depends on the formation of radical adducts forming a resonance stabilized carbon-centered radical[11]. However, the postulated intermediate product was not identified as yet. A variety of oxidation products have been detected upon ß-carotene autoxidation and upon interaction with radicals, including carotenoid epoxides and apocarotenoids of different chain length[12]. It should be noted that these compounds might possess biological activities as described for 4-oxoretinoic acid, an autoxidation product of canthaxanthin[13].

The antioxidant activity of carotenoids depends on the oxygen tension present in the system. At low partial pressures of oxygen (15 torr) such as those found in most tissues under physiological conditions, ß-carotene was found to inhibit the oxidation of artificial target molecules like tetralin or methyl linoleate[14]. In contrast, the initial antioxidant activity of ß-carotene is followed by a prooxidant action at high (760 torr) oxygen tension[11]. Similar results were obtained with rat liver microsomes[15] and lipid bilayer systems[16].

Carotenoids apparently interact with other lipophilic antioxidants such as vitamin E. In vitro, a synergism regarding the antioxidant activity has been observed in the inhibition of peroxyl radical induced lipid peroxidation, when ß-carotene was used together with α-tocopherol. Due to an increased consumption of tocopherol in the presence of ß-carotene, this effect was explained as a protection of ß-carotene by the vitamin E molecule[17].

2.2. Carotenoids and gap junctional communication

In cell culture carotenoids and retinoids reversibly inhibit the progression of carcinogen-initated fibroblasts to the transformed state. This inhibitory effect has been found to be related to an increased gap junctional communication induced by these compounds[18]. Gap junctions are composed of channels formed by connexins which connect the cytosol of one cell to that of another. They allow the free flux of molecules of mol. wt. < 1000 Da and are involved in the intercellular exchange of nutrients and signalling compounds. There is growing evidence that gap junctions play a role in the regulation of morphogenesis, cell differentiation, secretion of hormones, and growth control[19,20].

ß-Carotene, canthaxanthin, and lutein are efficient inducers of gap junctional communication, whereas α-carotene, and lycopene are less active. The influence of carotenoids on gap junctional communication does not correlate with their antioxidant activity. Both the carotenoid methyl-bixin and the noncarotenoid α-tocopherol are efficient inhibitors of lipid peroxidation but do not influence gap junctional communication. Since lycopene and canthaxanthin are non-provitamin A carotenoids, it has been suggested that their effects on cell communication are independent from retinoid pathways[21]. In contrast, it was demonstrated that 4-oxoretinoic acid is formed upon incubation of canthaxanthin in cell culture and that sufficient amounts of this compound are generated to induce gap junctional communication[13]. Thus, it is possible that a variety of retinoid analogs are formed from different parent carotenoids, which might possess interesting new biological properties.

The increase of intercellular communication by retinoids and carotenoids is accompanied by higher amounts of connexin m-RNA. This might be due to direct or indirect influences of retinoids and carotenoids on the gene expression of connexins. Indirect effects have been observed in various cell systems were retinoids regulate gap junctional communication by acting at the post-transcriptional level[22]. Retinoids function by binding to nuclear receptors which attach to specific regions of the DNA influencing the synthesis of specific proteins. It will be important to learn whether the connexin genes are directly influenced via this pathway and whether carotenoids are ligands or precursors of receptor ligands.

3 BIOKINETICS

The processes involved in carotenoid uptake and transport in the human organism are complex and are only partially understood. The biokinetics is influenced by various factors including fibre consumption, additional dietary fat components, the vitamin A status, or food processing[23]. After ingestion, carotenoids are incorporated into micelles which are formed from lipids and bile acids. Thus, a diminished bile flux impairs carotenoid availability. The carotenoid-loaded micelles are transferred into intestinal mucosa cells, where part of the carotenoids are metabolized to retinoids. Together with other lipophilic nutrients, carotenoids and retinol esters are assembled into chylomicrons and transported to the blood via the lymphatic system. The chylomicrons are enzymatically modified within the blood stream into chylomicron remnants which are taken up by the liver. In the liver the carotenoids are assembled into VLDL and released to the blood. Therefore, carotenoids first appear in the chylomicron fraction, then in VLDL and subsequently in LDL and HDL. Hydrocarbon carotenoids such as ß-carotene and lycopene are mainly found in the LDL fraction, whereas the more polar carotenoids are more equally distributed between LDL and HDL. Lipoproteins are the exclusive transport form of carotenoids in the human organism, and it has been shown that tissues equipped with lipoprotein receptors contain rather high amounts of these compounds. Carotenoid levels in liver or adrenals by far exceed the concentrations in other tissues such as kidney or fat. Thus, the lipoprotein association of carotenoids allows a more or less directed transport in the body[24].

3.1. Lycopene uptake

Lycopene is one of the major carotenoids in human blood and tissues and is almost exclusively provided by tomatoes and tomato products. Because of its high content of lycopene, tomato juice (50-120 mg/L) is a suitable source to study lycopene biokinetics[25].

After consumption of tomato juice equivalent to a single dose of 2.5 μmol lycopene/kg body weight, no increases in lycopene serum levels were detected within 7 days. Significant increases in lycopene serum levels were only observed when processed tomato juice was ingested. For processing, 10 ml of corn oil were added to 1 L of tomato juice and the mixture was heated to 100°C for 1h. This result clearly demonstrates that apart from the amount of carotenoid ingested, additional factors such as dietary fat, or release of the carotenoid from matrix components influence its bioavailability. The lycopene uptake from this dietary source varied with individuals, but peak serum concentration were always observed between 24-48h after ingestion. The apparent half-life of lycopene in human serum was about 2-3 days. The increases of lycopene serum levels upon ingestion of different amounts of tomato juice obviously depended on the dose but were not linear with the dose. Relatively more lycopene was absorbed when lower amounts were consumed. The repeated ingestion of tomato juice (4 times at 24h intervals) led to a continuous increase in lycopene serum levels for at least 5 days. Although the total dose of lycopene was lower as compared to a single-dose experiment with the same individual, the increases in lycopene serum levels were about the same. This finding is also indicative for improved absorption of small doses of carotenoids.

3.2. ß-Carotene uptake

Upon treatment with heat, light, or catalytic amounts of iodine carotenoids tend to isomerize and form a mixture of mono and poly cis isomers in addition to the all-trans form. The most important mono cis isomers of ß-carotene are the 9-, 13-, and 15-cis. Generally the all-trans form is predominant in nature, but organisms such as the halotolerant alga Dunaliella salina synthesize considerable amounts of cis isomers. In the human, the ß-carotene pattern of serum is dominated by the all-trans isomer, and only small amounts of cis-isomers (about 5% of total ß-carotene) are detected[26]. In contrast, considerable concentrations of different cis-isomers are found in human tissues such as liver, adrenals, or testes. With respect to cis isomers, interest was mainly focussed on the 9-cis form since it has been demonstrated that it is an isomer-selective precursor for 9-cis retinal[27]. 9-cis Retinal can be further oxidized in vivo to yield 9-cis retinoic acid which is a selective ligand for the retinoic X receptor (RXR). The RXR-ligand complex is a nuclear transcription factor likely involved in the regulation of specific genes[28]. It is capable to form heterodimers with other receptors of the same superfamily including the thyroid, vitamin D, and steroid receptors and therefore presumably interactive with a variety of signalling pathways.

In view of the importance of 9-cis retinoic acid in regulatory processes, more information is needed on the uptake, distribution, and metabolism of its precursor 9-cis ß-carotene. Betatene, an extract of the alga Dunaliella salina which contains up to 40% of the 9-cis isomer in its ß-carotene fraction, was used to study the biokinetics of this compound. Upon ingestion of a single dose of Betatene equivalent to 3.0 µmol of all-trans ß-carotene/kg body weight and 2.1 µmol of 9-cis ß-carotene/kg body weight, only the level of the all-trans isomer rose in human serum. No significant increase in the 9-cis ß-carotene level was observed when repeated doses of Betatene were applied[29], even after ingestion of high amounts of Betatene over a period of 150 days in the treatment erythropoietic protoporphyria[30]. Comparable results were obtained when analyzing the chylomicron fraction of human serum after Betatene ingestion[31]. The data suggest the existence of efficient discrimination mechanisms between the two isomers operative at the absorption step. Recently, a small increase in the concentration of 9-cis ß-carotene has been described after application of a 50:50 isomer mixture in the human[32]. However, the 9-cis concentrations were only a small fraction of the total plasma ß-carotene, confirming a strong discrimination between the all-trans and the 9-cis form. Interestingly, the 9-cis ß-carotene in human plasma also rose after ingestion of pure all-trans ß-carotene, possibly due to in vivo isomerization reactions.

3.3. Uptake of carotenoid esters

Xanthophylls carry at least one oxygen atom as keto- or hydroxy groups in their structure. The hydroxy group might be esterified with a variety of fatty acids, and an array of carotenol esters are detected in fruits and vegetables such as oranges, tangerines, or papayas[33]. Regarding dietary xanthophyll esters, those of ß-cryptoxanthin are of particular interest, since this carotenoid is a major source of vitamin A, often second only to ß-carotene. Cryptoxanthin and other carotenols like zeaxanthin and lutein are found in considerable amounts in human blood and tissues, but they occur only in their free and not in the esterified form. Thus, it has been suggested that carotenol esters are cleaved before entering the systemic circulation.

Upon ingestion of a tangerine concentrate rich in ß-cryptoxanthin esters, increasing amounts of free ß-cryptoxanthin were detected in human serum and in the chylomicron fraction of the lipoproteins[34]. The cryptoxanthin levels in chylomicrons rose from 0 to 6h after ingestion and returned towards basal levels at t = 9h. No cryptoxanthin esters were detected in chylomicrons or serum. The lack of carotenoid esters in the chylomicron fraction indicates that efficient cleavage enzymes are operative before the carotenoid is incorporated into other lipoproteins by the liver. Esters of lutein and zeaxanthin were present in small amounts in the tangerine juice concentrate. Like with ß-cryptoxanthin, no esters of these carotenoids were detected in serum or in chylomicrons. The data suggest that cleavage of carotenoid esters prior to release into the lymphatic circulation generally occurs in the human biokinetics of carotenoids.

4 CONCLUSION

Carotenoids are absorbed, transported, and metabolized in the human organism and serve as a source for vitamin A in the human organism. Thus, they are important micronutrients. Controversial data have been presented regarding the cancer protective properties of carotenoids, and data on the biochemical mechanisms underlying these effects are still missing. Since carotenoids occur widely in the human diet, further research is warranted to identify the biological properties of these micronutrients.

References

1. J. A. Olson, 1989, *J. Nutr.*, **119**, 94
2. N.I. Krinsky, 1993, *Annu. Rev. Nutr.*, **13**, 561
3. H. Sies, 1986, *Angew. Chem. Int. Ed. Engl.* **25**, 1058
4. C.S. Foote, and R. W. Denny, 1968, *J. Am. Chem. Soc.* **90**, 6233
5. W. Stahl, A.R. Sundquist, and H. Sies, 1994, in: Vitamin A in Health and Disease, Ed. R. Blomhoff, Marcel Dekker, NY, pp. 275
6. P.F. Conn, C. Lambert, E.J. Land, W. Schalch, and T.G. Truscott, 1992, *Free Rad. Res. Commun.* **16**, 401
7. P. Di Mascio, S. Kaiser, and H. Sies, 1989, *Arch. Biochem. Biophys.* **274**, 532
8. M. M. Mathews-Roth, M..A. Pathak, T.B. Fitzpatrick, L.C. Harber, and E.H. Kass, 1974, *JAMA* **228**, 1004
9. T.G. Truscott, 1990, *J. Photochem. Photobiol.* B., **6**, 359
10. J. Terao, 1989, *Lipids*, **24**, 659
11. G.W. Burton, and K.U. Ingold, 1984, *Science*, **224**, 569
12. G.J. Handelman, F.J.G.M. van Kuijk, A. Chatterjee, and N.I. Krinsky, 1991, *Free Rad. Biol. Med.* **10**, 427
13. M. Hanusch, W. Stahl, W.A. Schulz, and H. Sies, 1995, *Arch. Biochem. Biophys.*, **317**, 423
14. G.W. Burton, 1989, *J. Nutr.* **119**, 109
15. P. Palozza, and N.I. Krinsky, 1991, *Free Rad. Biol. Med.*, **11**, 407
16. T.A. Kennedy, and D.C. Liebler, 1992, *J. Biol. Chem.* **267**, 4658
17. P. Palozza, and N.I. Krinsky, 1992, *Arch. Biochem. Biophys.* **297**, 184
18. M.Z. Hossain, and J.S. Bertram, 1994, *Cell Growth & Differentiation.* **5**, 1253
19. H.A. Kolb, and R.S. Somogyi, 1991, *Rev. Physiol. Biochem. Pharmacol.* **118**, 1
20. J.W. Holder, E. Elmore, and J.C. Barrett, 1993, *Canc. Res.* **53**, 3475

21. A. Pung, J.E. Rundhaug, C. N. Yoshizawa, and J.S. Bertram, 1988, *Carcinogenesis*, **9**, 1533
22. V. Bex, T. Mercier, C. Chaumontet, I. Gaillard-Sanchez, B. Flechon, F. Mazet, O. Traub, and P. Martel, 1995, *Cell Biochem. Function* **13**, 69
23. J.W. Erdman, T.L. Bierer, and E.T. Gugger, 1993, *Ann. NY Acad. Sci.* **691**, 76
24. J.A. Olson, 1994, *Pure & Appl. Chem.* **66**, 1011
25. W. Stahl, and H. Sies, 1992, *J. Nutr.* **122**, 2161
26. W. Stahl, W. Schwarz, A.R. Sundquist, and H. Sies, 1992, *Arch. Biochem. Biophys.* **294**, 173
27. X.D. Wang, N.I. Krinsky, P.N. Benotti, and R.M. Russell, 1994, *Arch. Biochem. Biophys.* **313**, 150
28. D.J. Mangelsdorf, 1994, *Nutr. Rev.* **52**, 32
29. W. Stahl, W. Schwarz, and H. Sies, 1993, *J. Nutr.* **123**, 847
30. J. von Laar, W. Stahl, K. Bolsen, G. Goerz, and H. Sies, 1995, *J. Photochem. Photobiol. B: Biol.* in press
31. W. Stahl, W. Schwarz, J. von Laar, and H. Sies, 1995, *J. Nutr.* **125**, 2128
32. J.M. Gaziano, E.J. Johnson, R.M. Russell, J.E. Manson, M.J. Stampfer, P.M. Ridker, B. Frei, C.H. Hennekens, and N.I. Krinsky, 1995, *Am. J. Clin Nutr.* **61**, 1248
33. F. Khachik, G. R. Beecher, M.B. Goli, and W. Lusby, 1991, *Pure & Appl. Chem.*, **63**, 71
34. T. Wingerath, W. Stahl, and H. Sies, 1995, *Arch. Biochem. Biophys.* submitted

DESIRABLE VERSUS POTENTIALLY HARMFUL INTAKE LEVELS OF VARIOUS FORMS OF CAROTENOIDS

W. Stahl

Institut für Physiologische Chemie I
Heinrich-Heine-Universität Düsseldorf
P.O.Box 10 10 07
D-40001 Düsseldorf, Germany

1 INTRODUCTION

Epidemiological studies as well as in vitro and in vivo experiments support the idea that carotenoids are associated with protection against some common diseases such as cancer, cataract, or cardiovascular events[1,2]. The association of beneficial health effects with carotenoids initiated a discussion about optimal intake, protective serum levels, suitable carotenoid sources, and supplementation of carotenoids in addition to the diet. However, in a recent intervention trial with heavy smokers an increased risk for lung cancer was epidemiologically correlated with the intake of ß-carotene[3]. Although this study is still discussed controversially, the outcome raises questions about the safety of ß-carotene supplementation, especially in populations with high risk such as smoking, probably in late stages of malignant transformation.
   Apart from the provitamin A activity of some carotenoids, further biological properties of these compounds in mammals have been identified but as yet are poorly understood. Thus, the interpretation of biological effects remains a topic of study. The chemopreventive activities of carotenoids have been interpreted in context with their actions as antioxidants, enhancers of immune response, and inhibitors of mutagenesis and transformation, but an unequivocal confirmation of such relations is lacking[4].

2 VARIOUS CAROTENOIDS

Some of the difficulties regarding intake recommendations arise from the variety of carotenoids which are part of the human diet. It has been suggested that more than 40 different carotenoids are available from the diet and are absorbed, metabolized, or utilized in the human organism[5]. However, only the distribution of some major constituents such as lutein, zeaxanthin, cryptoxanthin, lycopene, α-, and ß-carotene have been analyzed more closely in human serum or tissues[6-8]. Most of the investigations on carotenoid biokinetics or biological effects were focussed on ß-carotene, while little is known about other carotenoids. Some studies, however, revealed specific properties of carotenoids other than ß-carotene which might also be relevant in human health.

Lycopene is a very efficient quencher of singlet molecular oxygen, likely involved in the lipophilic antioxidant defense system of the organism[9]. In contrast to other tissues where ß-carotene is predominant, lycopene was found to be the major carotenoid in human testes[7,10]. It is not known whether it has a specific function in this tissue or whether the shift in the carotenoid pattern towards lycopene is due to other causes, e.g. an enhanced degradation of ß-carotene to vitamin A in testes.

ß-Cryptoxanthin is present in human blood and tissues at levels generally exceeding the levels of α-carotene. Rather high amounts of this carotenoid are found in citrus fruits such as oranges and tangerines, major sources of cryptoxanthin in the human diet (Wingerath et al, unpublished data). The compound shows provitamin A activity but is not cleaved by the enzyme ß-carotenoid-15,15'-dioxygenase.[11].

Zeaxanthin and lutein are common oxocarotenoids (xanthophylls) and account for 15-30% of total carotenoids in the human blood. Both compounds are non provitamin A carotenoids, and they are efficient antioxidants scavenging singlet molecular oxygen and peroxyl radicals[12]. Lutein and zeaxanthin are the major carotenoids in the macula lutea in the retina of primates, and the optical density of the macular pigment correlates with the amount of lutein and zeaxanthin in the fovea[13,14].

A higher dietary intake of carotenoids is associated epidemiologically with a lower risk for advanced age related macular degeneration. Among various carotenoids the protection was most strongly associated with lutein and zeaxanthin. This also correlates with a higher frequency of intake of vegetables such as spinach and a diminished risk for macular degeneration[15]; spinach is especially rich in lutein. Antioxidant properties of carotenoids have been discussed in context with these protective effects[16]. However, the consumption of other lipophilic antioxidants, e.g. vitamin E, was not associated with a diminished risk for macular degeneration. The mechanisms underlying the selective uptake or incorporation of these xanthophylls into the macula are still unknown.

Very similar results were obtained in an epidemiological study on nutrients and cataract. The intake of carotene and vitamin A was inversely associated with the risk for cataract in women. Among specific food items, spinach rather than carrots was most consistently associated with a diminished risk. Therefore, it was speculated that carotenoids other than ß-carotene, e.g. lutein and zeaxanthin, might be protective in the genesis of this disease[17]. An increased consumption of other antioxidants such as tocopherols or ascorbic acid is also associated with a delayed development of cataract. Evaluating the role of nutrient intake, significant increases were noted in the risk of cataract for persons who consume fewer than three and a half servings of fruits and vegetables a day[18]. Depending on structural properties, carotenoids are differently orientated within a lipid bilayer matrix. Such physicochemical properties of different carotenoids might be of importance with respect to stabilization effects on cell compartments.

This short compilation shows that carotenoids of different structure likely exert different biological effects which should be taken into consideration for recommendations on carotenoid intake, especially when natural sources are recommended for carotenoid supply.

3 VARIOUS FORMS OF CAROTENOIDS

The natural sources of carotenoids are colored and green vegetables as well as various fruits such as oranges, tangerines, apricots, cantaloupe, or mango[19]. However, the uptake of carotenoids after the consumption of raw fruits and vegetables appears to be rather poor. In response to a single ingestion of raw carrots, broccoli, or tomato juice there was only little if any increase in human serum levels of ß-carotene or lycopene[20, 21]. Mild cooking and additional ingestion of dietary fat improves carotenoid absorption. This is likely due to the release of carotenoids from cellular compartments of the plants (socalled biological matrix) upon heating and the formation of carotenoid containing micelles from dietary fat which facilitate carotenoid uptake in the gut.

Adverse effects on carotenoid absorption have been associated with the intake of large amounts of fibre[22]. The increase in plasma ß-carotene was significantly restricted by pectin as compared to a control group. This might provide another explanation for diminished plasma ß-carotene response in humans after ingestion of carotenoid-rich food when compared with equivalent doses of ß-carotene supplements.

The particle size of unprocessed food is particularly important. The consumption of pureed or finely chopped vegetables yield considerably higher ß-carotene absorption as compared to the intake of whole or sliced raw vegetables [23]. Thus, recommendations on carotenoid supply from dietary sources should include advice on food-processing procedures suitable to improve the bioavailabilty of these compounds.

Within the last ten years there was a growing use of carotenoid supplements with either carotenoids alone or in combination with other vitamins and micronutrients. Most of these preparations contain synthetic all-trans ß-carotene, but there also are carotenoid mixtures extracted or enriched from natural sources. Several microorganisms produce considerable amounts of carotenoids, most likely for their own photoprotection[24]. The halophilic algae Dunaliella salina or Dunaliella bardawil contain different geometrical isomers of ß-carotene which amount up to 10% of the dry weight of this organism[25]. Lipophilic extracts of these algae are used in supplements and are utilized by the human organism[26, 27]. The influence of isomer composition on the bioavailability of ß-carotene in the human is discussed in a separate contribution (see: W. Stahl, H. Sies, this issue). Another natural source rich in carotenoids is palm oil which is mainly produced in Southeast Asia, Malaysia, and Indonesia. Its carotenoid fraction is composed of about 60% of ß-carotene, 35% of α-carotene, and some minor carotenoids as γ-, and δ-carotene[28]. Further enrichment procedures yield palm oil carotene containing 95% of carotenoids.

Specially cultivated lycopene-rich tomatoes are used as a natural source for this carotenoid. They also contain other compounds such as lycopene derivatives and ß-carotene.

4 INTAKE LEVELS

The intake levels of carotenoids are mainly influenced by the selection of food and thus dependent on individual preferences. Supplementation with specific carotenoids or increased consumption of a dietary component rich in specific carotenoids might shift the carotenoid pattern within an organism[29]. It has been suggested that plasma carotenoid concentrations are reflective of the dietary intake. The magnitude of correlation, however, varies [8,30] depending on the specific carotenoid and on the dietary assessment tool. The plasma concentrations of carotenoids in humans consuming a diet low in carotenoids continuously decrease[6]. Since the decline is not linear over a two week period, the presence of at least two different body pools was discussed, with one pool having a more rapid turnover. The mean residence time of carotenoids in tissues is unknown. Thus, little information is available about the correlation between plasma levels and tissue levels.

Broad variations in the individual response of serum carotenene after a single dose of ß-carotene have been observed in several studies[31, 32]. Some individuals were identified as consistently being nonresponders, probably due to genetically determined differences in carotenoid absorption or metabolism, as compared to the socalled responders. This is a further aspect to be considered for intake recommendations.

About 50 out of the 600 known carotenoids possess provitamin A activity [11] and thus are sources for vitamin A. It has been estimated that in the United States about 25% of vitamin A is delivered by carotenoids. Somewhat lower values were reported for Finland with 10-18% [33]. Especially in developing countries where vitamin A deficiency is one of the most serious and widespread nutritional disorders of children, carotenoids from fruits and vegetables represent a major source of vitamin A[34].

In addition to individual and regional differences, carotenoid intake levels appear to differ between seasons. In a Spanish study higher plasma levels of α-, and ß-carotene were detected in summer whereas ß-cryptoxanthin was present in greater proportions in winter. The seasonal variations did not affect zeaxanthin, lycopene, or retinol[35]. In Finland high ß-carotene serum concentrations were detected in spring and fall, likely reflecting the seasonality of dietary sources for this carotenoid in this Northern country[36]. However, no systematic trend on a seasonal influence on carotenoid serum concentrations were reported in a study performed in the United States[37].

5 DESIRABLE INTAKE LEVELS

In Western countries the intake of vitamin A from dietary sources including carotenoids is sufficient to prevent vitamin A deficiency, avoiding the classical ocular manifestations, such as xerophthalmia. An increased supply with carotenoids or carotenoid-rich fruits and vegetables should be suitable to improve the situation in developing countries. Such strategies include encouraging people to use the locally available plant food as a carotenoid source.

The recommended nutrient intakes (RNI) of vitamin A vary when looking at different references. The WHO and several individual countries refer to basal values which reflect an intake of vitamin A that just satisfies the needs of nearly all subjects of the population; 180 µg retinol equivalents for young infants, 300 µg equivalents for adult males, and 270 µg retinol equivalents for adult females per day. The socalled safe RNI of vitamin A reflects an intake of vitamin A that provides suitable reserves for nearly all subjects and are given with 350, 600, and 500 µg retinol equivalents per day, respectively. Unfortunately, considerable confusion exists on the equivalence of carotenoids and vitamin A. Between 2 and 6 µg ß-carotene are suggested to be equivalent to 1 µg of retinol[38].

Little information on the intake levels of carotenoids to achieve optimal protection against the degenerative diseases mentioned above is available. However, there seems to be consensus that the diet should contain various carotenoids including the provitamin A and non-provitamin A compounds. Such considerations are also reflected by the more general advice to consume five servings of fruits and vegetables per day [39, 40].

As a result of several epidemiological studies, serum concentrations of carotenoids have been correlated with reduced risks for several disease states. It should be mentioned, however, that the blood levels of ß-carotene may simply represent fruit and vegetable consumption as a marker and could reflect the ingestion of other, more important, protective factors that are also present in the diet [41].

An increased risk for ischemic heart disease has been associated with diminished plasma levels of antioxidants such as α-carotene and ß-carotene. The blood levels of these carotenoids necessary for protection have been estimated to 0.4-0.5 µmol/L (both compounds together)[42]. This value is close to the average concentration of α- and ß-carotene in healthy subjects reported in various studies.

The potential role of ß-carotene in the prevention of cardiovascular diseases has been mainly ascribed to their antioxidant properties in context with the inhibition of LDL oxidation[1]. Low ß-carotene plasma levels were also detected in Alzheimer's patients and in multi-infarct dementia patients. As compared to controls (0.13-1.53 µmol/L), the concentration of ß-carotene in the plasma of Alzheimer's patients varied from 0.13-0.42 µmol/L. In multi-infarct dementia patients ß-carotene values between 0.13 and 0.30 µmol/L were measured[43].

The results from several epidemiological studies suggest an increased risk for neoplasms such as lung cancer or stomach cancer to be correlated with lowered blood levels of carotenoids[2]. In a 12 year follow-up study conducted in Switzerland, the plasma levels of carotenes (α- plus ß-carotene) of controls were about 0.44 µmol/L whereas significantly lower concentrations were detected in lung cancer (0.29 µmol/L), and stomach cancer patients (0.28 µmol/L)[44].

Table 1: *Desirable plasma levels and dietary intake of antioxidant vitamins (modified from ref.: 40).*

	Optimal Plasma Level (µmol/l)	Dietary Intake (mg/day)
ß-Carotene	> 0.4	2 - 4
α-Tocopherol	> 30	15 - 30
Vitamin C	> 50	75 - 150

6 HARMFUL INTAKE LEVELS

One of the major concerns in context with high ß-carotene intake levels was associated with provitamin A activity, since vitamin A exhibits toxic and teratogenic properties. No hypervitaminosis A was observed upon ß-carotene ingestion even when extremely high doses were applied as e.g. in the treatment of erythropoietic protoporphyria (EPP)^{45}. This photosensitivity disease is treated with ß-carotene doses exceeding 100 mg/day^{46}. However, the vitamin A levels in EPP-patients are within the normal range. This suggests that effective mechanisms are operative in the control of the formation of retinoids from carotenoid precursors.

ß-Carotene appears to cross the placental barrier because carotenemic babies have been born from carotenemic mothers. With exception of the yellow color of the skin no adverse effects have been observed in these children. ß-Carotene also proved to be not active in mutagenicity and teratogenicity tests^{47,48}. Thus, the compound might be considered as safe even in high doses.

In contrast, adverse effects have been described upon the ingestion of high amounts of canthaxanthin. The compound appears to accumulate in the eye, forming canthaxanthin-rich deposits which impair vision, especially with a delayed dark adaptation period. Morphologically, lipid-soluble crystals appear in the inner layers of the entire retina and amounted to 42 µg of canthaxanthin/gram tissue $^{49, 50}$. After cessation of canthaxanthin treatment, the symptoms apparently disappeared, whereas crystalline deposits at the retina showed no signs of reversibility. No retinopathy was observed under the treatment with ß-carotene^{51}.

Recent results from experimental and epidemiological studies gave rise to questions about the treatment of specific risk groups with high amounts of ß-carotene. Upon application of high doses of ß-carotene to primates which simultaneously were treated with ethanol, an enhanced hepatotoxicity was observed as compared to alcohol or ß-carotene controls^{52}. In the animals fed alcohol and ß-carotene, multiple ultrastructural lesions appeared, including degeneration of the mitochondria accompanied by the release of mitochondrial enzymes into the blood. On cessation of ß-carotene treatment the plasma levels decreased more slowly in the alcohol-treated animals, indicating an impaired metabolism of this carotenoid.

In an intervention trial with Finnish smokers a significantly higher incidence of lung cancer was associated with the intake of 20 mg ß-carotene/day over a 6-year period^{3}. The latter finding raised concerns regarding the recommendation of ß-carotene in the prevention of lung cancer for this high-risk group. The study also initiated a renewed discussion about the biochemical mechanisms underlying potential adverse effects of carotenoids.

References

1. H. Gerster, 1991, *Internat. J. Vit. Res.*, **61**, 277
2. H. Gerster, 1993, *Internat. J. Vit. Res.*, **63**, 93
3. ATBC Cancer Prevention Study Group, 1994, *N. Engl. J. Med.*, **330**, 1029
4. N.I. Krinsky, 1993, *Annu. Rev. Nutr.*, **13**, 561

5. F. Khachik, G. R. Beecher, M. B. Goli, W.R. Lusby, and J. C. Smith, 1991, *Anal. Chem.*, **64**, 2111
6. C.L. Rock, M.E. Swendseid, R.A. Jacob, and R.W. McKee, 1992, *J. Nutr.*, **122**, 96
7. W. Stahl, W. Schwarz, A.R. Sundquist, and H. Sies, 1992, *Arch. Biochem. Biophys.*, **294**, 173
8. L.-C. Yong, M.R. Forman, G.R. Beecher, B.I. Graubard, W.S. Campbell, M.E. Reichmann, P.R. Taylor, E. Lanza, J.M. Holden, and J.T. Judd, 1994, *Am. J. Clin. Nutr.*, **60**, 223
9. P. Di Mascio, S. Kaiser, and H. Sies, 1989, *Arch. Biochem. Biophys.*, **274**, 532
10. L. A. Kaplan, J.M. Lau, and E.A. Stein, 1990, *Clin. Physiol. Biochem.*, **8**, 1
11. W. S. Blaner, and J.A. Olson, in: the Retinoids, eds.: M.B. Sporn, A.B. Roberts, and D.S. Goodman, Raven Press NY, 1994, p. 229
12. W. Stahl, A.R. Sundquist, and H. Sies, 1994, in: Vitamin A in Health and Disease, ed.: R. Blomhoff, Marcel Dekker NY, 1994, p. 275
13. G.J. Handelman, D..M. Snodderly, N.I. Krinsky, M.D. Russett, and A.J. Adler. 1991, *Invest. Ophthalmol. Vis. Sci.*, **32**, 257
14. R.A. Bone, J.T. Landrum, G.W. Hime, A. Cains, and J. Zamor, 1993, *Invest. Ophthalmol. Vis. Sci.*, **34**, 2033
15. Eye Disease Case-Control Study Group, 1994, *JAMA*, **272**, 1413
16. W. Schalch, in: Free Radicals and Aging, eds.: I. Emerit, and B. Chance, Birkhäuser Verlag, Basel, Switzerland, 1992, p. 280
17. S.E. Hankinson, M.J. Stampfer, J.M. Seddon, G.A. Colditz, B. Rosner, F.E. Speizer, and W. C. Willett, 1992, *Brit. Med. J.*, **305**, 335
18. A. Taylor, 1993, *J. Am. Coll. Clin. Nutr.*, **12**, 138
19. F. Khachik, G. R. Beecher, and M. B. Goli, 1991, *Pure Appl. Chem.*, **63**, 71
20. W. Stahl, and H. Sies, 1992, *J. Nutr.*, **122**, 2166
21. M.S. Micozzi, E.D. Brown, B.K. Edwards, J.G. Bieri, P.R. Taylor, F. Khachik, G.R. Beecher, and J.C. Smith, 1992, *Am. J. Clin. Nutr.* **55**, 1120
22. C.L. Rock, and M.E. Swendseid, 1992, *Am. J. Clin. Nutr.*, **55**, 96
23. J. W. Erdman, T. L. Bierer, and E. T. Gugger, 1993, *Ann. NY Acad. Sci.*, **691**, 76
24. H. J. Nelis, and A. P. De Leenheer, 1991, *J. Appl. Bacteriol.*, **70**, 181
25. A. Ben-Amotz, and M. Avron, 1990, *Trends Biotechnol.*, **8**, 121
26. W. Stahl, W. Schwarz, and H. Sies, 1993, *J. Nutr.*, **123**, 847
27. J.M. Gaziano, E.J. Johnson, R.M. Russell, J.E. Manson, M.J. Stampfer, P.M. Ridker, C.H. Hennekens, and N.I. Krinsky, 1995, *Am. J. Clin. Nutr.*, **61**, 1248
28. A. S.H. Ong, and E.S. Tee, 1992, *Meth. Enzymol.*, **213**, 142
29. M.R. Prince, J.K. Frisoli, M..M. Goetschkes, J.M. Stringham, and G.M. LaMuraglia, 1991, *J. Cardiovasc. Pharmacol.*, **17**, 343
30. A. Ascherio, M.J. Stampfer, G.A. Colditz, E.B. Rimm, L. Litin, and W. C. Willett, 1992, *J. Nutr.*, **122**, 1792
31. E.D. Brown, M.S. Micozzi, and N.E. Craft, 1989, *Am. J. Clin. Nutr.*, **49**, 1258
32. W. Stahl, W. Schwarz, J. von Laar, and H. Sies, 1995, *J. Nutr.*, **125**, 2128
33. M. Heinonen, 1989, *Internat. J. Vit. Nutr. Res.*, **61**, 3
34. K.P. West, in: Vitamin A in Health and Disease, ed.: R. Blomhoff, 1994, p. 585
35. B. Olmedilla, F. Granado, I. Blanco, and E. Rojas-Hidalgo, 1994, *Am. J. Clin. Nutr.*, **60**, 106

36. M. Rautalahti, D. Albanes, J. Haukka, E. Roos, C.-G. Gref, and J. Virtamo, 1993, *Am. J. Clin. Nutr.*, **57**, 551
37. L.R. Cantilena, T.A. Stukel, E.R. Greenberg, S. Nann, and D.W. Nierenberg, 1992, *Am. J. Clin. Nutr.*, **55**, 659
38. J.A. Olson, 1994, *Nutr. Rev.* **52**, s67
39. W.C. Willett, 1994, *Science*, **264**, 532
40. H. K. Biesalski et al, 1995, *Dt. Ärztebl.*, **92**, 1316
41. R. Ziegler, 1992, *Cancer Res.*, **52**, 2060s
42. K.F. Gey, U.K. Moser, P. Jordan, H.B. Stähelin, M. Eichholzer, and E. Lüdin, 1993, *Am. J. Clin. Nutr.*, **57**, 787s
43. Z. Zaman, S. Roche, P. Fielden, P.G. Frost, D.C. Niriella, and A.C.D. Cayley, 1992, *Age & Ageing*, **21**, 91
44. H.B. Stähelin, K.F. Gey, M. Eichholzer, and E. Lüdin, 1991, *Am. J. Clin. Nutr.*, **53**, 265s
45. M.B. Lewis, 1972, *Aust. J. Derm.*, **13**, 75
46. M. M. Mathews-Roth, M. A. Pathak, T.B. Fitzpatrick, L.C. Hatber, and E.H. Kass, 1974, *JAMA*, **228**, 1004
47. A. Bendich, 1988, *Nutr. Cancer*, **11**, 207
48. M. M.. Mathews-Roth ,1988, *Toxicol. Lett.*, **41**, 185
49. B. Daicker, K. Schiedt, J.J. Adnet, and P. Bermont, 1987, *Graefe's Arch. Ophthalmol.*, **225**, 189
50. U. Weber, and G. Goerz, 1985, *Dt. Ärztebl.*, **82**, 181
51. M.B. Poh-Fitzpatrick, and L.G. Barbera, 1984, *J. Am. Acad. Dermatol.*, **11**, 111
52. M.A. Leo, C.-I. Kim, N. Lowe, and C.S. Lieber, 1992, *Hepatol.*, **15**, 883

DESIRABLE VERSUS POTENTIALLY HARMFUL LEVELS OF VITAMIN E INTAKE

H. Korpela

Research Institute of Public Health, University of Kuopio,
P. O. Box 1627, FIN-70211 Kuopio, Finland

1 INTRODUCTION

Vitamin E occurs in the diet in eight different forms: d-α-, d-ß-, d-γ-, and d-δ-tocopherols and d-α-, d-ß-, d-γ- and d-δ-tocotrienols. Of the eight forms, d-α-tocopherol (RRR-α-tocopherol) has the highest biological activity.[1-3] On the other hand, d-α-tocopherol and d-α-tocotrienol have the highest antioxidant activity in lipoproteins.[3,4]

Although the antioxidant activity of γ-tocopherol (a major form of vitamin E in human diets) is about 50%, the biological activity is only about 10% of RRR-α-tocopherol.[1-3] The high biologic activity of RRR-α-tocopherol probably results from the activity of the hepatic α- tocopherol-binding protein which seems to regulate RRR-α-tocopherol transport into plasma very low-density lipoproteins (VLDL) and determines plasma concentrations of α-tocopherol[1-3] Other forms of vitamin E (such as γ-tocopherol or SRR-α-tocopherol) and excess RRR-α-tocopherol are excreted in the bile.

2 EFFECT OF VITAMIN E ON LDL OXIDATION

The oxidation resistance of LDL depends not only its vitamin E content but also on the ratio of PUFAs to saturated and monounsaturated fatty acids, and the amount of other antioxidants, cholesterol and endogeneous peroxide in LDL.[1,5-7] Human LDL rich in oleic acid and poor in linoleic acid is less susceptiple to oxidative modification. The PUFA and antioxidant content of LDL varies widely from person to person and the mean ratio of antioxidants to PUFA is about 1:170. Although food high in PUFA often contains high amounts of vitamin E, this is not always the case.[1] Vegetable oils (corn oil, soybean oil) are relatively rich in γ-tocopherol and therefore increasing PUFA in the diet may not increase RRR-α-tocopherol intake to maintain an ideal ratio of above 0.4 (mg RRR-α-tocopherol per g PUFA).

Each human LDL particle contains about 6 tocopherol molecules and LDL α-tocopherol contributes to about 30 - 50% of the oxidation resistance.[6-8] In humans low dose of α-tocopherol supplementation decreases the susceptibility of LDL to oxidation.[6-8] Natural α-tocopherol and α-tocotrienol are potentially beneficial antioxidants for the prevention of LDL oxidation.[1,3,4,7,9] The minimum dose of synthetic α-tocopherol (dl-α-tocopherol) needed to significantly decrease the susceptibility of LDL to oxidation in nonsmoking, normolipidemic men is 400 IU/day.[10]

3 INTAKE AND REQUIREMENT OF VITAMIN E

The optimal level of vitamin E intake for disease prevention is not known and it is difficult to determine the exact amount of vitamin E that would provide maximum protection against oxidative damage and degenerative diseases. The Recommended Daily Allowances (RDA) of vitamin E for men and women is 15 IU (10 mg) and 12 IU (8 mg), respectively.[1,11]

However, the RDA is designed to meet basic nutritional needs to prevent vitamin E deficiency in healthy people but it does not take into account factors that increase oxidative stress such as smoking, exercise, infection, alcohol intake, obesity and/or hyperlipidemia.[5,11-13] Oxidative stress promotes the development of oxidative damage. On the other hand, tissue damage can cause oxidative stress.[12] Several studies support the concept that the desirable intake of vitamin E should be at least 100 IU/day for individuals exposed to oxidative stress.[14-17]

Prospective epidemiological studies suggest that use of vitamin E supplements may reduce heart disease risk by up to 40%.[9,10] After adjusting for major cardiovascular risk factors, use of vitamin E supplements for more than two years at least 100 IU/day was associated with a 41% and 37% lower risk of coronary heart disease in women and men, respectively.[14,15]

These findings suggest that the optimal intake of vitamin E for disease prevention is preferably 100 IU/day rather than 15 IU/day. If further research confirms preliminary findings from protective effects of vitamin E on disease prevention it may be necessary to recommend intakes of vitamin E above the RDA for optimum protection against oxidative damage and disease prevention.

However, it may be difficult to obtain the desired intake of vitamin E from diet alone. Thus, in the future recommended nutrient allowances may need to be distinguished from recommended dietary allowances.

4 SAFETY OF VITAMIN E

A review of the literature concerning the safety of oral intake of vitamin E indicates that the toxicity of vitamin E is low.[18,19] Human studies report few adverse effects (side-effects) even at doses as high as 3200 mg daily. In addition, animal studies show that vitamin E is not mutagenic, carcinogenic or teratogenic. According to several human studies a daily dose of 100-300 mg vitamin E can be considered harmless and safe.[18,19]

However, oral intake of high amounts of vitamin E (>1000 IU/day) can exacerbate the blood coagulation defect of vitamin K deficiency caused by malabsorption or anticoagulant therapy (warfarin). Therefore, high levels of vitamin E intake are contraindicated in individuals receiving anticoagulant therapy.[18,19] Vitamin E has not been found to produce coagulation abnormalities in individuals who are not vitamin K deficient.

References

1. M. Meydani, Lancet 1995, 345,170.
2. H.J. Kayden and M.G. Traber, J. Lipid Res. 1993, 34, 343.
3. M. G. Traber, Free Radical Biol. Med. 1994, 16, 229.
4. C. Suarna, R.L. Hood, R.T. Dean and R. Stocker, Biochim. Biophys. Acta 1993, 1166, 163.
5. C. K. Chow, Free Radical Biol. Med. 1991, 11, 215.
6. H. Esterbauer, J. Gebicki, H. Puhl and G. Jürgens, Free Radical Biol. Med. 1992, 13, 341.
7. M. Dieber-Rotheneder, H. Puhl, G. Waeg, G. Striegl and H. Esterbauer, J. Lipid Res. 1991, 32, 1325.
8. H. M. G. Princen, W. van Duyvenvoorde and R. Buytenhek R, Arterioscler. Thromb. Vasc. Biol. 1995, 15, 325.
9. R.V. Acuff, S.S. Thedford, N.N. Hidiroglou, A.M. Papas and T.A. Odom Jr, Am. J. Clin. Nutr. 1994, 60, 397.
10. I. Jialal, C.J. Fuller and B.A. Huet, Arterioscler. Thromb. Vasc. Biol. 1995,15, 190.
11. P. Lachance and L. Langseth, Nutrition Rev. 1994, 52, 266.
12. B. Halliwell, Nutrition Rev. 1994, 52, 253.
13. F. B. Araujo, D.S. Barbosa, C.Y. Hsin, R.C. Maranhao and D.S.P. Abdalla, Atherosclerosis 1995, 117, 61.
14. M.J. Stampfer, C.H. Hennekens, J.E. Manson, G.A. Colditz, B. Rosner B and W.C. Willett, N. Engl. J. Med. 1993, 328, 1444.
15. E.B. Rimm, M.J. Stampfer, A. Ascherio, E. Giovannucci, G.A. Colditz and W.C. Willett, N. Engl. J. Med. 1993,328, 1450.
16. P. Knekt, A. Reunanen A and R. Järvinen, Am. J. Epidemiol. 1994, 139, 1180.
17. H. Hodis, W.J. Mack and L. LaBree, JAMA 1995, 273, 1849.
18. A. Bendich and L.J. Machlin, Am. J. Clin. Nutr. 1988, 48, 612.
19. H. Kappus and A.T. Diplock, Free Radical Biol. Med. 1992, 13, 55.

THE SUVIMAX STUDY: SCIENTIFIC JUSTIFICATION AND DESIGN OF A LARGE RANDOMIZED TRIAL FOR TESTING THE HEALTH EFFECTS OF A SUPPLEMENTATION WITH LOW DOSES OF ANTIOXIDANT VITAMINS AND TRACE ELEMENTS

A. FAVIER*, P. PREZIOSI°, A.M. ROUSSEL*, D. MALVY**, A. PAUL-DAUPHIN**, P. GALAN°, S. BRIANCON°°, S. HERCBERG °

*GREPO Research Group on Oxidative Diseases, Faculty of Pharmacy, University of Grenoble, 38700 La Tronche, France
° ISTNA Conservatoire des Arts et Métiers, Paris
°°Ecole de Santé Publique CHU Nancy-Brabois
**Laboratoire de Santé Publique, CHU de Tours

1. INTRODUCTION

Diseases presenting the major health problems in industrial countries have multifactorial etiology involving environmental, social, genetic and nutritional factors. An overproduction of free radicals named free radical stress is found in many such age related diseases as cancer, diabetes, rheumatisms, cardiovasular diseases, cataract, amyotrophic lateral sclerosis, Alzheimer disease[1]. Free radicals constitute a metabolic link between environmental factors, genetic of the free oxidant stress gene, and nutrition that brings as well antioxidant protective compounds that prooxidant contaminants. Antioxidant have been found protective against a great lot of experimental models of these diseases. Recent epidemiological data in industrial countries present a relationship between high risk of cancer or cardiovascular diseases and low intake of vegetables and fruits, or low intake of betacarotene, ascorbate, tocopherols, and selenium. However, the descriptive nature of these studies does not allow us to establish a causal relationship.
   To investigate the cause and effect relationship, we decided to carry out a controlled intervention trial with a combination of antioxidants at nutritional doses and evaluate the impact of such an intervention on the incidence of diseases in a large population. This study, called the SU.VI.M.AX study, is a large scale longitudinal study exploring the relationship between the intake of antioxidant vitamins and minerals and major health problems of the industrialised countries: cardiovasular diseases, cancers and cataract.

2. DESIGN OF THE SUVIMAX STUDY

2.1 Global view of the study

SU.VI.M.AX is an experimental epidemiologgical approach involving a double-blind supplementation protocol for 8 years, with a combination of antioxidant vitamins and minerals verus a placebo. A daily capsule containing 6 mg beta-carotene, 30 mg tocopherol, 120 mg ascorbate, 100 µg selenium as sodium selenite and 20 mg zinc is used. The efficiency of the supplement is tested on a randomly selected cohort of 15,000 adults (age range for men: 45-60 ; and for women: 35-60 years) across France, who will be periodically followed-up on nutritional intake, clinical events and biochemical status. The effects of the intervention trial will be evaluated in functional terms on the incidence of mortality and morbidity due to global and specific causes.

2.2 Selection of the participants

The study cohort of 15,000 adults has been selected from a national sample of 80,000 volunteers recruited for participating to the study via a multi-media campaign, involving the television, radio, newspapers and posters. The volunteers were asked to fill a questionnaire on demographic, health, life-style and nutritional variables, and to consume a daily placebo for a number of days to ensure their compliance. The study sample was selected from among those volunteers who return the information-sheets and a signed informed consent, by a weighted random sampling method. The sample is representative of the national population with regard to geographic density, socio-economic status and the distribution of various risk factors for diseases under study. The selected sample of 15000 adults was divided in a double-blind fashion by block randomization, stratified by sex, age-group and risk factors into two groups, one receiving the supplement, one receiving the placebo.

2.3 Following of the cohort

Subjects maintain monthly contact with the scientific data base at the SUVIMAX coordination site in Paris, using a particular French phone network named Minitel. This system functions as a small computer and permits each month to collect medical events, subjective perception of health and compliance to the supplementation. Every two months subjects fill, via Minitel, a dietary software reporting a 24-hours food record. To fill this questionnaire volunteers are helped by a manual coding foods and estimating portion size with photographic information.

Each subject undergo an annual check-up consisting of biochemical or laboratory test and a clinical examination every alternate year. We visit each year all the volunteers by using two mobile units, consisting of long semi-trailer with four rooms, moving all over France and equipped for medical investigation, electrocardiogram, mammography, radiology and biological sanpling. At the time of the visit participants received their pill for the next year as weekly blisters.

All the data obtained are sent to a scientific data bank in Paris and are analyzed on Alpha-VMS system using SAS and a specific database software.

2.4 Biological investigations in the SUVIMAX study

Three levels of biological investigations are realized on the subjects of the SUVIMAX study. A basic profile monitors health indicators such as glucose, hemoglobin, cholesterol, triglyceride, apolipoprotein A1, apolipoprotein B, serum ferritin and transferrin, urinary iodine extraction. A blood sample monitors their vitamins (E, C, carotene) and trace elements (selenium, zinc) status. The free radical stress is evaluated in all the participants by the determination of MDA and Cu-Zn Superoxide dismutase and on a sub-sample by determination of GPx, glutathione, plasma protein thiols and carbonyls and antibodies against oxidized proteins.

Beside these determinations blood is collected during the first sampling and stored in liquid nitrogen to realise a biological bank containing straws of plasma, erythrocytes and buffy coat. One of the biological banks is located in Lyon at the International Agency for Research in Cancer (IARC) (Dr Riboli). The second one is located in Annemasse (Biobank Merieux).

2.5 Organization and financial support

The central coordination is located in CNAM in Paris, but the follow-up of the cohort is monitor from 4 regional coordination, North East of France in Nancy, South Eastern in Grenoble, Western in Tours and North of France in Paris. Two laboratories have been built up for measuring vitamins in Tours and trace elements and radical damages in Grenoble.

The project is overseen by two committees: an Internationl Follow-up Committee which supervises the overall analysis and a Medical Committee which is responsible for the management, control and validation of individual clinical events. A sub-committee for ethical problems merges the heads and one member of these committees.

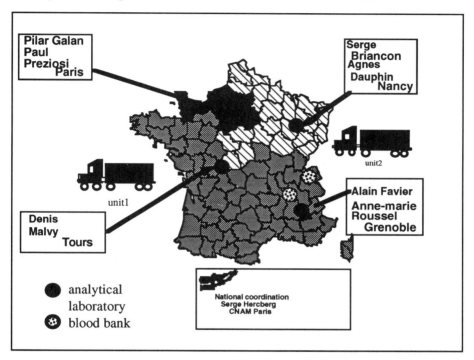

Figure 1 *Geographical organization of the SUVIMAX study in France*

Such a study is quite expensive with a large medical, technical and nurse team. The expenses are supported by the French Ministry of Health and Ministry of Agriculture and grants from industrial companies from the agro-food, pharmaceutical and clinical biology sectors. We thank our major sponsors for their trust and support; ASNAV Fondation Benjamin Delessert, Astra-Calve, Candia, Lipton, Kellog's, Orangina, Cereal, SODEXHO, L'Oreal, Estee Lauder, France-Telecom, Peugeot, Jet Service, Innothera, Merck-Clevenot, Lipha, Robapharm, Institut Danone, CERIN, Boerhinger Mannheim, Beckton Dickinson.

3. WHY USE A COMBINATION OF VITAMINS AND MINERALS AND NOT A HIGH DOSE OF A SINGLE ANTIOXIDANT ?

A few other trials have been carried out to test the effect of antioxidants. Some demonstrated a beneficial effect and a decrease in cancer mortality such as the Linxian study[2] supplementing with a combination of betacarotene, alfatocopherol and selenium, or the Chinese study[3] on liver cancer supplementing with selenium. Recently, the ABTC cancer prevention study[4] involving 29,000 male smokers in Finland supplemented with pharmacological doses of alfatocopherol and betacarotene observed a significant increase in lung cancer in the group receiving betacarotene and of cerebral hemorrhages in the group receiving alfatocopherol. We decided thus to test the effect of a combination of antioxidants at a lower, closer to nutritional level of dose.

3.1 Risk induced by high doses of antioxidant

Antioxidants as well as free radicals are ambiguous compounds with a range of beneficial and toxic intake. They often exert a biphasic relationship between dose and activity.

3.1.1.Usefulness of a normal production of free radicals. Low level of activated oxygen increases antioxidant capacity of cells by stimulating oxidative response genes. Cells and organisms will adapt to resist oxidative stress by increasing their antioxidant capacity. Cell lines treated by low doses of H_2O_2 increase 40 time their resistance to a cytotoxic dose of H_2O_2[5]. Rats treated by TNF alpha, an oxidative cytokine increase their resistance to hyperoxia[6], and reciprocally an overexpression of HSP by heat or arsenate increase the resistance to TNF[7].

In bacteria an organized system involving regulon oxy R and sox R and named oxidative stress gene has been demonstrated[8]. In mammals such an organized system does not exist, but various genes encoding for antioxidants are stimulated by free radicals, genes of antioxidant enzymes MnSOD, heme oxygenase[9], genes of antioxidant proteins such as metallothioneins or lactoferrin[10], but also genes of repair systems, proteases[11] or heat shock proteins[12]. Free radicals can switch the expression of genes of transferrin receptor and gene of ferritin[13].

With aging capacity of adaptation to oxidative stress decreases as demonstrated in aging Caenorhabditis elegans[14].

Free radicals can not only kill bacteria but at a reasonable level they are beneficial stimulating T lymphocytes. B lymphocytes possess a NADPH oxidase inside their membrane stimulated by IgM, IgD , IgG, HLA-DR, et CD19 and important in their physiology[15]. H_2O_2 at a micromolar concentration is a potent activator of T lymphocyte functions by activating transcription factors as NF-KB[16], even if it will kill or damage lymphocytes at a high level[17].

Stimulation of phagocytic cells such as PMN or macrophages result in an overproduction of oxygen radicals named oxidative burst[18]. Such a process is variable according to cell type and mediated by NADPH oxidase, myeloperoxidase, iNO synthase, and is important in bacterial killing by producing a great variety of toxic species[19] damaging bacteria: superoxide anion, hydrogen peroxide, hypochlorite, chloramines, nitroperoxide[20], simultaneously with enzymes.

3.1.2. Free radical induced-apoptosis may be a beneficial process protecting against cancer and immune disturbances. Cells infected by a virus or damaged by carcinogenic agents and promoted to become cancerous are dangerous. Fortunately they possess a genetic system named apoptosis that leads to an altruistic suicide of these cells. H_2O_2 or NO can induce or facilitate apoptosis, and various factors known to be apoptotic inducers, such as TNF, generate free radicals. An oncogenic factor bcl2 produces an antioxidative mitochondrial protein[21]. N-acetyl cystein inhibits TNF-induced apoptosis in HIV-chronically infected U937 cells[22]. But suppression of apoptosis has been described as increasing viral production and maintaining a high level of infection[23]. We thus have to be very careful and never forget that apoptosis is supposed to be a beneficial mechanism. For instance, the effect of antioxidants in sunscreen or topicals can be beneficial when applied before exposure and tumor promotion and dangerous when applied after tumor promotion by preventing apoptosis. A supplementation with high doses of beta carotene or tocopherol realized in Finnish adults smoking from many years had deleterious effects, beta carotene increasing the number of lung cancer by 18 % and tocopherol the number of brain hemorrhage[24]. This may be due to an unknown statistical bias or hazard but we speculate that these voluntaries smoking for a very long time, had carcinogenic cells in their lungs whose destruction had been prevented by beta carotene.

3.1.3. Pro-oxidant and antagonistic effects of high doses of antioxidants. By scavenging, free radical molecules form a rather reactive radical. This is the case even for vitamin E, vitamin A, vitamin C, glutathion-$RO°_2$ + Tocopherol ---> RO_2H + Tocopheryl°.

Recently an antioxidative effect by betacarotene has been found at low oxygen pressure and a prooxidative effect at very high oxygen pressure, thus increasing lipid peroxidation and protein carbonyl [25].

Other mechanisms are involved in explaining the oxidative effect by excessive antioxidants. Ascorbate increases iron absorption and forms a strong pro-oxidative system with iron or copper. Ascorbate increases the redox cycling of naphtoquinone mediated by the DT-diaphorase and thus the production of superoxide anions [26]. High levels of ascorbate in the diet of the guinea pig decreased glutathione reductase in tissues [27]. Selenium, in the form of selenite, produces free radicals mainly by interacting with glutathione [28]. The reaction produces superoxide anions as demonstrated by chemiluminescence [29] or by ESR [30]. This effect, which occurs mainly with selenite and not selenate or selenomethionine, has been observed in animals fed selenium rich diets [31], and in children with phenylketonuria supplemented with a high level of selenite for one year [32].

Copper and particularly iron, which are beneficial at normal levels in the diet as cofactors of the antioxidant enzyme SOD and catalase, are strong oxidants inducing the Fenton reaction in tissues when they exist in the "free form" in excess or in intoxication as in Wilson's disease or hemochromatosis. Manganese, the cofactor of an other SOD inhibits copper-zinc SOD at excessively high intake levels. An excess of zinc increased cell death after an oxidative stress in cultured fibroblasts [33]. A zinc deficiency disturbed selenium metabolism and GPx activity mainly in male rats [34]. Supplementation with zinc elevated GPx activity in diabetic patients [35]. Ascorbate supplementation of normal adult women also improved selenium status [36]. Experimental zinc deficiency decreases tocopherol status[37] and glutathione transferase activity [38].

Competition for absorption exists among the trace elements copper, zinc, manganese [39], and iron [40]. It can be beneficial when zinc supplementation decreases free iron [41] or copper overload. In normal subjects a low dose supplementation with zinc increased Cu-Zn SOD [42] when an excessive long-term intake of the same metal led to a reduction in activity, by decreasing copper absorption[43]. Iron supplementation generally has a deleterious effect on antioxidant status. A similar competitive phenomenon in absorption or transport can exist between the various vitameric forms of some vitamins. Equimolar doses of beta-carotene and lutein exert a reciprocal inhibition of their absorption in humans. Betacarotene reduced the area below the curve for lutein from 54 to 61%, whereas lutein lowered it for beta-carotene in the majority of subjects but enhanced it in other[44]. Such an antagonism can be observed at a cellular level between cis-and trans-betacarotene[45]. The occurrence and the fact that isomeric forms can have different antioxidant activities and biological properties favors the use of fruits and vegetables, or extracts thereof instead of synthetic supplements.

Even antioxidant enzymes such as superoxide dismutases can worsen DNA damage induced by oxygen radicals [46]. Overexpression of SOD by transvection of its gene decreases GSH and promotes cell death in mice epidermal cells [47], and increases lipid peroxidation in the brain of transgenic mice [48]. However, superoxide not only can initiate lipid peroxidation but also terminate the chain reaction as demonstrated in ischemia-reperfusion of isolated heart [49].

3.2 Metabolic interrelationships exist between antioxidant nutrients with a beneficial mutual protection and regeneration

When exerting their protective action, antioxidants can be damaged by transforming into radicals or oxidation. However, some can be protected or recycled by another antioxidant. For instance, Cu-Zn SOD is inactivated by H_2O_2 [50], and its lifetime prolonged by peroxidase, when various peroxidases are inactivated by the superoxide anion[51].

Ascorbate has a synergistic action with alfatocopherol [52]. The tocopheryl quinone radical formed when breaking the chain reaction of peroxidation inside the membrane is transferred to cytosolic ascorbate (Figure 2) which becomes an ascorbyl radical. The ascorbyl radical evolves to a semidehydroascorbate regenerated to ascorbate by an ascorbate reductase consuming NADPH. Ubiquinone is involved in the recycling process of tocopherol within the mitochondrial membrane. Ubiquinone is able to reduce the vitamin E radical leading to a semi-ubiquinone regenerated by the mitochondrial or microsomial electron transport [53]. A synergistic action may also exist between tocopherol and beta-carotene within the membrane as they have a synergy in protecting against mutagenesis [54].

Such a synergistic effect between alfatocopherol and ascorbate exists also at the plasma level within lipoproteins. Following an oxidative stress in plasma, the ascorbyl radical rapidly peaks then disappears immediately before the appearance of the tocopheroxyl radical [55]. Supplementation of volunteers with ascorbate decreasde in vitro LDL oxidation by 15 %, with tocopherol by 50 %, while simultaneous supplementation with these two vitamins decreased LDL oxidation by 78 % [56].

In peroxide reduction GPx oxidizes reduced glutathione that is regenerated by glutathione reductase with riboflavin and NAPDH as cofactors. Riboflavin protected the rat from lipid peroxidation and had a synergistic effect with alfatocopherol [57,58].

3.3 Heterogeneity in the origin and frequency of antioxidant deficiencies among the French population

Some years ago we conducted a large survey of the dietary intake and nutritional status of the French population. This study, called the Val de Marne 88 study, was conducted on 1200 healthy males and females aged 1 to 90 years and living at home. Great heterogeneity was observed in the origin of nutritional deficiencies. According to serum levels, the elderly were more prone to ascorbate deficiency. Tocopherol deficiency was more frequent among the young [59]. Similarly low levels of zinc were more frequent simultaneously in children and elderly, but copper levels were more often below the cut off level in young male adults [60]. The pattern of nutritional deficiency is also different between males and females. For instance, zinc deficiency was more frequent in females.

These differences all derived from the variations in the concentration and bioavailability of vitamins and trace elements in foods. Meat is a good source of zinc but contains very low levels of ascorbate or carotenoids. Fruits are a rich source of carotenoids or ascorbate but a very poor source of zinc, selenium or tocopherols. It is well known that diversity in diet is the only way of guaranteeing an adequate intake of all nutrients.

Therefore supplementation of only one nutrient, for instance, a high dose of betacarotene, will only normalize the nutritional antioxidant status of a small segment of the population. By supplementation with five different antioxidants, we hope to achieve a five-fold increase in the efficiency of our study.

3.4 Antioxidant synergy

A synergistic effect by antioxidants, reciprocal protection, has been observed by means of various mechanisms for the scavenging of different oxygen species. For instance, a mixture of betacarotene, alfatocopherol and selenium in the diet of rats had a better antioxidative effect in tissues than supplementation with each of these nutrients alone [61]. The enhanced cell protective antioxidative effect was obtained by a combination of glutathione peroxidase, Cu-Zn SOD and alfatocopherol [62]. A combination of riboflavin and alfatocopherol exert a synergistic effect in protecting tissues against lipid peroxidation induced by a diet rich in oxydized fatty acids [63].

3.4.1. Complementary scavenging of free radicals. The efficiency of each antioxidant varies according to the oxygen reactive species. Thus, the efficiency of an antioxidant differs according to the type of oxidative stress. Rats were better protected against a CBrCL3-induced stress by alfatocopherol than by selenium against a hydroperoxide-induced stress. In the same rat model, a synergistic effect of the addition of water-soluble antioxidants (ascorbate, N-acetyl cystein, catechine, trolox) and liposoluble (canthaxanthine, betacarotene, coenzyme Q) was observed only with an iron-induced stress, and a stronger effect was observed with an addition of all antioxidants in a a copper-induced stress [64]. Other factors beside oxygen pressure can alter nutrient efficiency. Betacarotene was a more efficient scavenger than alfatocopherol in tissues with low oxygen pressure whereas a reverse effect was observed in other tissues [65]. Each antioxidant is part of a chain that tends to divert the flux of various radicals to form inert water molecules. They complete the action of the upstream antioxidant. Hence, the addition of SOD to cultured fibroblasts had a weak effect, however, a concomitant addition of GPx exerted a strong protective effect against cytotoxicity [66].

3.4.2 Complementary effect in protecting various cellular compartments or tissues.

In tissues antioxidant system is quite different in each tissue of the organism. For example, catalase is abundant in the liver but does not occur in the thyroid or the brain [67]. Lung muscle and the eyes are well protected by glutathione and glutathione system. Body fluids are poorly protected by lipophilic scavengers or enzymes but possess high levels of iron-sequestering proteins, such as transferrin or lactoferrin. Plasma is protected by albumin, uric acid and scavengers.

Each subcomponent within the cellular compartments has a specific antioxidant system. Catalase is mainly found inside peroxisomes where copious amounts of H_2O_2 are produced by fatty acid oxidation, glycollate oxidase, urate oxidase. Only the mitochondria contain the Mn superoxide dismutase (MnSOD), clearly because a substantial quantity of cytosol is produced in the electron transfer chain. Cytosol contains ascorbate and ascorbate reductase, selenium glutathione peroxidase (SeGPx), GTr, metallothionein, and cytosolic Cu-Zn superoxide dismutase. Membranes are mainly protected by scavengers such as alfatocopherol, betacarotene and ubiquinone. However, also specific iso-enzymes such as selenium hydroperoxide glutathione peroxidase or extracellular Cu-Zn superoxide dismutase (EC-SOD) protect its exterior surface. Within cellular membrane molecules, tocopherol molecules are oriented to their phenol part on the surface and their phytyl chain in fatty acids. Therefore, tocopherols are not efficient in scavenging deep seated free radicals within the membrane. Ubiquinone, which does not have such a localization, can scavenge in the middle part of membranes.

4. PRELIMINARY STEPS OF THE STUDY AND THEIR RESULTS

Before implementing our study we built up and tested various tools during two preliminary years. Effortless logging of dietary records by volunteers via the Minitel computer was an entirely new technique which had to be set up and the precision and reliability followed-up

The nutritional efficiency of our formula (containing the mixture of antioxidants or placebo) was tested on the nutritional status of 400 volunteers supplemented during six months. Figure 2 shows that our antioxidant doses facilitated an increase in the level of each nutrient in blood, thus decreasing the risk of biological deficiency in normal French adults.

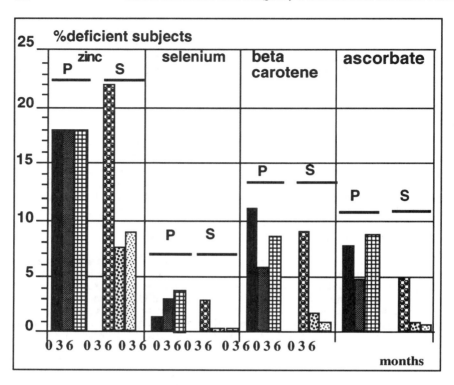

Figure 2 *Percentage of French adult volunteers found under a cut-off level for plasma zinc, selenium, betacarotene, and ascorbate, before and after s supplementation with a placebo (P) or the mixture of antioxidants used for the SUVIMAX study (S).*

5 ACTUAL DEVELOPMENT AND FUTURE EUROPEAN ENLARGEMENT OF THE STUDY

The original idea for such a study came into being in 1990, but the recruitment of 80 000 volunteers was carried out only four years later in March 1994, with the selection of 15 000 representative subjects. Since October 1995, all subjects have been sampled for biology and blood bank and are taking their daily pill with as low as a 1% withdrawal rate. We have determined 15 000 samples for ferritin, TIBC, cholesterol, triglyceride, apo A1, apo B and glucose concentration. Selenium, zinc, and vitamins have all been measured. These very recent data are not yet available for evaluation, but we have observed a large variation in lipid status in terms of the subjects' geographical location in France. A paradoxical increase in cholesterol level in the South of France, where a low incidence of cardiovascular disease exists, was observed, thus confirming the Monica European study. To enlarge the statistical reach of the SUVIMAX study we are about to organize identical studies in some other European countries. A trial of 3 000 subjects will begin in Catalonia during 1996, coordinated by professor Luis Serra in Barcelona. Other similar studies in Belgium (professor Neve) and in Italy are being prepared. We hope that scientists from the Nordic coutries will join us and initiate similar trials in their countries, thus enabling comparison of the efficiency of antioxidant enrichment of the diets of more diverse European populations.

References

1 J. Pincemail, 'Analysis of free radicals in biological systems', Birkhauser, Favier A. et al. editors Basel , 1995, p. 83
2 W. Blot, Jy Li., P. Taylor ,W. Guo, S. Dawsey , G. Wang , CS.Yang , S.Zheng, M. Gail , G.Li , Y.Yu , B. Liu , J.Tangrea , Y.Sun, F. Liu, JF.Fraumenti , Y.Zhang, B. Li *J.Nat Cancer Inst.* 1993, **85**, 1483
3 S.Y Yu .,Y. Zhu, W.Li, Q.Huang, C.Huang, Q. Zhang, C. Hou, *Biol.Trace El. Res.*, 1991, **29**, 289
4 The Alpha-tocopherol Beta carotene Cancer Prevention Group, *New. Engl. J. Med.* , 1994, **330**, 1029
5 A.Wiese, R. Pacifici, K.J. Davies, *Arch. Biochem. Biophys.* 1995, **318**, 231
6 C.W.White, P.Ghezzi *Biotherapy*, 1989 , **1**, 361
7 D. Kusher, C.Ware, L.Gooding , *J.Immunol.*1990, **145**, 2925
8 D.Touati ,B.Tardat, I.Compan, Free Rad. Biol.Med. 1990, **9**, 1
9 Tyrrel R. M.,Keyse S.M.,Lutier D.,Applegate L., *Free Rad.Biol. Med.* 1990, **9**, 1
10 J. Lash, T. Coates, R.Lafuze, R.Baehner, L.Boxer , *Blood* , 1983,**61**, 885
11 P.E.Starke, C.Oliver, E.R.Stadtman , *Faseb J.* ,1987 ,**1**, 36
12 J.Becker, V.Mezger , AM.Courgeon, M.Best-Belpomme, *Free. Rad. Res. Comm.* , 1991,**12-13**, 455
13 Rouault T, Klausner R., in 'Trace Elements In Oxidative Diseases', A.Favier, J.Neve, P.Faure editors, AOCS Press ,Champaign, 1994, p 8
14 D.Darr , I.Fridovitch , *Free Rad. Biol. Med.*, 1995,**18**, 195
15 S.Kobayashi , S.Imajoh-Ohmi , F.Kuribayashi, H.Nunoi , M.Nakamura, S.Kanegasaki *J. Biochem.*, 1995,**117**,758
16 M.Los, W.Droge, K.Stricker, P.Baeuerle, K.Schulze-Osthoff, *Eur.J.Immunol.* 1995, **25**, 159
17 A.El-Hag , P.E.Lipsky , M.Benett , R.A.Clark , *J. Immun.* , 1986,**136**, 3420
18 C. Winterbourn , in 'Ogygen Radicals:Systemic Events And Disease Process', D.Das, W.Essman editors , Karger , Bale, 1990 p31
19 J.A.Badwey , M.L.Karnovsky, *Ann.Rev.Biochem.*, 1980,**48**, 695
20 J.Hibbs, Z.Vavrin, R.Taintor , *J.Immun.*,1987, **138**, 550
21 T.Buttke, P.Dandstrom , *Immun. Today* , 1994, **15**, 7
22 W.Malorni , R.Rivabene , M.T.Santini , G.Donelli, *FEBS* , 1993, **327**, 75
23 B.A.Antoni , P.Sabbatini , A.Rabson , E.White, *J.Virol.*, 1995, **69** ,2384
24 The Alpha-Tocopherol Beta Carotene Cancer Prevention Group, *New. Engl. J. Med.* , 1994, **330**, 1029
25 P.Palozza , G.Calviello , M.G.Bartoli , *Free Rad. Biol. Med.*, 1995, **19**, 887
26 R.Jarabak , J.Jarabak , *Arch. Biochem. Biophys.*, 1995, **318,** 418
27 G.Barja, M.Lopez-Garcia, R.Perez-Campos, C.Rojas, S.Cadenas, R.Pamplona, *Free Rad. Biol. Med.* , 1994,**17**, 105
28 Y.Seko, Y.Saito, J.Kitahara, N.Imura ,in 'Selenium In Biology And Medicine', Wendel A editor, Springer-Verlag , Berlin, 1983, p70
29 L.Yan, JE.Spallholz, *Biochem. Pharmacol.*, 1993, **45**, 429
30 H.Xu , Z.Feng , C.Yi , *Huzahong Longong Daxue Xuebao* ,1991, **19**, 13
31 JJ.Dougherty, WG.Hoekstra , *Proc. Soc. Exp. Biol. Med.*, 1982, **169**, 209
32 A.Favier, B.Wilke, J.Arnaud, MJ.Richard, V.Ducros, M.Vidailhet , in 'Trace Elements In Man And Animals VII', B.Momcilovic Editor, Published by IMI, Zagreb. 1991, p 7.15
33 MJ.Richard , P.Guiraud , MT.Leccia , JC.Beani , A.Favier , *Biol. Trace Elem. Res.*, 1993, **37**, 187
34 D. Behne, A.Kyriakopoulos, H.Gessner, J.Vormann, T.Gunther, *J.Trace. Elem. Electrol. Health Dis.* , 1992, **6**, 21
35 P.Faure, AM.Roussel, S.Halimi, A.Favier ,in 'Tema 7', Anke, Meissner, Mills editors. Verlag Media Touristik, 1994, p 888

36 M.Mutanen , H.M.Mykkanen , *Hum Nutr Clin Nutr* ,1985 , **39**, 221
37 M.Bunk, A.Dnistrian, M.Schwartz, R.Rivlin , *P.S.E.B.M.*,1989, **190**,379
38 I. S. K.Lee, L.Y.Y.Fong, , *Carcinogenesis*,1986, **7**, 1111.
39 C.Davis, J.L.Greger, *Am.J.Clin.Nutr.*, 1992, **55**, 747
40 P.J.Aggett, R.W.Crofton, C.Khin, S.Gvozdanovic, D.Gvozdanovic, in 'Zinc Deficiency In Human Subjects', Alan R. Liss, New-York, 1983. p. 117
41 C.Coudray, S.Rachidi , A.Favier, *Biol. Trace Elem. Res.*, 1993, **38**, 273
42 C.Coudray, MJ.Richard, F.Laporte, P.Faure, AM.Roussel, A.Favier, *Journal Nutr. Med.*, 1992, **3**,13
43 P.Fisher , A.Giroux , M.L'Abbe, *Am.J.Clin.Nutr.*, 1984, **40**, 743
44 D.Kostic, W.White, J.Olson, *Am. J. Clin.Nutr.* 1995, **62**, 604
45 M.Oarada, W.Sthal , H.Sies, *Biol. Chem. Hoppe Seyler* ,1993, **374**, 1075
46 L.Daniel, Y.Mao, U.Saffiotti , *Free Rad.Biol. Med.*, 1993,**14**, 463
47 P.Amstad, A.Peskin, G.Shah, M.Mirault, R.Moret, I.Zbinden, P.Cerutti , *Biochemistry* , 1991, **30**, 9305
48 A.Nicole,M.Portier, F.Agid, M.Lafond, M.Thevenin, P.M.Sinet, I.Ceballos-Picot , in 'Free Radical And Aging', Paris,1991
49 S.K.Nelson, S.K.Bose, J.M.Mac Cord , *Free Rad. Biol. Med.* , 1994, **16**, 195
50 P.M.Sinet , P.Garber, H.Jerome, *Bull. Europ. Physiopath. Resp.* 1981, **17**, 91
51 J. Blum, I.Fridovich , *Arch. Biochem. Biophys.* 1985, **240**, 500
52 E. Niki, A.Kawakami, Y.Yamamoto, Y.Kamiya, *Bull Chem Soc Jpn* , 1985, **58**, 1971
53 V.Kagan, L.Packer , in Free Radicals And Antioxidants In Nutrition, F.Corongiu , S.Banni , M.Dessi , C.Rice-Evans,editors, Richelieu Press , London, 1993, 27
54 R.Pizzala , L.Bianchi , R.Melli , L.Rehak , R.Rossi , L.A.Stivala , V.Vanini, *Med. Biol. Environn.* 1995, **23**, 89
55 M.Sharma , G.Buettner , *Free Rad. Biol. Med.*, 1993,**14**, 649
56 V.Rifici , A.Khachadurian , *J.Am. Col. Nutr.*, 1993, **12**, 631
57 DE.Halquist , F.Xu , KS.Quandt , et al. , *Med.J. Hematol.* 1003, **42**, 13
58 T.Miyazawa , K.Tsuchiya , T.Kaneda , *Nutr. Rep. Int.*, 1984, **29**, 157
59 S.Hercberg , P.Preziosi , P.Galan , M.Devanlay , H.Keller , C.Bourgeois , G.Potier De Courcy, F.Cherouvrier , *Int. J. Vit. Nutr. Res.* 1994, **64**, 220'
60 A.Favier , S.Hercberg , J.Arnaud , P.Preziosi , P.Galand , in Trace Elements In Man And Animals VII', B. Momcilovic editor, IMI, Zagreb, 1991, P13-5
61 H.Chen , L.Pellett , H.Andersen , A.L.Tappel , *Free Rad.Biol.Med.*, 1993, **14**, 473
62 C.Michiels ,M.Raes ,A.Houbion ,J.Remacle , *Free Rad. Biol.Med.*, 1991, **14**, 323
63 T.Miyazawa , K.Tsuchiya , T.Kaneda , *Nutr. Rep. Int.* 1984, **29**, 157
64 H.Chen , A.Tappel , *Free Rad. Biol. Med.*, 1994, **16**, 437
65 P.Palozza ,N.Krinsky , *Free Rad.Biol.Med.*,1991,**11**, 407
66 C.Michiels ,M.Raes ,A.Houbion ,J.Remacle , *Free Rad. Biol.Med.*, 1991,**14**, 323
67 S. Marklund, Ng. Westman, E. Lundgreen, G.Roos, *Cancer Research* 1982, **42**, 1955

ANTIOXIDANTS, DIET AND MORTALITY FROM ISCHAEMIC HEART DISEASE IN RURAL COMMUNITIES IN NORTHERN FINLAND

P.V. Luoma, S. Näyhä, K. Sikkilä and J. Hassi

Regional Institute of Occupational Health, and Department of Public Health Science and General Practice, University of Oulu, FIN 90220 Oulu, Finland

1 INTRODUCTION

The evidence that antioxidants may play a role in the prevention of atherogenesis has been increasing rapidly in recent years.[1-3] Several studies have suggested that antioxidants such as alpha-tocopherol,[4-7] retinol,[4] albumin[8] and selenium[9] may reduce cardiovascular mortality. They have also suggested that variations in serum antioxidant levels may explain the cross-cultural differences in the incidence of ischaemic heart disease (IHD) better than the classic risk factors such as raised serum cholesterol, high blood pressure and smoking.[10] In an earlier study we found a favourable serum antioxidant status in men living in Mountain Lapland or the Saami area, Northern Finland.[11] The present study was undertaken to investigate how antioxidants are associated with IHD mortality in rural communities, i.e. in small geographical areas in northern Finland. Serum antioxidants such as alpha-tocopherol, albumin, retinol and selenium together with serum cholesterol and other classic risk factors were evaluated in relation to diet and IHD mortality in the seven northernmost communities of the country.

2 SUBJECTS AND METHODS

2.1 Subjects and protocol

This study is a part of a health survey performed in northern Finland.[12] In 1989, 574 men engaged in reindeer herding in Mountain Lapland, i.e. the three northernmost communities of Finland, or the four neighbouring communities to the south (the reference area) (Figure 1) were invited to participate in a medical examination. This study included the 334 participants whose characteristics and serum lipid values were known.

The participants attended the local health centre for a medical checkup, interview and the drawing of blood samples for the determination of serum alpha-tocopherol, retinol, albumin, selenium, total cholesterol, HDL cholesterol and triglycerides. Information was gathered on age, body weight and height, blood pressure, smoking and dietary habits. The participants were asked about their consumption of reindeer meat (meals/week), fish (meals/week), milk (glasses/day) and alcohol,[11] and smoking status (current smoker/ex-smoker/never smoked). A sample of 118 men were subjected to a dietary interview performed by trained nurses using the 24-hour recall method.[12]

Subjects with both parents of Finnish ancestry were considered Finns, those with both parents of Saami origin Saamis (also known as Lapps, i.e. members of the ethnic minority living in the northern parts of Norway, Sweden, Finland, and the Kola Peninsula, Russia), and those with one parent or at least one of their grantparents of Saami origin as mixed.

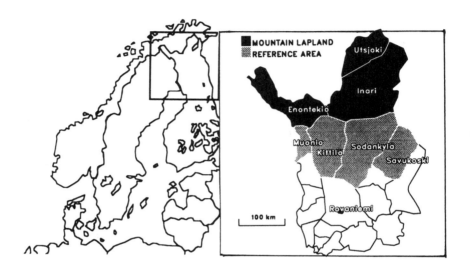

Figure 1 The location of the seven northernmost communities of Finland.

2.2 Methods

Venous blood samples for the determination of serum lipids and antioxidants were drawn after an overnight fast. The concentrations of antioxidants and lipids were determined by methods used earlier.[11] Serum LDL cholesterol was calculated by the formula of Friedewald et al.,[13] and according to the method the triglyceride values higher than 4.5 mmol l^{-1} (400 mg l^{-1}) (n=8) were excluded from the determination of LDL. Alcohol consumption was calculated as g/ day according to a method used earlier.[11]

Male mortality from IHD in the general population in the two regions and the seven communities was determined for the years 1961-1990, the information being obtained from the cause-of-death files prepared by the Central Statistical Office of Finland.[14] The mortality was expressed as a standardised mortality ratio (SMR).[14]

The differences in the parameters were tested by analysis of variance and covariance, adjusting for age, body mass index (BMI, kg/m^2), smoking status, systolic and diastolic blood pressure, and also for LDL cholesterol in the case of alpha-tocopherol and retinol. Spearman correlation analysis was used to compare antioxidants, lipids and IHD mortality.

3 RESULTS

3.1 Subject characteristics

The average age of the 334 men was 46.0 (range 20-85) years and BMI 26.2 (17.3-41.6) kg/m^2. Twenty-two percent of the men were Saami, 37 % were smokers, and alcohol consumption averaged 12 (range 0 - 122) g/ day. In Mountain Lapland (n=123) and in the reference area (n=211) the average ages were 46.1 and 46.0 years, the BMIs 25.6 and 26.6 kg/m^2 and the proportion of smokers 41 and 34 %, respectively.

Table 1 shows the characteristics of the men living in the seven northernmost communities of the country. All the Saami men except one were living in Mountain Lapland. All the males living in Utsjoki, and none of those in Savukoski, Muonio and Kittilä, were Saami.

Table 1 *Characteristics (means) of the men living in the seven northernmost communities of Finland*

	Utsjoki (19)	Enontekiö (32)	Inari (72)	Savukoski (45)	Sodankylä (109)	Muonio (12)	Kittilä (45)
Age, years	46.5	46.1	46.0	47.5	46.1	43.8	44.8
BMI kg/m^2	25.2	26.3	25.4	27.3	26.7	24.8	26.4
Men of Saami origin,%[1]	100	39	53	0	1	0	0
Smokers, %	32	56	38	36	32	42	33

The number of the subjects in brackets.
[1]The ethnic origin of 317 men was known.

3.2 The mortality from IHD and antioxidants and lipids

3.2.1 Regions. The mortality from IHD and the concentrations of serum antioxidants and lipids varied according to the place of residence. IHD mortality (SMR) was 20 % lower in Mountain Lapland than in the reference area (80 vs. 100, 95 % confidence interval (CI) for the difference -28 to -13) (Figure 1), the concentrations of alpha-tocopherol (18.0 vs. 15.6 µmol l^{-1}, 0.9 to 3.9, P<0.001), albumin (46.8 vs. 46.1 g l^{-1}, 0.1 to 1.3, P<0.05), selenium (1.59 vs. 1.48 µmol l^{-1}, 0.00 to 0.22, P<0.05), cholesterol (6.76 vs. 6.37 mmol l^{-1}, 0.09 to 0.69, P<0.01) and LDL cholesterol (4.81 vs. 4.50 mmol l^{-1} 0.04 to 0.58, P<0.05) showing reverse trends. The HDL cholesterol/ cholesterol ratio tended to be lower in Mountain Lapland than in the reference area (0.20 vs. 0.21, -0.02 to 0.00, P=0.06), but retinol (2.90 vs. 2.86 µmol l^{-1}, P>0.1) and HDL cholesterol (1.30 vs. 1.29 mmol l^{-1}, P>0.1) were similar in the two regions. Serum triglyceride values were skewed, but the median values (1.17 vs. 1.11 mmol l^{-1}) showed little difference.

3.2.2 Communities. IHD mortality (SMR = 100 in Finland) varied from 78 to 140 and was lowest in the communities of Mountain Lapland (Table 2, Figure 1). Serum albumin and lipid adjusted alpha-tocopherol levels were high in communities in Mountain Lapland and lowest in the community of Kittilä (Table 2). Serum selenium was determined in six communities. It was highest in Utsjoki (Table 2) and higher in the two communities of

Table 2 *Serum antioxidants, lipids (means) and the mortality from ischaemic heart disease (IHD) in the seven northernmost communities of Finland*

	No. of analyses	Utsjoki (19)	Enontekiö (32)	Inari (72)	Savukoski (45)	Sodankylä (109)	Muonio (12)	Kittilä (45)	P-value
Alpha-tocopherol (μmol l^{-1})[1]	322	17.2	17.1	18.5	17.1	15.6	16.0	13.8	<0.005
Retinol (μmol l^{-1})[1]	322	2.64	2.93	2.96	2.73	2.83	3.02	3.00	>0.1
Albumin (g l^{-1})[2]	329	47.3	47.6	46.3	47.2	46.4	45.3	44.6	<0.001
Selenium (μmol l^{-1})[2]	203	1.63	-	1.57	1.37	1.53	1.56	1.50	<0.1
Cholesterol (mmol l^{-1})[2]	334	6.36	6.75	6.88	5.99	6.51	7.11	6.20	<0.005
LDL cholesterol (mmol l^{-1})[2]	326	4.61	4.74	4.92	4.07	4.66	4.95	4.40	<0.01
HDL cholesterol (mmol l^{-1})[2]	334	1.21	1.32	1.30	1.40	1.24	1.51	1.21	<0.01
HDL cholesterol/ cholesterol[2]	334	0.20	0.20	0.20	0.24	0.20	0.21	0.21	<0.005
General male population, standardised mortality ratio, SMR, IHD 1961-1990[3,4]		78	83	103	110	110	111	140	

The number of the subjects in the brackets.
[1] Adjusted for age, BMI, smoking, systolic and diastolic blood pressure and LDL cholesterol
[2] Adjusted for age, BMI, smoking and systolic and diastolic blood pressure
[3] SMR of Finland = 100
[4] Male population in 1989: Utsjoki 796, Enontekiö 1292, Inari 3882, Savukoski 982, Sodankylä 5475, Muonio 1463, Kittilä 3158.

Mountain Lapland than in the four reference communities (see preceding paragraph). Serum retinol showed no association with the place of residence. Adjusted serum albumin (Spearman correlation coefficient, r=-0.85, P<0.01) and alpha-tocopherol (r=-0.79, P<0.05) (Figure 2) correlated inversely with IHD mortality in the communities, whilst other antioxidants showed no significant association with it.

Serum cholesterol and other lipids (Table 2) and also systolic and diastolic blood pressure (data not shown) varied between communities, but they showed no significant correlation with IHD mortality. The individual triglyceride values were skewed, but the median values in the communities showed little difference (data not shown).

3.3 Diet, antioxidants and lipids

The principal differences in the diet between Mountain Lapland and the reference area were in the consumption of reindeer meat, milk and dietary fat. A high consumption of

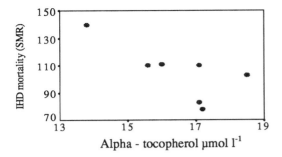

Figure 2 Serum alpha-tocopherol adjusted for age, BMI, smoking status, systolic and diastolic blood pressure and LDL cholesterol and mortality from IHD (SMR) in 1961 - 1990 in the seven northernmost communities of Finland (Spearman r = -0.79, P<0.05).

reindeer meat (at least three meals/week) was most common in the communities of Mountain Lapland and Savukoski (Figure 1). The consumption of milk (3.4 vs 4.5 glasses/day, 95 % CI for the difference -2.0 to -0.2, P<0.02) and dietary fat (34.0 vs 49.1 g/day, -25.2 to -4.9, P<0.01), particularly butter (17.3 vs 28.2 g/day, -20.5 to -1.4, P<0.05), was lower in Mountain Lapland than in the reference area. The serum alpha-tocopherol level increased with the consumption of reindeer meat, the association being most significant in the Finns who showed greater variation in consumption than the Saamis. The adjusted alpha-tocopherol levels in Finns eating reindeer meat at least twice a week and less frequently were 15.8 and 11.8 µmol/l, respectively (95 % CI for the difference 0.6 to 7.3, P<0.05). The adjustment for fish or alcohol consumption had no significant effect on this difference. Serum albumin (Table 2) was highest in the three communities with the highest consumption of reindeer meat (data not shown).

Serum selenium levels were higher in men eating fish at least twice a week than in those who did so less frequently (1.57 vs. 1.45 µmol/l, 95 % CI for the difference 0.03 to 0.21, P=0.01). The finding was not affected by the consumption of reindeer meat or alcohol. Serum antioxidants and lipids in Finns did not deviate from those in Saamis both living in the same area and eating the same amount of reindeer meat or fish.

3.4 Serum antioxidants and lipids in Saamis and Finns

Since antioxidants and lipids were related to the place of residence, the values for the Saamis and Finns living in the same area, i.e. Mountain Lapland, were compared. In this region the Saamis showed higher selenium levels than the Finns (1.60 vs 1.47 µmol l^{-1}), but the difference was statistically insignificant (P>0.1). Adjusted serum alpha-tocopherol (17.8 vs. 18.3 µmol l^{-1}, P>0.1) and of cholesterol (6.74 vs. 6.63 mmol l^{-1}, P>0.1) levels were high both in Saamis and Finns. The differences in other antioxidants and lipids were also insignificant.

4 DISCUSSION

Our study showed variations in the diet, serum concentrations of antioxidants and lipids and IHD mortality according to the place of residence. Serum alpha-tocopherol, albumin and selenium were high in the men who lived in the rural communities with low IHD mortality, whilst low serum cholesterol, alpha-tocopherol and albumin were characteristic of men living in the community with the highest IHD mortality. These findings conform with those relating alpha-tocopherol,[5-7,10] albumin[8] and selenium[9] to differences in cardiovascular mortality, but not with earlier observations in this country which have shown no association between regional variation of alpha-tocopherol and IHD.[5]

In this study the classic risk factors, when considered together, showed no clear association with mortality, and hence other factors, possibly the antioxidants, may have significance. The differences in serum alpha-tocopherol and albumin levels between the communities were wide enough to be reflected in the mortality.[4,8] Alpha-tocopherol is the major chain-breaking antioxidant in human tissues and the LDL particle, and albumin is an important extracellular antioxidant.[15] Albumin is synthesized in the liver and antioxidants such as alpha-tocopherol[16] and selenium,[16] which protect hepatocellular structures against oxidative damage, may contribute to its serum level. Albumin itself acts as a powerful inhibitor of peroxidation and free radical generation.[15] Selenium is a constituent of glutathione peroxidase, an enzyme vital to the protection of cellular structures against oxidative damage.[17] It acts as a peroxide removing enzyme. The antioxidants may retard the atherosclerotic vascular process and the manifestation of IHD by preventing the oxidative, atherogenic modification of LDL.[1-3]

Serum antioxidant levels were related to dietary factors. Alpha-tocopherol increased with the consumption of reindeer meat and selenium with fish consumption. Furthermore, serum alpha-tocopherol and albumin were high in the communities where the consumption of reindeer meat was high. These associations suggest that the local diet contributes to the antioxidative defence. Reindeer meat and fish are the main constituents of the diet in the north, and important sources of antioxidants.[18,19] Reindeer meat, which is rich in protein, alpha-tocopherol and selenium and low in fat content,[18] may reduce the atherogenity and thrombogenity of the diet.[19] A diet rich in reindeer meat may also beneficially increase the intake of monounsaturated oleic acid.[18,20] It is highly resistant to oxidation and may slow the progression of atherosclerosis by protecting LDL against atherogenic modification.[21] In addition the low consumption of milk and dietary fat, particularly butter, in Mountain Lapland may have significance. The present selenium data conform with those showing an inverse association of serum selenium and cardiovascular mortality.[9] An improvement of selenium status with fish consumption may contribute to the antioxidative defence, and the prevention of IHD.[19,22]

About one third of the people living in Mountain Lapland are of Saami origin. Hence it is possible that genetic factors may affect antioxidant and lipid levels and mortality there. In our previous studies we determined the distribution of apolipoprotein E phenotype, one of the many genetically determined factors that may affect the risk of IHD, and found the frequency of the type E4/4 to be high particularly in Mountain Lapland[23] and more common amongst the Saamis than amongst the Finns.[24] The association of this phenotype with the high serum cholesterol and IHD risk[25] may contribute to the high cholesterol level in the north, but it is unusual to associate it with the low death date from IHD observed there. Interestingly, however, Saamis in northern Norway, despite high serum cholesterol, also have a low occurrence of IHD compared with the Finns or Norsemen in the area.[26,27]

References

1. J.T. Salonen, R. Salonen, K. Seppänen, M. Kantola, S. Suntioinen and H. Korpela. *Br Med J* 1991, **302**, 756.
2. H. Esterbauer, J. Gebicki, H. Puhl and G Jurgens. *Free Radic Biol Med* 1992, **13**, 341.
3. D. Steinberg and workshop participants. *Circulation* 1992, **85**, 2338.
4. K.F. Gey KF and P. Puska. *Ann NY Acad Sci* 1989, **570**, 268.
5. R.A.Riemersma, M. Oliver, R.A. Elton, G. Alfthan, E. Vartiainen, M. Salo, et al. *Eur J Clin Nutr* 1990, **44**, 143.
6. E.B. Rimm EB, M.J. Stampfer, A. Ascherio, E. Giuovannucci, G. Colditz and W.C. Willett. *N Engl J Med* 1993, **328**, 1450.
7. P. Knekt, A. Reunanen, R. Järvinen, R. Seppänen R, M. Heliövaara and A. Aromaa. *Am J Epidemiol* 1994, **139**, 1180.
8. A. Phillips, A.G. Shaper and P.W. Whincup. *Lancet* 1989, **334**, 1434.
9. J.T. Salonen, G Alfthan, J.K. Huttunen, J. Pikkarainen and P. Puska. *Lancet* 1982, **ii**, 175.
10. K.F. Gey, H.B. Stähelin and B.E. Ballmer. *Ther Umsch* 1994, **51**, 475.
11. P.V. Luoma, S. Näyhä, K. Sikkilä and J. Hassi. *J Intern Med* 1995, **237**, 49.
12. S. Näyhä and J. Hassi, eds. *Publications of the Social Insurance Institution ML, Helsinki,* 1993;127.
13. W.T. Friedewald, R.I. Levy and D.S. Fredrickson. *Clin Chem* 1972, **18**, 499.
14. S. Näyhä. *Scand J Soc Med* 1989, suppl 40.
15. B. Halliwell and M.C. Gutteridge. *Arch Biochem Biophys* 1990, **280**, 1.
16. H. Sies, W. Stahl and A.R. Sundquist AR. *Ann NY Acad Sci* 1992, **699**, 7.
17. J.T. Rotruck, A.L. Pope, H.E. Ganther, A.B. Swanson, D. Hafeman and W.G. Hoekstra. *Science* 1973, **179**, 588.
18. M. Nieminen. "Proceedings of the 3rd International Wildlife Ranching Symposium," van Hoven W, Ebedes H, Conroy A, eds. University of Pretoria, Pretoria 0002, South Africa, 1992, p.196.
19. T.L.V. Ulbricht and D.A.T. Southgate. *Lancet* 1991, **338**, 985.
20. G. Garton and W.R. Duncan. *J.Sci Fd Agric* 1971, **22**, 29.
21. S. Parthasarathy, J.C. Khoo, E. Miller, J. Barnett, J.L Witztum and D. Steinberg. *Proc Natl Acad Sci USA* 1990, **87**, 3894.
22. D. Kromhout, E. Bosschieter E. and C. De Lezenne Coulander. *N Engl J Med* 1985, **312**, 1205.
23. P.V. Luoma, T.L. Lehtimäki, S. Lehtinen, S. Näyhä, J. Hassi J, C. Ehnholm et al. "Proceedings of the Laboratory Medicine 95" 11th IFCC European Congress of Clinical Chemistry, Tampere, Finland. H Adlercreutz, I Penttilä, R. Tenhunen, T. Weber, eds. Forssan kirjapaino, Finland, 1995, 595, abstr.
24. S. Lehtinen, T. Lehtimäki T, P. Luoma, S. Näyhä, J. Hassi, C. Ehnholm et al. Differences in genetic variation of apolipoprotein E in Lapps and Finns. *Eur J Lab Med* 1995; in press.
25. J. Davignon, R.E. Gregg RE and C.F. Sing. *Arteriosclerosis* 1988, **8**, 1.
26. The Cardiovascular Study in Finnmark 1974-75. *Nordic Council for Arctic Medical Research Report 25*, 1979, pp.1-191.
27. D.S. Thelle and O.H. Førde. *Am J Epidemiol* 1979, **110**, 708.

CALCIUM PANTOTHENATE AND ANTIOXIDANT VITAMINS IN PREVENTION OF SURGICAL STRESS

O.I. Dubrovshchik [*], I.T. Tsilindz [*], I.Y. Makshanov [*], A.G. Moiseenok [**]

[*] Department of General Surgery, Medical Institute,
80 Gorky str., 230015, Grodno, Belarus
[**] Laboratory of Coenzymes, Institute of Biochemistry of the Academy of Sciences of Belarus, 50 BLK, 230017, Grodno, Belarus

1 INTRODUCTION

The authors suggest that lipid peroxidation (LPO) products play an important role in the mechanisms of development of stress reactions in the organism subjected to operative trauma. Their manifestations depend on the resistance of the organism and the extent of the adaptation to the extreme factor.[1,2] Preclinical experiments showed that Ca pantothenate (CaP) had antistress activity.[3] The evident manifestation of CaP activity is the ability to prevent LPO in organs and tissues, which was demonstrated by the models of hypobaric hypoxia, galactosamine hepatitis, diabetes mellitus and profound hypothermia. In particular, one injection of the metabolite, Phosphocalciumpantothenate, completely prevented the development of hyperfermentomia and accumulation of LPO products in the blood of experimental animals, which underwent profound hypothermia.[4] This finding has had application in cardiosurgical practice in which prevention of hypothermia complications is of prime importance.

2 SUBJECTS AND METHODS

We examined 51 patients with ulcer of the pyloroduodenal region and 55 patients with cancer of the stomach admitted to a surgical clinic. The age of the patients varied between 26 and 70 years. To enhance the organism resistance to the operative trauma, the 7-day standard preoperative preparation was used, including anabolic steroids, infusions of protein and carbohydrate substrats, vitamin therapy (ascorbic acid, cyancobalamin).

To investigate study the CP influence on the prevention of surgical stress at the preoperative period, a test was made to control the increase in the level of blood 11–oxycorticosteroids (OCS) within 7-8 hours after the single CP intramuscular injection at a dose of 3.0 to 3.5 mg/kg body weight. If the increase of OCS level was over 10% of the initial content, the test was considered to be positive. It is not advisable to apply the test imitating a stress reaction to patients with a negative CP response since it is impossible to realize the adaptational syndrome without increased OCS level. [6] As the earlier investigations showed the majority of the examined patients displayed a positive CP test.

A positive test was displayed by 91% of the healthy individuals examined, by 66.7 % of the patients with peptic ulcer, by 53.8 % of patients with cancer of the stomach. The CP (20% solution, 1-2 ml, daily) was administered starting from the third day. The total number of the injections was 4 or 5. The last one being 2 h before the operation.

The levels of blood plasma were investigated fluorimetrically, [7] malonedialdehyde (MDA)- by a thioborbituratel method, and diene conjugates (DC) spectrophotometrically.[9] The activity of acid phosphatase was measured a photometric technique. Schiff bases content (SB) was determined by a phosphorescent method.[10]

3 RESULTS AND DISCUSSION

As Figure 1 shows the supplementary CP prescribtion for the patients with peptic ulcer had a moderate antistress effect and did not influence the content of free fatty acids in the blood plasma. For the patients with cancer of the stomach, CP stimulated during the preoperative period, the OCS release without essential changes in the latter over the further observation periods. Little of any CP effect on the LPO parameters and the acid phosphatade activity was found in the patients with peptic ulcer, as for the oncological patients, the effect of the vitamins turned out to be rather significant (Figure 2) at the early postoperative period. This effect was confirmed by the investigation of SB content and by the changes in enzyme activity.

Judging by the clinical data (Table 1) CP application in both groups of the patients stimulated the shortening of the treatment period, the decreasing of the complications number and the lowering of the lethal outcomes.

Table 1. *Effect of supplementary pantothenate administration on treatment results*

Group of patients	Treatment period (days)	Number of complications (%)	Lethal outcomes (%)
Patients with peptic ulcer (conventional standard preoperative preparation)	110 ± 5.7	26.6	6.9
Pantothenate-treated patients with peptic ulcer	83.1 ± 4.9 *	4.4	0
Patients with cancer of the stomach (conventional preoperative preparation)	128.0 ± 6.3	22.9	8.3
Pantothenate-treated patients with cancer of the stomach	94.2 ± 3.4 *	12.6	5.4

Note: * - $p < 0.05$ as compared to control group.

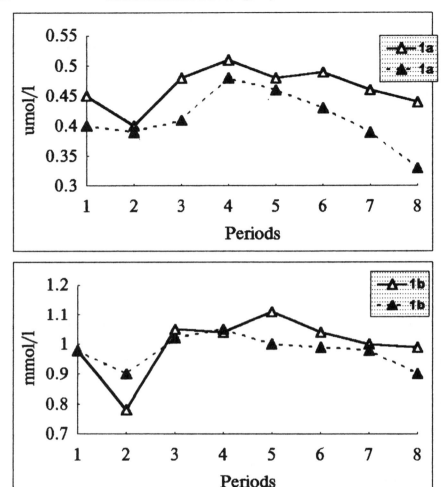

Figure 1 *Changes in the levels blood plasma of oxycorticosteroids (1a) and free fatty acids (1b) in patients with peptic ulcer, standard receiving preoperative preparation (-△-) or supplementary pantothenate (-▲-). Here and further the periods of observations are designated in the figures as: 1 — on admission to hospital; 2 — on completion of preparation for surgery; 3 — before initial narcosis during the operation; 4 — on completion of the operation; 5 — after 1 day following the operation; 6 — after 3 days following the operation; 7 — after 5 days following the operation; 8 — before discharge from hospital*

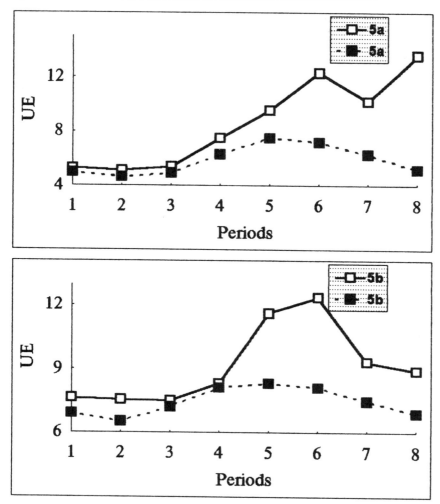

Figure 2 *Changes in the concentrations of diene conjugates (5a) and malonedialdehyde (5b) in the blood plasma of patients with cancer of the stomach receiving standard preoperative preparation (-□-) or additional pantothenate (-■-). Note: see the legend to Figure 1*

It may be suggested that CP administration leads to an increase in corticosteroid levels and therefore serves as a trigger for the general adaptation syndrome development. Repeated simulation of this reaction (4 or 5 CP injections) makes it possible to increase the effectiveness of the adaptational mechanisms to the operative irritant, i.e. to elevate the potential of stress - limiting systems. A very important element is also the CP ability to prevent LPO activation processes which is obviously related to general metabolic properties of this compound,as a precursor of coenzyme A biosynthesis - a powerful factor of homoeostasis of energy and phospholipid metabolism.[3,4] On the whole, the approach providing for training influence on the stress-limiting systems and enhancement of body resistance (including prevention of LPO activtation) seems to be extremely promising in modern surgery. [14-16]

The above data testify to expediency of antistress defence of the body and prevention of the development of oxidative stress by repeated administration of Ca pantothenate at preoperative period. In this situation the patients have repeated functional tension of the adrenal system, which produces training effect and causes their functional rearrangement and as a result optimizes the hormonal response to the operative trauma.

References

1. I.Y. Makshanov. Theoretical and practical organiam resistance aspects in Medicine, (Rus), Grodno Medical Institute, Grodno, 1991.

2. O.I. Dubrovshchik, D.Sci. Thesis, (Rus.), Moscow, 1992.

3. A.Moiseenok, Pantothenic acid. Biochemistry and application of the vitamins, (Rus.), Nauka i Technika, Minsk,1980.

4. A.Moiseenok, W.Gurinovich, S.Omeljanczyk, *News Acad. Sci. BSSR*, ser. biol. sci., 1988, **N 1**, 36.

5. L. Utno, M. Girgensone, D. Tribe, *J. Md. and Cell Cardiology*, 1989, **N 5**, 55.

6. F. Meyerson. Adaptation to stress situations and physical loading, (Rus), 1985, **3**, 137.

7. Y. Pankov and I. Uszatova, *Works on new apparatus and methods, (Rus)*, 1985, **3**, 137.

8. I. Asakawa and S. Hatushita, J. *Lipids*, 1980, **15**.137.

9. V. Costyuk, A. Potapovitch and Y. Lunets. *Problems in Medical Chemistry, (Rus.)*, 1984, **30**, 125.

10. Yu. A. Vladimirov, A.I. Archakov. Lipid peroxidation in biological membranes. Moscow, Nauka, 1972, 252 p.

11. E.D. Hall and J.M. Pangler, J. *Neurosurg.*, 1982, **57**, 247.

12. S. Bruskov, D. *Med. Sci. ,Thesia (Rus.)*, Kcharkov, 1981.

13. R. Mitchel, G. Smith, *Brit. J. Anaesth.*, 1989, **63**, 147.

14. E. Tomascsyk, *Vestnik kchirurgie (Rus.)*, 1985, **N 9**, 20.

15. E. Tomascsyk, D. *Med. Sci. Thesis (Rus)*, Vilnius, 1989.

SERUM ß-CAROTENE RESPONSE TO CHRONIC SUPPLEMENTATION WITH RAW CARROTS, CARROT JUICE OR PURIFIED ß-CAROTENE

R. Törrönen[1], M. Lehmusaho[1], S. Häkkinen[1], O. Hänninen[1] and H. Mykkänen[2]

[1]Department of Physiology and [2]Department of Clinical Nutrition
University of Kuopio
P.O. Box 1627, FIN-70211 Kuopio, FINLAND

1 INTRODUCTION

Epidemiological studies on the relationship between diet and cancer show that consumption of large amounts of fruits and vegetables reduces the risk of certain cancers[1]. The protective effect of fruits and vegetables has been attributed to their content of carotenoids, especially ß-carotene[2-4].

Several intervention trials using synthetic ß-carotene are currently in progress[5], and so far the results of three studies on cancer incidence have been published. A five-year placebo-controlled trial showed no effect of ß-carotene on the occurrence of new skin cancers in persons with previous non-melanoma skin cancers[6]. A five-year trial in China showed a significant reduction of mortality due to cancer among persons whose diets were supplemented with the combination of ß-carotene, vitamin E, and selenium[7]. In a five to eight-year randomized, double-blind, placebo-controlled trial in Finland (the ATBC trial), no reduction was found in the incidence of lung cancer or cancers of other sites among male smokers after dietary supplementation with ß-carotene[8]. On the contrary, a higher incidence of lung cancer was observed among the men who received ß-carotene than among those who did not. Also total mortality was higher among the participants who received ß-carotene than among those who did not. This study raised the possibility that ß-carotene supplements may have both harmful and beneficial effects.

Although dietary intake and blood level studies strongly suggest the cancer preventive potential of ß-carotene, the results published so far on the human intervention trials using purified ß-carotene supplements are surprisingly negative. To raise the blood ß-carotene level, increased intake of carotenoid-rich foods would be a natural alternative to synthetic ß-carotene preparations. In Finland, as well as in other Western countries, carrot is the most important source of dietary ß-carotene.

The purpose of the present study was to compare the serum ß-carotene responses in free-living healthy volunteers on a low-carotenoid diet supplemented for 6 weeks either with raw carrots, carrot juice or capsules containing purified ß-carotene. We also wanted to see if the wide variation in the serum response to purified ß-carotene preparations[9-16] is detected also in individuals consuming raw carrots or carrot juice.

2 MATERIALS AND METHODS

Thirty eight healthy non-smoking adult female volunteers were recruited from the personnel and students of the university of Kuopio, Finland. They were assigned into three groups so that the mean habitual serum ß-carotene concentrations of the groups were not statistically different.

The study consisted of a 10-day depletion period and a 6-week supplementation period. The subjects were instructed to consume a low-carotenoid diet for 10 days before the study and throughout the study period. During this time they ate a self-selected diet low in carotenoids. They received detailed instructions about the dietary modifications they had to make. They were given a list of fruits and vegetables to avoid and those that were allowed. They were told to follow their usual diet in every other respects.

The low-carotenoid diet was then supplemented for the next 6 weeks either with raw carrots (120 g/day), carrot juice (1 dl/day) or capsules containing purifed ß-carotene (2 capsules/day), each providing 12 mg of ß-carotene/day. The carrots (Nantes Narland; 10 mg ß-carotene/100 g) were purchased from a local cultivator. The carrot juice (Biotta; 140 mg ß-carotene/l) and ß-carotene capsules (Karocaps®; 6 mg ß-carotene/capsule) were kindly provided by A. Vogel Oy (Vantaa, Finland) and Leiras Oy (Turku, Finland), respectively. The subjects were instructed to take the supplements with meals.

Blood samples for serum ß-carotene measurements were collected by venipuncture after an overnight fast before the start of the low-carotenoid diet (habitual level), at the beginning of the supplementation period (initial level) and after 3 and 6 weeks of supplementation. The blood samples were protected from light and centrifuged within one hour after being drawn. The sera were stored at -70 °C. Serum ß-carotene concentrations were analysed using the reversed phase HPLC method of MacCrehan and Schönberger[17].

Dietary intakes of energy, energy nutrients and fibre were determined using 4-day food records by household measures, including one weekend day. The supplements were not included in the records. Intake of ß-carotene was estimated using 1-month food frequency questionnaires. Nutrient intakes were computed by the Nutrica computer program (Social Insurance Institution, Finland) using the Finnish foods database[18].

Statistical significance of the differences was analysed using the StatView 512+™ software (BrainPower, Inc., Calabasas, CA, USA). Data of each test group at different time points were first analysed by analysis of variance (ANOVA) for repeated measures, followed by the Scheffé multiple comparison test. Data of the different test groups at each time point were compared by factorial ANOVA. Relation between the serum ß-carotene responses after 3 weeks and 6 weeks of supplementation was tested by calculating Spearman's rank correlation coefficients. Pearson's correlation coefficients were calculated to describe the relationship between the serum ß-carotene response and age, body mass index, serum cholesterol and triglyceride levels, habitual ß-carotene intake and blood level, or nutrient intake. Values of $p < 0.05$ were considered statistically significant.

3 RESULTS

The mean age of the 38 participants was 30 years (range 20-53). Their mean body mass index was 23.2 kg/m^2 (16.9-37.5), and the serum cholesterol and triglyceride concentrations were 5.2 mmol/l (3.8-7.6) and 0.94 mmol/l (0.54-2.12), respectively. There were no significant differences in the mean age, body mass index or serum lipid concentrations between the test groups. No change in the body mass index was observed during the supplementation period.

The mean habitual intake of ß-carotene was 4.5 mg/day (range 1.3-10.1), with no significant difference between the groups. During the test period, the dietary intake of ß-carotene (without the supplements) decreased to 0.4 mg/day (0.06-1.0, one subject 1.8). During the supplementation period, there were no significant differences between the test groups in the intakes of energy, energy nutrients nor fibre.

The mean habitual serum ß-carotene concentration of all the participants was 536 ± 228 µg/l. There was a large interindividual variation in serum concentrations of ß-carotene throughout the study. The habitual ß-carotene concentration ranged from 103 to 1182 µg/l, and decreased after 10 days on the low-carotenoid diet to 53 % of the habitual level, ranging from 94 to 574 µg/l. Despite the interindividual variation in the serum values, the differences between the mean initial ß-carotene concentrations of the test groups at the beginning of the supplementation period were less than 5 % (Table 1).

Table 1 Serum β-Carotene Concentrations (μg/l)

Group	Habitual (Normal Diet)	Initial (Low-carotenoid Diet)	After 3 Weeks of Supplementation	After 6 Weeks of Supplementation	ANOVA
Raw carrots	570 ± 214	284 ± 133	471±209 [b,c]	452±232 [a,d]	$p < 0.001$
Carrot juice	512 ± 213	282 ± 128	448±226 [c]	570±378 [a,c]	$p = 0.004$
Capsules	524 ± 269	294 ± 134	851±463 [a]	937±456 [b]	$p < 0.001$
ANOVA	$p = 0.800$	$p = 0.972$	$p = 0.004$	$p = 0.006$	

Values are means ± SD.
Statistics within the groups (habitual values not included in the analysis): ANOVA (repeated measures) + Scheffe F test ([a] $p < 0.01$, [b] $p < 0.001$, as compared to the initial values; no significant differences between weeks 3 and 6). Statistics between the groups: ANOVA (factorial) + Scheffe F test ([c] $p < 0.05$, [d] $p < 0.01$, as compared to the capsule group; no significant differences between raw carrots and carrot juice).

The compliance of each participant was checked weekly. The daily doses of the supplements (120 g carrot, 1 dl carrot juice and 2 capsules) were well tolerated, and the subjects were able to ingest these doses during the whole 6-week test period. Serum ß-carotene concentrations increased significantly in all test groups (Table 1). Carrot juice doubled and raw carrots nearly doubled the serum ß-carotene level; no statistically significant difference in the effect was found between them. Supplementation with purified ß-carotene elicited mean serum ß-carotene concentrations that were 3-fold compared to the initial level and nearly 2-fold compared to the respective levels in the groups consuming carrots or carrot juice. There were no statistically significant differences between the mean values after 3 and 6 weeks of supplementation in any of the test groups.

There was a wide variation in the individual response of serum ß-carotene to supplementation, regardless of the source (Figure 1). Variation was largest – ranging from a lack of response to over 8-fold increase in the serum ß-carotene concentration – in the group receiving ß-carotene capsules. In the groups receiving carrots or carrot juice, the variation ranged from a lack of response to nearly 3-fold and to over 4-fold increases in the serum ß-carotene levels, respectively. Figure 1 presents the individual serum ß-carotene responses after 3 and 6 weeks of supplementation. There was a statistically significant correlation ($r = 0.615$; $p < 0.05$) between the 3-week and 6-week responses in the group ingesting raw carrots, but not in those ingesting carrot juice or ß-carotene capsules. There were no consistent correlations of serum ß-carotene response with age, body mass index, serum lipid levels, habitual ß-carotene intake nor blood level, nor nutrient intake during the supplementation.

4 DISCUSSION

In this study we have investigated the bioavailability (assessed as the increase of serum ß-carotene concentration) of natural ß-carotene obtained from raw carrots and carrot juice in healthy non-smoking women. The serum response was lower after ingestion of natural ß-carotene than after ingestion of synthetic ß-carotene from capsules. Similarly to synthetic ß-carotene, there was a large interindividual variation in the serum response also among the subjects who received natural ß-carotene.

The subjects were adult female volunteers with no apparent illnesses. They did not use any supplements containing ß-carotene or vitamin A. Only non-smokers were included since smoking affects the serum response to dietary ß-carotene. Their habitual intake ranged from 1.3 to 10.1 mg/day and serum ß-carotene concentration from 103 to 1182 μg/l. To deplete the body ß-carotene stores before the supplementation, the subjects were instructed to

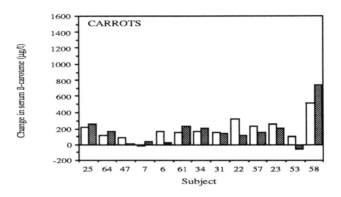
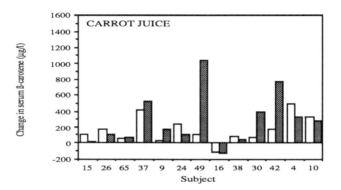
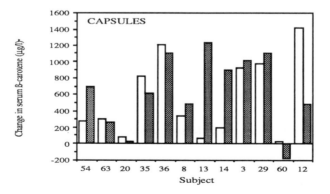

Figure 1 *Individual responses of serum β-carotene concentration after supplementation with 12 mg of β-carotene either from raw carrots, carrot juice or capsules containing purified β-carotene. Serum values were measured after 3 (white columns) and 6 (grey columns) weeks of supplementation. Subjects in each test group are arranged in the ascending order of their habitual serum β-carotene concentration.*

consume a low-carotenoid diet for 10 days before the study (and throughout the study period). The depletion diet provided 0.4 mg ß-carotene/day. After 10 days of depletion, serum ß-carotene concentrations had decreased drastically, to 53 % of the habitual level. The supplementary dose of ß-carotene was relatively small, 12 mg/day, corresponding to 120 g of carrot and 1 dl of juice. These doses of carrot and juice were well tolerated. Ingestion of larger amounts of carrot daily for 6 weeks would probably have been difficult for the subjects, leading to poor compliance.

The serum ß-carotene response was lower in the subjects receiving 12 mg of ß-carotene from raw carrots than from the capsules. After 6 weeks of consuming carrots, the mean serum ß-carotene concentration was approximately 1.5 times the initial level. The response was only 26 % of that elicited by the capsules. The difference in the serum ß-carotene response to supplementation with purified ß-carotene and with carrots observed in the present study is similar to that observed in other studies. Brown et al.[11] compared plasma ß-carotene responses in men after a single dose (30 mg) of ß-carotene administered as cooked carrots or as purified ß-carotene in capsules. Ingestion of carrots produced maximum plasma response that was only 21 % of that produced by the capsules. In a chronic supplementation study of Micozzi et al.[19], healthy men received a controlled diet with either cooked carrots or purified ß-carotene capsules providing 30 mg of ß-carotene daily for 6 weeks. The plasma response to ß-carotene from carrots was only 18 % of that to an equal amount of ß-carotene from the capsules. In children, administration of 6 mg of ß-carotene daily for 20 days in cooked carrots resulted in no significant change in plasma ß-carotene concentration, despite the approximately 3-fold increase in response to purified ß-carotene[15].

The serum ß-carotene response in the subjects receiving 12 mg of ß-carotene from carrot juice was 45 % of that from the capsules. Carrot juice doubled the mean serum ß-carotene level in 6 weeks, and no statistically significant difference in the effect was found between carrot juice and raw carrots. Kim et al.[20] have reported that after consuming approximately 20 mg of ß-carotene from carrot juice for 7 days, serum ß-carotene concentration was approximately 2.5 times the baseline level.

The present results together with previous results suggest poorer bioavailability of natural ß-carotene than of purified ß-carotene. The inhibitory effect of fibre on ß-carotene absorption may provide one explanation for the reduced serum response after the ingestion of carrots and carrot juice compared to the equal dose of purified ß-carotene from the capsules. The increase in plasma ß-carotene concentration after ingestion of ß-carotene in a capsule was found to be significantly reduced by pectin[21].

Substantial interindividual variation in the extent of serum response after supplementation with purified ß-carotene preparations has been documented[9,16], and was observed also in the present study. The variation ranged from a lack of response to over 8-fold increase in the serum ß-carotene concentration. Moreover, there was no statistically significant correlation between the individual serum ß-carotene responses after 3 and 6 weeks of supplementation, suggesting that the steady state of the serum concentration was not reached during the first 3 weeks.

The interindividual variation in the serum response was observed also in the subjects supplemented with natural ß-carotene from carrots and carrot juice. The reasons behind this remain undetermined, since none of the background characteristics (age, body mass index, serum lipid levels, habitual ß-carotene intake and blood level, or nutrient intake) displayed any consistent correlation with the serum ß-carotene response. Brown et al.[11] have reported a 3- to 4-fold difference between the lowest and highest responses in subjects ingesting a single dose (30 mg) of ß-carotene from cooked carrots, suggesting interindividual differences in the absorption of ß-carotene. However, efficiency of absorption may not be the only determinant of the serum response to oral ß-carotene. Serum concentrations are affected also by such factors as destruction of ß-carotene in the gastrointestinal tract, rate of metabolism and rate of tissue uptake.

In conclusion, smaller average increases of serum concentration were observed in healthy non-smoking females supplemented for 6 weeks with 12 mg of ß-carotene from raw carrots or carrot juice than in those supplemented with an equal dose of purified ß-carotene from capsules, suggesting poorer bioavailability of natural ß-carotene than of purified ß-carotene. Substantial interindividual variation in the serum response was detected in subjects ingesting purified ß-carotene from capsules, as well as in those ingesting natural ß-carotene from raw carrots or carrot juice.

References

1. K. A. Steinmetz and J. D. Potter, *Cancer Cause Control*, 1991, **2**, 325.
2. R. Peto, R. Doll, J. D. Buckley and M. B. Sporn, *Nature*, 1981, **290**, 201.
3. R. G. Ziegler, *Am. J. Clin. Nutr.*, 1991, **53**, 251S.
4. R. G. Ziegler, A. F. Subar, N. E. Craft, G. Ursin, B. H. Patterson and B. I. Graubard, *Cancer Res.*, 1992, **52**, 2060S.
5. G. van Poppel, *Eur. J. Cancer*, 1993, **29A**, 1335.
6. E. R. Greenberg, J. A. Baron, T. A. Stukel, M. M. Stevens, J. S. Mandel, S. K. Spencer, P. M. Elias, N. Lowe, D. W. Nierenberg, G. Bayrd, J. C. Vance, D. H. Freeman Jr, W. E. Clendenning, T. Kwan and The Skin Cancer Prevention Study Group. *N. Engl. J. Med.*, 1990, **323**, 789.
7. W. J. Blot, J. Y. Li, P. R. Taylor, W. D. Guo, S. Dawsey, G. Q. Wang, C. S. Yang, S. F. Zheng, M. Gail, G. Y. Li, Y. Yu, B. Q. Liu, J. Tangrea, Y. H. Sun, F. S. Liu, J. F. Fraumeni, Y. H. Zhang and B. Li, *J. Natl Cancer Inst.*, 1993, **85**, 1483.
8. The Alpha-Tocopherol, Beta Carotene Cancer Prevention Study Group, *N. Engl. J. Med.*, 1994, **330**, 1029.
9. J. C. Meyer, H. P. Grundmann, B. Seeger and U. W. Schnyder, *Dermatologica* 1985, **171**, 76.
10. N. V. Dimitrov, C. Meyer, D. E. Ullrey, W. Chenoweth, A. Michelakis, W. Malone, C. Boone and G. Fink, *Am. J. Clin. Nutr.*, 1988, **48**, 298.
11. E. D. Brown, M. S. Micozzi, N. E. Craft, J. G. Bieri, G. Beecher, B. K. Edwards, A. Rose, P. R. Taylor and J. C. Smith Jr, *Am. J. Clin. Nutr.*, 1989, **49**, 1258.
12. L. M. Canfield, J. Bulux, J. Quan de Serrano, C. Rivera, A. F. Lima, C. Yolanda Lopez, R. Perez, L. Kettel Khan, G. G. Harrison and N. W. Solomons, *Am. J. Clin. Nutr.*, 1991, **54**, 539.
13. D. W. Nierenberg, T. A. Stukel, J. A. Baron, B. J. Dain, E. R. Greenberg and The Skin Cancer Prevention Study Group, *Am. J. Clin. Nutr.*, 1991, **53**, 1443.
14. E. J. Johnson and R. M. Russell, *Am. J. Clin. Nutr.*, 1992, **56**, 128.
15. J. Bulux, J. Quan de Serrano, A. Giuliano, R. Perez, C. Yolanda Lopez, C. Rivera, N. W. Solomons and L. M. Canfield, *Am. J. Clin. Nutr.*, 1994, **59**, 1369.
16. A. Carughi and F. G. Hooper, *Am. J. Clin. Nutr.*, 1994, **59**, 896.
17. W. A. MacCrehan and E. Schönberger, *Clin. Chem.*, 1987, **33**, 1585.
18. M. Rastas, R. Seppänen, L.-R. Knuts, R.-L. Karvetti and P. Varo, 'Nutrient Composition of Foods'. Publications of the Social Insurance Institution, Helsinki, Finland, 1990.
19. M. S. Micozzi, E. D. Brown, B. K. Edwards, J. G. Bieri, P. R. Taylor, F. Khachik, G. R. Beecher and J. C. Smith Jr, *Am. J. Clin. Nutr.*, 1992, **55**, 1120.
20. H. Kim, K. L. Simpson and L. E. Gerber, *Nutr. Res.*, 1988, **8**, 1119.
21. C. L. Rock and M. E. Swendseid, *Am. J. Clin. Nutr.*, 1992, **55**, 96.

EFFECTS OF DIETARY N-3 FATTY ACIDS AND VITAMIN E ON LIPID TRANSPORT AND GLUCOREGULATION IN HYPERTRIGLYCARIDAEMIA.

A. Vrána, L. Kazdová, V. Nováková, and M. Matějčková,

Institute for Clinical and
Experimental Medicine Prague, Czech Republic

1 INTRODUCTION

The marked hypotriglyceridaemic action of fish oil rich in n-3 fatty acids is well documented both in human hypertriglyceridaemia and in an animal models of this disorder.[1,2]

Moreover, in hypertriglyceridaemic states, which are frequently associated with disturbances of carbohydrate metabolism, improvement of glucoregulation was observed after n-3 fatty acids administration.[3,4,5] Improved glucose tolerance was mediated by increased insulin action, while insulinemia and glucose-stimulated insulin release from isolated rat pancreatic islets were depressed as a result of fish oil administration.[5,6]

On the other hand, high intake of these fatty acids known as potent prooxidant, depletes body antioxidant capacity and increases concentrations of cellular and body fluid lipoperoxidation products.[7]

It was reported, that in hypertriglyceridemic patients, high doses of vitamin E elicited a mild improvement in glucoregulation as well as a beneficial effect on plasma lipids and lipoproteins, including triglyceridaemia.[8]

The aim of the present study was therefore to investigate effects of combined administration of n-3 fatty acids and of vitamin E, a natural antioxidant, on some parameters of lipid transport in animal model of hypertriglyceridaemia, in hereditary hypertriglyceridaemic rats. The line selected from rats of the Wistar strain by inbreeding exhibits in addition to genetically transmitted and high carbohydrate-diet potentiated hypertriglyceridemia, a cluster of other symptoms of metabolic syndrome X, i.e.hyperinsulinemia, insulin resistance, impaired glucose tolerance, elevation of blood pressure, and hyperuricemia.[9,10] Serum concentrations of triglycerides, free fatty acids, alpha-tocopherol and conjugated dienes, and oral glucose tolerance were monitored in rats fed diets containing either fish oil or olive oil without or with vitamin E supplementation.

2 MATERIALS AND METHODS

Male rats of a hereditary hypertriglyceridaemic line selected in our laboratory and originating from the Wistar strain weighing 250-270g were used in all experiments. A high carbohydrate diet[11] containing 70cal% sucrose, 20cal% protein and 10cal% fat to which 30g/kg of fish oil (FO) or olive oil (OLO) was added without or with 500mg/kg of vitamin E was fed ad libitum for 4 weeks except for the last 12 h prior to the glucose load. The following additions to diets were used: EPAX 5500, a fish oil preparation EPAX 5500 (Pronova a.s., Norway) containing, according to the manufacturer information, 55 per cent of n-3 fatty acids, olive oil of medicinal quality and vitamin E (alpha-tocopherol acetate, Slovakofarma, Slovak Republic).

Blood for serum triglyceride determination of several intervals of the feeding experiment was sampled from tail veins under light ether anaesthesia and serum separated by centrifugation at 4°C.

Glucose tolerance tests were performed after 2 weeks of diets feeding. Glucose (3g/kg b.wt) was administered intragastrically as 30% solution in saline by gastric gavage, and blood was collected without anaesthesia by cutting of tail veins before and 60 mins after the

glucose load. Concentrations of blood glucose, serum triglycerides and free fatty acids were determined by commercial enzymatic kits (LACHEMA, Czech Republic).
Serum alpha-tocopherol were determined HPLC as described before,[12] serum concentrations of conjugated dienes (precursors of cytotoxic hydroperoxides) were detected by UV spectrophotometry at 235 nm.[13]
Statistical significance of differences was calculated using Student's t-test.

3 RESULTS AND DISCUSSION

Fish oil administration alone mildly, but statistically significantly, increased the serum alpha-tocopherol concentration. On the contrary, alpha-tocopherol supplementation led to the distinctly smaller increase in its serum concentration in the FO diet fed group compared with controls fed OLO diet (Table 1). A high sucrose diet in combination with OLO led to markedly increased triglyceridemia compared with the group fed FO diet and, also, with baseline concentrations found chow-diet fed (Table 2). In addition, vitamin E supplementation in rats fed FO diet led to a further mild, but statistically significant decrease of serum triglycerides.
Serum free fatty acids concentration was lowered by FO administration, no effect of vitamin E supplementation on serum free fatty acids was observed (Table 3).

Table 1. *Serum alpha-tocopherol concentrations in rats given olive oil diet (OLO) or fish oil diet (FO) without or with vitamin E (E) supplementation.*

	Serum alpha-tocopherol, $\mu mol/L$		
OLO	OLO+E	FO	FO+E
11.2 ± 0.3^a	42.4 ± 2.7	14.2 ± 0.3	22.1 ± 0.8

aMeans $\pm$ S.E.M. (n = 8-9)
Statistical significance: OLO vs OLO + E $p < 0.001$, OLO vs FO $p < 0.001$, OLO + E vs FO + E $p < 0.001$

Table 2. *Serum triglycerides concentrations of chow diet fed controls and of rats given olive oil diet (OLO) or fish oil diet(FO) without or with vitamin E (E) supplementation.*

	Serum triglycerides, mmol/L			
	Days on diets			
	0 (chow diet)	2	7	21
OLO		6.71 ± 0.86	4.28 ± 0.54	3.48 ± 0.26
OLO+E		7.57 ± 1.06	4.44 ± 0.59	3.30 ± 0.13
	1.81 ± 0.10^a			
FO		1.78 ± 0.22	1.24 ± 0.11	1.45 ± 0.13
FO+E		1.55 ± 0.15	0.90 ± 0.09	1.03 ± 0.10

aMeans $\pm$ S.E.M. (n=8-9)
Statistical significance: OLO and OLO + E vs FO and FO + E $p < 0.001$ at all measurements, FO vs FO + E (7 and 21 days) $p < 0.05$.

Blood glucose concentrations before and after the oral glucose load are given in Table 4. After overnight fasting (0 min) blood glucose was lower in the FO diet fed group compared with olive oil diet controls. Post-load glycaemia was also lower in the FO group than in OLO controls. In rats fed OLO diet, vitamin E supplementation decreased both initial and post-load glycaemia. In FO group no further effect of vitamin E supplementation on already low blood glucose concentration was observed.

The above results indicate that, in rats fed FO and not supplemented with vitamin E, the serum alpha-tocopherol levels were higher compared with controls fed OLO diet. This difference probably reflects the higher content of the vitamin in FO preparation than in OLO. On the other hand, vitamin E supplemetation of FO diet led to a smaller increase in its serum concentration than in OLO and vitamin E supplemented group. This is in keeping with recent findings demonstrating a decreased serum antioxidant capacity after the administration of an n-3 fatty acids rich fish oil preparation.[7] The increased serum conjugated dienes after FO administration compared with OLO given controls are compatible with the above results.

As expected, FO administration, markedly reduced triglyceridaemia - this effect might be interpreted as the previously described reduction of liver lipogenesis, triglyceride synthesis and their export from the liver to the plasma in the form of very low density lipoproteins.[5,14] One possible explanation of mechanism/s leading to a small, yet statistically significant decrease in triglyceridaemia due to the vitamin E administration might be an increment of tissues clearance of plasma triglycerides.

Table 3. *Serum concentrations of free fatty acids (FFA) and of conjugated dienes (CD) in rats given olive oil diet (OLO) or fish oil diet (FO) without or with vitamin E (E) supplementation.*

	Serum FFA, $\mu mol/L$	Serum CD, abs.U
OLO	590 ± 24^a	0.150 ± 0.02
OLO+E	530 ± 26	0.134 ± 0.04
FO	460 ± 35	0.221 ± 0.02
FO+E	460 ± 20	0.187 ± 0.01

[a]Means $\pm$ S.E.M. (n = 8-9).
Statistical significance: Serum FFA, OLO and OLO + E vs FO and FO + E $p < 0.05$, serum CD, OLO vs FO $p < 0.01$, OLO + E vs FO + E $p < 0.01$.

Table 4. *Blood glucose concentrations before (0 min) and 60 mins after intragastric administration of glucose in rats given olive oil diet (OLO) or fish oil diet (FO) without or with vitamin E (E) supplementation.*

	Blood glucose, mmol/L	
	0 mins	60 mins
OLO	5.44 ± 0.23^a	6.97 ± 0.17
OLO+E	4.20 ± 0.18	6.38 ± 0.19
FO	4.22 ± 0.15	6.19 ± 0.15
FO+E	4.50 ± 0.20	6.25 ± 0.23

[a]Means $\pm$ S.E.M. (n = 8-9).
Statistical significance: 0 min, OLO vs OLO + E $p < 0.01$, OLO vs FO $p < 0.05$, 60 mins, OLO vs OLO + E $p < 0.01$, OLO vs FO $p < 0.01$.

Previously, we found that FO administration to hereditary hypertriglyceridaemic rats, while increasing basal and epinephrine-stimulated release of free fatty acids from isolated adipose tissue, decreased serum free fatty acids concentration.[15] The mild, but statistically significant reduction of serum free fatty acids concentrations in the FO diet given group was confirmed in the present study, and no effect of vitamin E supplementation of either diet on serum free fatty acids levels was detected. The reduction of serum free fatty acids levels might be due to their increased utilization as an energy substrate, since availability of blood triglycerides is markedly decreased by FO administration and, in addition, glycaemia is also mildly lowered by such a dietary intervention.

Glucoregulation was also affected by the type of the prevalent source of dietary fat. Improvement of glucose tolerance in the FO diet fed group after intragastric glucose load confirms previous results.[2,3,4] These studies indicated a dichotomy of the effect dietary FO on glucoregulation - depressed insulinaemia due to decreased insulin release from the islets of Langerhans[6] and, on the other hand, increased insulin action on isolated adipose tissue.[3]

In conclusion, FO compared with OLO administration profoundly decreased triglyceridaemia in an hereditary animal model of hypertriglyceridaemia. Vitamin E supplementation in FO diet fed rats led to an additional mild decrease in serum triglyceride concentration. No such an effect of the vitamin was observed in OLO diet fed rats. As regards the effects of the type of dietary oil and vitamin E administration on glucose homeostasis, both fasting and post-load glycaemia was lower in the FO groups compared with OLO controls. Vitamin E supplementation lowered fasting and post-load glycaemia, but only in the OLO diet fed groups.

REFERENCES

1. W. E.Connor, C. A. DeFrancesco and S. L. Connor, *Ann. N. Y. Acad. Sci.*, 1993, **683**, 16.
2. I. Klimeš, E. Seboková, A. Vrána, and L. Kazdová, *Ann. N.Y. Acad. Sci., 1993*, **683**, 69.
3. A. Vrána, A. Žák and L. Kazdová, *Nutr. Rep. int.*, 1988, **38**, 687.
4. A. Žák, M. Zeman, P. Hrabák, A. Vrána, H. Švarcová, and P. Mareš, *Nutr. Rep. int.*, 1989, **39**, 235.
5. A. Vrána, L. Kazdová,I. Klimeš, M. Ficková, and A. Žák, in I. Klimeš, B. Howard and R.C. Kahn (eds), Insulin and the Cell Membrane, 1990, p. 397, Gordon and Breach Sci. Publ., New York.
6. A. Vrána and L. Kazdová, *Diabetologia*, 1990, **33** (Supplement 1), 484.
7. M. Meidani, F. Natiello, B. Goldin, N.Free, M. Woods, E. Schaefer, J.B. Blumber, and S.L. Gorbach, *J.Nutr.*1991,**121**,484.
8. G. Paolisso, A. D'Amore, D. Galzerano, V. Balbi, D. Giugliano, M. Varricchio and F.D'Onofrio, *Diabetes Care*, 1993, 16, 1433.
9. A. Vrána and L. Kazdová, *Transpl. Proc.*, 1990, **22**, 2579.
10. A. Vrána, L. Kazdová, Z. Dobešová, J. Kuneš, V. Křen, V. Bílá, P.Štolba and I. Klimeš, *Ann. N.Y.Acad. Sci.*, 1993, **683**, 57.
11. P. Fábry, R. Poledne, L. Kazdová and T. Braun, *Nutr. Dieta*, 1968, **10**, 81.
12. G.L. Catignani, in Methods in Enzymology, 1985, **123**, 215.
13. P.A. Ward, G.O Pill, J.R. Hatherill, T.M. Hannesly, and R.J. Kunkel, *J. clin. Invest.*, 1985, **76**, 517.
14. A. Žák, K. Hátle, A. Vrána, and J. Skořepa, in H. Haller, M. Hanefeld, W. Jaross and W. Leonhard (eds), Proc. 5th Dresden Lipid Symposium, 1985, p. 463, VEB Chemie, Berlin.
15. A. Vrána and L. Kazdová, *Diabetologia*, 1994, **37**, 145.

ACKNOWGLEDGEMENT

This study was supported by grants 2024-3/94 and 3097-2/95 from the Internal Grant Agency of the Ministry of Health of Czech Republic.

ENHANCED ANTIOXIDANT STATUS IN LONG-TERM ADHERENTS OF A STRICT UNCOOKED VEGAN DIET ("LIVING FOOD DIET")

A-L. Rauma[1], R. Törrönen[2], O. Hänninen[2], H. Verhagen[3] and H. Mykkänen[1]

[1]Departments of Clinical Nutrition and [2]Physiology, University of Kuopio, P.O. Box 1627, FIN-70211 Kuopio, Finland, and [3]TNO Nutrition and Food Research Institute, P.O. Box 360, 3700 AJ Zeist, The Netherlands

1 INTRODUCTION

The vegetarian diets may play a preventive role in many chronic diseases due to their antioxidant content.[1,2] Antioxidants derived from plants include antioxidant vitamins (vitamin C, vitamin E, beta-carotene) and non-nutrient antioxidants such as flavonoids and other polyphenols, and the degradation products of glucosinolates, i.e. isothiocyanates and indoles.[3-5]

Most of the studies concerning the antioxidant vitamin status of vegetarians have involved dietary assessment without biochemical measurements. The purpose of the present cross-sectional study was to investigate the effects of a strict uncooked vegan diet composed of fermented and sprouted foods ("living food diet") on the antioxidant status of the long-term adherents of this diet. The dietary data are presented together with the concentrations of antioxidant vitamins and activities of antioxidant enzymes in the blood.

2 SUBJECTS AND METHODS

The subjects were 20 females long-term adherents of a strict uncooked vegan diet ("living food diet") were recruited via an advertisement in a living food eaters' newsletter in Finland. The average duration of the vegan diet was 5.2 years (SD 3.9, range 0.7-14). Each vegan subject was matched with an omnivorous control subject for sex, age, social status and residence. The mean values for age and height were similar within the pairs, but the vegans had significantly lower body weights and body mass indices.

The clinical blood chemistry of the subjects was within the reference ranges, except of low hemoglobin and ferritin concentrations in one vegan and two control subjects. Five of the vegans had originally started this diet due to an illness, and four of the control subjects suffered from a chronic disease.

The participants were asked to report the consumption of alcohol and health supplements, physical and sport activities, menstrual cycle and compliance to the vegan diet (vegans only). None of the participants smoked or used any medication regularly. Informed consent was obtained from each subject before the study, and the study protocol was accepted by the Ethical Committee of Kuopio University and Central Hospital.

Food intake was estimated using 5-day dietary records. Nutrient intakes were calculated by the Nutrica computer program (Social Insurance Institution, Finland) using the Finnish nutrient database[6] supplemented with the nutrient values of the vegetable foods commonly used by the vegans.

Blood samples were collected by venipuncture after an overnight fast. Serum cholesterol concentration was analyzed by an enzymatic method using Boehringer Mannheim GmbH test kit (Cat. No. 104083, Mannheim, Germany). Serum beta-carotene and alpha-tocopherol concentrations were analyzed by high-performance liquid chromatography, as described by Miller et al.[7] and Catignani et al.[8], respectively, and whole blood vitamin C by the method of Speek et al.[9] Glutathione peroxidase activity in whole blood was measured according to Paglia and Valentine[10], and superoxide dismutase using the Ransod kit (No. SD 125, Randox Laboratories Ltd, Crumlin, N. Ireland).

24-h urine samples were collected for the determination of sodium concentration using an ion-selective electrode in Hitachi 717 analyzer (Hitachi, Ltd., Tokyo, Japan).

The SPSS/PC computer program was used in the statistical analyses. Statistical significance of the differences between the matched pairs (vegan vs omnivorous control) was assessed by the Wilcoxon test for paired samples. Spearman's rank correlation coefficients were used to establish the correlations between the observed blood levels and calculated dietary intakes of different antioxidant nutrients.

3 RESULTS

On the basis of the urinary sodium excretion the compliance to the vegan diet was good (mean 50 mmol/d, SD 25, range 14-88 in the vegans, vs. 146 mmol/d, SD 71, range 78-391 in the controls, $p<0.001$).

According to the dietary records, the vegans consumed more root crops, pulses, nuts, fruits and berries than the omnivorous controls. Based on the frequency questionnaire, 10 vegans consumed no alcohol at all and 10 vegans used alcohol less than once a month or once a month. Eight controls used alcohol less than once a month or once a month, 12 controls used alcohol on weekends or daily.

The vegans received significantly more dietary antioxidant vitamins (beta-carotene, vitamin C and vitamin E) and significantly less selenium than their omnivorous controls. The mean daily intakes of dietary antioxidants by the vegans, expressed as percentages of the RDA[11], were: vitamin C 305 %, vitamin A 247 %, vitamin E 313 %, zinc 92 %, copper 120 %, selenium 49 %. The mean dietary intakes of antioxidant vitamins and selenium by the omnivorous controls exceeded the RDA (176, 167, 138 and 133%, respectively for vitamins C, A and E, and selenium), while those of copper and zinc were 80 and 83 % of the RDA.

The number of subjects using supplements (including seaweed) was high in both groups (vegans 16, controls 11). However, a lower number of the vegans consumed antioxidant supplements (4 vs. 11), and only vitamin C intake by the controls was considerably increased by supplementation.

The vegans had significantly higher blood levels of beta-carotene ($P<0.001$), vitamin C ($P<0.01$) and vitamin E expressed per millimole of plasma cholesterol ($P<0.01$) than the controls (Figure 1). The activity of erythrocyte superoxide dismutase was also significantly higher ($P<0.01$) in the vegans than in the controls. Despite the lower dietary selenium intake, the activity of blood glutathione peroxidase in the vegans was not different from that in the omnivorous controls. The differences in blood levels of antioxidant vitamins and erythrocyte superoxide dismutase activity between the groups were seen also in pairs using no antioxidant supplementation. Serum beta-carotene in the vegans correlated significantly with the dietary intake (r=0.52, p<0.01).

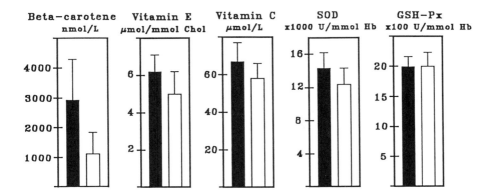

Figure 1 *Serum concentrations of antioxidant vitamins and activities of superoxide dismutase (SOD) and glutathione peroxidase (GSH-Px) in the blood of female vegans and their matched omnivorous controls.*

4 DISCUSSION

The present study revealed significantly higher intakes of the antioxidants from food sources, and a greater antioxidant status in the long-term users of a strict vegan diet ("living food diet") as compared to matched omnivorous controls. Since these differences remained after controlling the effects of several confounding factors (use of antioxidant supplements, smoking, alcohol consumption, serum cholesterol concentration), the present data indicate that the greater antioxidant status of the long-term adherents of this strict vegan diet is due to the higher intakes of dietary antioxidants.

The high serum beta-carotene concentration among the "living food" eaters may be due, at least partly, to the daily use of carrot juice and wheatgrass juice, which are good sources of carotenoids[12]. The use of these juices is strongly recommended in the "living food" regimen. High beta-carotene status has been reported earlier in studies on other vegetarian groups.[13,14] In the present study, some vegans had serum beta-carotene concentrations exceeding those achieved by supplementation with 20-30 mg of pure beta-carotene per day.[15,16]

The findings of high dietary vitamin E intakes by the long-term adherents of the "living food diet", and high serum vitamin E concentration, as adjusted for serum cholesterol concentration, are in agreement with that of Pronzcuk et al.[17] on other vegetarian groups. Similarly, the higher vitamin C concentration in the blood of the long-term adherents of the "living food diet" as compared to their omnivorous controls is in accordance with the findings on other vegetarian groups,[13,18] and probably due to the higher dietary intake of vitamin C containing fruits (other than citrus fruits) and berries.

The intakes of copper and zinc from the "living food diet" were within the current recommendations, and in agreement with the earlier reports on other vegetarians,[19,20] but the intake of selenium by the vegans was only one third of the average selenium intake by the Finnish population.[21] However, the latter result is in agreement with earlier reports on low dietary selenium intakes by vegans and mixed vegetarian groups.[20,22] Bioavailability of dietary zinc, copper and selenium from the "living food diet" appears to be good based on the observed activities of superoxide dismutase and glutathione peroxidase in the vegans. Although the vegetarian diets typically have a high content of phytic acid, a well-known inhibitor of mineral absorption,[23] the intake of phytic acid on a "living food" regimen (412 mg/day by analysis[24]) is less than half of that reported in Indian and in Caucasian vegetarians.[25]

The "living food eaters" usually adopt their diet with the purpose of curing an illness or maintaining their health, and they avoid antioxidant supplements because they believe that their diet will provide sufficient amounts of all needed antioxidants. Based on results of the present study, this practice seems justified. The "living food diet" provides significantly more dietary antioxidants than the omnivorous diet, and the long-term adherents of this diet have a greater antioxidant status than the omnivorous controls.

References

1. W. Willet, Am. J. Clin. Nutr., 1994, **59** (suppl), 1162S.
2. G. F. Frazer, Am. J. Clin. Nutr., 1994, **59** (suppl), 1117S.
3. M. G. L. Hertog, E. J. M. Feskens, P. C. H. Hollman, M. B. Katan and D.Kromhout, Lancet, 1993, **342**, 1007.
4. K. A. Steinmetz and J. D. Potter, Cancer Caus. Control, 1991, **2**, 427.
5. H. Verhagen, H. E. Poulsen, S. Loft, G. van Poppel, M. I. Willems and P. J. van Bladeren, Carcinogenesis, 1995, (in press).
6. M. Rastas, R. Seppänen, L. R. Knuts, R. L. Karvetti and P. Varo, Nutrient Composition of Foods. Publications of the Social Insurance Institution, Helsinki, 1993.
7. K. W. Miller, N. A. Lorr and C. S. Yang, Anal. Biochem., 1984, **138**, 340.
8. G. L. Catignani and J. G. Bieri, Clin. Chem., 1983, **29**, 708.
9. A. J. Speek, J. Schrijver and W. H. P. Schreurs, J. Chromatogr., 1984, **305**, 53.
10. D. E. Paglia and W. N. Valentine, J. Lab. Clin. Med., 1967, **70**, 158.
11. Committee on Dietary Allowances. Food and Nutrition Board, National Research Council. Recommended Dietary Allowances. 10th ed. Washington, DC:National Academy Press,1989.
12. J. Laakso, I. Ruokonen, T. Helve, M. Nenonen and Hänninen O. Nutrient status of rheumatoid arthritis patients on an extreme vegan diet. 25th Scandinavian Congress on Rheumatology, June 1994. Lillehammer.
13. P. Millet, J. C. Guilland, F. Fuchs and J. Klepping, Am. J. Clin. Nutr., 1989, **50**, 718.
14. M. R. H. Löwik, J. Schrijver, J. Odink, H. van den Berg and M. Wedel, J. Am. College. Nutr., 1990, **6**, 600.
15. D. Albanes, J. Virtamo, M. Rautalahti, J. Haukka, J. Palmgren, C. G. Gref and O. P. Heinonen, Eur. J. Clin. Nutr., 1992, **46**, 15.
16. E. D. Brown, M. S. Micozzi, N. E. Craft, J. G. Bieri, G. Beecher, B. K. Edwards, A. Rose, P. R. Taylor and J. C. Smith, Am. J. Clin. Nutr., 1989, **49**, 1258.

17. A. Pronczuk, Y. Kipervarg and K. C. Hayes, J. Am. Colleg. Nutr., 1992, **1**, 50.
18. J. G. Bergan and P. T. Brown, J. Am. Diet. Assoc., 1980, **76**, 51.
19. R. S. Gibson, A. Heywood, C. Yaman, L. U. Thompson and P. Heywood, Ecol. Food Nutr., 1991, **25**, 69.
20. M. Abdulla, I. Andersson, N-G. Asp, K. Berthelsen, D. Birkhed, I. Dencker, C-G. Johansson, M. Jägerstadt, K. Kolar, B. M. Nair, P. Nilsson-Ehle, Å. Norden, S. Rasnner, B. Åkesson, P-A. Öckerman, Am. J. Clin. Nutr., 1981, **34**, 2464.
21. G. Alfthan, A. Aro, H. Arvilommi and J. K. Huttunen, Am. J. Clin. Nutr., 1991, **53**, 120.
22. U. M. Donovan, R. S. Gibson, E. L. Ferguson, S. Ounpuu and P. Heywood, J. Trace Elem. Electrol. Health Dis., 1992, **128**, 39.
23. O. Hänninen, M. Nenonen, W. H. Ling, D. S. Li and L. Sihvonen, Appetite, 1993, **19**, 243.
24. D. Oberleas and B. F. Harland, J. Am. Diet. Assoc., 1981, **79**, 433.
25. S. K. Pushpanjali and G. R. Fenwick, Proceedings of the International Conference Euro Food Tox IV. 'Bioactive Substances In Foods Of Plant Origin', (H. Kozlowska, J. Fornal, Z. Zdunczyk, eds. Centre for Agrotechnology and Veterinary Sciences, Olsztyn,Poland, vol 2, pp. 316-21, 1994.

FRUIT AND VEGETABLE SUPPLEMENTATION - EFFECT ON EX VIVO LDL OXIDATION IN HUMANS

M. Chopra[1], U. McLoone[1], M. O'Neill[1] N. Williams[2] and D.I. Thurnham[1]

[1]Human Nutrition Research Group, School of Biomedical Sciences
University of Ulster, Coleraine BT52 1SA
and [2]COAG Laboratory, Papworth Hospital
Papworth Everard, Cambridge CB3 8RE
UK

1 INTRODUCTION

Oxidation of cholesterol-rich, low-density lipoproteins (LDL) is now widely accepted as an important factor contributing to the pathogenesis of atherosclerosis and hence the risk of cardiovascular disease. Role of antioxidant nutrients in the protection of LDL from oxidation is of interest as their levels can be manipulated without any clinical side effects. Among these antioxidants are vitamin C, vitamin E, ubiquinone, flavanoids and carotenoids. A healthy diet of fruits and green leafy vegetables is believed to be protective against the risk of cardiovascular disease because it improves the lipid levels and provides high level of antioxidants.[1,2] Since carotenoids are important constituents of fruits and vegetables, it is suggested that they provide part of the protection against cardiovascular disease.

The present study was undertaken to investigate if the carotene component of fruit and vegetable diet protects LDL from oxidation.

2 EXPERIMENTAL

2.1 Effect of fruit and vegetable supplementation on LDL oxidation ex vivo

2.1a. In a preliminary study 11 healthy volunteers were recruited. Ten volunteers (5 men, 5 women) were asked to increase their fruit and vegetable consumption and one volunteer was asked to reduce his fruit and vegetable consumption for two weeks. Volunteers on the supplementation study were advised to take fruits and vegetables in proportions which would have given them approximately 30 mg of specific carotenes/day. Food rich in three carotenoids, viz ß-carotene (mainly carrots, mango), lutein (mainly spinach, spring greens, broccoli, green beans) and lycopene (mainly fresh tomatoes, tomato soup/juice and watermelon) was given and to avoid possible interaction between carotenes for absorption, volunteers were asked to take food rich in one particular carotenoid each day. EDTA (1mg/ml) blood was taken at baseline and at the end of week 1 and week 2 after supplementation or depletion. Volunteers were asked to keep a dietary record of their daily food consumption and weekly intake of carotenes was calculated using the carotenoid database Comp-Eat (Royal Society of Chemistry, 1993) supplemented with additional information on lycopene and lutein content of vegetables.[3] Mean weekly carotene intake is shown in table 1a.

Carotene analysis was carried out on plasma using reverse phase HPLC.[4] LDL were isolated by differential ultracentrifugation of plasma at 10°C for 22 hours at 40,000 rpm in a Beckman ultracentrifuge using a swing out rotor.[5] On the following day LDL layer was removed and its cholesterol content was determined using a CHOD-PAP kit from Boehringer. LDL was stored under argon at 4°C until further analysis. Prior to LDL oxidation excess EDTA was removed after dialysing the LDL samples for 24 hours at 4°C in a pH 7.4 buffer containing 0.01 mol sodium phosphate, 0.16 mol sodium chloride, 10 µmol EDTA and 0.1 gram of chloramphenicol per litre of the buffer. Cholesterol content was determined as above and copper-initiated oxidation of LDL was monitored by measuring diene conjugate formation at 234 nm. Oxidation was carried out in phosphate buffered saline pH 7.4 at 37°C. Final concentration of LDL was adjusted to 0.25 mg total LDL/ml (this is equivalent to 0.1 µmol LDL/ml) and of copper was 11.7 µmol/l. Change in plasma carotenes and lagphase is shown in table 1b.

Levels of all carotenes were halved in the individual who was on fruit and vegetable depletion diet (data not shown). Lag phase in this volunteer was reduced from 25 minutes to 16 minutes after one week and 17 minutes after 2 weeks on the depletion diet.

2.1b Fruit and vegetable study was repeated in another 11 volunteers (6 men, 5 women). Everything was same as in study 2.1a except that dietary records for the week prior to supplementation were also obtained (table 2a).

2.2 Effect of in vivo supplem ntation with lutein on LDL oxidation ex vivo

We have shown previously that lutein, a xanthophyll carotenoid, can inhibit oxidation of LDL in vitro.[6] Our first fruit and vegetable supplementation study showed an increase in all plasma carotenes after supplementation but only the increase in plasma lutein was significant ($p<0.05$). However, the increase in the resistance of LDL to oxidation was not associated with any change in any carotene concentration. In the second fruit and vegetable supplementation study all carotenes were increased except lutein, also, there was no change in the resistance of LDL to oxidation, but, there was a statistically significant correlation between plasma lutein concentration and lagphase.

To investigate if lutein was an important component of fruits and vegetables for protection of LDL against oxidation, we recruited another 12 volunteers. Six volunteers were given 30 mg/day of lutein in sunflower oil (10 mg in 5 ml oil, three times a day) orally with a meal and six were given placebo i.e. sunflower oil only. LDL isolation and oxidation were carried out in the same way as above. Carotene analysis of both plasma and LDL fractions were carried out using reverse phase HPLC as above.

3 RESULTS AND DISCUSSION

3.1 Effect of fruit and vegetable supplementation on Cu-initiated oxidation of LDL

Results of this study show that fruit and vegetable supplementation increases the plasma carotene levels after supplementation for two weeks. Table 1a and 1b show the weekly dietary intake values and plasma carotene for the first supplementation study.

Table 1a   *Dietary intake values of carotenes {mean (range)in mg/week}*

	Lutein	ß-Carotene	Lycopene
Baseline	----------	-------------	-------------
Day 7	47.3 (10-66)	47.6 (10-65)	113.5 (62-174)
Day 14	51.8 (12-66)	35.7 (12-67)	170.4 (81-245)

Baseline - Calculated intakes week prior to supplementation

Table 1b   *Plasma antioxidant concentration after increased consumption of fruit & vegetable for two weeks.*

	ß-carotene	Lutein	Lycopene	ß-crypto-xanthin	α-tocopherol	Lag
Baseline (n=9)	0.54±0.50	0.26±0.17	0.79±0.35	0.17±0.15	24.4±5.9	31±7
Day 7 (n=9)	0.65±0.48	*0.35±0.18	0.95±0.28	0.21±0.19	23.9±5.9	***46±7
Day 14 (n=8)	0.70±0.60	0.35±0.20	1.00±0.47	0.30±0.28	23.1±5.1	*44±11

Results of antioxidants are expressed in µmol/l and lagphase (lag) in minutes as mean ± S.D. Significance of change when compared to the baseline value was calculated using paired t-test, $*p<0.05$, $***p<0.002$.

Table 2a and 2b show the weekly dietary intake values of carotenes and their plasma levels after supplementation for two weeks (second study).

Table 2a   *Dietary intake values of carotenes { mean (range) in mg/week}*

	Lutein	ß-Carotene	Lycopene
Baseline	4.5 (0-6)	10.8 (0-25)	7.3 (0-17)
Day 7	42.0 (15-79)	70.5 (36-99)	112.4 (10-164)
Day 14	57.7 (27-151)	86.2 (35-120)	87.2 (7-100)

Baseline - Calculated intakes week prior to supplementation

Table 2b   *Plasma lipophilic antioxidants and the effect on LDL oxidation after two weeks supplementation with fruits and vegetables.*

	ß-carotene	Lutein	Lycopene	ß-crypt	α-toco	Lag
Baseline (n=11)	0.29±0.23	0.38±0.11	0.26±0.12	0.14±0.10	36.3±6.9	48±5
Day 7 (n=11)	***0.79±0.54	0.41±0.09	**0.67±0.33	**0.28±0.16	*32.7±8.8	49±3
Day 14 (n=11)	*0.80±0.63	0.38±0.15	**0.93±0.24	**0.28±0.14	**29.3±8.6	49±2

Results for antioxidants are expressed in µmol/l and lag phase in minutes as mean ± S.D. ß-crypt = ß-cryptoxanthin, α-toco = α-tocopherol and lag = lagphase. Significance of change when compared to the baseline value was calculated using paired t-test, *p<0.05, **p<0.005, ***p<0.001.

Table 3   *Effect of fruit and vegetable supplementation on the plasma cholesterol*

	Baseline mean±S.D.	**Day 7** mean±S.D.	**Day 14** mean±S.D.
Study 1 (n=9)	5.45±0.95	**3.11±0.60	*4.75±0.88
Study 2 (n=11)	5.06±1.20	4.79±1.03	5.18±1.39

Results are expressed as mmol/l and show the significance of differences when compared to baseline values calculated using paired t-test, *p<0.05, **p<0.001.

Results of the fruit and vegetable supplementation studies suggest that plasma carotenes can be raised after supplementation for one week with carotene rich diet (table 1 and 2). Supplementation produced a significant decrease in plasma total cholesterol but only in the first supplementation study (table 3). Since information on dietary fat intake during the supplementation period was not obtained, it is difficult to interpret whether this decrease was due to a reduced fat intake to compensate for the increased fruit and vegetable consumption.

In the first study although all plasma carotenes appeared to increase after supplementation, only plasma lutein was significant ($p<0.05$, paired t-test) (table 1b). There was an increase in the resistance of LDL to oxidation, this increase was however, not associated with the change in plasma carotenes. Plasma α-tocopherol did not change, which suggests that the increase in the lagphase observed was not due to the changes in α-tocopherol. In the second fruit and vegetable study, all carotenes except lutein were raised significantly after supplementation (table 2b). No change in lag phase was observed and no correlation was found between the change in carotenes and lag phases. Lutein and lag phase ($r=0.66$, $p<0.03$) were correlated in the day 14 samples. Although the number of subjects in these studies are small and the results variable the results of our study suggest that fruit and vegetable enriched diet increased the plasma carotenes, reduced the plasma cholesterol and may increase the resistance of LDL to oxidation. Also, changes in the resistance of LDL to oxidation may be related to lutein.

3.2 Effect of in vivo supplementation with lutein on LDL oxidation ex vivo.

Results of the in vivo lutein supplementation study are shown in table 4. Both placebo and supplemented group had similar mean LDL lutein levels at the beginning of the study. On supplementation, there was a 4 fold increase in LDL lutein concentration after one week and ≈ 6 fold after 2 weeks supplementation. Levels of LDL lutein remained unchanged in the placebo group.

Table 4 *Effect of in vivo supplementation with lutein on LDL oxidation ex vivo*

Group	*Placebo (n=5)		*Lutein supplemented (n=6)	
	Lutein LDL mol/mol	Lagphase mins	LDL lutein mol/mol	Lagphase mins
Before supplementation	0.049 ± 0.019	47 ± 7	0.055 ± 0.019	43 ± 7
Week 1 after supplementation	0.048 ± 0.01	41 ± 7	0.223 ± 0.12	40 ± 6
Week 2 after supplementation	0.053 ± 0.02	45 ± 3	0.317 ± 0.15	41 ± 5

*Results are expressed in mean ± S.D.

No significant change in the lag phase was observed in either group after the 2 weeks study period. This suggests that either even higher concentrations of lutein are required before any effect on LDL oxidisability will be observed or other component/s in green vegetables either on their own or in combination with carotenoids, contribute to the protective effect of fruits and vegetables against disease.

More studies are required to clarify the proposed antioxidant role of carotenes. Studies to investigate the effect of long term supplementation with carotenes ß-carotene, lutein and lycopene are already underway in our laboratory.

Acknowledgements:

This work was funded by AAIR2-CT93-0888 DG XII. The lutein in vivo supplementation studies were done in collaboration with The Howard Foundation in Cambridge, UK. Fruit and vegetable study was done parallel with Prof Anne-Marie Roussell and Dr Isabelle Hinninger from Laboratoire de Biochimie Micronutrients-Radicaux Libres, UFR des Sciences, Domaine de la Merci, 38700 La Tronche, Grenoble, France, who will be reporting their results separately.

References:

1. R.B. Singh, S.S. Rastogi, M.A. Niaz, S. Ghosh, R. Singh, S. Gupta, Am. J. Cardiol 1992, 70:869.
2. K.F. Gey. J. Nutr. Biochem 1995, 6:206.
3. D.J. Hart and K.J. Scott. Food Chemistry 1995, 54(1):101
3. D.I. Thurnham, W. E. Smith and P. S. Flora, Clin. Chem. 1988, 34:377.
4. H. Puhl, G. Waeg, H. Esterbauer, Methods in Enzymology 1994, 233:425.
5. M. Chopra and D. I. Thurnham, Proc Nutr Soc 1994, 53(2):18A.

VITAMIN-C REDUCES UPTAKE OF MERCURY VAPOR IN RATS

W. A. Rambeck, B. Lohr, S. Halbach

Institut für Physiologie, Physiologische Chemie und Tierernährung, Universität München, 80539 München, and
Institut für Toxikologie, GSF-Forschungszentrum, 85758 Oberschleissheim

1 INTRODUCTION

Certain parts of the population like workers in the chloralkali industry or dentists might be exposed to mercury vapor (Hg^0) in concentrations that are toxicological relevant.
Binding of inhaled mercury to tissues and organs will depend to a certain extent on the oxidative capacity of the H_2O_2-consuming enzyme catalase. In order to investigate a possible effect of a variation of the oxidative metabolism on the uptake of mercury, vitamin C was added to the drinking water of rats, which were exposed to various concentrations of mercury vapor.

2 MATERIAL AND METHODS

Design and operation of the exposure system [1] allowed Hg^0-concentrations down to 0.05 mg/m^3 (MAK-Value 0.1). Treated rats obtained 1 g ascorbic acid/l in the drinking water for 8 weeks. Treated and control rats inhaled pairwise radioactive Hg-Vapor (0.05, 0.1, 0.5 and 1 mg/m^3) for 30 and 60 min. Hg-concentration was measured in heart, lung, liver, spleen, kidney, brain, erythrocytes, plasma and in the whole body.

3 RESULTS

The vitamin-C concentration in plasma increased significantly after supplementation from 7.9±1.2 to 13.6±3.1 µg/ml. The following significant reductions of Hg-uptake were observed in the vitamin-C group. Hg-retention in kidney and lung was lowered by more than 30%, plasma values by more than 80%. The additive effect of these partial reductions of Hg-retentions in certain organs and tissues was obviously large enough to lower the whole body concentration, i.e. the Hg body burden, by up to 45% (Figure 1). Control experiments with i.p. injection of $HgCl_2$ excluded the possibility of a reduction of Hg^{++} to Hg^0 by vitamin-C and subsequent exhalation.

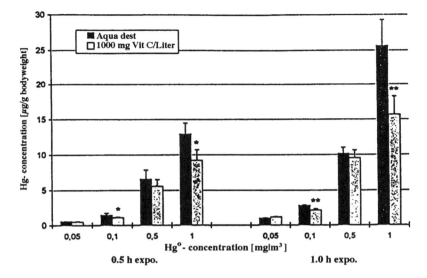

Figure 1. *Whole body mercury retention with and without vitamin C in the drinking water of rats*

4 DISCUSSION

The antioxidant properties of vitamin-C confirm the role of oxidative processes for the in-vivo retention of inhaled Hg-vapor. Possible explanations of the vitamin-C effect include formation of a less reactive complex with catalase (compound II) or an enhanced production of GSH followed by a reduced availability of H_2O_2 to generate sufficient catalase-compound-I for the oxidation of Hg^0. These results indicate that vitamin-C not only protects against accumulation of the toxic metal cadmium [2], but possibly also against the effects of mercury vapor uptake.

References

1. S. Halbach and R. Fichtner, Toxicol. Methods, 3, 25-.36, 1993.
2. W.A. Rambeck, Übers. Tierernähr. 22, 184-190, 1994

Selenium Intake and Status of Various Populations

SELENIUM INTAKES AND PLASMA SELENIUM LEVELS IN VARIOUS POPULATIONS

Georg Alfthan[1] and Jean Neve[2]

[1]Department of Nutrition, National Public Health Institute, FIN-00300 Helsinki, Finland and [2]Free University of Brussels, Institute of Pharmacy, Campus Plaine 205-5, B-1050 Brussels, Belgium

1 INTRODUCTION

The selenium status of humans is mainly determined by the intake of selenium from the diet. The major determinants of dietary selenium are of plant origin: grains, pulses and tubers. The selenium concentration of plants may vary greatly depending on the fraction of soluble selenium in the soil. For example, low-selenium grains are found in areas having a humid climate and soils with a low pH, high iron and aluminum concentration[1].

2 INTAKE OF SELENIUM IN VARIOUS COUNTRIES

Comparative studies have shown that the selenium concentration of e.g. wheat varies between different areas from < 10 µg/kg to 720 µg/kg[2,3]. Therefore it is not surprising that this is reflected in the daily intake of selenium of populations living in different areas. Table 1 shows a compilation of mean intakes from some populations worldwide[4,5]. The intake ranges from 10 to 250 µg/day. The intake of selenium in only a few countries reaches the Recommended Dietary Allowance of USA, 70 µg/day for men and 55 µg/day for women. For China, Canada and USA a wide range is shown illustrating that within these countries there are both high and low selenium endemic areas.
   Estimation of intake is laborious, costly and difficult. An alternative approach is to use biomarkers to describe the selenium status of an individual or population.

3 TOENAIL SELENIUM AS BIOMARKER OF INTAKE

As biomarkers of selenium status different human tissues have been used such as whole blood, plasma, hair, finger- and toenails, liver and heart tissue[4]. Each of these reflect intake on a different time scale. Toenails are a long-term indicator reflecting the previous intake over a time span of 6-12 months[6]. Figure 1 shows data on both intake and toenail selenium concentration for groups of healthy subjects from studies where both of these variables were available. The nail selenium data is arranged according to ascending selenium values. The lowest mean toenail selenium concentrations, 0.17 mg/kg and 0.26 mg/kg, have been reported from China[7] and New Zealand[8], respectively, and the highest from China[12], 2.72 mg/kg. The corresponding mean intakes do not follow the same ascending order, Figure 1, but the subjects can nevertheless be classified into three intake categories according to toenail selenium concentration, low 11-56 µg/day, medium 66-125 µg/day and high 196-250 µg/day.

Table 1. *The daily Intake of Selenium in Various Countries*

COUNTRY	INTAKE, µg/day
China	10 - >200
New Zealand	20 - 30
Turkey	30
Sweden	38
United Kingdom	30 - 40
France	40 - 55
Belgium	45 - 55
Italy	47
Thailand	47
Brazil	57
Iran	57
Spain	57
Netherlands	67
Switzerland	70
Finland 1994	80
Canada	80 - 180
USA	60 - 250

Compiled from ref. 4 and 5.

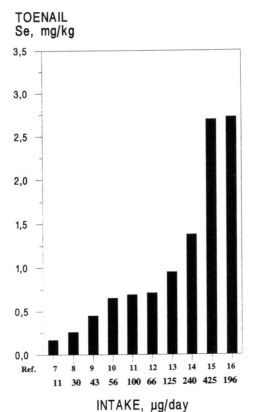

Figure 1 *Selenium Intake and Toenail Selenium Concentration in Various Countries*

4. PLASMA SELENIUM AS BIOMARKER OF INTAKE

Plasma is the most popular biomarker of selenium intake in human studies. It reflects the mean intake over a time period of a few weeks. Figure 2 shows mean plasma selenium values of some healthy subpopulations in ascending order. The data for New Zealand[17] is low, 0.69 µmol/l as is to be expected from the intake and toenail data. In Europe, levels below 1 µmol/l are found generally in the eastern European countries and the highest have been reported from Norway[32], 1.58 µmol/l. The highest mean values have been reported from the seleniferous regions of South Dakota and eastern Wyoming, USA[14], 2.50 µmol/l. In China[34] both extremes are to be found, 0.22 µmol/l in selenium deficient areas and 6.2 µmol/l in seleniferous areas.

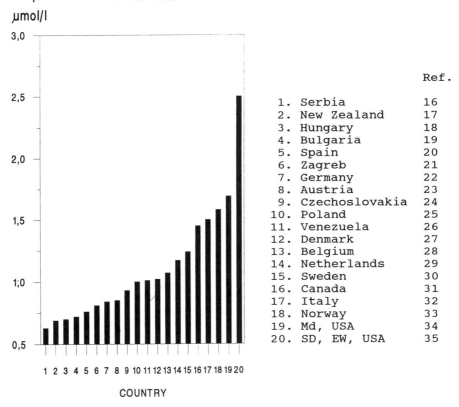

Figure 2 *Plasma Selenium Concentration in Various Countries*

5 FACTORS WHICH AFFECT PLASMA SELENIUM LEVELS

There are limitations as to the use of plasma as a biomarker of selenium status because several factors may influence plasma selenium levels[37]. Among lifestyle-related factors which may affect plasma selenium concentrations are area of residence, time of sampling, occupation and smoking. Factors which may influence plasma selenium concentrations on a metabolical level include diet, supplements, medication, alcohol and fasting. Documentation of these factors is of importance when reporting reference values for healthy subjects[38]. Determination of plasma selenium does not usually suffer from problems related to external contamination of samples. Accuracy of the results requires quality assurance either by analyzing certified reference sera, participating in interlaboratory trials or verifying the result by another method.

6 TEMPORAL VARIATION OF SELENIUM INTAKE AND PLASMA SELENIUM IN FINLAND

Addition of sodium selenate to artificial fertilizers was started in Finland in 1985. This is a nationwide program which has increased the mean intake from 38 µg/day in 1984 to a maximum level of 125 µg/day in 1989-91. The amount of added selenate was decreased from 16 mg/kg fertilizer to 6 mg/kg in 1991 resulting in an estimated plateau of about 80 µg/day[37].

Figure 3 illustrates the effect of the increased intake on mean plasma selenium concentrations of urban subjects in Helsinki (n=40) and rural Leppävirta county in eastern Finland (n=45). The mean plasma concentrations were 0.89 µmol/l before the selenium addition started and 1.52 µmol/l in 1989-91. The latest results from May 1995 are 1.25 µmol/l. The Figure serves as an example showing the importance of documenting the time of sampling. Variations of plasma selenium levels of this magnitude are not confined to Finland alone. An analogous situation has been reported from New Zealand due to occasional importation of high-selenium wheat which doubled the blood selenium concentrations in some areas[38]. In Belgium[28] and Scotland[39] systematic studies have also revealed that plasma selenium concentrations vary with time, i.e. have decreased in these countries.

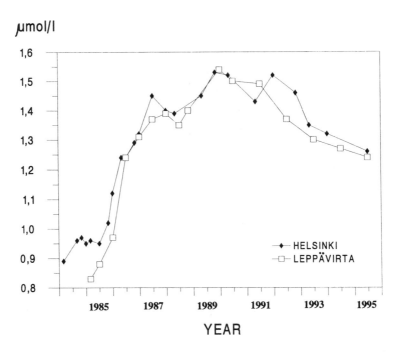

Figure 3 Temporal Variation of Plasma Selenium Concentration Before and During Addition of Selenium to Fertilizers in Finland

7 CONCLUSIONS

It is well documented that the selenium concentration of grains, especially wheat, varies many fold between areas contributing to a wide variation in the intake of populations. To assess intake, biomarkers of selenium status are used widely. There is still limited data on toenails as a biomarker, but they can be used to classify subjects into groups of low, medium and high intake. Plasma selenium of healthy subjects varies from 0.6 µmol/l to 2.5 µmol/l. There has been a temporal variation of plasma selenium due to addition of selenium to fertilizers in Finland and in other countries due to importation of high-selenium wheat. Preanalytical factors may influence plasma selenium concentrations and analytical quality assurance to verify accuracy is recommended.

References

1. T. Yläranta, Norwegian J. Agric. Sci.,1993,suppl.**11**,141.

2. P. Varo and P. Koivistoinen, Internat. J. Vit. Nutr. Res., 1981,**53**,62.

3. M. Eurola, P. Ekholm, M. Ylinen, P. Koivistoinen and P. Varo. Cereal Chem. 1990,**67**,334.

4. J. Neve, J. Trace Elem. Electrolytes Health Dis., 1991,5,1.

5. J.T. Kumpulainen, J. Trace Elem. Electrolytes Health Dis,1993,**7**,107.

6. M.P. Longnecker, M.J. Stampfer, J.S. Morris, V. Spate, C. Baskett, M. Mason and W.C. Willett, Am. J. Clin. Nutr., 1993, **57**,408.

7. G. Yang, L.Z. Zhu, S.J. Liu, L.Z. Gu, P.C. Qian, J.H. Huang and M.D. Lu, 1987. In: G.F. Combs, J.E. Spallholz, O.A. Levander and J.E. Oldfield eds., 'Selenium in Biology and Medicine', Van Nostrand Reinhold Co, New York,1987,p. 589.

8. J.S. Morris, M.J. Stampfer and W. Willett, Biol. Trace Elem. Res.,1983,**5**,529.

9. M.L. Ovaskainen, J. Virtamo, G. Alfthan, J. Haukka, P. Pietinen, P. Taylor, J.K. Huttunen, Am. J. Clin. Nutr., 1993;,**7**,662.

10. P. van't Veer, P.J. van der Wielen, F.J. Kok, R.J.J. Hermus and F. Sturmans, Am. J. Epidem.,1990,**131**,987.

11. G. Alfthan, A. Aro, H. Arvilommi and J.K. Huttunen, In:'Biomarkers of Dietary Exposure', 3rd Meeting Nutritional Epidemiology. Proceedings. Rotterdam 23-25th January 1991,p.110.

12. G. Yang, R. Zhou, S. Yin, B. Gu, Y. Liu, Y. Liu and X. Li, J. Trace Elem. Electrolytes Health. Dis., 1989,**3**,77.

13. G. Alfthan, J. Kumpulainen, A. Aro and A. Miettinen, In: Proceedings of the 5th International Symposium on Selenium in Biology and Medicine, Nashville, Tennessee, July 20- 23,1992.

14. M.P. Longnecker, P.R. Taylor, O.A. Levander, S.M. Howe,C. Veillon, P.A. McAdam, K.Y. Patterson, J.M. Holden, M. J. Stampfer, J.S. Morris and W.C. Willett, Am. J. Clin. Nutr., 1991,**53**,1288.

15. V.E. Negretti de Brätter, P. Brätter, D. Gawlik, W.L. Jaffe, International Symposium on Selenium in Biology and Medicine, Nashville, Tennessee, July 20-23, 1992.

16. Z. Maksimovic, I. Dujic, V. Jovic and M. Rsumovic, Biol. Trace Elem. Res., 1992, **33**, 187.

17. C.D. Thomson, M.F. Robinson, J.A. Butler, P.D. Whanger, Br. J. Nutr., 1993, **69**, 577.

18. G. Alfthan, G. Bogye, A. Aro and J. Feher, J. Trace Elem. Electrolytes Health Dis., 1992, **6**, 233.

19. K. Tzachev, Z. Georgiou, G. Gentchev, In: J. Anastassopoulou J, P. Collery, J.C. Etienne, T, Theophanides, eds., 'Metal Ions in Biology and Medicine', John Libbey Eurotext, Paris, 1992, **1**, 420.

20. F. Fernandez-Banares, C. Dolz, M. Mingorance, E. Cabre', M. Lachica, A. Abad-Lacruz, A. Gil, M. Esteve, J. Gine' and M. Gassull, Eur. J. Clin. Nutr., 1990, **44**, 225.

21. M. Mikac-Devic, N. Vukelic and K. Kljaic, Biol. Trace Elem. Res., 1992, **33**, 87.

22. O. Oster, G. Schmiedel and W. Prellwitz, Biol. Trace Elem. Res., 1988, **15**, 47.

23. B. Tiran, A. Tiran, W. Petek, E. Rossipal and O. Wawschinek, Trace Elem. Med., 1992, **9**, 75.

24. V. Korunova, Z, Skodova, J. Dedina, Z. Valenta, J. Parizek, Z. Pisa and M. Styblo, Biol. Trace Elem. Res., 1993, **37**, 91.

25. W. Wasowicz and B. Zachara, J. Clin. Chem. Clin. Biochem., 1987, **25**, 409.

26. J. Burguera, M. Burguera, M. Gallignani, O. Alarcon and J. Burguera, J. Trace Elem. Electrolytes Health Dis., 1990, **4**, 73.

27. P. Grandjean, G. Nielsen, P. Jorgensen and M. Horder, Scand J. Clin. Lab. Invest., 1992, **52**, 321.

28. J. Neve, F. Vertongen, A. Peretz and Y. Carpentier, Ann. Biol. Clin., 1989, **47**, 138.

29. G. Michaelsson, B. Berne, B. Carlmark and A. Strand, ActaDerm. Venerol., 1989, **69**, 29.

30. R. S. Gibson, O. Martinez and A. Mac Donald, J. Gerontol., 1985, **40**, 296.

31. G. Sesana, A. Baj, F. Toffoletto, R. Sega, I. Ghezzi, Sci. Total Environ., 1992, **120**, 97.

32. J. Ringstad, B. Jacobsen, S. Tretli and Y. Thomassen, J. Clin. Pathol., 1988, **41**, 454.

33. O.A. Levander and V.C. Morris, Am. J. Clin. Nutr., 1984, **39**, 800.

34. Y. Xia, X, Zhao, L. Zhu and P.D. Whanger, J. Nutr. Biochem., 1992, **3**, 211.

35. H. Robberecht and H. Deelstra, J. Trace Elem. Electrolytes Health Dis., 1994, **8**, 129.

36. G. Alfthan and J. Neve, submitted 1995.

37. Working Group on Selenium. 'Annual Report on Selenium', Report of Finnish Ministry of Agriculture and Forestry, 1994, 2.

38. J.H. Watkinson, Am. J. Clin. Nutr., 1981, **34**, 936.

39. M.N.I. Barclay and A. MacPherson, Br. J. Nutr., 1992, **68**, 261.

VARIOUS FORMS AND METHODS OF SELENIUM SUPPLEMENTATION

Antti Aro

Department of Nutrition
National Public Health Institute
FIN-00300 Helsinki, Finland

1 INTRODUCTION

In certain areas of the People's Republic of China selenium (Se) intake is extremely low, around or below 10 µg/d. In these areas an endemic, often fatal cardiomyopathy has affected people, particularly children and young women. The disorder, Keshan disease, can be prevented by prophylactic administration of sodium selenite[1]. It is assumed that the condition is caused by deficiency of Se, combined with a viral infection or possibly with other nutritional factors.

There is no direct evidence on harmful health effects of mild Se deficiency. Nevertheless, epidemiological evidence suggests that low Se intakes, reflected in low serum Se levels, may increase the risk of cardiovascular disease and cancer[2]. In prospective case-control studies from Finland[3] and Denmark[4] increased risk of cardiovascular death or ischemic heart disease was observed in men whose serum Se values were within the lowest tertile of the respective population. No such association has been evident in populations where low serum selenium values are not found[2]. With respect to cancer, studies from Finland and the USA have suggested that the risk of certain types of cancer, particularly that of upper gastrointestinal tract and lungs is increased in men but not in women[5,6]. On the other hand, in several studies, most of them from the USA, no association has been found between serum Se and the risk of cancer[2].

The Se intake of people can be increased either indirectly by supplementing fertilizers or animal feeds with selenium compounds or directly by supplementation of foods or by using Se containing dietary supplements.

2 SUPPLEMENTATION OF FERTILIZERS WITH SELENIUM

The Se content of foods is low when the Se concentration in the soil is low or Se exists in chemical forms that are poorly available for plants. Selenates are generally water-soluble and do not form poorly soluble adsorption complexes whereas selenites are mostly insoluble. In alkaline soils and under aerobic conditions Se exists mostly in the form of selenates which are readily available for plants[7]. In soils with low pH and high humidity, reduced forms of Se are found such as selenites and selenides which form insoluble complexes with ferric and aluminium compounds[8]. This is a condition typical for Finland where only ca 5 % of total Se in soils is in a soluble form and available for plants[9]. In studies in the 1970s in Finland, low Se concentrations were found in locally produced foods, and the total Se intake of people was estimated to be about 25 µg/d only[10].

Studies in Denmark and Finland indicated that supplementary selenate, in contrast to selenite, stayed in the soil in a form that was available for plants for several months[12, 13]. It was also found that the amount of selenate which was necessary to increase the Se content of grass is smaller than the amount needed for increasing the Se concentration of grains[13]. Being aware of the low Se intake of the Finnish population and the suggestions about the possible harmful health effects of this, the Ministry of Agriculture and Forestry of Finland decided in 1984 to start supplementing fertilizers with Se[11]. The amount of Se added to fertilizers for grain production was 16 mg/kg and that added to fertilizers for the production of hay and fodder was 6 mg/kg. This corresponded, on average, to 8 g/ha for cereals and 3 g/ha for pastures annually. Although Se is not an essential element for plants these take up the mineral and incorporate it into organic compounds. It was predicted that the supplementation of fertilizers with selenate would affect the Se intake of both domestic animals and human people. An expert group was appointed by the Ministry of Agriculture and Forestry to monitor the effects of Se supplementation of fertilizers on soils and waters, fertilizers, animal feeds, foods and the Se status of people, and to give annual reports on the situation. The amounts of Se added to fertilizers were kept constant until 1990 when it was decided to reduce the supplemented amount to 6 mg/kg for all fertilizers[11].

2.1. Effects on human nutrition in Finland

Since 1969, animal feeds have been enriched with selenite in Finland. This practice had very little effect on the dietary Se intake of people which remained low throughout the 1970s. Serum Se concentrations of healthy people were around 50-60 µg/l[14]. In 1979 and 1982 there were poor domestic crops and high-Se wheat was imported from the USA. This resulted in considerable fluctuations in the Se intake of people, and between 1980 and 1984 the mean serum Se concentrations showed annual variations between 70 and 100 µg/l[14].

When the effects of Se supplementation became evident in 1985, milk was the first food item which showed an almost immediate increase in Se content. This was followed by meat products, cereal products and other dairy products within a period of 6 months[15]. The mean Se intake was increased by 2-3 -fold during the first year, and during the subsequent 3 years the intake was stabilized to the level of 110-120 µg/10 MJ[11, 15]. The impact of meat products increased most, to more than 40 % of total Se intake, followed by that of dairy products (25 %) and cereal products (20 %). Se intake from fish remained quantitatively unchanged but its share of total Se intake declined from 30 % before Se supplementation to less than 10 % afterwards[11].

2.2. Effect on serum Se and platelet GSHPx

The serum Se concentration of a small reference group of healthy individuals has been monitored since mid 1970s[14]. In 1985 this group, consisting of urban people from Helsinki, was expanded by recruiting another group of healthy individuals who were rural people, mainly members of farmer families from central Finland. The serum Se concentrations of both groups increased concomitantly, from levels of 65-70 µg/l in 1985 to 120 µg/l in 1989-1991. Similar effects were also observed in other groups of people including children, adults, elderly men and women, and pregnant women[11].

After 1990, due to the reduction in the amount of Se supplemented to fertilizers, the mean Se intake was reduced gradually to 85 µg/10 MJ in 1993. A concomitant decline was observed in mean serum Se concentrations to the level of 100 µg/l in both reference groups of healthy subjects[16].

Selenocysteine is an integral part of the enzyme glutathione peroxidase (GSHPx) which catalyzes the reduction of hydrogen peroxide to water and fatty acid hydroperoxides to their respective alcohols. GSHPx activity is dependent on Se intake up to a certain saturation level which is different for different tissues: relatively low (about 40 µg/d) for serum and higher (about 120 µg/d) for platelets[17]. When a group of healthy Finnish men was supplemented with Se, 200 µg/d, in 1981 before the supplementation period, platelet GSHPx activity was doubled[17]. In a similar supplementation study in the same men in 1988, the GSHPx activity was increased only by 0-30 % by different Se compounds showing that the increased Se intake had stimulated the GSHPx activity in platelets to near-maximal levels[17]. GSHPx activity in serum did not respond to supplementation in either study[17, 18].

3 SUPPLEMENTATION OF ANIMAL FEEDS

Rapidly growing domestic animals need dietary supplementation with Se and vitamin E in order to provide optimal growth and to avoid deficiency diseases such as muscular dystrophy in ruminants[19] and hepatosis, mulberry heart disease and muscular dystrophy in other domestic animals[20]. It has been customary to supplement animal feeds with inorganic selenite. However, inorganic Se has very little effect on the Se content of milk and meat whereas organic Se is more available for secretion into milk[21, 22]. Similarly, organic Se compounds are more effectively incorporated into meat products than inorganic selenite, with the exception of kidneys[23, 24].

The marked effect on the Se content of meat, milk and eggs by the supplementation of fertilizers with sodium selenate in Finland depends on conversion of selenate into organic Se compounds, mainly seleno-methionine by the plants. The Se requirements of the animals were met by the supplementation as well[24].

4 OTHER METHODS

Because of the toxicity of Se, direct supplementation of foods may be hazardous. In Finland it was considered too dangerous, and supplementation of fertilizers was preferred. In the People's Republic of China, supplementation of table salt with selenite has been used in trials[1]. Individual self-supplementation with Se-containing pills has been common in some countries. In Finland 17% of women and 9% of men reported that they used Se supplements in 1990[25]. Dietary supplements are expensive and not feasible for increasing the Se intake of populations.

5 SUMMARY

The selenium concentration of foods can be increased by supplementing fertilizers with soluble Se compounds or by supplementing the feeds of domestic animals with organic Se compounds. Both methods involve a biological barrier which increases the safety of the intervention for humans. The ongoing supplementation of fertilizers with Se as sodium selenate has been successful in increasing the Se intake of people in Finland[11, 16]. It has increased the Se content of almost all foods, doubled the serum Se concentration of people and increased the activity of GSHPx in blood platelets to near-maximal levels. In addition, it has reduced the need to supplement animal feeds with inorganic Se. Plants transform the selenate they take up into organic Se compounds. This increases animal intake of organic Se which is incorporated in tissues and increases the Se content of meat, milk and eggs. It is possible to modify the Se intake of the population and to bring it to the desired level by adjusting the amount of supplementary Se. Supplementation of fertilizers with Se is feasible in countries with relatively uniform geochemical conditions. Simple supplementation of animal feeds with organic Se is an alternative approach which may be an easier way of improving human Se nutrition for some countries with low dietary Se intakes.

References

1. X. Chen, G. Yang, J. Chen, X. Chen, Z. Wen and K. Gen, Biol. Trace Element Res. 1980, **2**, 91.
2. A. Aro, Norwegian J. Agric. Sci., 1993, suppl **11**, 127.
3. J. T. Salonen, G. Alfthan, J. Pikkarainen, J. K. Huttunen and P. Puska, Lancet 1982, **2**, 175.
4. P. Suadicani, H. O. Hein, and F. Gyntelberg, Atherosclerosis 1992, **96**, 33.
5. P. Knekt, A. Aromaa, J. Maatela, G. Alfthan, R-K. Aaran, M. Hakama, T. Hakulinen, R. Peto and L. Teppo, J. Natl. Cancer Inst. 1990, **82**, 864.
6. W. C. Willett, M. J. Stampfer, D. Hunter and G. A. Colditz, in: Trace Elements in Health and Disease. ed. A. Aitio, A. Aro, J. Järvisalo, H. Vainio, The Royal Society of Chemistry, Cambridge, 1991, pp. 141-155.
7. A. M. Fan, S. A. Book, R. R. Neutra and D. M. Epstein, J. Toxicol. Environ. Health 1988, **23**, 539.
8. G. F. Combs and S. B. Combs, The role of selenium in nutrition. Academic Press, New York, 1986
9. J. Sippola, Ann. Agric. Fenn. 1979, **18**, 182.
10. P. Varo and P. Koivistoinen, Acta Agric. Scand., 1980, suppl **22**, 165.
11. P. Varo, G. Alfthan, J. K. Huttunen and A. Aro, in Selenium in biology and human health. ed. Burk, R. F., Springer, New York, 1994, pp. 198-218.
12. G. Gissel-Nielsen and B. Bisbjerg, Plant and Soil 1970, **32**, 382.
13. T. Yläranta, Increasing the selenium content of cereal and grass crops in Finland, Dissertation, University of Helsinki, 1985.
14. G. Alfthan, Nutr. Res., 1988, **8**, 467.
15. P. Varo, G. Alfthan, P. Ekholm, A. Aro and P. Koivistoinen, Am. J. Clin. Nutr. 1988, **48**, 324.
16. A. Aro, G. Alfthan and P. Varo, Analyst, 1995, 120, 841.
17. G. Alfthan, A. Aro, H. Arvilommi and J. K. Huttunen, Am. J. Clin. Nutr., 1991, **53**, 120.
18. O. A. Levander, G. Alfthan, H. Arvilommi, C-G. Gref, J. K. Huttunen, M. Kataja, P. Koivistoinen and J. Pikkarainen, Am. J. Clin. Nutr. 1983, **37**, 887.
19. B. Pherson, Norwegian J. Agric. Sci., 1993, suppl 11, 79.
20. L. Jönsson, Norwegian J. Agric. Sci., 1993, suppl 11, 95.
21. H. R. Conrad and A. L. Moxon, J. Dairy Sci., 1979, 62, 404.
22. P. Aspila, J. Agric. Sci. Finland, 1991, 63, 1.
23. P. Ekholm, P. Varo, P. Aspila, P. Koivistoinen and L. Syrjälä-Qvist, Br. J. Nutr., 1991, 66, 49.
24. L. Syrjälä-Qvist and P. Aspila, Norwegian J. Agric. Sci., 1993, suppl 11, 159.
25. A. Aro and A. Halttunen. Suom. Lääkäril. 1990, 45, 751.

LOW DIETARY SELENIUM INTAKES IN SCOTLAND AND THE EFFECTIVENESS OF SUPPLEMENTATION

A MacPherson, M N I Barclay and J. Molnár

SAC
Auchincruive, Ayr,
Scotland KA6 5HW

1 INTRODUCTION

Human daily dietary intakes of selenium in Scotland were first estimated in the mid seventies[1] and were reported as 60µg/d. Further studies in 1985 and 1990[2,3] revealed a progressive decline in dietary selenium supply such that calculated intakes were reduced to 43 and 30µg per person per day respectively. This decline was identified as being due to a changeover in source of breadmaking flour from North America to Europe following the UK's accession to the Common Market. Thus by 1990 the daily dietary intake of selenium was comparable to that in Finland prior to the introduction of the selenium fertilization policy of all cereal and other crop growing ground.

This paper describes the results of a further and more comprehensive survey of the selenium concentration of UK foods undertaken in 1993-94. It also examines the effectiveness of selenium supplementation in raising the plasma selenium concentration of human subjects. Finally it examines the possibility of increasing human dietary selenium intakes by raising the selenium content of carcass meat by means of pre-slaughter supplementation of farm animals.

2 MATERIALS AND METHODS

2.1 Food Survey Samples

Over 700 food samples representing 100 different types of food were collected. These were purchased from a wide range of outlets and across the nine different regions of mainland Britain: South East/East Anglia; South West; East Midlands; West Midlands; Wales; Yorkshire and Humberside; North West; North of England and Scotland. For foods such as milk where seasonal variations might be expected purchases were made throughout the year. Each set of products was combined, milled, homogenised or macerated in a food processor to a uniform material which was then sub-sampled and dried(at $60^{\circ}C$ or in a freeze drier as appropriate) for selenium analysis.

2.2 Selenium Supplementation of Human Subjects

Two studies have been undertaken as follows:

    2.2.1 *Young Males.* Fifty young men were allocated at random to either a placebo or selenium supplementation treatment and maintained on these for 3 or 6 months. Blood samples were collected prior to treatment and at the end of the supplementation period. In some cases a further sample was collected 3 months after the end of treatment.

    2.2.2 *Mixed Age Males* Forty eight adult males aged between 20 and 61 years were allocated at random to a placebo or selenium treatment group and maintained on these treatments for 4 months. At the end of the supplementation period blood samples were collected from forty subjects(20 in each group) and analysed for selenium. In both studies selenium was given as L-selenomethionine and at a rate of 100µg/day.

2.3 Selenium Supplementation of Sheep and its Effect on Concentration in Carcass Meat

Sixteen Scottish Blackface lambs were allocated at random to one of 4 treatment groups and given 0, 3.5, 7.0 or 10.5mg selenium per week as sodium selenite for a period of 14 weeks prior to slaughter. Post mortem samples of liver, kidney and shoulder and thigh muscle were collected and analysed for selenium.

2.4 Analytical Methods

Food and post mortem tissue samples were digested in an acid mixture[4] and the selenium content was determined following production of the hydride using a PS Analytical continuous flow hydride generator by means of a PS Analytical Excalibur atomic fluoresence detector using a high powered selenium lamp to initiate the fluorescence. Plasma selenium concentrations were determined similarly but without the requirement for the initial digestion. The accuracy of the method was verified by the use of Community Bureau Reference samples and validated for plasma analysis with a Nycomed standard serum. The results of these analyses are given in Table 1.

Table 1 *Selenium concentration of standard samples as determined and compared to their reference values*

BCR No.	Material	Result (mg/kg)	Reference Value(mg/kg)
B1-89	Wholemeal Flour	0.140 ± 0.003	0.132 ± 0.010
C85-04	Cabbage leaves	0.080 ± 0.007	0.083 ± 0.008
B1-84	Bovine muscle	0.188 ± 0.019	0.183 ± 0.012
B2-78	Mussel	1.715 ± 0.035	1.660 ± 0.040
	Plasma	98.3 ± 4.3 µg/l	100.0 ± 6.0 µg/l

3 RESULTS

3.1 Selenium content of UK foods and dietary intake

The highest selenium concentrations were found for brazil nuts(254 µg/100g), kidney (145 µg/100g) and crab meat(84 µg/100g) while most vegetables and dairy products tested were less than 2 µg/100g fresh weight. A wide range of breads and bread rolls were tested with wholemeal bread being highest at 9 µg/100g and currant bread lowest at 3.5 µg/100g. Table 2 presents the selenium concentration of the foods categorised by their selenium content.

Table 2 *Foods with selenium concentrations of >20, 10-20 and <10 μg/100g fresh weight*

Food	Se Concentration	Range
Brazil nuts	254	86-686
Kidney	145	78-200
Crab	84	28-126
Lobster	54	41-76
Liver	42	18-135
Cashew nuts	27	17-39
Prawns(tiger)	26	8-42
Duck	22	18-24
Grouse	20	11-29
Prawns(regular)	19	16-23
Turkey leg	15	12-18
Pork	14	12-15
Pheasant	14	12-15
Rice(long grain)	13	7-23
Parmesan	12	11-14
Cheddar(reduced fat)	11	9-12
Turkey breast	10	9-12
Rice(brown)	10	4-14
Venison	9	5-13
Gouda	8	7-9
Stilton	8	7-9
Beef	8	7-8
Camembert	7	5-8
Processed cheese	7	6-9
Danish Blue	6	5-7
Edam	6	6-7
Brie	4	3-4
Chocolate(Dark)	4	2-11
Peas dried	3	1-6
Fromage frais	2	1-3
Soya milk	2	1-5

Using the data provided by this study where possible and that from standard tables[5] where not the average daily intake of selenium was calculated by using the estimated quantities of food eaten each day for the UK as a whole and also for its constituent regions[6]. This gave a value of 34 μgSe/day for the UK and values ranging from 30 μg/day in the North of England to 41 μg/day in North West England. Scotland was at the lower end of the range at 32 μg/day while the South East and East Anglia were close to the top at 40 μg/day.

3.2 Selenium Supplementation Studies

The effectiveness of the selenium supplementation of the young males can be seen in Figure 1. It raised plasma selenium concentrations from 83 to 124 μg/l while control subjects experienced a decline from 80 to 71 μg/l. Selenium supplementation of the mixed age males significantly($P<0.001$) elevated their plasma selenium concentrations to 120 μg/l compared to 77 μg/l in their unsupplemented controls(Figure 2).

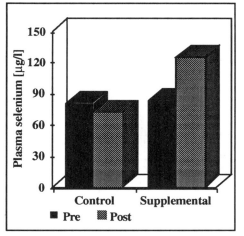

Figure 1. *Response in mean plasma selenium concentration of control and selenium-supplemented subjects.*

Figure 2. *Mean plasma selenium concentration of subjects on placebo or 100 µg selenium daily.*

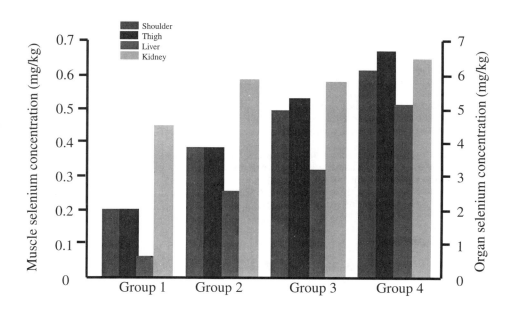

Figure 3. *Response in muscle and organ selenium concentrations to pre-slaught supplementation of lambs with selenium.*

3.3 Selenium Supplementation of Sheep

A dose response relationship was found between the level of supplementation and the selenium concentration of shoulder and thigh muscle and liver(Figure 3). This was less evident in the case of kidney where there was some response to the lowest treatment level(3.5mgSe/week) but little change thereafter.

4 DISCUSSION

Prior to the analysis of the food samples in this survey the calculated daily selenium intake based on data in the compositional tables was 62 µg. The marked fall to the present value of 34 µg/day has been due to a drop in the selenium intake from bread and cereals from 40 to 10 µg/day. This has been brought about by the change in source of our breadmaking flour from Canada to Europe following our accession to the European Union. Small increases in the selenium concentration of meat, eggs and dairy products prevented the fall from being even more drastic. This level of intake is fairly comparable to that in Finland prior to their supplementation programme and puts the UK intake among the lower national levels and well below the RNI's of 75 and 60 µg/day for adult males and females respectively.

The supplementation studies showed that it is possible to raise below normal plasma selenium concentrations to within the normal range with 100 µgSe/day. Withdrawal of the supplement, however, meant that concentrations returned to their previous level within the 3 months interval between samples. It was interesting to note that the highest plasma selenium concentration(161 µg/l) in the mixed age study was for an unsupplemented subject. It transpired that he regularly consumed brazil nuts which suggests that their high selenium content is in an available form. It also suggests that normal plasma selenium status(>90 µg/l) can be maintained by dietary means by a careful selection of foods. On the other hand anyone selecting against the relatively few good sources of dietary selenium will end up with a very low intake indeed.

The sheep supplementation study was undertaken to investigate a possible alternative approach to raising human dietary selenium intakes to the fertilization policy of crop growing ground adopted in Finland. It showed that muscle and organ meat selenium concentrations can be markedly increased by pre-slaughter supplementation. A 100g portion of muscle meat from lambs in groups 2-4 would provide 16, 31 or 44 µg selenium and thus contribute significantly to the daily requirement.

This paper has shown a marked decline in UK human dietary selenium intakes which has resultd in below normal plasma selenium concentrations. These can be effectively increased by selenium supplementation or by careful dietary selection. A way of enhancing the selenium concentration of meat has been evaluated and could contribute significantly to meeting our dietary requirements.

References

1. Thorn, J., Robertson, J., Buss, D. H. and Bunton, N. G., *Brit. J. Nutr.*, 1978, **39**, 391.
2. Barclay, M. N. I. and MacPherson, A., *J. Sci. Fd and Agric.* 1986, **37**, 1133.
3. Barclay, M. N. I. and MacPherson, A. *Brit. J. Nutr.*, 1992, **68**, 261.
4. Hershey, J. W., and Oostdyke, T., 1988, *JAOAC,* **71**, 1090.
5. Holland, B., Welch, A.A., Unwin, I. D., Buss, D. H., Paul, A.A. and Southgate, D.A.T., 'McCance and Widdowson's The Composition of Food, 5th Edit.' London, Cambridge: RSC/MAFF. 1991.
6. Ministry of Agriculture, Fisheries and Food, 'The British Diet, Finding the Facts 1989-1993, appendix vi Total Diet Study' HMSO, London, 1994.

SELENIUM DEFICIENT STATUS OF INHABITANTS OF SOUTH MORAVIA

J. Kvíèala*, V. Zamrazil* and V. Jiránek**

*Institute of Endocrinology, Národní 8, 116 94 Praha 1, Czech Republic
**Immunotech, Radiová 1, 102 27 Praha 10, Czech Republic

1 SUMMARY

To estimate status and intake of selenium in inhabitants of country side of South Moravia, 264 sera, 28 hair, and 356 urine samples from random selected inhabitants of both sexes in the age from 6 to 65 years were analyzed. Quality of the neutron activation analysis (sera and hair) and fluorimetry (urines) was checked by coanalyses of reference materials. Overall values (serum - 42 µgSe/l; hair - 0.228 µgSe/g; urine - 8.2 µgSe/l or 7.2 µgSe/g creatinine) demonstrate severe Se deficiency with the extremely low intakes of selenium in the locality (mean assess 14-21 µg/day). Se levels for boys, girls, men and women as well as for individual sex and age groups are mentioned together with recommended or optimal values. Frequency distributions of Se concentrations in sera, urines, and hair are presented together with the estimation of selenium intake in individual sex and age groups. Age-dependence both for serum and urine levels was found in the region, while sex dependence was found only for 10 years old children and 36-49 years old adults. Reasons for preferential use of urine Se concentration (µg/l) over its recalculation for creatinine (µg/g creatinine) are explained. Se status of South Moravia inhabitants is one of the lowest statuses not only in Europe but all the world round and only the countries with the epidemiological occurrence of Se-dependent diseases have lower Se indices of their inhabitants.

1 INTRODUCTION

Selenium is one of the most beneficial trace elements for human organism. Connection of its deficiency with so called 'oxidative stress' diseases (cardiovascular diseases, cancer, and others) by the decrease of enzyme Glutathione Peroxidase (Se-GSH-Px)[1,2], and with alterations of thyroid hormone metabolism by the decrease of enzyme Iodothyronine 5'-Deiodinase (IDI)[3,4] has been proved by the epidemiological studies as well as in vitro and in vivo experiments. Se affects detoxification of both heavy metals and mutagenic, teratogenic, and carcinogenic organic substances[5,6] and influences immunity of the organism[7].

From supplementation studies has been deduced optimal concentration of serum Se (at about 100-135 µg/l)[8-10] and intake (55-75 µgSe/day)[11,12]. Lower Se serum concentrations are said to be deficient, with various effects upon biochemical and physiological pathways of the organism. Various severity of deficiency (20-100 µg/l) leads to exhaustion of Se pools, decrease and exhaustion of Se-GSH-Px, decrease of Selenoprotein P. Deiodinase I (IDI) starts to be depressed in various organs during severe deficiency. Higher occurrence of cancer, cardiovascular and other 'oxidative' diseases was revealed by epidemiological studies for lower Se percentiles. Concentration of serum Se below 20 µg/l - vital deficiency - may result in various clinical difficulties leading to death or nonreparative changes of functions of organism[13,14] (Keshan Disease, Kashin-Beck Disease, Cretinism) without Se supplementation.

Status of selenium in the population is influenced by various circumstances, but the most important is its intake by food, which is determined by the local geochemical condition and possibility of plants to absorb Se to the nutritional chain. Low selenium belts seem to affect Se nutrition status of the inhabitants from north of Europe (Scandinavian countries) to south (former Yugoslavia, Greece, Italy) and from east (Belorussia) to west (Germany, France). Czech Republic is situated just in the cross of the belts. Preliminary studies have shown indeed very low concentrations of Se in serum, hair, and urine, the most often used indices of Se status. Because of important influence of Se on the health of population as well as individuals, Se status of population of chosen regions is searched in detail.

2 MATERIAL AND METHODS

2.1 *Materials.* Inorganic standards were prepared by dissolving of SeO_2(99.999%) (Alfa Products, Karlsruhe, Germany) in deionized (17.5-18 M) water (apparatus NANOpure Barnstead, Wilhelm Werner GmbH, Berg, Germany). Suprapur HNO_3, $HClO_4$, HCL, (Merck, Darmstadt, Germany) were used for wet-ashing of biological material for measurement. EDTA p.a. was purchased from Lachema, Brno, CR. 2,3-diaminonaphtalene hydrochloride 99% (DAN) and cyclohexane HPLC (99,94%) from Aldrich, Steinheim, Germany were used for fluorimetric determination of Se. Human Serum of the Second Generation (kind gift of prof. Versieck from Gent, Belgium), LYPHOCHEK Urine Metals Control Level 1 from BioRad, Anaheim, CA, USA, and IAEA reference material H-4 (animal muscle), were used as biological standards. Wet-ashing was done in heated steel block from Liebisch, Bielefeld, Germany. Fluorimetric measurements were done by fluorimeter LS 50 from Perkin-Elmer, Norwalk, CT, USA. Gamma-spectroscopy was performed with multichannel analytical system from Silena, Milano, Italy, which consisted of multichannel analyzer MB 7329/s-16K with coaxial HPGe detector (rel.efficiency 34,6% and FWHM 1,87 keV for 1,33 MeV). Quantitative analyses of peaks were done by the programme Silgamma.

2.2 *Subjects.* Sera from 264, urines from 356 random selected inhabitants of both sexes in the age from 6 to 65, and hair from 28 men in the age of 36 - 49 years of South Moravia were analyzed for selenium. Creatinine was measured in corresponding urines for the purpose of recalculation of Se in urine to creatinine[15].

2.3 *Methods.* Blood serum and hair were analyzed by neutron activation analysis (NAA) as described elsewhere[16,17]. Human serum of the second generation and IAEA standard reference material H-4 were coanalyzed for quality assurance of the Se detection with the results of 1.03 - 1,07 µg/g of dry serum (declared value 1,05 µg/l), and 0,273 µg/g of dry muscle (certified 0,28 µg/g). Within batch precision 8.5% (n=10) and between batch precision 8.7% (n=19) was reached for human serum.

Urine selenium was measured by fluorimetry as complex with DAN after one-tube wet-ashing with a mixture of acids (HNO_3,$HClO_4$;4+1) and reduction of Se(VI) to Se(IV) by HCl according Sheehan and Gao[18], with the minor modifications[19]. Precision of urine selenium determination was checked by Lyphocheck Urine analysis. Our values were 53-59 µg/l, declared values were 50 µg/l with acceptable values 40-60 µg/l. Within batch precision of analyses of Lyphocheck urine Se was 6.4% (n=6), between batch precision of apparently healthy adult was 6.5% (n=9; mean 11.18 µgSe/l urine).

Creatinine was analyzed by bio-test Lachema[20] - picric acid reacts with creatinine in an alkaline medium to form orange-red colored adduct suitable for the photometric determination.

Declared reproducibility is 6%.

2.4 Satistics Analytical results were evaluated by the usual statistical methods.

3 RESULTS AND DISCUSSION

360 randomly selected inhabitants of the lowland region Znojemsko in South Moravia were searched for status of selenium by the analyses of serum, urine, and in case of group of 36-49 years old men also hair selenium content. Se in urine was also recalculated to urine creatinine. Overall arithmetic means of the results of the whole population of the region are presented in Table 1. There are values obtained as µgSe/g lyophilized tissue (serum, hair) and values recalculated to µgSe/l serum, µgSe/g hair. (15% of weight of hair is H_2O).

Table 1 - Materials and Results

MATERIAL	SEX	AGE	No. cases	MEAN ± S.D. µgSe/g dry tissue		MEAN /l liquid /g tissue
SERUM	m,f	6-65	264	0.457	0.117	42
HAIR	m	36-49	28	0.269	0.045	0.23
URINE*	m,f	6-65	356	8.19*	3.75	8.2
URINE+	m,f	6-65	352	7.17+	3.82	7.2+

Reculcation of µgSe/g dry serum into µgSe/l serum by multiplication by 90.9; recalculation of µgSe/g dry hair to µgSe/g hair by multiplication by 0.85.
*µgSe/l;   +µgSe/g creatinine

Serum (or plasma) selenium is the best known and the most often used indicator of Se status[8,13,21,22]. It is significantly correlated both to some biochemical activities and to intake, especially under low or moderate exposure to Se. The results of serum Se concentrations for groups of children (boys and girls) as well as for men and women are shown in Figure 1. Optimal concentration is also added for comparison. Differences between adults and children and between measured and optimal levels of serum Se are statistically significant ($p<0.01$) ; measured values are at about one third of optimal ones.

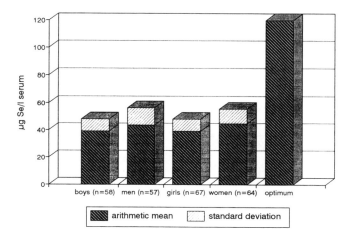

Figure 1 - Serum Se Concentration

Differences between sexes were not significant. Table 2 documents again dependence of serum Se on age. Differences between groups according the age are statistically significant (p>0.05 and higher) both for males and females, namely between 6 years old boys and 36-49 years old men, for females between 36-49 years old women and all groups of girls. Differences between 10 and 6 years old boys as well as 10 and 13 years old boys are also significant. These results are very close to those from Prague district[23]. Sex differences were obtained between men and women at the age of 36-49 years. Very close numbers were obtained for arithmetic means, geometric means, and medians (data are not presented), which indicates normal distribution of the values and possibility to compare arithmetic means.

Table 2 - Se Serum Concentrations of Individual Groups According Sex and Age (µg/g dry serum)

age	6	10	13	18-35	36-49	50-65 years
males						
n	20	19	19	19	19	19
X	0.37	0.50	0.42	0.45	0.50	0.48
S.D.	0.08	0.10	0.07	0.09	0.16	0.16
females						
n	23	22	22	21	21	22
X	0.42	0.45	0.42	0.47	0.52	0.49
S.D.	0.09	0.08	0.11	0.09	0.11	0.15

Recalculation into µgSe/l sera by multiplication by 90.9
n - Number of cases; X - Arithmetic mean; S.D. - Standard deviation.

Critical levels for various stages of Se status according its indexes are only rough estimates obtained from comparison of epidemiological studies and supplementation trials in the countries and regions with various intakes of Se. Their mechanistic application to individuals may lead to diagnostic errors. On the other hand inverse associations postulated between Se status and incidence of cardiovascular diseases, some types of cancer, and other oxidative stress illnesses raise the necessity to define optimum Se levels as well as various stages of Se deficiency both for populations and individuals. The level of 100-135 µgSe/l of serum[8] was estimated as optimum from the Se saturation experiments with the analyses of the concentrations and biochemical functions of selenoproteins[9,10,24,25]. Discussions on minor disturbances in Se status and their health effects raised the question of marginal deficiency (the range 70-100 µgSe/l serum), which might have deleterious effects on human health during the extreme load of the organism. Epidemiological occurrence of some diseases (Keshan disease, Kashin-Beck disease, myxedematous cretinism) in very low Se regions (China, Russia, Zaire) was proved to be associated with decrease of Se and activity of Se-enzymes Se-GSH-Px or IDI. Upper limit of this vital deficiency was detected between 7 and 20 µgSe/l serum (or adequately 10-25 µgSe/l whole blood)[8,21,22,26]. More stages of Se deficiency (mild, moderate, severe) may be classified between optimal status (or marginal deficiency) and vital deficiency and are connected with higher occurrence of cardiovascular diseases and some types of cancer. Their limits are connected not only with absolute Se concentrations but also with long-term levels of the local population[13,25], e.g. development of its reparative possibilities and exhaustion of Se pools in the case of individual. It is supposed[13] that in low Se populations like in New Zealand[27] or China[28] (and also in Czech Republic) the limits are lower then in the regions with higher Se status (USA)[29].

Frequency of serum Se concentrations in the population of South Moravia according various stages of Se deficiency is shown in Figure 2. There are not values in the limits of vital deficiency (minimum was 21 µgSe/l serum), but more than three quarters are under the limit of 55 µg/l and maximal value is 81 µg/l. These numbers show poor Se status of the region. Mean concentration of serum Se obtained from South Moravia

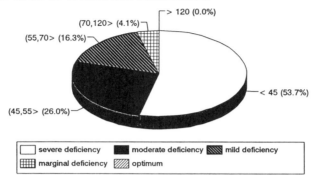

Figure 2 - Frequency of Serum Se Concentration [µg/l]

inhabitants is one of the lowest in Europe[30] (together with a few regions in Serbia)[31] and only inhabitants of the areas with epidemiological occurrence of Se-dependent diseases have lower values[13].

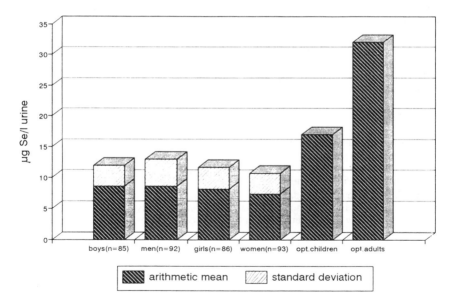

Figure 3 - Urine Se Concentration

Concentrations of Se in urines of boys, girls, men, and women are depicted in Figure 3 together with means of optimum for children and adults. Concentrations for individual age and sex groups are tabulated in Table 3. Geometric means and medians are again tight to arithmetic means which justifies calculations with and comparison of arithmetic means. We were not able to find significant differences between adults and

Table 3 - Se Urine Concentrations of Individual Groups According Sex and Age (μg/l)

age	6	10	13	18-35	36-49	50-65 years
males						
n	27	28	30	29	33	30
X	8.2	8.5	9.2	9.9	8.0	7.9
S.E.	3.2	3.9	3.3	4.7	3.4	5.0
females						
n	31	28	27	33	31	29
X	8.0	7.4	9.1	9.5	6.0	6.7
S.E.	4.3	2.3	3.5	3.1	3.3	2.7

n - Number of cases; X - Arithmetic mean; S.D. - Standard deviation.

children as was the case in the Prague region[23]. Difference between men and women was the opposite one to that from Prague - men have by 17% higher mean Se urine level ($p<0.05$). The reasons are probably different nutritional habits in the regions. In this way it is possible to explain different literary results[32]. Significant differences between age groups (Tab. 3) were found only for females (10 - 13, 10 - 18-35, 18-35 - 36-49, and 18-35 - 50-65 years old women; $p<0.05$ and $p<0.01$); sex differences were found only for 10 years old children and middle aged adults ($p<0.05$).

The possibility to use urine Se concentrations for estimation of Se status of population[8,13,32,33] is a consequence of the comparison of urine levels in areas with various Se statuses and intakes[32,33,34], and also from studies with correlations between urine and serum Se[32,33].

Needs for Se intake differ according the body weight[8,13,27]. It must be considered also in case of children and all recommended daily intakes count for it in their limits[11,12]. Because urine Se levels are dependent on Se intake, they should be much lower, but no extended studies have been carried out on selenium excretion in childhood[8]. Because of this theoretical consideration, halves of the limits used for adults are used in Figure 4 for children between 6 and 13. In fact, urine Se concentrations for children are the same or even higher then those for adults in both our studies (Prague region[23] and this one). It seems that in our country nutritional habits much better satisfy Se requirements of children then those of adults.

According to the literature[32,33], mean Se values do not exceed 30 μg/l for Europeans generally, but their concentrations increase to 100 μg/l in areas with high Se content in soil. On the other hand, mean urine Se concentration of 7 μg/l urine (n=49) was published in region with very poor soil Se content and epidemiological occurrence of Keshan disease[35]. From this point of view, at about one half of the population searched for this study has the same levels of urine Se concentration as the population with epidemiological occurrence of the disease associated with Se deficiency.

Urine Se is used as body Se index[8,13,22] but it is connected with Se intake[8,13,32,33], too. Metabolism of Se is tightly regulated by renal retention and excretion, especially in medium or low intakes; in this case at about 60% of long-term Se absorption is excreted by urine[14,36,37,].

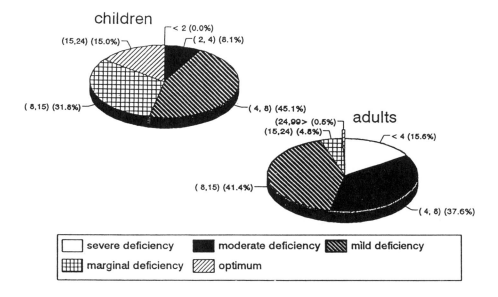

Figure 4 - Frequency of Urine Se Concentration [µg/l]

It is possible to assess medium Se absorption from daily urine Se on these grounds. Palmquist and co-workers[38] claimed that the daily urinary Se excretion could be estimated from a morning single voided sample. Oster and Prellwitz[33] have found a significant correlation between Se excretion in 24 h (µg/24 h) and Se excretion in single voided urine (µg/l) for 24 healthy subjects (r=0.553; p<0.01). Moreover, day-to-day variations of excreted Se are in the low Se population depressed by homeostatic renal retention of Se and random fluctuations of urine volume would tend to nullify one another in epidemiologic studies. Even when use of 24 h urine is highly recommended, the possibility to use single void (morning) urine is very important, because it is nearly impossible to obtain 24 h urine during epidemiological collection of samples, especially when also children are involved. That is why we have used morning single void urine Se for estimation of Se intake for individual groups of inhabitants according age and sex. Assumptive 1 to 1.5 l of daily urine were used for estimation of lower and upper limit of long-term mean of daily ingested Se. Comparison of obtained values with both WHO

Table 4 - Estimation of Daily Intake of Selenium

AGE	ASSESSED VALUES MALES	FEMALES	RECOMMENDED INTAKE MALES	FEMALES*	BOTH+
6	14-20	13-20	25		20
10	14-21	12-19	35		30
13	15-23	15-23	40		35
18-35	16-25	16-24	70	55	55
36-49	13-20	10-15	70	55	55
50-65	13-20	11-17	70	55	55

Assumptions: 1. 60% of Se intake is excreted by urine
2. Mean urine volume is 1 - 1.5 l.
*American Academy of Sciences (1989)[11] + WHO;
+Scientific Committee for Food, EC[12].

and EC recommendations in Table 4 shows very low intakes of Se in the country of South Moravia.

To improve information on Se status from single void urine, some authors proposed to use recalculation of urine Se to urine creatinine[15]. But, according Oster and Prellwitz results [33], coefficient of correlation between 24 h Se urine and Se/creatinine is better than between 24 h Se urine and Se/l only for patients. Both coefficient of correlation and significance were better in case of correlation between µgSe/24 h and µgSe/l of 24 healthy subjects. Our values for groups of children and adults recalculated to µgSe/creatinine are shown in Figure 5. Table 5 presents the same calculation for individual groups according the sex and age. It is obvious from the comparison of the Figures 3 and 5 and Tables 3 and 5 that the relations between age groups were shifted by the use of creatinine. The reason is that the protein and Se metabolism are quite different in childhood and maturity. Requirements for Se in developing organism are connected much more with

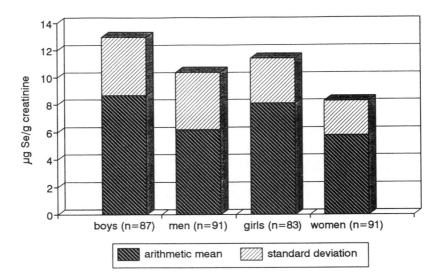

Figure 5 - Urine Se in Relation to Creatinine

anabolic metabolism and, on the other hand, muscle mass with its greatest Se content is not so great as that in adults and, moreover, hormone status is quite different. That is why we think that Se/creatinine might be a useful indicator only for comparative studies of the same age and sex

Table 5 - Se Urine Concentrations of Individual Groups According Sex and Age Recalculated to Urine Creatinine (µg/l)

age	6	10	13	18-35	36-49	50-65 years
males						
n	28	29	30	28	33	30
X	9.9	8.4	7.9	4.5	5.4	8.7
S.E.	5.1	3.2	4.2	1.6	2.1	6.1
females						
n	31	25	27	31	31	29
X	9.7	7.7	6.7	5.2	4.7	7.6
S.E.	4.3	2.4	1.5	2.1	2.4	2.0

n - Number of cases; X - Arithmetic mean, S.D. - Standard deviation.

groups and one must be very careful with the use of this type of data even in the medical studies with various types of diseases.

Another often discussed Se indice is hair Se[8,13,21,22]. It has been related to the long-term selenium status and intake in epidemiological studies of large population groups[26], and also in studies with Se supplementation[14,26]. Its applicability was documented by the comparison of hair Se concentrations in areas with very low Se

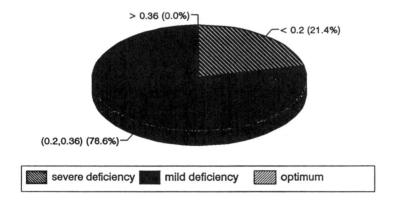

Figure 6 - Frequency of Hair Se Concentration [µg/g]

status (Keshan disease)(0.1 µgSe/g hair), low (deficient areas) (0.2 µg/g), medium (Se-sufficient areas) (0.36 µg/g), high (without selenosis) (0.6 µg/g) and very high (with selenosis)(>0.6 µg/g) in China[35,39]. Hair has advantage in noninvasive obtaining of the samples and simple transport and manipulation. Some questions have been connected with its correlation with content of Se in other tissues[8], but there are more papers on the correlations both with Se in organs and with its intake[39-41]. Possible contamination from anti-dandruff shampoos with the content of Se may arise in western countries, but this problem has probably not been actual in our country till now. Hair Se content was used in this study in the group of middle aged men (36-49 years old men). Mean concentration is in the Table 1, frequency of concentrations compared to stages of Se deficiency is depicted on the Figure 6. None of 28 men had this indice in the vital deficiency limits, on the other hand the highest obtained concentration was 0.3 µgSe/g hair, so that all values were in the limits for severe (21%) or mild deficiency. Also this long-term indice confirms very low Se status of the region.

4 CONCLUSIONS

A. All three selenium indices employed show a very low status of the inhabitants of the country side of South Moravia (Znojemsko). The levels are the lowest reported in Europe and in the global scale only the areas with the endemic occurrence of Se-dependent diseases have lower Se indices. Se status of children was better than that of adults.

B. The same as above was true for intake of selenium assessed from the urinary Se level.

C. Different serum Se levels were settled for children and adults, the same was true for urinary Se levels of females.

D. Statistically significant gender differences were found for urine Se (men-women) but not for serum Se.

E. Preferential use of concentration data (µgSe/l urine) instead of recalculation against urine creatinine concentration (µgSe/g creatinine) is concluded both on the grounds of the literary data and of our results for comparison studies, if it is impossible to obtain 24 h urine samples, which is preferable according to both former parameters. This is true especially for compararative purposes for groups with various age and/or sex.

ACKNOWLEDGMENT

This work was supported partly by grant IGA MZ ÈR No. 1428-3, and grant GA ÈR No. 311\93\0986. Prof. Versieck (Rijksuniversiteit Ghent,Belgium) is gratefully acknowledged for the kind gift of the reference material of the second generation human serum.

REFERENCES

1. *Trace Elements and Free Radicals in Oxidative Diseases*, Proceedings of the 4th International Congress on Trace Elements in Medicine and Biology, Chamonix (France), April 5th-9th, 1993, eds.Favier,A., Neve,J., and Faure,P.,AOCS Press, Champaign (Il,USA),1994
2. Hocman,G.(1988) Int.J.Biochem.,1988,**20**,123.
3. Arthur,J.R., Nicol,F., Beckett,G.J., Am.J.Clin.Nutr., 1993,**57**,236S
4. Vanderpas,J.B., Contempré,B., Duale,N.L., Decks,H., Bebe,N., Longombé,A.O., Thilly,C.H., Diplock,A.T., Dumont,J.E. Am.J.Clin.Nutr.,1993,**57**,271S
5. Whanger,J.D. (1981), Selenium in Biology and Medicine, Spallholz,J.E., Martin J.L., Ganther,H.E., eds., Avi Publishers, Westport,CT,1981, p.230.
6. Pence,B.C., and Budding,F., J.Nutr.1985,**115**,1198.
7. Petrie,H.T., Klassen,L.W., O'Dell,J.R., Kay,H.D., J.Leu- kocyte Biol.,1989,**45**,210
8. Neve,J.,J.Trace Elem.Electr.Health.Dis.,1991,5,1.
9. Alfthan,G.,Aro,A.,Arvilommi,H.,Huttunen,J.K., Am.J.Clin. Nutr.,1991,**53**,120
10. Varo,P., Alfthan,G., Huttunen,J.K., Aro,A., in Selenium in Biology and Human Health, ed. Burk,R.F., Springer-Verlag New York,Inc.,1994,199
11. US National Academy of Sciences, Recommended Dietary Allowances, 10th

edition, National Academy Press, Washington,D.C.,1989.
12. Report SCF-EU, Nutrition Reviews,1993,**51**,209
13. Levander,O.A.,Federation Proc.,1985,**44**,2579
14. Xia,Y., Zhao,X., Zhu,L., Whanger,P.D., J.Nutr.Biochem., 1992,**3**,211
15. Hojo,Y., Bull.Environm.Contam.Toxicol.,1982,**29**,37.
16. Kvíèala,J., Havelka,J., J.Radioanal.Nucl.Chem. 1987,**121**, 261.
17. Kvíèala,J., Havelka,J., J.Radioanal.Nucl.Chem. 1987,**121**, 271.
18. Sheehan,M.T., and Min Gao,Clin.Chem.,1990,**36**,2124
19. Kvíèala,J., Zamrazil,V., Soutorová, M., Tomíka,F., Analyst, 1995,**120**,959
20. Van Pilsen,J.F., and Boris,M.,Clin.Chem.,1957,**3**,90.
21. Diplock,A.T., Am.J.Clin.Nutr.Suppl.,1993,**57**,256S,
22. Van Dael,P., and Deelstra H., Flair Concerted Action No 10 Status Papers,1993,312
23. Kvíèala,J., Havelka J., Zamrazil,V., Èeøovská,J., Èermák S. in: Trace Elements in Man and Animals, eds. Anke,M., Meissner,D., Mills,C.F., Verlag Media Touristik, Gersdorf, 1993, p.233
24. Néve,J., Chamart,S., Van Erum,S., Vertongen,F., Dramaix, M., in: Selenium in Medicine and Biology, eds. Néve,J., Favier,A., Walter de Gruyer, Berlin-New York,1989,315
25. Levander,O.A., Alfthan,G., Arvilommi,H., Gref,C.G., Huttunen,J.K., Kataja,M., Koivistoinen,P., Pikkarainen, J., Am.J.Clin.Nutr., 1983,**37**,887
26. Ge,K., Yang,G., Am.J.Clin.Nutr.,1993,**57**,259S
27. Thomson,C.D., Ong,L.K., Robinson,M.F., Am.J.Clin.Nutr., 1985,**41**,1015
28. Luo,X., Wei,H., Yang,H., Xing,J., Liu,X.,Quiao,C., Feng, C., Liu,Y., Liu,J.,Wu, W., Liu,X., Guo,J., Stoecker,B., Spallholz,J.E., Yang,S.,.J.Clin.Nutr.,1985,**42**,439
29. Valentine, J.L., Kang,H.K., Dang,P., Schluchter,M, J.Toxicol.Environ.Health, 1980,**6**,731
30. Van Cauwenberg, R., Robbrecht,H., Deelstra,H., Picramenos,D., Kostakopoulos,A., J.Trace Elem. Electrolytes Health Dis.,1994,**8**,99
31. Maksimovic,Z.J., Kidney Int., 1991,**40**,S12
32. Robberecht,H.J., Deelstra,H.A., Clinica Chimica Acta, 1984,**136**,1984
33. Oster,O., Prellwitz,W., Biol.Trace Elem.Res.,1990,**24**,119
34. Robinson,J.R., Robinson,M.F., Levander,O.A., Thomson, C.D., Am.J.Clin.Nutr., 1985,**41**,1023
35. Yang,G., Wang,S., Zhou,R., Sun,S., Am.J.Clin.Nutr., 1983,**37**,872
36. Martin,R.F., Janghorbani,M., Young,V.R., Am.J.Clin. Nutr., 1989,**49**,854
37. Levander,O.A., Sutherland,B., Morris,V.C., King,J.C., Am.J.Clin.Nutr., 1981, **34**, 2662
38. Palmquist,D.L., Moxon,A.L., Cantor,A.H., Fed.Proc., 1979,**38**,391
39. Chen,X., Yang,G., Chen,J., Chen,X., Wen,Z., Ge,K., Biol. Trace Elem.Res., 1980, **2**,91
40. Cheng,Y.D., Zhuang,G.S., Tan,M.G., Zhi,M., Zhou,W., Biol.Trace Elem.Res., 1990, **26-7**, 737
41. Salbe,A.D., Levander,O.A., J.Nutr., 1990,**120**,200 42. Chatt,A., Holzbecher,J., Katz,S.A., Biol.Trace Elem. Res., 1990,**25**,513

SELENIUM STATUS AND CARDIOVASCULAR RISK FACTORS IN POPULATIONS FROM DIFFERENT PORTUGUESE REGIONS

A.M. Viegas-Crespo*, M.L. Pavão**,V. Santos***, M.L.Cruz****, O.Paulo*,J. Leal*, N. Sarmento*, M.L. Monteiro*****, M.F. Amorim *****, M.J. Halpern***** and J. Néve******

* Dep. Zoology and Anthrop., Univ. Lisbon, Bloco C2 - 3º piso, 1700 Lisbon,Portugal; **Dep. Technological Sciences, Univ. Azores, 9500 - Ponta Delgada, Portugal;***Hosp. P. Delgada, 9500 - P. Delgada, Portugal; **** Hosp. S. de Magos, 2120 - Salvaterra de Magos, Portugal; *****Fac. Medicine, New Univ. Lisbon, 1100 Lisbon, Portugal; ****** Dep. of Pharmac. Org. Chem., Free Univ. Brussels, B-1050 Brussels, Belgium

1 INTRODUCTION

Selenium as a cofactor of glutathione peroxidase, which prevents lipid peroxidation in mammals [1], takes part in the direct protection of endothelial cells against reactive oxygen species that have been implicated in atherogenesis [2-4]; moreover, it is involved in the biosynthesis of arachidonic acid derivatives in platelets [5,6] and in the regulation of lipoprotein cholesterol metabolism in human beings and in animal models [7-10]. These aspects are relevant enough to conclude that low selenium status may be related to atherosclerosis and, consequently, to the occurrence of cardiovascular diseases [11].

Clinical studies showed a decrease in plasma selenium of patients with congestive cardiomyopathy and/or myocardial infarction [12,13]. A significant inverse correlation between plasma selenium and severity of coronary atherosclerosis was also reported in man [14]. However, prospective epidemiological studies on the relationship between selenium and cardiovascular disease are rather controversial [15,16]

The aim of this work was to compare the selenium status by determining serum levels of this element in inhabitants of two urban and one rural portuguese regions.The relationship between serum selenium levels and generally accepted cardiovascular risk factors was also an objective. In this context. serum selenium and serum lipid parameters (total cholesterol, HDL cholesterol, LDL cholesterol and triglycerides) were evaluated. Age and sex as well as alcohol and tobacco consumption were also considered.

2 SUBJECTS AND METHODS

2.1. Subjects

The studied groups consisted of 101 (39 women and 62 men), 98 (50 women and 48 men) and 35 (19 women and 16 men) volunteer portuguese subjects, aged 20 to 60 years and living in Lisbon-Mainland (urban region), Ponta Delgada - Azores´ Archipelago (urban region) and Salvaterra de Magos - Mainland (rural region), respectively.

The donors were non-alcoholic persons and they did not abuse drugs. Age and sex as well as the date of sampling were registered.The existence of chronic diseases and a history of any cardiovascular condition or stroke were also considered. The subjects were asked to begin to fast 12 h before blood sampling, which occurred in the morning and was carried out in 1994 from April to July.

2.2. Methods

2.2.1. Analytical procedures. Blood was collected in polyethylene tubes by venipuncture. Serum was removed after centrifugation without addition of anticoagulants and an aliquot kept frozen at -20° C until analysed for selenium.

HDL proteins were obtained by adding polyethylene glycol to fresh samples to precipitate other lipoproteins. Their cholesterol content as well as the serum total cholesterol were determined enzymatically by the cholesterol CHOD-PAP method (Boehringer, Mannheim, FRG). Serum triglycerides were determined enzimatically by the triglycerides GPO-PAP method (same manufacturer). LDL cholesterol concentration was calculated using the Friedewald formula.

Serum selenium was quantified by a direct electrothermal atomic absorption spectrometric procedure with Zeeman background correction [17]. Accuracy of the procedures was checked with standard reference material.

2.2.2. Statistics. Normality of the distribution was evaluated by the Kolmogorov-Smirnov test. Distribution was studied by drawing frequency polygons after division into class intervals.

Individual mean comparisons were tested for significance by the Student's t-test or by the Mann-Whitney test.

The correlations of serum selenium with age and lipid parameters were analysed by linear regression or correlation coeficient. For the discrete variables, smoking and alcohol consumptions associations with other parameters were made by the Student´s t-test.

3 RESULTS AND DISCUSSION

3.1. Results

3.1.1. Serum selenium concentrations - Intrapopulational and interpopulational differences. A significant difference ($p<0.01$) of average serum selenium concentrations was found between the two sexes in the Azores population. A less important variation ($p<0.05$) was also observed in the population of Lisbon (Table 1).

Women and men of the rural population from Salvaterra de Magos exhibited significant differences ($p<0.01$) in the average serum selenium concentration when compared with both the urban populations (Lisbon and Azores). An exception was found for the women of the Azores, which had a serum selenium concentration similar to the women of Salvaterra de Magos (Table1).

Table 1. *Serum selenium concentration ($\mu g\ l^{-1}$) of populations from Lisbon, Ponta Delgada - Azores and Salvaterra de Magos.*

Sex	Lisbon	P.Delgada-Azores	Salvaterra Magos
W	94.4±16.7	88.1±14.5	84.2±14.4
M	99.0±20.5	97.9±16.2	84.5±16.5

Values represent the mean ± S.D..
W - women (n = 39, Lisbon ; n = 50, P.Delgada - Azores; n = 19, Salvaterra);
M - men (n = 62, Lisbon; n = 48, P.Delgada - Azores; n = 16, Salvaterra).

For the azorean population, the results revealed an increase (p<0.02) of serum selenium levels with age, when considering all samples irrespective of sex (Figure 1).

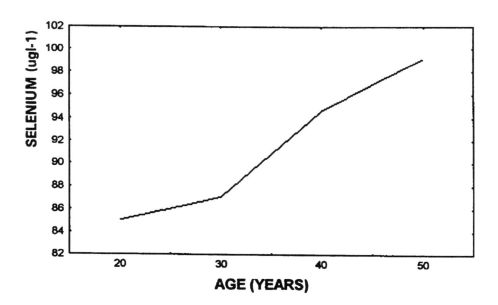

Figure 1. *Change of serum selenium levels with age of subjects from the Azorean population.*

3.1.2.Association between the serum selenium concentration and the serum lipid parameters. In order to study the relationship between the lipid parameters and serum selenium levels, the subjects were divided according to respective total cholesterol and triglycerides contents.The normolipidaemic group consisted of healthy subjects having serum total cholesterol and triglycerides concentrations < 200 mg dl -1;the hiperlipidaemic group consisted of subjects with impairment of lipid status, having the serum total cholesterol and /or triglycerides concentrations > 200 mg dl -1.The HDL cholesterol and the HDL cholesterol/total cholesterol ratio is within the normal range, for the normolipidaemic group, but this ratio is lower (p<0.01) for the hiperlipidaemic individuals when compared with normal values for the three populations (Tables 2,3 and 4).

Mean serum selenium was not significantly different for the two groups of normolipidaemic and hiperlipidaemic subjects in the two populations from Azores and Salvaterra de Magos (Tables 3 and 4).However, a weak increase (p<0.05) in that parameter was observed, for the hiperlipidaemic group of Lisbon, which exhibited also the highest values in cholesterol and triglycerides concentrations (Tables 2,3 and 4).

Table 2. *Lipid parameters and selenium concentration in serum of the subjects from Lisbon population.*

Parameter	Sex	Normolipidaemic	Hiperlipidaemic
Cholesterol	W	181.1±27.6	271.2±54.2*
(mg d l^{-1})	M	193.6±22.7	298.3±100.6*
HDL cholest.	W	56.1± 9.5	46.0±12.7**
(mg d l^{-1})	M	46.3±13.5	39.6±10.1**
HDL cholest./	W	31.4±6.0	17.0±5.8*
cholesterol,%	M	23.8±7.0	14.4±5.4*
LDL cholest.	W	117.2±20.0	175.5±54.4*
(mg d l^{-1})	M	118.6±20.9	182.7± 54.8*
Triglycerides	W	77.0±34.4	276.4±82.8*
(mg d l^{-1})	M	103.5±35.8	400.6±120.0*
Se	W	89.1±13.6	99.7±18.4**
(μg l^{-1})	M	94.7±23.2	103.3±17.0**

Values represent the mean ± S.D..
W - women (n=18 in normolipidaemic group; n=21 in hiperlipidaemic group);
M - men (n=28 in normolipidaemic group; n=34 in hiperlipidaemic group).
Asterisks denote the signifiance of the t test of differences between means for the hiperlipidaemic and nor-molipidaemic groups(*p<0.01;**p<0.05).

Table 3. *Lipid parameters and selenium concentration in serum of the subjects from the population of Ponta Delgada - Azores.*

Parameter	Sex	Normolipidaemic	Hiperlipidaemic
Cholesterol	W	175.2±20.1	243.8±27.7*
(mg d l^{-1})	M	174.4±26.4	237.6±44.2*
HDL cholest.	W	52.4±11.3	46.1± 7.3**
(mg d l^{-1})	M	37.5±10.8	40.7±11.4
HDL cholest./	W	29.9±7.1	19.2±2.8*
cholesterol,%	M	21.6±7.2	17.1±4.9*
LDL cholest.	W	108.2±22.8	170.2±33.0*
(mg d l^{-1})	M	114.0±27.2	156.0±42.0*
Triglycerides	W	82.3±39.0	165.6±91.9*
(mg d l^{-1})	M	114.0±32.3	255.0±83.9*
Se	W	85.6±15.2	93.1±12.9
(μg l^{-1})	M	98.4±16.0	98.1±16.3

Values represent the mean ± S.D..
W - women (n=38 in normolipidaemic group; n=12 in hiperlipidaemic group);
M - men (n=20 in normolipidaemic group; n=28 in hiperlipidaemic group).
Asterisks denote the significance of the t test of differences between means for the hiperlipidaemic and normolipidaemic groups(*p<0.01;**p<0.05).

Table 4. *Lipid parameters and selenium concentration in serum of the subjects from the population of Salvaterra de Magos.*

Parameter	Sex	Normolipidaemic	Hiperlipidaemic
Cholesterol	W	166.9±26.2	223.9±15.2
(mg d l^{-1})	M	173.1±21.4	204.1±30.6
HDL cholest.	W	53.1±17.0	43.1±4.0
(mg d l^{-1})	M	44.8±5.2	24.3+4.2*
HDL cholest./	W	32.4±10.1	19.3±2.3*
cholesterol,%	M	26.4±5.4	11.8±3.2*
LDL chol.	W	103.1±15.4	138.3+8.3*
(mg d l^{-1})	M	112.0±13.6	132.1±19.8
Triglycerides	W	109.9±25.8	110.0±37.6
(mg d l^{-1})	M	90.7±30.1	143.5+64.0
Se	W	82.6±14.3	84.7±14.4
(µg l^{-1})	M	83.8±17.4	87.1+15.1

Values represent the meanS.D..
W - women (n=10 in normolipidaemic group;n=9 in hiperlipidaemic group);
M-men (n = 13 in normolipidaemic group;n= 3 in hiperlipidaemic group).
Asterisks denote the significance of the t test of differences between means for the hiperlipidaemic and normolipidaemic groups(*p<0.01).

For the three populations no significant correlations were found between the several lipid parameters and the serum selenium concentrations,either considering the sexual difference or not.

3.1.3.Association between serum selenium concentration and drinking and smoking habits. Concerning alcohol consumption, no statistical difference in the serum selenium concentration was found among individuals for the three studied populations.

However, there is a strongly significant statistical difference (p<0.001) of serum selenium levels between male smokers (82.924.4gl-1,n=15) and non-smokers (102.114.0 gl-1,n=41) in the Lisbon population.

No differences were found in other populations, as well as between smoking and non-smoking women within the same population.

3.2. Discussion

3.2.1.Comparison of the obtained serum selenium concentrations with data from other portuguese regions and other european countries. The similarity of serum selenium concentrations in the urban populations of this study (one from the portuguese mainland and the other from the Azores' Archipelago) with data from fishing portuguese populations (104.2 21.3 µg l-1, n= 59 men - Câmara de Lobos - Madeira Island) [18] is observed.

The most striking result of this comparive study is the significantly lower mean serum selenium concentration of the rural population of Salvaterra de Magos when compared to the serum selenium concentration found in fishing [18] and in the two studied urban populations of the portuguese territory.This fact, added to the very low serum selenium concentration found in another rural population (59.8 17.0 µgl-1, n= 16 men - Curral das Freiras - Madeira Island) [18], suggests that the selenium status is related to the feeding habits, with the serum selenium concentration being inversely related to the consumption of animal proteins.

A deeper study on the diet and its nutrient composition in these populations is essential to answer these questions.

However, present data for the three portuguese populations are in the same range of values than those obtained in other countries of Europe [19,20]. Nevertheless, they seem to be higher than the values obtained in southern european countries, according to data observed in Greece [20] and Yugoslavia [21], including the ones from Barcelona - Spain [22], but they are similar to those found in some populations from Italy according to Cauwenbergh et al [20].

3.2.2. Association of the serum selenium concentration with other factors. The increase of selenium serum levels with age observed in the azorean population, taking into account the both sexes, is in accordance with results obtained by other authors for the same range of age [23,24]. However, conclusions about age-dependency are questionable, because the studied groups are sometimes poorly defined and the age-range is too small or too large [23].

The tendency observed in portuguese populations concerning the sexual differences in serum selenium levels agrees with data reported by some authors [23,24]. However, most of them have found no significant variations in selenium correlated with sex [23,24]. According to Robberecht et al [23] the race and hormonal status may jeopardize the conclusions.

The finding of no correlation between serum selenium and lipid parameters agrees with data of Crespo et al in normolipidaemic Portuguese individuals [25] and with the data of Bukkens et al in healthy Dutch subjects [26]. But it disagrees with results presented by Salonen et al. [27], who reported a weak positive correlation between serum selenium and HDL cholesterol in Eastern Finnish men.

The weak increase of serum selenium levels observed in most hiperlipidaemic individuals from the population of Lisbon had not been reported so far. So, a more detailed study about hiperlipidaemic subjects considering serum lipid parameters and antioxidant indicators, as well as nutritional, metabolic and genetic factors should be faced in further studies.

A significant difference between smokers and non smokers was observed in men from the population of Lisbon. This result agrees with the referred by some authors [26], but differs of data reported by others [23]. Probably lack of registration of type and amount of cigarettes smoked is partly responsible for the discrepancies in literature data [23].

3.2.3. Final remark.The present results encourage further investigations on the selenium dietary intake as well as on the environmental selenium in portuguese regions. On the other hand, a further study considering not only selenium, but also other parameters, namely indicators of oxidative stress and their relationship with cardiovascular risk factors should be made for the portuguese populations.

References

1. C. Little and P.J. O'Brien, *Biochem. Biophys. Res. Commun.*, 1968, **31**, 145.
2. K. Yagi, *Bioessays*, 1984, **19**, 58.
3. B. Hennig and C.K. Chow, *Free Radical Biol. & Med.*, 1988, **4**, 99.
4. P.V. Luoma, J. Stengard, H. Korpela, A. Rautio, E.A. Sotaniemi, E. Suvanto and J. Marniemi, *J. Internal. Medicine*, 1990, **227**, 287.
5. M.A. VanRij, C. Kirk, I. Wade, C. Thomson and M.F. Robinson, *Clin. Science*, 1987, **73**, 525.

6. M.J. Panham, E. Graf, E. Hoff and R. Niemann, *Agents and Actions*, 1987, **22**, 353
7. W.L. Stone, M.E. Stewart, C. Nicholas and S. Pavuluri, *Ann. Nutr. Metab.*, 1986, **30**, 94.
8. A.M. Crespo, Ph. D. Thesis, University of Lisbon, 1990.
9. P.V. Luoma, E.A. Sotaniemi, H. Korpela and J. Kumpulainen, *Res. Commun. Chem. Pathol. Pharmacol.*, 1984, **46** (3), 469.
10. M.E. Haberland, A.M. Fogelman and P.E. Edwards, *Proc. Natl. Acad. Sci. U.S.A.*, 1982, **79**, 1712.
11. J. Néve, *Path. Biol.*, 1989, **37** (10), 1102.
12. Ph. Auzépy, M. Blondeau, Ch. Richard, D. Pradeau, P. Thérond and T. Thuong, *Acta cardiologica*, 1987, **42** (3), 161.
13. O. Oster, M. Drexler, J. Schenk, W. Meinertz, W. Kasper, C.J. Schuster and W. Prellwitz, *Ann. Clin. Res.*, 1986, **18**, 36.
14. J.A. Moore, R. Noiva and I.C. Wells, *Clin. Chem.*, 1984, **30**, 1171.
15. J. Virtamo and J.K. Huttunen, *Ann. Clin. Res.*, 1985, **17**, 87.
16. J.T. Salonen, G. Afthan, J.K. Huttunen, J. Pikkarainen and P. Puska, *Lancet*, 1982, **2**, 175.
17. J. Néve, S. Chamart and L. Molle, "Trace Element Analytical Chemistry in Medicine and Biology", Vol. 4, Walter de Gruyter, Berlin-New York, 1987, p. 349.
18. A. M. Viegas-Crespo, I. Torres, M.L. Mira and J. Néve, "First Int. Meet. of Trace Elem. and Vit. in Medicine", Monastir, Tunisie, PA B12, 1995, p.60.
19. J. Neve, *J.Trace Elem. Electrolytes Health Dis.*, 1991, **5**, 1.
20. R. Van Cauwenbergh, H. Robberecht, H. Deelstra, D. Picramenos and A. Kostakopoulos, *J. Trace Elem. Electrolytes Health Dis.*, 1994, **8**, 99.
21. Z. Maksimovic, V. Jovic, I. Djujic and M. Rsumovic, *Environmental Geochem. Health*, 1992, **14**, 107.
22. F. Fernandez-Banares, C. Dolz, M.D. Mingorance, E. Cabré, M. Lachica, A. Abad-Lacruz, A. Gil, M. Esteve, J.J. Giné and M.A. Gassul, *Eur. J. Clin Nutr.*, 1990, **44**, 225.
23. H. Robberecht and H. Delstra, *J. Trace Elem. Electrolytes Health Dis.*, 1994, **8**, 129.
24. J. Versieck and R. Cornelis, " Trace Elements in Plasma or Serum" ,C.R.C. Press, Boca Raton,Florida, U. S., 1989.
25. A. M. Viegas-Crespo, J. Néve, M.L. Monteiro, M.F.Amorim, O. Paulo and M. J. Halpern, *J. Trace Elem. Electrolytes Health Dis.*, 1994, **8**, 119.
26. S.G.F. Bukkens, N. de Vos, F.J. Kok, E.G. Schouten, A.M. Bruijn and A. Hofman, *J. Am. Coll. Nutrition*, 1990, **9** (2), 128.
27. J.T. Salonen, R. Salonen, K. Kantola, M. Parviainen, G. Alfthan, P.H. Maenpaa, E. Taskinen and R. Rauramaa, *Atherosclerosis*,1988, **70**, 155.

ALTERATION IN PLASMA SELENOPROTEIN P LEVELS AFTER SUPPLEMENTATION WITH DIFFERENT FORMS OF ORAL SELENIUM IN HEALTHY MEN

M. Persson-Moschos*, G. Alfthan** and B. Åkesson*

*Department of Applied Nutrition and Food Chemistry, University of Lund, Lund, Sweden, and **National Public Health Institute, Helsinki, Finland

1 INTRODUCTION

The major functional forms of selenium in mammalian tissue are selenocysteine-containing proteins. Their concentrations in tissue are dependent on the dietary intake of selenium. In mammals at least eight such selenoproteins have been identified (Table 1). So far only few selenoproteins have been used as biochemical markers of selenium status, e.g. glutathione peroxidase activity in plasma, erythrocytes and platelets. Recently plasma glutathione peroxidase was also assayed using a radioimmunoassay[1].

Selenoprotein P (SeP) is the major selenoprotein in human plasma[2]. Its function is at present unknown, but it may act as a selenium carrier or as a free radical scavenger[3,4]. In contrast to most other selenoproteins it contains more than one selenocysteine residue in the polypeptide chain. Recently SeP was isolated from human plasma, and an immunoassay was developed[2,5]. The present communication reports preliminary results from studies of SeP levels in plasma from healthy men who were given different forms of selenium as oral supplements for several months.

2 SUBJECTS AND METHODS

As described in previous reports[6,7] the study group consisted of 45 Finnish men who participated in two supplementation trials (Trial I and Trial II). The first was performed in 1981 and the second one in 1987. In 1985 a nation-wide supplementation of fertilizers with selenium was started in Finland[8]. As a consequence, the mean dietary intake of adults increased from approx. 40 to 100 μg/d in 1987. The subjects were divided into one placebo group (n=14-20) and three treatment groups that received 200 μg selenium per day as selenium-enriched yeast (group 1, n=10), sodium selenate (group 2, n=9-10) and selenium-enriched wheat (Trial I) or sodium selenite (Trial II) (group 3, n=10). Plasma samples were obtained before the interventions and at 11 weeks after the start of the supplementation in both trials.

SeP levels were measured with a radioimmunoassay[2,5]. Its concentration was expressed in arbitrary units. The intra-assay coefficient of variation (c.v.) was 4.4 % and the inter-assay c.v. was 7.8%. Differences between SeP levels in different groups

Table 1. *Selenoproteins in mammals*

Cellular glutathione peroxidase (cGSHPx)
Extracellular glutathione peroxidase (eGSHPx)
Phospholipid hydroperoxide glutathione peroxidase (phGSHPx)
Gastrointestinal glutathione peroxidase (giGSHPx)
Type I 5'-iodothyronine deiodinase (5'-I-DI)
Selenoprotein P (SeP)
Selenoprotein W (SeW)
Mitochondrial capsule selenoprotein (MCS)

changes in SeP levels after selenium supplementation were assessed by ANOVA followed by Bonferroni's test and Duncan's test.

3 RESULTS

Previous studies showed that after supplementation with 200 µg selenium per day in Trial I plasma selenium concentrations increased more when selenium-rich wheat or yeast was used than when selenate was used, whereas selenate and wheat selenium were the most effective supplements to stimulate platelet glutathione peroxidase activity[6]. No significant changes in plasma glutathione peroxidase activity were observed. In the later supplementation trial[7] (Trial II) the same study group was studied again when the subjects had acquired a higher selenium status through country-wide selenium supplementation[8]. Plasma selenium increased markedly after supplementation with yeast, less so after selenite and not significantly after selenate. Selenate and selenite but not yeast resulted in increased platelet glutathione peroxidase activity, but no effects on plasma glutathione peroxidase activity were observed[8].

The new data on plasma SeP levels were expressed in two ways. In Figure 1 and

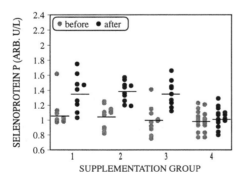 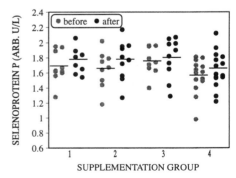

Figure 1 *Trial I: Selenoprotein P before and after selenium supplementation. 1 = yeast, 2 = selenate, 3 = wheat and 4 = placebo. Horizontal line indicates mean.*

Figure 2 *Trial II: Selenoprotein P before and after selenium supplementation. 1 = yeast, 2 = selenate, 3 = selenite and 4 = placebo. Horizontal line indicates mean.*

Figure 2 the results from Trial I and Trial II are given in arbitrary units relative to a serum standard. Furthermore, to facilitate comparison of SeP levels at different occasions its concentration for each individual prior to the first supplementation (when the subjects had low selenium status) was set to 1.00. In Trial I the relative SeP levels (mean (SD)) after supplementation with selenium as yeast (1.28 (0.19)), wheat (1.39 (0.19)) or selenate (1.33 (0.22)) were not significantly different among supplementation groups but differed from that in the placebo group (1.04 (0.09)), $p<0.05$ in Bonferroni's and Duncan's tests.

Before Trial II at high selenium status in 1987 the relative plasma SeP level had increased to 1.62 (0.32) (n=43). The SeP levels at 11 weeks after the start of the supplementation with yeast (1.74 (0.28)), selenate (1.71 (0.43)) and selenite (1.81 (0.41)) were not significantly different from each other or from that in the placebo group (1.62 (0.28)).

4 CONCLUDING REMARKS

The responses in SeP levels after supplementation with different forms of selenium are different compared with other indices of selenium status. After several months of selenium supplementation the different forms of selenium resulted in similar plasma SeP levels, although total plasma selenium varied markedly among the groups. The mean increase in plasma SeP of all supplemented subjects in Trial I was 33%, whereas the corresponding increase in Trial II was only 5%. The small increase in Trial II suggests that plasma SeP had reached a plateau as also recently reported by Hill et al[9]. The underlying mechanisms for varying responses in different parameters of selenium status remain to be established. Moreover, the time course of changes in SeP levels needs further study.

5 ACKNOWLEDGEMENTS

The study was supported by the Swedish Council for Forestry and Agricultural Research, the Påhlsson Foundation and the Swedish Nutrition Foundation.

6 REFERENCES

1. W. Huang and B. Åkesson B, *Clin. Chim. Acta*, 1993, **219**, 139.
2. B. Åkesson, T. Bellew and R. F. Burk, *Biochim. Biophys. Acta*, 1994, **1204**, 243.
3. R. F. Burk and K. E. Hill, *J. Nutr.*, 1994, **124**, 1891.
4. R. F. Burk, K. E. Hill, J. A. Awad, J. D. Morrow, T. Kato, K. A. Cockell and P. R. Lyons, *Hepatology*, 1995, **21**, 561.
5. M. Persson-Moschos, W. Huang, T. S. Srikumar, S. Lindeberg and B. Åkesson, *Analyst*, 1995, **120**, 833.
6. O. A. Levander, G. Alfthan, H. Arvilommi, C. G. Gref, J. K. Huttunen, M. Kataja, P. Koivistoinen and J. Pikkarainen, *Am. J. Clin. Nutr.* 1983, **37**, 887.

7. G. Alfthan, A. Aro, H. Arvilommi and J. K. Huttunen, *Am. J. Clin. Nutr.* 1991, **53**, 120.
8. P. Varo, G. Alfthan, P. Ekholm, A. Aro and P. Koivistoinen P, *Am. J. Clin. Nutr.*, 1988, **48**, 324.
9. K. E. Hill, Y. Xia, M. E. Boeglin, P. R. Lyons, B. Åkesson and R. F. Burk, *FASEB J.*, 1994, **8**, A436.

SELENIUM IN FOOD AND POPULATION OF SERBIA

I. S. Đujić

Center of Chemistry
Institute of Chemistry, Technology and Metallurgy
Njegoševa 12, Beograd

1 INTRODUCTION

Selenium (Se) is an essential trace elements for animals and humans, but not for plants. The quantity of Se in food and feed is strongly dependent on the concentration in plants, that is the main source of selenium, for humans and animals. The uneven distribution and availability of soils Se imply that different areas of the world are characterized, from the point of human and animal nutrition, as Se-deficient, Se-adequate or Se-toxic areas.

A geographical pattern of Se situation in Europe, showing Scandinavia as a natural low Se-area, while central Europe is balanced between deficiency and sufficiency. Results from southern Europe, as well as Serbia indicaties that Se status ranges between adequate and inadequate (1,2).

Main biologically active components with Se in their active sites are enzymes glutathion peroxidase (GSHPx) (3), phospholipid hydroperoxide glutathion peroxidase (PLGSHPx) (4) and iodothyronine 5'-deiodinase (ID) (5) and selenoprotein P (6,7,8). Beside these compounds, many other Se-proteins have been identified in animal tissues. Antioxidant function is the major role of Se in human and animals, because Se is a required factor for antioxidant enzymes that control tissue peroxide levels in cells by degrading hydroperoxide.

The tissue concentrations of the Se-enzymes appear to be homeostatically controlled (9). Intakes of Se above that which is required for optimal growth, may slightly or moderately increase activities of Se-enzymes. Therefore, elevated concentrations of tissue Se found after supplementation, especially with Se-methionine, is a consequence of nonspecific Se incorporation into other proteins. Dietary Se-deficiency causes non-uniform decrease of Se-dependent proteins. Factors that effect the fall of some Se-proteins with significant fraction of the Se (eg. cGSHPx in liver) under deficiency conditions probably serve to increase the Se available for synthesis of selenoproteins that are more important to the survival of the animal (10).

Numerous studies with animals have demonstrated that Se at nutritional levels is a potent inhibitor of virally- and chemically induced tumorogenesis (11,12). Some epidemiological results (13,14,15,16,17,18) have demonstrated that in some cases strong, significant inverse association exists between Se levels in blood or daily intake by food and risk of cardiovascular death, cancer and other degenerative diseases (19,20,21). Another study indicate that supplementation with Se may diminish the damage induced by peroxidation in patients with increased requirement for antioxidant compounds (22).

Because Se may be considered only at nutritional levels as a prophylactic agent, knowledge on Se content in human diet and biochemical values are important (23). Only in this case we can, if it is necessary correct intake of Se and provide recommended daily intakes (RDA).

In order to obtain more information on safe and adequate dietary Se levels in Serbia, we have studied the Se levels in food, average intake by food, and Se status in population.

2 EXPERIMENTAL

2.1 Food collection and Se determination

Samples of wheat, wheat flour, potatoes, milk and pork were collected according to the sampling protocol designed by the FAO European Research Network on Trace Elements (24) during spring and winter of 1991 in Serbia (Table 1) and then pooled, to obtain nationally representative samples. The content of Se in pooled samples was, after wet digestion, determined by hydride generation AAS (25). For analytical quality control, the following certified reference materials were used: NBS SRM-1567 (for wheat), IAEA A-11 and NBS SRM-1549 (for milk), ARC/Cl secondary reference material (for potatoes) and IAEA H-4 (for animal muscle).

Table 1 *Sampling protocol for nationally representative food products collected during spring and winter of 1991 in Serbia*

	Vojvodina	*Central Serbia*	*Kosovo*
Whole wheat			
Production, %	55	38	7
No of silo-samples	18	14	2
Wheat flour			
No of mill-samples	108	92	4
No of supply-samples	52	43	7
Potatoes			
Production,%	26	67	7
No of samples	14	56	6
Milk			
Production,%	17	68	15
No of dairy samples	60	150	30
Meat			
Production,%	46	53	1
No of slaughterhouse samples	96	56	-

Other food item samples that do not represent entire national production were also collected during 1991 in Serbia and Se content determined in them.

On the basis of mentioned data intake of Se by food in Serbia was calculated, using household budget survey for 1991 (26) as source of data for daily weight of food eaten. Obtained value for daily dietary Se intake in Serbia was then diminished by 10 % for average cooking loss.

Average value for cooking loss of 10% was obtained on the bases of measurement losses of Se during bread beaking, meat roasting, mixed vegetable cooking, potatos frying and 3 national cooked dishes (sarma, musaka and djuvec).

2.2 Blood and scalp hair collection and Se determination

Samples of venous blood and hair were collected from healthy individuals originally enrolled in epidemiological programs and blood donors. Four ml of venous blood were taken by syringe heparinized with 0.1 ml of heparin and stored at -20^0C for Se determination in whole blood, or immediately centrifuged for 5 min. at 3500 r/min. to separate blood plasma. The erythrocytes were washed 3 time's with 0.9 % NaCl and then resuspended in it (1:1). Plasma and suspension of erythrocytes were stored at -20^0C. Scalp Hair was clipped with stainless steel scissors and collected in polyethylene bags. To remove external adherent contaminations, the hair was washed following IAEA procedure (27), and stored after drying.

All stored samples were wet ashed with concentrated $HNO_3/HClO_4$ mixture in borsilicate test tubes placed in an aluminium block heater. The digested material, after addition of 6M HCl, was heated at 80^0C for 30 min. and transferred to the volumetric flask. The Se concentrations were than determined by gaseous hydride generation AAS, using a Perkin Elmer 5000 atomic absorption spectrometer equipped with MHS-10 vapour generator accessory.

Analytical procedures were validated using certified reference material IAEA A-13 - Animal blood, Contox 0148 level I - Human serum and GBW 09101 - Chinese hair).

3 RESULTS AND DISCUSION

3.1 Se in food

The results obtained for Se concentration in nationally representative food products from Serbia collected 1991 (Table 2) were same or lower than in other countries (28).

3.1.1 Whole wheat . Comparison of values obtained for Se in wheat from Sweden with our value indicate that similar naturally low level of Se existed in Sweden. Addition of Se to multimineral fertilizers increased significantly Se in Finnish wheat after 1985 (Table 2).

3.1.2 Wheat flour. The value for Se in wheat flour, an important source of trace elements, in our country was very low, lower than in all countries except Sweden that has a similar value (Table 2.). The high Se content in samples from Finland from 1984 and Switzerland was due to mixing of 30% of imported high-Se wheat from North America or due to Se fertilization of Finish soils after 1984.

3.1.3 Potatoes. Variation in contents of Se in potato samples among countries are less significant (Table 2.). Se content in Finnish potatoes, which was similar to those in other countries, in the 1984 crop has increased over tenfold due to Se fertilization or, in the case of Denmark and Netherlands, due to import of potatoes.

3.1.4. Milk. In general, variation in the content of Se in milk was relatively small among the countries (Table 2). The Se content in Serbian milk is higher than in Finnish for 1984, but does not represent only national production, because some milk was imported from other countries. Se fertilization increased the selenium content of Finnish milk from 80 µg/kg to 300 µg/kg.

3.1.5 Pork. Se content was very similar in the pork samples of all countries (Table 2.). Strong increase in the Se content of Finnish pork later on was due to Se fertilization of Finnish soils.

Table 2. *Contents of Se (µg/kg) in food products from different countries on a dry weight basis obtained after 1985.*

Country	Wheat	Wheat Flour	Potatoes	Milk	Pork
Finland	9*	72*	7*	75*	409*
	174	139	88	305	706
Austria	34	25	-	-	-
Sweden	26	18	9	115	371
Scotland	23	-	6	115	-
FRG	30	25	-	122	-
France	-	-	-	111	379
Switzerland	23	213	-	-	-
Turkey	72	-	-	-	-
Norway	-	-	6	-	-
Denmark	-	-	39	-	-
Netherlands	-	-	35	-	-
Serbia	23	17	8	98	357

* Data for 1984

Values obtained for Se in pooled samples of food items collected during 1991 that do not represent entire national productions are presented in Table 3. They indicate that all analyzed food items are more or less Se deficient (29).

3.2 Dietary Se intake

Data obtained for daily weight of food eaten in Serbian diet by household budget survey in 1991 were taken by calculating average daily intake of Se and are presented in Table 3.

Average daily dietary intake of Se by analyzed food groups in Serbia obtained on the basis of house hold budget survey for 1991 were as follows: dairy products -- 8.07 µg (27.15 %), meat and fish (without meat products) -- 12.46 µg (41.92 %), cereals -- 7.26 µg (24.43 %), vegetable -- 0.85 µg (2.86 %), fruits -- 0.11 µg (0.37 %) and alcoholic beverage -- 0.87 µg (2.93 %).

Contribution of analyzed food groups in dietary Se intake in Serbia was different from other countries. Main sources of Se in them are cereals (about 50 %), meat and fish (about 40 %) and dairy products (about 6 %). Participation of other food groups is negligible. In our case, participation of cereals in dietary Se intake is low, only about 25 %, although the quantity of consumed cereals in Serbia is higher than in most other countries.

Table 3. *Average daily intake of selenium by food in Serbian diet*

Food items	Weight of food eaten (g/day)	Se content, µg/kg	Intake of Se (µg/person per day)
Dairy products			
Milk	270.4	12.2	3.29
White cheese	27.9	52.5	1.46
Processed cheese	3.6	78.2	0.28
Egg	22.2	138.0	3.04
Meat & fish			
Pork	48.2	85.6	4.12
Beef	26.8	92.0	2.47
Chicken	32.1	122.4	3.93
Lamb	8.5	86.2	0.73
Fresh water fish	8.8	90.3	0.79
Sea fish	1.9	221.4	0.42
Cereals			
Bread	221.3	16.4	3.63
Flour	194.8	17.0	3.31
Corn	24.1	13.5	0.32
Vegetables			
Potatoes	92.1	0.1	<0.01
Beans	17.8	0.46	<0.01
Onion	4.7	13.7	0.06
Carrots	9.3	14.4	0.13
Cabbage	41.6	12.3	0.51
Tomatoes	35.1	2.5	0.09
Paprika	29.9	1.2	0.04
Fruits			
Apples	38.6	1.2	0.05
Pears	9.9	2.3	0.02
Melons	23.6	1.1	0.03
Grapes	9.4	0.9	<0.01
Alcoholic Beverage			
Beer	48.2	3.2	0.15
Wine	15.1	1.4	0.02
Brandy	15.3	52.4	0.80

Total daily intake of selenium in Serbia obtained for 1991 by analyzed food items was 29.72 µg/day, without correction for cooking loss. With this correction daily intake of selenium by food was 26.75 µg/day. Although we did not analyze all of the most consumed food items that may be significant sources of dietary selenium in the Serbian diet, our results indicate that average daily intake was very low compared with the US

RDA, as well as other countries (Table 4), and similar to those found in the Finnish diets before soil fertilization with selenium (30).

Table 4. *Average dietary intake of selenium in various countries expressed as a percentage of the relevant RDA* (1) for adults (0.07 mg/day)*

Country	Study group(s)	Sampling methods	No of samples**	Se
Brazil	Low social class, Manaus	DD	20	85
Iran	Urban & Rural	DR	27	85
Italy	Industrial area, traditional diet & high intake of seafood	DD	35	70
Japan	Typical (but different) diet	DD	5	190
Spain	Urban, higher & lower middle class	DD	30	85
Sudan	Urban, lower middle class	DD	20	160
Thailand	Rural, different nutritional habits	DD	20	70
Turkey	Rural	DD	16	45
USA	General US population (composite total diets)	MB	5	140

*DD = duplicate diets; DR = Dietary records; MB = Market basket
**No. of total diet samples analyzed

3.3 Se in population

Results of our measurements obtained for Se in blood, plasma, erythrocytes and scalp hair of adult healthy population from Serbia during last 8 years are presented in Table 5.

Table 5. *Se levels in blood, plasma, erythrocytes and scalp hair of Serbian population (1987-1994)*

Sample	n	Se, µg/l
Whole blood	462	58.2 ± 18.3
Plasma	1456	52.4 ± 12.9
Erythrocytes	1228	61.8 ± 19.6
Scalp hair	472	98.0 ± 15.7

The mean value determined for Se in whole blood, plasma, erythrocytes and scalp hair in our population are lower than the values reported in other countries (Table 6).

A high correlation was found between Se levels in blood and scalp hair ($r = 0.87$, $p<0.001$), plasma and scalp hair ($r = 0.82$, $p<0.001$) and erythrocytes and scalp hair ($r = 0.77$, $p<0.001$).

Data obtained for blood and scalp hair Se indicates that low dietary Se intakes are reflected on Se level in our population. Data given for a few selected countries by Gissel-Nielsen (45) indicate a similar correlation between the Se-intake and blood Se.

Our studies of the changes induced by oxidative stress in Se and other trace elements content, as well as GSHPx and other antioxidant enzymes activity (45,46,47) indicate that oxidative stress may induce significant changes in Se level and tissue distribution. Data obtained by supplementation with Se-yeast on rats exposed to oxidative stress (48,49,50) showed that increased Se intake in our conditions diminish changes in Se and other essential trace elements in tissues, as well as in antioxidant enzymes activity.

Table 6 *Se concentrations in healthy population from different countries*

Country	*n*	*Mean ± SD*	*Reference*
Whole blood			
USA	210	206	31
Belgium (Ghent)	109	122 ± 17	32
Finland (Hartola)	109	115 ± 43	33
Greece		150 ± 50	34
Sweden	104	148 ± 2	35
Italy	40	70 - 114	36
Serbia	462	58 ± 18	
Plasma			
USA (Georgia)	206	104 ± 21	37
USA (Florida)	212	131 ± 22	38
Belgium (Ghent)	110	97 ± 12	32
Finland	6	82 ± 6	39
France (Strasburg)	3500	75 ± 15	40
Sweden	104	106 ± 2	35
Italy	4201	87 - 93	41
Venezuela (Portugesa)	49	207	42
Serbia	1456	52 ± 13	
Erythrocytes			
France (Bordeaoux)	57	108 ± 20	43
Venezuela	49	360	42
Serbia (Portugesa)	1228	62 ± 20	
Scalp hair			
USA		493	44
China (Endemic area)		64.0	44
China (Nonendemic area)		107	44
Venezuela (Portugesa)	49	1460	42
Serbia	472	98 ± 16	

Considering our data and those reported for Se in food, diet and population groups in other countries it appears that insufficient Se concentration in food, from the point of human and animal are possible to correct by Se addition. As use of supplements with Se to increase Se intake is only a partial solution of the Se deficiency problem in Serbia, we have concluded to test, in a series of experiments, different ways of Se supplementation of field crops in Serbia. All available data indicates that well-defined Se supplementation of crops is a cheap, sure and easy way of ensuring a desirable Se intake by animals and humans (1).

References

1. G. Giessel Nielsen, Proc. of *STDA's Fifth International Symposium*, 1994, 103.
2. I.Đujić, V. Đermanović, M. Milovac, M. Ševaljević, Proc. *I Regional Symposium: Chemistry and the Environment*, 1995 (in press)
3. Chada, C Whitney and P. E. Newburger, *Blood*, 1989, **74**, 2535.
4. Q F. Ursini, Proc. of *International Symp. on Selenium*, Belgrade, 1991, 12.
5. J. L. Leonard and T. J. Visser In: *Thyroid Hormone Metabolism* (G. Hennemann ed.). Marcel Dekker, Neww York and London, 1986, 189.
6. Burk and K. E. Hill, *Biol. Trace Elem. Res.*, 1992, **33**, 151.
7. Hill, R. S. Lloyd and R. F. Burk, , Proc. *Natl Acad Sci U S A*, 1993, **90**, 537.
8. R. F. Burk and K. E. Hill, *J. Nutr.*, 1994, **124**, 1891.
9. C. Michiels, M. Raes, O. Toussaint and J. Remacle, *Free Radic. Biol. Med.*, 1994, **17**, 235.
10. R. F. Burk and K. E. Hill, *Annu. Rev. Nutr.*, 1993, **13**, 65.
11. Shuamberger, *Mutat. Res.*, 1985, **154**, 29.
12. Fishbein, *Arch Geschwulstforsch*, 1986, **56**, 53-78.
13. Salonen, G. Alfthan, J. K. Huttunen, J. Pikkarainen and P. Puska, *Lancet*, 1982, **2**, 175.
14. Salonen, G. Alfthan, J. K. Huttunen and P. Puska, *Am. J. Epidemiol.*, 1984, **120**, 342.
15. Willet, B. F. Polk and J. S. Morris, *Lancet*, 1983, **2**, 130.
16. Birt, *Magnesium*, 1989, **8**, 17.
17. Knekt, A. Aromaa, J Maatela, G. Alfthan, R. K. Aaran, M. Hakama, T. Hakulinen, R. Peto and L. Teppo, *J.Natl. Cancer Inst.*, 1990, **82**, 86.
18. Z. Pavlowicz, B. A. Zachara, U. Trafikowska, A. Maciag, E. Marchaluk and A. Nowicki, *J. Trace Elem. Electrolites Health Dis.*, 1991, **3**, 275.
19. U. Tarp, *Dan. Med. Bull.*, 1994, **41**, 264.
20. G.J. Beckett, F.E. Peterson, K. Choudhury, P. W. H. Rae, F. Nicol, P. S-C. Wu, A. D. Toft and A. F. Smith, *J. Trace Elem. Electrolites Health Dis.*, 1991, **5**, 265.
21. A. Favier, C. Sappey, P. Leclerc, P. Faure and M. Micoud, *Chem. Biol. Interact.*, 1994, **91**, 165.
22. H. B. Stahelin, *Support Care Cancer*, 1993, **6**, 295.
23. Yang, R. Zhou, S. Yin, L. Gu, B. Yan, Y. Liu and X. Li, *J. Trace Elem. Electrolites Health Dis.*, 1989, **3**, 77.
24. J. Kumpulainen, *Protocol for the 1986-1988 activities of the FAO European Cooperative Network on trace elements Sub-network E: Trace Element Status in Food*. Agricultural Research Center of Finland, Jokoinonen, Finland, 1985, 30.
25. J. Kumpulainen and M. Paakki, *Fresenius Z Anal Chem.* 1987, **326**,684.
26. Anketa o utrosenoj kolicini artikala licne potrosnje u domacinstvima u SRJ u 1991 godini, Savezni zavod za statistiku, (Survey of the amount of food items consumed in householg in FRJ in the year 1991, Federal Bureau of Statistics), Beograd,1992, 85.
27. IAEA, *Scientific Committee Report on Food Sampling*, IAEA-82-2, IAEA, Wiena, 1989, 9.
28. J. Kumpulainen, *Report of the 1989 Consultation of the European Cooperative Research Network on Trace Elements*, Lausanne, Switzerland, 1989, VI - 19.
29. O. A. Levander , B. Sutherland, B. A. Morris, J. C. King, *Am. J. Clin. Nutr.* 1981, **34**, 2662

30. J. Kumpulainen, M. Sinisalo, M. Paakkaki and R. Tahvonen, *Kemia-Kemi* 1987,**14**, 10.
31. W. H. Allaway, J. Kubita, F. Losee and M. Roth, *Arch. Environ. Health.*, 1968, **16**,342.
32. M. Verlinden, M. Van Sprudel, J. C. Van der Auvera and W. Eylenbosch, *Biol. Trace Elem. Res.*, 1983, **5**, 91.
33. M. Tolonen, M. Halne and S. Sarna, *Biol. Trace Elem. Res.*, 1985, **7**, 161.
34. T. Paradellis, *Eur. J. Nucl. Med.*, 1977, **2**, 277.
35. G. Michaelson, B. Berne, B. Carlmark and A. Strand, *Acta Derm. Venerol. (Stickholm)*, 1989, **69**, 29.
36. G. F. Clementi, L. C. Rossi and G. P. Santaroni, *J. Radioanal. Chem.*, 1977,**37**, 549.
37. P. A. McAdam, D. K. Smith, E. B. Feldman and C. Holmes, *Biol. Trace Elem. Res.*, 1984, **6**, 3.
38. P. A. Pella and R. C. Dobbyn, *Anal. Chem.*, 1988, **60**, 684.
39. M. Mutanen, *Inter. J. Vitam. Nutr. Res.*, 1986, **56**, 297.
40. E. A. Mailer, M. L. Sargentini-Maier, F. Rastegar, C. Christophe, C. Ruch, R. Heimburger and M. J. F. Leroy, *Fresenius Z. Anal. Chem.*, 1988, **331**, 58.
41. G. Morisi and M. Patriarca, *Acta Clinica Hungarica*, 1991, **128**, 581.
42. Brätter, V. E. Negretti de Brätter, W. G. Jaffé and H. Mendez Castellano, *J. Trace Elem. Electrolites health Dis.*, 1991, **5**, 269.
43. M. Simonoff, C. Conri, B. Fleury, B. Berden, P. Moretto, G. Ducloux and Y. Liabador, *Trace Elem. Med.*, 1988, **5**, 64.
44. D. Combs and S. B. Combs, *The Role of Selenium in Nutrition*, Academic, Orlando, Fl, 1986.
45. Djujić, O. Jozanov-Stankov, M. Mandić, M. Demajo, M. Vrvić: *Biol. Trace Elem. Res.*, 1992, **33**, 197.
46. Maksimović, I. Djujić, M. Ršumović, V. Jović, *Biol. Trace Elem. Res.*, 1992, **33**, 187.
47. Maksimović, I. Djujić, V. Jović, M. Ršumović, *Bulleten T. CV de l'Academie Serbe des Sciences et des Arts*, Sciences naturelles, 1992, **33**, 65.
48. Djujić, M. Demajo, M. Mandić, M. Spasić, Z. Sajčić, *Anticancerogenesis and Radiation Protection* 2, Ed. M. Simic, Plenum Press, New York 1991, 323.
49. Djujić, J. Vučetić, V. Matić, V. Milić, M. Vrvić, *Trace Elements in Health and Disease*, Eds. G.T. Yureger, O. Donna & L. Kayrin, Adana, 1991, 579.
50. Demajo, I. Djujić, *Low level Radiation Acheivements, Concerns and Future Aspects*, Eds.D.Horvat & P. Stegnar, Ljubljana,1991, 40.

SELENIUM LEVELS IN LEAD EXPOSED WORKERS

B. Giray, A. Gürbay, N. Basaran, F. Hincal

Hacettepe University, Faculty of Pharmacy
Department of Pharmaceutical Toxicology, Ankara, Turkey

1 INTRODUCTION

Selenium has been suggested to interact with heavy metals by forming chemical complexes and frequentlyto prevent their toxicity [1-4]. Although the mechanism, has not been clearly identified, lead is also considered to be one of these interacting metals. As in the case of cadmium or mercury, formation of a lead- selenium complex, an increase in selenium excretion and/or the antagonistic effects of the two metals on each other have been suggested by some investigators [4,5]. Very few human data are available about possible interaction between selenium and lead [6] and this study has been undertaken to investigate this relationship in subjects heavily exposed to lead.

2 SUBJECTS AND METHOD

2.1 Subjects

The study group was composed of 20 male, lead- exposed workers from a storagebattery plant. They were aged 22-55 years (mean age 33±9), and mean exposure period was 6±5 years. 20 healthy male subjects, with no history of occupational lead exposure and matched for age were also examined for their lead and selenium status as a control group. The dietary and smoking habits of both groups were similar.

2.2 Sampling and Analytical Methods

Blood samples were collected in the morning and blood lead levels measured by AAS (Hitachi model 28100) [7]. Serum was obtained after centrifugation and kept at - 35 °C until measurements. All samples were analyzed for their selenium content by a spectrofluorometric method as described by Lalonde et al [8], using a Hitachi model 650-40 spectrofluorometer.

3 RESULTS

As seen in Table 1, the mean blood lead level of exposed workers was much higher, 74 ± 5 µg/dl, than in controls, 16 ± 1.7 µg/dl, suggesting a heavy exposure of lead. However, mean serum selenium levels were not found to be different in exposed group (76 µg/L) and controls (78 µg/L), and there was no correlation between neither the mean nor the individual levels of selenium and lead.

Table 1 Serum Selenium and Blood Lead Levels in Lead Exposed Workers

	Selenium Conc (µg/L)	Blood Lead Conc (µg/dl)
Control	78 ± 13	16 ± 1.7
Workers	76 ± 13	74 ± 5.0 *

Blood lead concentration of the workers is statistically significantly higher compared with that of the control group * p<0.001

4 DISCUSSION

Lead and selenium have long been considered environmental pollutants, although selenium has also been shown to be an essential element in human and animal nutrition. Even though some heavy metals like mercury, cadmium and arsenic have been shown to interact with selenium and reduce their toxicity [3], studies on lead and selenium interaction are limited and the underlying mechanism is not clear. The protective effect of selenium against mercury toxicity and vice versa in a number of different organisms [2,3] has long been recognized, but the exact mechanisms of interaction between mercury and selenium are not well understood either. Redistribution of mercury in the presence of selenium, competition for binding sites between mercury and selenium, formation of a mercury - selenium complex, conversion of toxic forms of mercury to other forms, and prevention of oxidative damage induced by mercury possible due to an inhibitory effect on glutathione peroxidase activity are the possible mechanisms being considered [3,9]. Similar mechanisms have also been discussed for lead and selenium interaction [4], and although they were observed to be antagonistic to each other in animal experiments, the beneficial effects of selenium against the toxic effects of lead seems to be limited [1,10]. Among the proposed mechanisms, protection against oxidative damage may deserve much attention due to the essential role of selenium as an intrinsic component of the antioxidant defence system of the cells. In fact, Khan et al [11] reported that toxic levels of both lead and selenium had profound effect on the concentration of iron and copper in tissues of chicks, and the mortality due to concomitant exposure to lead and selenium was lowered by the presence of vitamin E, in the diet. Earlier studies of Levander et al [10] also suggest the protective roles of natural antioxidants, selenium and/or vitamin E, on the toxicity of lead.

However, there is little human data on the interaction or the relation of these two elements. Gustafson's data [6] indicate a minor interaction, in humans, between moderate occupational lead exposure and selenium status. Our data, however, does not support their results indicating that there is no difference between the selenium status of subjects who are heavily exposed to lead and those with no history of exposure. Thus, our data might suggest that lead exposure does not affect the bioavailability and/or utilization or elimination of selenium. However, animal experiments focusing on the methodological aspects of possible interaction among selenium and lead are needed in order to allow meaningful planning of possible further human studies.

References

1. M. Cikrt and V. Bencko, *Toxicol. Letter*, 1989, **48**, 159.
2. C. Leonzio, S. Focardi and C. Fossi, *Sci. Total Environ.*, 1992, **119**, 77.
3. M. Lourdes, A. Cuvin- Aralar and R. W. Furness, *Ecotoxicol. Environ. Safety*, 1991, **21**, 348.
4. S. C. Rastogi, J. Clausen and K. C. Srivastava, *Toxicology*, 1976, **6**, 377.
5. F. L. Cerklewski and R. M. Forbes, *J. Nutr.*, 1976, **106**, 778.
6. A. Gustafson, A. Schütz, P. Andersson and S. Skerfving, *Sci. Total Environ*, 1987, **66**, 39.
7. D. J. Hodges and D. Skelding, *Analyst*, 1983, **108**, 813.
8. L. Lalonde, Y. Jean, K. D. Roberts; A. Chapdelaine and G. Bleau, *Clin. Chem.*, 1982, **28**, 172.
9. J. Parizek and 1. Ostadalova, *Experientia* 1967, **23**, 142.
10. 0. A. Levander, V. C. Morris and R. J. Ferretti, *J. Nutr.*, 1977, **107**, 378.
11. M. Z. Khan and K. Markiewicz, *J. Vet. Med.*, 1993, **40**, 652.

Vitamins and Sesame Seed Lignans in Foods and Nutrition

EFFECTS OF PROCESSING, PACKAGING AND STORAGE ON THE LEVELS OF VITAMINS IN FOOD AND DIETS.

Margareta Hägg

Agricultural Research Centre of Finland, Laboratory of Food Chemistry
FIN-31600 Jokioinen, Finland

1 INTRODUCTION

Adequate levels of vitamin intakes are normally supplied by a balanced diet. Nutritional changes takes place during processing, the extent of which varies with the type of food and the process employed. Among the variables requiring consideration are the time period and temperature during processing and storage, the concentration and temperature dependences of the degradation reaction, vitamin concentation, enzyme activity in the food system, pH, concentration of oxygen, metal ions and various reducing and oxidizing agents, stability of the various forms of the vitamin, water activity in the food system. Chichester[1] has summarized the effects of pH, air, light and heat on vitamins. Vitamin C is unstable at pH 7, in alkaline solutions, in air, light and heat and is thus easily oxidized. Light also affects many other vitamins, such as vitamin A, D, B_1, B_2, B_6, B_{12}, and E. Niacin is the most stable of all vitamins and remains stable in acid and alkaline solutions, at pH 7 and in air, light and heat. Vitamin B_6 is also fairly stable, but unstable in light and heat.

2 PROCESSING

2.1 Canning

As citrus fruits are good sources of vitamin C, considerable research has been carried out on the retention of vitamins during canning. All investigations have reported high retentions during canning of grapefruit juice; 97-99 %[2]. Vitamin C contents of citrus products appear to be marginally affected by canning.

In garden peas canning results in losses of at least 50 % of the ascorbic acid contents at harvest on dry weight basis, although these were partly compensated by moisture losses during canning. Ascorbic acid is also lost during the reheating of canned garden peas.[3]

Canned mango slices were studied by Cano and de Ancos[4]. Thermal treatment during the canning process caused the degradation of xanthophyll carotenoids. After the canning, ß-carotene is the major carotenoid left, being approximately 90 % of the remaining total carotenoid concentration. Canned kiwi fruit slices have also been studied by Cano and Marin[5]. ß-carotene contents remained unchanged during the canning process. The chlorophyll a and b were distroyed during the canning process, however.

2.2. Freezing

Freezing processes include prefreezing treatments, freezing, frozen storage and thawing. Losses during freezing can be due to physical separation, leaching and chemical degradation. Most vegetables are blanched prior to freezing in order to inactivate the enzyme which would cause changes in sensory properties and nutritional value during freezing. According to studies performed in Campden, England[3], blanching and freezing garden peas caused about a 33 % loss of ascorbic acid, but the loss was likely due to blanching, not freezing.

After two years of frozen storage, strawberries retained 66 ± 3.9 % of the vitamin C contents[6], but the 9 individual samples studies had different contents, ranging between 58.6 % and 71.4 %. Black currant stored frozen for one year retained about 72 % of their vitamin C contents[6]. The individual samples (N=5) differed from 64.8 % to 77.3 %.

Excellent thiamine retention was found in frozen bread. About 100 % was retained in fresh and frozen French breads calculated on a dry weight basis, while mixed wheat and rye bread retained 95 %, coffee breads 92 % and rolls 100 %[7]. In addition, riboflavin retention was found to be very good in frozen breads, being 92 % in French bread, 97 % in mixed wheat and rye bread, 96 % in coffee bread and 92 % in rolls.[7]

The vitamin B_6 contents of food decrease substantially during frozen storage, according to a 1994 investigation[8]. Storage at -18 °C for 5 months decreased the vitamin B_6 contents of vegetables about 20-40 %, and by about 60 % in beef and fish.

Cano and de Ancos[4] have also determined the carotenoids in frozen, raw and canned mango. Extracts from raw and frozen mango slices showed the same qualitative carotenoid composition, but significant quantity differences. Much lower carotenoid concentrations were found in frozen than in raw mango slices. Frozen kiwi slices stored at -18 °C for 6 months also showed a pigment pattern similar to that of the fresh kiwi fruit, but in much lower concentrations[5].

2.3 Cooking

A recent study on the effects of preparation procedures and packaging on the nutrient retention in different vegetables obtained the following results. The three different potato cultivars, Van Gogh, Bintje and Nicola were investigated. The potato samples were hand-peeled and washed in different washing solutions. The solutions were plain water, or water containing different amount of citric acid, ascorbic acid, sodiumbenzoate and 4-hexylresorcinol. Samples were vacuum-packed in 20 % CO_2 and 80 % N_2. All trials showed that vitamin C content was destroyed during cooking and further when the potatoes were stored hot for one hour (Figure 1). On average cooking destroyed about 30 % of the vitamin C content and a further 10 % was lost when potatoes were kept hot. These results are from a one day storage of packaged, prepeeled potatoes. Solms and Genner-Titzmann[9] also found that prepeeled, cooked potatoes retained about 76 % of the vitamin C content and unpeeled potaoes retained about 89 %. The corresponding retention rate for thiamine was 100 %. After cooking, unpeeled potatoes retained 85 % of their riboflavin contents and peeled potaoes 63 %. Wills and Sutkilucksanavanish[10] also found retention rates of about 72 % of the vitamin C contents and 100 % of the riboflavin contents.

The vitamin C content of fresh, steamed broccoli decreased by 6 % and after storage for 30 minutes at 63 °C by 39 %, according to Carlson and Tabacchi[11]. Vitamin C content of steamed, frozen broccoli decreased by 17 % and 29 % after a 30 minutes storage at 63 °C, resulting in a similar total loss for fresh and frozen broccoli.

Awonorin and Ayoade[12] investigated thiamine degradation in minced meat cooked at different temperatures. Loss of thiamine was lower when meat was cooked at 70 °C than at 90 °C. When cooked at 70 °C, about 50-60 % of the thiamine content was retained. Thiamine retention in the minced samples was approximately 23-33 % higher at 70 °C than at 90 °C. Uherova et al.[13] determined thiamine retention in microwaved meat. After microwaving, thiamine retention rates were about 85-95 % for pork and about 90 % for chicken. Retention rates of vitamin B_6 (pyridoxine) were about 62-87 % for pork and about 60 % for chicken. These investigators used two different microwave ovens which thus resulted in different retention rate.

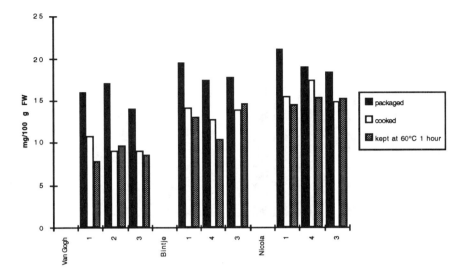

Figure 1 *Vitamin C contents of potatoes after various processing and washing treatments. (1=0.5% citric acid and 0.1 % ascorbic acid, 2=0.1% citric acid and 0.1 % ascorbic acid and 0.1% sodiumbenzoate, 3=water, 4=0.1 % citric acid and 0.005 % 4-hexylresorcinol)*

Piironen et al.[14] has determined vitamin E contents of various foods. Stewing reduced the α-tocopherol content of Baltic herring by about 5 %. After cooking, the contents of different tocopherols and tocotrienols in rolled oats decreased about 4-10 %.

2.4 Packaging and processing

The following results are from a recent study on the processing, packaging and nutritional value of raw vegetables carried out in Finland. The Ajax white cabbage cultivar was first mechanically cut into 5 mm strips and washed with plain water, or in water containing 100 ppm chloride once or twice at different temperatures. The 1 kg cabbage samples were then packaged in orientated polyproplylene film bags (40 μm, 300 x 280 mm) in normal air. The samples were stored for 8 days at +4 °C. Figure 2 gives the results for vitamin C contents after 2, 4 and 8 days of packaged storage. A fresh shredded white cabbage sample was also included for reference. After 8 days vitamin C contents were surprisingly higher than after 2 days. Vitamin C contents of the packaged samples decreased slightly after the two days storage compared to the fresh sample. After 8 days storage the vitamin C content of the packaged samples roughly equalled that of fresh cabbage. The different washing methods did not affect the vitamin C content. decreased slightly after the two days storage compared to the fresh sample. After 8 days storage the vitamin C content of the packaged samples roughly equalled that of fresh cabbage. The different washing methods did not affect the vitamin C content.

Vitamin C contents were additionally determined in shredded white cabbage kept in serving dishes for three hours at room temperature after a packaged storage of two days (Figure 3). A slight decrease in vitamin C contents was again detected in all packaged samples and also in the fresh sample.

Trials were conducted on the Chinese cabbage cultivars. The cabbage cultivars studied were the summer cultivar Kasumi and the winter cultivar Kingdom. After removal of all external parts, the cabbage was mechanically shredded into 6 mm strips and packaged in

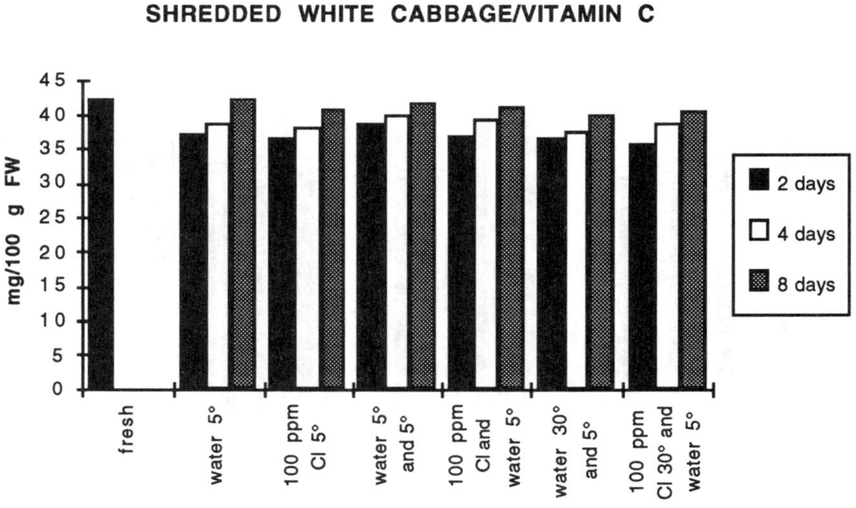

SHREDDED WHITE CABBAGE/VITAMIN C

Figure 2 *Vitamin C contents of shredded white cabbage over time after various washing treatments.*

1 kg lots into polypropylene bags in normal air. Vitamin C contents in fresh samples of the summer cultivar Kasumi were 25 mg/100 g fresh weight (FW) after one day storage and after 4 d storage only 16 mg and remained the same after 7 d storage at +4 °C. The 7 d packaged storage decreased the vitamin C content to 13 mg. Trials were also performed after the shredded Chinese cabbage had been kept in serving dishes at room temperature or at +10 °C for 3 hours. Vitamin C content in fresh shredded cabbage after 3 hours was 17 mg, having decreased by 8 mg. The vitamin C content of packaged cabbage was about the same as before the 3 h period when kept at room temperature.

The ß-carotene content of Kasumi cultivar was low, about 90 μg/100 g FW in fresh cabbage and decreasing further during the 7 d packaged storage. ß-carotene content decreased when fresh cabbage was left for 3 hours in serving dishes. No decrease was detected in packaged cabbage. The vitamin C content of the winter cultivar Kingdom was 17 mg/100 g FW and remained about the same after 7 days packaged storage. Again, 3 h storage in serving dishes at room temperature did not affect the vitamin C contents. The ß-carotene content of the Kingdom cultivar was even lower than that of the Kasumi cultivar, or only about 15 μg/100 g FW. The low ß-carotene content was reflected in the slight changes in ß-carotene contents during storage.

Various washing methods were also studied in the winter cultivar, Kingdom. The quantity of water used for washing was 3 l/kg, once or twice, depending on the trials. The temperature of the washing solution was +6 or +30 °C and the trials were performed either with chlorinated water (0.01 % free chlorine) or with citric acid (1g/ l). The vitamin C content of fresh shredded Chinese cabbage found in these trials was 14 mg/100 g FW and in packaged cabbage about 11-13 mg/100 g FW after 1 day, depending on the washing method employed. Vitamin C content decreased during 7 d packaged storage if the cabbage had been washed in plain water at +6 °C. The vitamin C content of cabbage washed in water at +30 °C and in chlorine was about the same after 1 and 7 d packaged storage. Trials were also performed after a 6 weeks winter storage. Vitamin C content of fresh cabbage was about 13 mg/100 g FW, and only 1 mg of vitamin C had been destroyed during the 6 weeks winter storage. The spring trials also used bags containing 10 air holes 350 μm in diameter. The vitamin C contents of packaged shredded cabbage stayed higher in

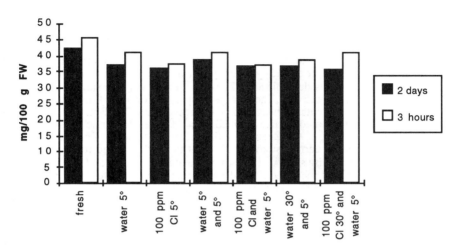

Figure 3 *Vitamin C contents of shredded white cabbage over time after various washing treatments.*

perforated bags. There has been one Danish investigation on shredded Chinese cabbage[15]. The vitamin C content in fresh Chinese cabbage was 49.5 mg /100g and, after shredding, it was 32.1 mg. After 12 d packaged storage the vitamin C content was 26.8 mg. Hence about 54 % of the vitamin C content was retained.

In the above study on processing, packaging and nutritional value of raw vegetbles, the Navarre cultivar was used for the trials on carrots. Carrots were washed, peeled and thereafter only some were washed in various washing solutions employing plain water, water containing 100 ppm chloride or 0.5 % citric acid. Carrots were grated into 3 mm strips and packaged into polypropylene bags in normal air. The ß-carotene content decreased during 8 d packaged storage. ß-carotene content decreased the most if the water temperature had been +30 °C. Washing the samples in water containing chloride best preserved the ß-carotene contents. α-carotene content also decreased slightly during the 8 d packaged storage period. After the winter storage the ß-carotene content of the carrots had decreased by about 10 %, but the α-carotene content was the same that in autumn. When the trials were carried out using perforated bags the ß-carotene content increased during the 8 d packaged storage and decreased in the other trials employing unperforated bags. Trials were also performed after grated carrot were kept in serving dishes for 3 hours at room temperature. The ß-carotene content of a fresh, grated carrot sample decreased by 50 %. The corresponding decrease in the packaged samples varied between 10 to 40 %, depending on the washing method. Washing with chloride, which earlier best preserved the carotene content during packaged storage, caused the highest decrease in the ß-carotene content of carrots kept in serving dishes.

The three potato cultivars, Van Gogh, Bintje and Nicola, were also studied in the trials. Whe prepeeled, packaged potatoes were stored 7 days at +4 °C vitamin C contents were retained and the decreases were slight.

Gosselin and Mondy[16] carried out a study on packaging materials, including paper mesh bags, double wall, wet strenght Kraft paper and polyethylen bags. Potatoes were stored for 1, 4 and 8 weeks in darkness at room temperature. In all the bags the ascorbic acid decreased during 8 weeks storage. Potatoes packaged in Kraft paper best preserved the ascorbic acid content.

2.5 Storage

In a Danish study[17] the ascorbic acid contents in Chinese cabbage was determined during storage at different temperatures and relative huminity. The various experimental conditions were +1 and 3 °C at the relative huminities of 99 %, 90 % and 80 %. The ascorbic acid content was best preserved at +1 °C and 80 % reletive huminity. Retention was 93.5 % after a 15-wk storage. Ascorbic acid contents decreased the most at +3 ℃ resulting in 67 % retention after a 15-wk storage. Vitamin C retention of cabbage during a 30-50 d storage period was determined by Zee et al.[18] employing the storage temperatures +3, +8, and +23 °C, respectively. The storage temperature of +23 °C or +8 °C caused a decrease in vitamin C contents. Cabbage stored in darkness at + 23 ℃ retained 61 % of the vitamin C, and samples stored in light retained 88 %. The retention rate for storage at +8 °C was over 90 %. Storage at +3 °C increased the vitamin C contents of cabbage.

Carotene contents of carrots stored at different temperatures and relative humidities were determined in a Danish study[17]. The experimental temperatures and relative humidity were +1 °C and 3 °C and 99 %, 90 % and 80 %, respectively. After 18 wk storage the carotene content was about the same as in fresh carrots. Slight changes in the carotene contents (both increases and decreases) occurred during the storage period. Lee[19] also found changes in the carotene content in both directions during storage. During 155 days contents of α- and ß-carotene initially increased and then decreased slightly.

In another Danish study[20] vitamin C, thiamine and pyridoxine contents of four different potato cultivars were studied during 32 wk of storage. Vitamin C retention was 50 % for the Asparages cultivar, 45 % for Bintje, 46 % for Hansa and only 37 % for Sava after a 32 wk storage at +3 °C and 90 % relative huminity. Surprisingly, the rates of thiamine retention found were 112 % for Bintje, 109 % for Asparages, 111 % for Sava and 101 % for Hansa, respectively. The corresponding retention rates for pyridoxine were 89 % for Hansa, 83 % for Bintje and 80 % for both Asparages and Sava. Wills et al.[21] determined the vitamin C contents of potatoes during a 12 months period. Every 2 weeks they determined the vitamin C contents in 1 kg of each type of potato on sale at five retail outlets. The level of vitamin C was relative constant from March almost to July (early autumn-midwinter) then decreased markedly to minimum levels in August to September (late winter-early spring). Rapid increases occurred in October (spring) with maximum levels occuring from Deceember to January (mid summer). Keijbets and Ebbenhorst-Seller[22] determined ascorbic acid contents in six potato cultivars stored for 8 months at +5-6 °C. The ascorbic retention was about 50 %. Although there were different initial amount of ascorbic acid in the various cultivars the amount was roughly the same after storage.

Shaw and Moshonas[23] studied the vitamin C contents of orange juices differently packaged, processed and stored after opening. Average retention after 5-7 days was 88 %. The average retention of juice left in an open plastic container for a second week was 66.5 %. The range of ascorbic acid retention after two weeks for all samples varied widely from 47 to 96 %.

Riboflavin degradation in milk can occur slowly in dark, during storage in a refridgerator, according to Munoz et al.[24]. The resulting loss is a function of both the packaging material and the penetration of light into the product. The loss of riboflavin in three milk samples stored 6 days in darkness at +8 °C ranged from 16.5 to 23.4 %.

Skrede[25] studied the ascorbic acid contents of syrup stored in different packages at +20 °C in both light and dark conditions. After an 18 months storage the ascorbic acid content varied from 125 to 150 mg in black currant syrup packaged in glass, PET or PVC, and about 200 mg in the beginning. Polyethylene packaging destroyed ascorbic acid during a 12 months storage. Practically the same was applicable for orange syrup. Johansson and Marmoni[26] studied the ascorbic acid content of syrups after different preparation methods and after a 4 month storage period. The ascorbic acid content of fresh red currant was 24 mg/100 g and that of syrup ranged from 5 to 15 mg after preparing. After 4 month storage the ascorbic acid contents decreased to 0-5 mg/100 g. Ascorbic acid content were best preserved in stirred syrup. The ascorbic acid content of fresh black currant was 140 mg/100 g ranging between 62 and 110 mg in prepared syrup. After 1 month the ascorbic acid content was about 60 mg and after 4 months from 36 to 65 mg/100 g. Ascorbic acid contents were best preserved in stirred syrup. The ascorbic acid content of fresh strawberries was 42 and in syrup 5-24 mg/100 g. After 4 months the content varied from 0 to 14 mg/100 g. In this case, steamed syrup best preserved the ascorbic acid contents. The ascorbic acid content of fresh raspberry was 31 mg/100 g and syrup contained 6-20 mg/100 g. After 4 months the content was 0-5 mg/100 g. Stirred syrup best preserved ascorbic acud contents.

3 DISCUSSION

A report[27] on nutrient losses and gains during food preparation published by the Swedish National Food Administration found excellent vitamin E retention, 100 % after boiling vegetables, potatoes and eggs and in prepared fish. Vitamin E retention of meat varied from 60 to 100 %. Vitamin A retention ranged from 30 to 100 %, with the main range of 80-95 %. Vitamin C retention varied from 20 to 80 %, thiamine retention 45-95 % and riboflavin retention 50-100 %. Such differences are very substantial. The present report was based on data obtained in Denmark, Germany, Russia, Sweden, the UK and the USA and on papers published chiefly between 1970 and 1990. These results are in agreement with newer results reviewed in the present paper.

New methods to extend shelf-live of foods have been developed during recent years. Minimal processing (for instance mild heating methods, sous vide), modified atmosphere packaging, active packaging (for instance using oxygen absorbers or scavengers), and vacuum packaging have been developed, where vitamins are better preserved. Thus, in future the retention of vitamins in food will be even higher then today.

References

1. C.O. Chichester, World Review of Nutrition and Dietetics, 1973, **16**, 318.
2. F.C. Lamb, R.P. Farrow, E.R. Elkings, Handbook of Nutritive value of processed food, volume 1, Ed M. Rechcigl 1986.
3. M.N. Hall, M.C. Edwards, M.C. Murphy, R., Pither, Campden Food and Drink Research Association Technical Memorandum no. 553, 1989.
4. M.P. Cano, B. de Ancos, J. Agric. Food Chem., 1994, **42**, 2737.
5. M.P. Cano, M.A. Marin, J. Agric. Food Chem. 1992, **40**, 2141.
6. M. Hägg, S. Ylikoski, J. Kumpulainen, J. Food Comp. Anal., 1995, **8**, 12.
7. M. Hägg, J. Kumpulainen, J. Food Comp. Anal. 1994, **7**, 94.
8. I. Vedrina-Dragojevic, B. Sebecic, Z Lebensm Unters Forsch, 1994, **198**, 44.
9. J.Solms, R. Genner-Titzmann, 9th Triennial Conference of the European Association for Potato Research 1984, 51.
10. R. Wills, K. Suthilucksanavanish, Food Australia, 1991, **43(1)**, 19.
11. B. Carlson, M. Tabacchi, J. Amer. Diet. Assoc., 1988, **88(1)**, 65.
12. S. Awonorin, J. Ayoade, Int. J. Food Sci. Techn., 1993, **28**, 595.
13. R. Uherova, B. Hozova, V. Smirnov, Food Chem., 1993, **46**, 293.
14. V. Piironen, P. Varo, P. Koivistoinen, J. Food Comp. Anal., 1987, **1**, 53.
15. H. Aabye, Bioteknisk institut, Beretning nr **107**, 1983 (in Danish).
16. B. Gosselin, N. Mondy, J. Food Sci. 1989, **54(3)**, 629.
17. T. Leth, Levnedsmiddelstyrelsen, Publikation **126**, 1986 (in Danish).
18. J. Zee, L. Carmichael, D. Codere, D. Poirier, M. Fournier, J. Food Comp. Anal. 1991, **4**, 77.
19. C. Lee, Food Chem., 1986, **20**, 285.
20. T. Leth, A. Christensen, K. Jakobsen, Levnedsmiddelstyrelsen Publikation **122**, 1986 (in Danish).
21. R. Wills, J. Lim, H. Greenfield, J. Sci. Food Agric., 1984, **35**, 1012.

22. M. Keijbets, G. Ebbenhorst-Seller, Potato Research, 1990, **33**, 125.
23. P. Shaw, M. Moshonas, J. Food Sci., 1991, **56(3)**, 867.
24. A. Munoz, R. Ortiz, M. Murcia, Food Chem., 1994, **49**, 203.
25. G. Skrede,Norsk Institutt for næringsmiddelforskning Rapport **2**, 1981 (in Norwegian).
26. H. Johansson, E. Marmoni, Vår Föda, 1986, **7**, 407 (in Swedish).
27. L. Bergström, Livsmedelsverket Rapport **32**, 1994.

ANALYTICAL METHODOLOGY FOR ANTIOXIDANT VITAMINS IN INFANT FORMULAS

J. M. Romera, A. J. Angulo, M. Ramirez, A. Gil.

R&D Departament Abbott Laboratories Camino de Purchil 68, 18004 Granada, Spain.

1 INTRODUCTION

Antioxidant vitamins limit oxidation process in dietetic products. Especial attention has been paid to search analytical methodology for the quality control of adapted infant formulas because those formulas are the only source of nutrients for the majority of infants artificially fed during the first months of life. Vitamin analysis of infant formulas is therefore a method for Quality Assurance during manufacturing and is necessary to ensure compliance with the legislative requirements for these products [1,2].

Several methods (tritiations, colorimetry, UV spectrometry, gas chromatographic techniques, etc.) have been extensively utilized to quantify retinol, tocopherols, ascorbic acid and its esters. The official methods for these vitamins are colorimetrics but these analytical methods are often time-consuming and have poor reproducibility [1]. HPLC now offers the analyst a relatively easy way of determining these vitamin in infant formulas [3-7].

This study describes a HPLC method for the simultaneous determination of vitamin A and E and a HPLC method for quantification of ascorbic acid in infant formulas. These methods were applied to six kind of products especially designed for infant nutrition: starting, follow-up and preterm infant formulas and special diets based on protein hydrolysates, lactose free and soy protein from several manufacturers. We have optimized antioxidant vitamin analysis including extraction, clean-up and end-methods of assay. These methods are highly specific, precise and less time-consuming.

2 MATERIALS AND METHODS

2.1 Reagents

Acetonitrile and methanol (LC grade); phosphoric acid, sodium phosphate, metaphosphoric acid, potassium hydroxide, pyrogallol (reagent grade) and filtered deionized water were the materials used.

An Ascorbic acid L (+) working standard solution of 250 µg mL^{-1} (A) was prepared by dissolving 25 mg ascorbic acid in 1% (w/v) metaphosphoric acid solution and diluting to 100 ml with 1% (w/v) metaphosphoric acid solution. Vitamin A and E standard solutions were prepared with Retinol acetate and a-tocopherol acetate in 20 µg/ml concentrations.

0.1 M potassium phosphate buffer, was prepared by diluting 6.78 ml o-phosphoric acid 85% (v/v), adjusting the pH to 3.5 with potassium hydroxide 50% (w/v) and completing to 1 litre. This solution was filtered and used as mobile phase for ascorbic acid determination. Vitamin A and E mobile phase was a mixture of methanol and acetonitrile (80:20).

2.2 Apparatus

The liquid chromatographic system (Waters) consisted of double piston pump, autosampler, diode array detector, fluorescence detector and 5-μm reverse-phase Supersphere C-18 column (250mm x 4.6mm id) (Merck). The diode arrays detector were set at 325 nm for retinol. Tocopherols were detected using fluorescence detector with excitation 290 nm and emission 325 nm. Vitamin C was analized with the same liquid chromatographic system but in this case UV detector was set at 245 nm.

2.3 Sample preparation and HPLC assay

Vitamin A and E were performed by saponification and diethyl ether-petroleum ether extraction. A sample aliquot (1 g) was homogenized in 5 ml of warm water and added to methanol (5 ml) with 1% pyrogallol. Potassium hydroxide solution (50%;W/V) (2ml) were added and maintained at 70 °C during 20 min with shaking each 5 min. Retinol and tocopherols were extracted with diethyl-ether/petroleum ether (50/50;V/V). Organic phase was evaporated under nitrogen and insaponificable compounds extracted were dissolved in methanol (5ml). 20 μl of this solution was injected into the reverse phase HPLC system (mobile phase: methanol-acetonitrile (80:20).

Ascorbic acid was determined according to Behrens and Madere with the following modifications: An aliquot of infant formula (5 g) was brought to a final volume of 100 ml with 1% metaphosphoric acid solution. Homogenates were submitted to 2.000 g in a refrigerated centrifuge for 10 min. The clear supernatant were filtered through Whatman 501 paper. A portion of 1 ml was diluted to 5 ml with 1% metaphosphoric acid and a 10 μl aliquot of this solution was injected into the HPLC system to determine the concentration of ascorbic acid in the sample. Aliquots of standards (1.25 to 10.00 μg/ml) were used to do a calibration curve by plotting peak area vs ascorbic acid amount after HPLC analysis.

3 RESULTS AND DISCUSSION

We have optimized an HPLC method for vitamin A which permits the simultaneous analysis of vitamin A and E, reducing analysis time (run time = 12 min). Ascorbic acid was determined according to Behrens and Madere[4] method with several modifications. These methods accurately measure concentrations as low as a few μg of retinol, tocopherol and ascorbic acid. The linearity between standard concentrations and detector responses was good: r = 0.9992 for retinol, r = 0.9996 for α-tocopherol and r = 0.9997 for ascorbic acid. This high sensitivity makes the HPLC method with UV detection suitable for analysis of infant formulas containing low amounts of vitamin A and C. Tocopherols show poor UV absorbance and were monitored by fluorescence detection. Sample preparations are necessary to eliminate a number of interferences, mainly from proteins and lipids, during the chromatographic process [8-10]. Sample treatments can impair vitamin stability. Thus, standard solutions were processed to evaluate the influence of sample treatment. Each standard solution was assayed in quadruplicate to determine the precision of the HPLC method. The recoveries were 97.9 for retinol, 101.4 for α-tocopherol and 95.6 for vitamin C; coefficients of variation were 4.50, 2.79 and 1.41, respectively. Augmented samples were treated obtaining excellent recoveries: 100.2% for vitamin A, 102.8% for vitamin E and 97.4% for ascorbic acid, on average.

No significant differences were obtained between vitamin levels of different batches, indicating that vitamins are added in a proper way. Most of all infant formulas analyzed contained antioxidant vitamins concentrations higher than those that were declared on their nutritional labels (Tables 1 and 2), indicating that losses of these vitamins during formula manufacture were overestimated. The oversupplementation is a usual practice in these products because antioxidant vitamin contents (mainly vitamin C contents) decrease during shelf-life. Nevertheless, vitamin concentrations measured in all samples complied with international recommendations.

Table 1. *Quantitative determination of antioxidant vitamins in infant formulas.*

Infant formula	Vitamin A [1]		Vitamin E [2]		Vitamin C [2]	
	D [3]	A [4]	D	A	D	A
Starting	550	639 ± 25	5.5	6.5 ± 0.3	50	30.5 ± 1.1
	465	1468 ± 16	6	10.4 ± 0.1	42	78.8 ± 0.5
	465	1834 ± 3	5.7	9.4 ± 0.2	46	89.0 ± 0.3
	450	481 ± 44	2.7	6.6 ± 0.7	45	37.3 ± 0.1
follow-up	550	668 ± 30	5.5	7.4 ± 0.3	50	48.2 ± 2.6
	429	1026 ± 27	5.7	8.3 ± 0.3	39	62.9 ± 0.1
	404	621 ± 2	4.9	9.5 ± 0.2	40	66.8 ± 0.9
	450	883 ± 24	2.7	9.3 ± 0.1	125	134.7 ± 0.4
preterm	480	584 ± 48	5.5	7.3 ± 0.4	100	45.5 ± 1.1
	444	820 ± 5	10	12.1 ± 0.3	77	96.4 ± 1.2
	612	614 ± 4	17.8	19.8 ± 0.5	75	73.7 ± 1.3
	450	711 ± 37	2.7	7.4 ± 0.3	195	213.7 ± 7.5

[1] Mean of four samples ± SEM expressed as µg / 100g of sample.
[2] Mean of four samples ± SEM expressed as mg / 100g of sample.
[3] Declared in nutritional label
[4] Analyzed following the present methodology

Table 2. *Quantitative determination of antioxidant vitamins in special infant diets.*

Infant formula	Vitamin A [1]		Vitamin E [2]		Vitamin C [2]	
	D [3]	A [4]	D	A	D	A
Hydrolysates	550	503 ± 26	5.5	5.0 ± 0.3	50	34.8 ± 0.4
	456	712 ± 6	6	7.2 ± 0.2	41	30.5 ± 0.1
	500	623 ± 2	6.4	16.5 ± 0.4	64	66.6 ± 0.1
	450	565 ± 51	4	6.9 ± 0.4	45	43.3 ± 0.5
Lactose free	550	570 ± 31	5.5	7.9 ± 0.2	50	47.6 ± 1.1
	453	867 ± 9	6	8.6 ± 0.1	40	43.0 ± 0.5
	450	566 ± 18	12	17.6 ± 1.1	75	64.2 ± 1.0
	450	666 ± 27	2.7	10.3 ± 0.3	45	48.5 ± 1.4
Soya protein	550	586 ± 35	5.5	8.5 ± 0.1	50	51.2 ± 1.8
	453	669 ± 2	6	12.9 ± 0.1	80	116.5 ± 0.5
	450	627 ± 19	12	14.5 ± 0.5	75	82.8 ± 0.5
	450	809 ± 20	2.7	7.65 ± 0.5	45	35.1 ± 0.9

[1] Mean of four samples ± SEM expressed as µg / 100g of sample.
[2] Mean of four samples ± SEM expressed as mg / 100g of sample.
[3] Declared in nutritional label
[4] Analyzed following our methodology

References

1. AOAC. Official Methods of Analysis. 15th Ed. *Association of Official Analytical Chemists*, Washington, DC. 1990.
2. D. C.Woollard, *Food technology in Australia*, 1987, **39**, 250.
3. S. H. Ashoor, W. C. Monte, J. Welty, *J. Assoc. Off. Anal. Chem.*, 1984, **67**, 78.
4. W. A. Behrens, R. Madere, *J. Food Comp. Anal.*, 1989, **2**, 48.
5. J. T. Tanner, S. A. Barnett, *J. Assoc. Off. Anal. Chem.*, 1986, **69**, 777.
6. J. N. Thompson, G. Hatina, W. B. Maxwell, *J. Ass. Off. Analyt, Chem.*, 1980, **63**, 894.
7. J. N. Thompson, P. Erdody, W.B. Maxwell. *Biochem. Med.* 1973, **8**, 403.
8. J. N. Thompson, *J. Assoc. Off. Anal. Chem.*, 1986, **69**, 727.
9. D. I. Thurnham, E. Smith, P. S. Flora, *Clin. Chem.*, 1988, **4**, 377.
10. D. C. Woollard, A. D. Blott, *J. Micronutr. Anal.* 1986, **2**, 97.

VITAMIN E CONTENTS OF BREAKFAST CEREALS AND FISH

Margareta Hägg and Jorma Kumpulainen

Agricultural Research Centre of Finland, Laboratory of Food Chemistry
FIN-31600 Jokioinen, Finland

1 INTRODUCTION

Interest in vitamin E as an antioxidant has greatly increased recently. However, the antioxidative potense of tocopherols and tocotrienols is different. Thus it is inportant to determine the content of various tocopherols and tocotrienols in foods in order to estimate the total vitamin E content. Vitamin E contents of imported and domestic breakfast cereals and fish products were determined at the Laboratory of Food Chemistry, Agricultural Research Centre of Finland as part of a 5-year project on the quality of domestic and imported foods.

2 MATERIALS AND METHODS

Breakfast cereal samples consisting of common domestic brands and some imported brands were collected in autumn 1990. Ten retail size packages of the domestic brands were sampled from every raw material batch. In 1994 imported and domestic fish samples were optained from fishermen and wholesalers. The sardine, tuna, and mackerel samples comprised of 5 - 6 cans. The sample size of fresh fish samples was about 0.5-1 kg. Figure 1 shows a typical HPLC diagram of a tuna sample.

The collected subsamples were pooled and carefully homogenized and then analyzed in duplicate. Vitamin E contents of fish were determined by the HPLC method according to Syväoja et al.[1]. About 10 mg of the samples were weighed in duplicate. After that 12 g ascorbic acid, 100 ml ethanol, and 40 ml water were added and mixed. After 30 min 30 ml 30 % potassium hydroxide was added and saponified at 70 °C for 30 min. The sample was then filtered and extracted with 3 x 100 ml hexane for 2 min, rinsed with 3 x 100 ml water and evaporated to dryness. The sample was then diluted in methanol and determined fluorimetrically at 292:324 nm. The mobile phase was hexane:di-isopropylether (93:7) at a 2 ml/min flow rate. An Hibar Lichrosorb Si 60 5 µm (250 x 4 mm) (Merck) column was employed. The HPLC instrumentation employed an HP 1090 M liquid chromatograph, an HP 1046, a fluorimetric detector (Hewlett-Packard) or an Hewlett-Packard 1084 B equipped with a fluorimetric detector (Perkin-Elmer 3000). All standards were Merck tocopherols (Art. 15496) and tocotrienols (Art 8224) throughout. Vitamin E contents of cereals were determined according to Piironen et al.[2].

3 RESULTS AND DISCUSSION

Tocopherol and tocotrienol contents in müslies are shown in Table 1. Calculation of vitamin E contents was made according to McLaughlin & Weighrauch[3] and was based on the formula α-tocopherol equivalent (TE) = mg α-tocopherol + 0.3 (mg α-tocotrienol) +

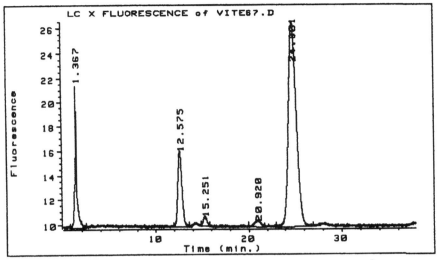

Figure 1 HPLC diagram of tocopherols and tocotrienols in a tuna sample.

0.4 (mg β-tocopherol) + 0.1 (mg γ-tocopherol) + 0.01 (mg γ-tocotrienol) + 0.01 (mg δ-tocopherol). Vitamin E contents differed substantially. Of the various tocopherols, α-tocopherol was found in the greatest abundance in most müslis and δ-tocopherol was not detected in müsli samples. The müsli samples contained α-tocotrienol and γ-tocotrienol.

The tocopherol and tocotrienol contents of Finnish rolled oats and rolled mixed grain are shown in Table 2. Marked differences in the tocopherol and tocotrienol contents of the different rolled oat samples were found, but not in the total vitamin E contents. Rolled oats contained α-, β- and γ-tocopherols and α-tocotrienol. The same tocopherols and tocotrienols were also detected in domestic rolled mixed grain samples. Total vitamin E contents of rolled oats and rolled mixed grain samples were about equal.

Table 3 shows the tocopherol and tocotrienol contents of fish. δ-tocopherol and β-, γ- and δ-tocotrienols were not encountered in the present fish samples. Vitamin E contents were highest in Baltic herring, but contents of the three samples differed substantially, ranging from 3.2 - 4.3 mg TE/100 g fresh weight. Vitamin E contents of fish canned in oil were also high, but part of the vitamin E content originated from oil. Oil mainly contributed to the γ-tocopherol content which was abundant only in these samples. Sardines canned in oil contained 2.7 mg vitamin E/100 g compared to sardines canned in tomato sauce, 1.90 mg. Vitamin E contents of tuna canned in oil were 1.83 mg, compared to tuna canned in water with only 0.6 mg.

Acknowledgments

Many thanks to Mrs. Tuula Kurtelius for skillful technical assistance.

Table 1 Tocopherol and tocotrienol contents and total vitamin E as α-tocopherol equivalents (TE) in müslies (mg/100 g fresh weight).

Müslies	tocopherols			tocotrienols		TE
	α–	β–	γ–	α–	γ–	
United Kingdom						
1	0.53	0.11	1.08	0.30	0.10	0.8
2	0.60	0.07	0.74	0.62	0.15	0.9
3	1.84	0.14	0.78	0.40		2.1
4	0.47	0.14	0.63	0.27		0.7
5	0.61	0.21	0.99	0.62		1.0
6	0.25	0.12	0.58	0.22		0.4
7	0.41	0.12	0.61	0.31		0.6
8	0.76	0.13	0.83	0.66		1.1
9	0.85	0.13	0.67	0.28		1.1
10	1.76	0.24	0.88	0.21		2.0
11	0.40	0.09	0.51	0.70	0.15	0.7
Germany						
1	1.53	0.13	0.32	1.43	0.07	2.1
2	1.31	0.20	1.19	0.79		1.8
3	1.03	0.15	0.83	0.61	0.11	1.4
Sweden						
1	0.45	0.21	0.73	0.57	0.16	0.8
2	0.98	0.16	0.68	0.49	0.26	1.3
Finland						
1	1.33	0.09	0.71	0.74		1.7
2	0.70	0.11	0.21	1.11		1.1
3	0.77	0.14	0.61	0.65		1.1
Average	0.87 ± 0.5	0.14 ± 0.1	0.71 ± 0.2	0.58 ± 0.3	0.14 ± 0.1	1.2 ± 0.5

Table 2 Tocopherol and tocotrienol contents and total vitamin E as α-tocopherol equivalents (TE) in rolled oats and rolled mixed grain (mg/100 g fresh weight).

	tocopherols			tocotrienols	TE
	α–	β–	γ–	α–	
rolled oats					
1	0.87		0.14	2.05	1.5
2	0.85		0.11	2.12	1.5
3	0.92		0.52	2.07	1.6
4	1.00		0.09	1.97	1.6
5	0.80	0.05	0.12	1.71	1.4
6	0.81		0.16	2.16	1.5
7	0.93	0.05	0.17	2.42	1.7
8	1.27	0.06	0.17	1.33	1.7
9	1.20	0.06	0.13	1.27	1.6
10	0.61	0.02	0.11	0.74	0.9
11	0.71	0.06	0.15	1.95	1.3
12	0.94	0.08	0.17	2.55	1.8
average	0.82 ± 0.3	0.05 ± 0.02	0.17 ± 0.1	1.86 ± 0.5	1.5 ± 0.2
rolled mixed grain					
1	1.10	0.30	0.95	1.75	1.8
2	1.01	0.17	0.82	1.50	1.6
3	0.73	0.11	0.55	1.43	1.3
4	0.61	0.13	0.70	1.18	1.1
average	0.80 ± 0.2	0.18 ± 0.1	0.76 ± 0.2	1.47 ± 0.2	1.5 ± 0.3

Table 3 Tocopherol and tocotrienol contents and total vitamin E as α-tocopherol equivalents (TE) in fish (mg/100 g fresh weight).

Fish	Country	tocopherols α-	β-	γ-	tocotrienols α-	TE
Baltic Herring	Finland	3,63 ± 0,525		0,22 ± 0,007	0,13 ± 0,015	3.7
Sardine, oil	Marocco	2,19	0,22	4,71	0,24	2.8
Sardine, oil	Spain	2,18	0,11	3,79	0,07	2.6
Salmon	Finland	2,49			0,06	2.5
Vendace	Finland	2,29			0,04	2.3
Sardine, tomato	Thailand	2,64	0,05	0,03	+	2.7
Sardine, tomato	Slovakia	1,07		0,11	0,10	1.1
Tuna, oil	Philippines	1,21	0,12	0,26	1,03	1.9
Tuna, oil	Indonesia	1,29	0,08	3,56	0,41	1.7
Whitefish	Canada	1,98		2,81	0,17	2.1
Whitefish	Finland	1,41 ± 0,583		0,45	0,05 ± 0,031	1.4
Rainbow trout	Finland	1,55 ± 0,570		+	0,24 ± 0,171	1.6
Perch	Finland	1,41		0,22 ± 0,267	0,03	1.4
Mackerel, tomato	Denmark	1,59 ± 0,202	0,18 ± 0,049	0,09 ± 0,032	0,06 ± 0,006	1.4
Herring, pickled	Iceland	1,16 ± 0,789		0,12 ± 0,202	0,07 ± 0,044	1.2
Saithe	Faeroe Islands	1,16			0,06	1.2
Saithe	Norway	1,07				1.1
Cod	Norway	0,75			0,05	0.8
Tuna, water	Thailand	0,55 ± 0,189			0,10 ± 0,064	0.6
Tuna, water	Philippines	0,75			0,06	0.8
Flounder	Netherlands	0,56 ± 0,221	+	0,03 ± 0,043	0,06 ± 0,015	0.6
Sole	Netherlands	0,51			0,07	0.5
Mackerel	Ireland	0,08			0,07	0.1

References

1. E.-L. Syväoja & al, JAOCS, 1985, **62(8)**, 1245.
2. V.Piironen & al, Intern. J. Vit. Nutr. Res., 1984), **53**, 35.
3. McLaughlin & Weilrauch, J. Amer. Diet Assoc., 1979, **75**, 647.

THE IN VIVO ANTIOXIDANT PROPERTIES OF SESAME SEED LIGNANS

A. Kamal-Eldin, D. Pettersson and L.-Å. Appelqvist

Department of Food Science, Swedish University of Agricultural Sciences,
Box 7051, 750 07 Uppsala, Sweden

1 INTRODUCTION

The sesame plant (*Sesamum indicum*, Linn., Order: Tubiflora, Family: Pedaliaceae) is well known for its edible seeds and oil but also for some medicinal uses[1]. Sesame seeds are characterized by the presence of two oil-soluble lignans: sesamin and sesamolin [2].

Sesamin   Sesamolin

Lignans have been reported to be antioxidants, anticancer agents, platelet activating factor receptor antagonists, antiviral (and anti-HIV) agents, bactericides, viricides, disinfectants, moth repellants and anti-tubercular agents. [3-12]

2 EXPERIMENTAL

Thirty-six Sprague-Dawley rats (6X6) were fed experimental diets containing (g/kg): starch (570), casein, (200), ground nut oil (100), cellulose powder (40), sucrose (40) and mineral and vitamin premixes *ad libitum*. The rats were fed equal amounts of tocopherols (13 mg alfatocopherol and 50 mg gammatocopherol per kg diet) and sesamin was the only variable in diets 1-6 at 0.0, 0.25, 0.50, 1.0, 2.0 and 4.0 g/kg, respectively. After 4 weeks, blood was collected from 24 h. fasting animals in tubes containing EDTA as anticoagulant. The tocopherols were extracted from the plasma, livers and lungs and analyzed by HPLC. Linear regression analyses of tocopherol concentrations (Y) against sesamin level (X) were done by using the regression procedure (PROC REG) and analysis of variance by using the general linear model (GLM) with level of sesamin as the only main effect. For experimental details, the reader is referred to. [13]

3 RESULTS

Positive regression slopes were obtained for both alfatocopherol (a-T) and gammatocopherol (g-T) showing an increasing effect with increasing sesamin levels in the diets (Figure 1). The increase was significant for g-T and g-/a- tocopherol ratios in the plasma ($P < 0.05$), liver ($P < 0.001$) and lungs ($P < 0.001$). Although sesamin induced almost equal increases in the a-T levels in rat plasma, liver and lungs, these increases were not statistically significant, perhaps due to the large within-group variation in the levels of this tocol in the plasma and tissues. Thus, sesamin spared g-T in rat plasma and tissues, and this effect persisted even in the presence of alfatocopherol, a known competitor to g-T.

4 DISCUSSION

Although the human intake of g-T in Sweden is approximately twice that of a-T, the bioavailability of g-T is only about 10-35% of that of a-T. Both homologues are absorbed to the same extent from the gastrointestinal tract but g-T is cleared from the tissues at a considerably faster rate than a-T [14,15]. This study showed that sesamin-feeding increases the total tocopherol level and significantly the g-T levels, in rat plasma, liver and lung [13]. Similar results were previously reported by Yamashita *et al.* [16].

Sesame oil produced a better carotene utilization in the rat than did soybean and peanut oils[17]. Sesamin effectively supressed 7,12-dimethylbenz[a]anthracene induced rat

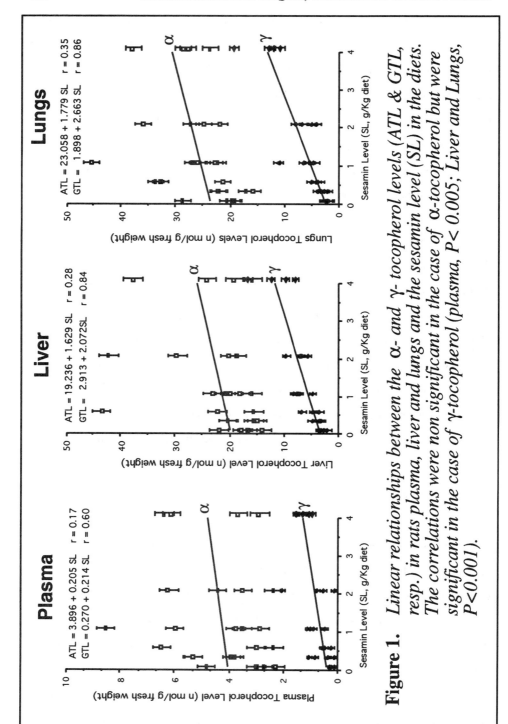

Figure 1. *Linear relationships between the α- and γ- tocopherol levels (ATL & GTL, resp.) in rats plasma, liver and lungs and the sesamin level (SL) in the diets. The correlations were non significant in the case of α-tocopherol but were significant in the case of γ-tocopherol (plasma, $P < 0.005$; Liver and Lungs, $P < 0.001$).*

mammary carcinogenesis.[18] and had protective effects against liver damage caused by alcohol or carbon tetrachloride in rodents[19]. The antioxidant effects of sesamin and its sparing of vitamin E may explain the observations that some antioxidants, other than tocopherols, in sesame suppressed senescence in mice[19] and that daily injection of sesame oil to adrenalectomized female rats increased the number of successful pregnancies, prolonged the survival time of those animals having a successful gestation and increased their ability to rear youngs[20].

The mechanism by which sesamin acts to spare the tocopherols is not known but it seems to involve competition with g-T oxidation and/or clearance. There is only one report on the absorption and metabolism of sesamin in rats. Only 0.15% of the fed sesamin was recovered in the lymph during 24 hours, *ca* 15-19% was excreted in the feces in rats fed a purified diet and about 34% in rats fed a non-purified diet[21]. Lignans are known to undergo demethylation and transform into mammalian lignans (mainly 1,2-dihydroxy compounds) by microbial action in the intestine[22-24]

Lignans having the methylenedioxyphenyl (1,3-benzodioxole) function are also known to be metabolized by the cytochrome P-450-dependent mixed function oxidases (mfo) associated with the endoplasmatic reticulum of the microsomes to the corresponding catechol (1,2-dihydroxy) derivatives[25]. The metabolic intermediates are capable of binding to reduced cytochrome P-450 to give relatively stable complexes correlated with non-competitive inhibition of xenobiotic metabolism. The metabolism of the methylenedioxyphenyl group of safrole was studied[26] and it seems to cause non-competitive inhibition to the mfo. Since many mfo are active in the metabolism of lipids, steroids and other compounds foreign to the metabolic network[27] the sesamins may act as competitive or non-competitive inhibitors of the metabolism of these lipophilic substances by the mfo.

Acknowledgements: The results presented in this paper were reproduced from the reference 13 after kind permission from the American Oil Chemist's Society.

References

1. Watt, J. M. and Breyer-Brandwijik, M. G. (1962) *The medicinal and poisonous plants of southern and western Africa*, 2nd. edition, E. and S. Livingstone Ltd., Edinburg & London, pp 831-33.
2. Kamal-Eldin, A. and L. Å. Appelqvist (1994) *J. Am. Oil Chem. Soc.* **71** (2): 149-156.
3. McRae, W. D. and Towers, G. H. (1984) *Phytochemistry* **23**(6): 1207-1220.
4. McRae, W. D., Hudson, J. B. and Towers, G. H. (1989) *Planta Med.* **55**(6): 531-535.
5. Gringrich, D. E., Hussaini, I. and Shen T. Y. (1990) Abstracts, 1990's Am. Chem Soc Meeting, MEDI 8, Boston, MA.
6. Schroder, H. C., Merz, H., Steffen, R., Muller, W. E., Sarin, P. S., Trumm, S., Schultz, J. and Eich, E. (1990) *Z. Naturforsh* **45**(11/12): 1215-1221.
7. Shen, T. Y. (1991) *Lipids* **26**: 1154-56
8. Adlercreutz, H., Mousavi, Y. and Hockerstedt, K. (1992a) *Acta Oncol.* **31**(2): 175-181.
9. Adlercreutz, H., Mousavi, Y., Clark, J., Hockerstedt, K., Hamalainen, E., Wahala, K., Makela, T. and Hase, T. (1992b) *J. Steroid Biochem. Mol. Biol.* **41**(3-8): 331-337.
10. Caragay, A. B. (1992) *Food Technol.* **46**(4): 65-68.
11. Castro-Faria-Neto, H. C., Martins, M. A., Silva, P. M., Bozza, P. T., Cruz, H. N., de-Quiroz-Paulo, M., Kaplan, M. A. and Cordeiro, R. S. (1993) *J. Lipid Mediat.* **7**(1): 1-9.
12. Thompson, L. U. (1993) *Food Res. Intern.* **26**(2): 131-149.
13. Kamal-Eldin, A., Pettersson, D. and L. Å. Appelqvist (1995) *Lipids* **30** (6): 499-505.
14. Gloor, U., Wursh, J., Schwieter, U., and Wiss, O. (1966) *Helv. Chim. Acta* **49**: 2303-2312.
15. Peake, I. R., and Bieri, J. G. (1971) *J. Nutr.* **101**: 1615-1622.
16. Yamashita, K., Kawagoe, Y., Nohara, Y., Namiki, M., Osawa, T., and Kawagishi, S. (1990) *J. Jpn. Soc. Nutr. Food Sci.* **43**: 445-449.
17. Chou, T. C., and Marlatt, A. L. (1953) *J. Nutr.* **51**: 305-315.
18. Hirose, N., Doi, F., Ueki, T., Akazawa, K., Chijiiwa, K., Sugano, M., Akimoto, K., Shimizu, S. and Yamada, H. (1992) *Anticancer Res.* **12**: 1259-1266.
19. Akimoto, K., Kitagawa, Y., Akamatsu, T., Hirose, N., Sugano, M., Shimizu, S. and Yamada, H. (1993) *Ann. Nut. Metabolism.* **37**: 1218-224.
20. Tobin, C. E. (1941) *Endocrinology* **28**: 419-25.

21. Hirose, N., Inoue, T., Nishihara, K., Sugano, M., Akimoto, K., Shimizu, S., and Yamada, H. (1991) *J. Lipid Res.* **32**: 629-638.
22. Axelson, M., Sjoevall, J., Gustafsson, B. E. and Setchell, K. D. R. (1982) *Nature* **298**(5875): 659-660.
23. Thompson, L. U., Robb, P., Serraino, M., Cheung, F. (1991) *Nutr. Cancer* **16**(1): 43-52.
24. Nose, M., Fujimoto, T., Takeda, T., Nishibe, S., Ogihara, Y. (1992) *Planta Med.* **58**(6): 520-523.
25. Cassida, J. E. (1970) *J. Agric. Food Chem.* **18**: 753-772.
26. Ioannides, C.; Delaforge, M. and Parke, D. V. (1981) *Food Cosmet. Toxicol.* **19**: 657-666.
27. Mason, H. S., North, J. C., and Vanneste, M. (1965) *Fed. Proc.* **24**: 1172-1180.

SESAME SEED AND ITS LIGNANS PRODUCE MARKED ENHANCEMENT OF VITAMIN E ACTIVITY IN RATS FED A LOW α–TOCOPHEROL DIET

K. Yamashita, Y. Iizuka, T. Imai and M. Namiki

Department of Food and Nutrition, School of Life Studies,
Sugiyama Jogakuen University, Nagoya 464 Japan

1 INTRODUCTION

Sesame seed is rich not only in oil (about 50%) and protein (about 20%), but also in lignans (such as sesamin and sesamolin). Empirically, sesame oil is stable against the oxidative deterioration in cooking, which appears to be related to its characteristic lignan components[1,2]. On the other hand, vitamin E is recognized as a food component having an anti-aging effect. Gamma-tocopherol has a potent antioxidative activity comparing to α-tocopherol in in vitro tests, whereas its vitamin E activity is only about 5-16% of α-tocopherol [3]. Sesame seed contains only γ-tocopherol accompanying a trace amount of α-tocopherol, indicating it is a relatively poor source of vitamin E. If sesame seed is truly an anti-aging food, a question occurs whether or not it acts without an inherent vitamin E function. We have thus tested effects of sesame seed on vitamin E activities in rats fed diets containing various levels of α-tocopherol. For full details of this investigation see papers published by Yamashita et al[4, 13].

2. VITAMIN E ACTIVITIES OF SESAME SEED IN RATS FED DIETS WITHOUT ALFA-TOCOPHEROL (VITAMIN E-FREE DIET).

To compare the relative vitamin E activities of sesame seed with those of α-tocopherol and γ-tocopherol, rats were fed four different diets for 8 weeks: (a) vitamin-E free diet, (b) α-tocopherol-containing diet (51.7 mg/kg diet), (c) γ-tocopherol-containing diet (51.7 mg/kg diet) and (d) the same diet as C + 20% powdered sesame seed. The content of lipid peroxides in the liver was higher in the control animal group fed vitamin E-free diet than in the α-tocopherol-fed group. The γ-tocopherol fed group contained a significantly higher amount of peroxides than in the α-tocopherol-fed group, whereas the sesame seed-fed animal group contained the same level as the α-tocopherol-fed group. In parallel both red blood cell hemolysis and pyruvate kinase activities were measured as indices of vitamin E status using the same rats. The results obtained supported the data on peroxide analysis. Then α-and γ-tocopherol concentrations in plasma were determined. A high concentration of α-tocopherol was seen only in the α-tocopherol-fed group, whereas in the other groups only a trace amount was detected. A substantial amount of γ-tocopherol was found only in the sesame seed-fed animal group, while it was very low in the γ-tocopherol-fed group, despite the fact that the sesame seed diet and γ-tocopherol supplemented diet contained equal amounts of γ-tocopherol. We also measured the level of tocopherols in liver; results were shown to support the data obtained on the plasma. Overall results appear to indicate the possible presence of components in sesame seed which may raise the content of γ-tocopherol in both plasma and liver and possibly promote their suppressive effect on the lipid peroxidation.

In order to examine the role of components in sesame seed which may raise the blood and liver level of γ-tocopherol we tested sesaminol and sesamin as sesame lignans (Figure 1), by giving them to rats.

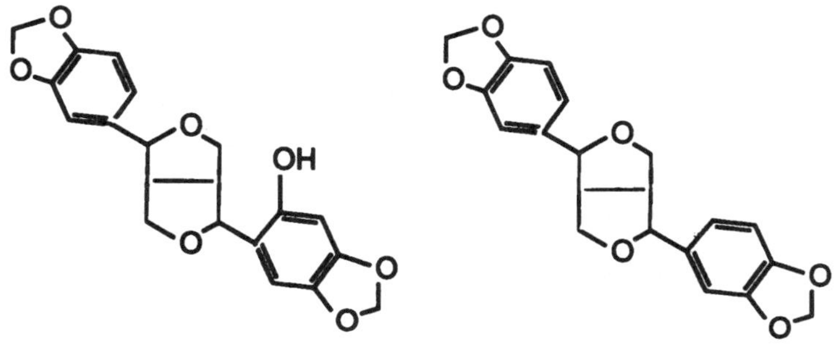

Sesaminol   **Sesamin**

Figure 1 *Chemical structure of sesaminol and sesamin*

It is shown that sesaminol is an antioxidant but sesamin is inactive. A series of experiments were performed using (a) (γ-tocopherol+sesaminol) diet and (b) (γ-tocopherol+sesamin) diet instead of the sesame seed diet. Results obtained were essentially the same as those in the former experiment. The level of lipid peroxide in liver in both animal groups was shown to be nearly the same as that in the α-tocopherol-fed group. Although γ-tocopherol was not detectable in the γ-tocopherol-fed animal, it was shown to be present in the (γ-tocopherol + sesamin)-fed animal group and more significantly in the (γ-tocopherol + sesaminol)-fed group. It appears evident that both sesaminol and sesamin exhibit the effect of enhancing the level of γ-tocopherol in vivo, antioxidant sesaminol being superior to sesamin (non-antioxidant). We thus conclude that sesame seed lignans and γ-tocopherol may act synergistically to produce vitamin E activities in rats [4].

3 SESAME SEED AND ITS LIGNANS PRODUCE MARKED ENHANCEMENT OF VITAMIN E ACTIVITIES IN RATS FED DIET CONTAINING LOW ALFA-TOCOPHEROL.

Synergistic effect between sesame seed lignans and γ-tocopherol was observed in rats fed diet without α-tocopherol supplementation. However, usually both α- and γ-tocopherol are present in an ordinary meal. A number of previous studies have demonstrated that α- and γ-tocopherol are absorbed from intestine in a similar manner, although α-tocopherol is preferentially bound with the transporting protein in liver, and most of γ-tocopherol is excreted through bile without combining with the transporting protein [5-9]. Therefore, it is evidently of importance to clarify whether or not sesame seed lignans may act with α-tocopherol in synergistical manner to raise vitamin E activities. Consequently, we examined how to improve the vitamin E status of the low α-tocopherol-fed animal group by supplementing either sesame seed or its lignans to the basal diets. Rats were fed five different diets: (a) control vitamin E-free diet, (b) low α-tocopherol diet, and three low α-tocopherol diets containing (c) 5%, (d)10%, and (e)15% sesame seed. The level of liver peroxides was found to be extremely low in rats fed the sesame seed and even supplementation of 5% sesame seed to the low α-tocopherol-containing diet caused nearly complete suppression of lipid peroxides formation. Although the low α-tocopherol-fed animal group (b) contained lesser amount of α-tocopherol, it was greater than in the vitamin E-free animal (a); supplementation of 5% sesame seed to the low α-tocopherol-containing diet (c) caused the formation of a significantly higher level of both α- and γ-tocopherol. The enhancement of tocopherol formation was more marked in the 10% supplemented animal group (d), but no further increase was seen in the 15% supplemented group (e). Overall results indicate that supplementation of 5-10% sesame seed to the basal diet appears to be sufficient to improve the status of vitamin E deficiency in the low α-tocopherol-fed animals.

Two different types of diet containing either sesame lignans (sesaminol or sesamin) or low α-tocopherol were tested. The results obtained were essentially the same as in the former experiment. The high level of lipid peroxide observed in the low α-tocopherol-fed group was completely suppressed by supplementing sesame lignans. Concentrations of α-tocopherol in plasma and liver were low in the low α-tocopherol-fed group but slightly higher than those in the vitamin E-free group. In both (low α-tocopherol + sesaminol)- and (low α-tocopherol +sesamin)-fed rat groups significantly higher levels of α-tocopherol were detected in comparison with those in the low α-tocopherol diet without either sesaminol or sesamin, although it was higher in the sesaminol group than in the sesamin group; lower in both groups comparing to the normal α-tocopherol fed group. Therefore, it appears evident that both sesaminol and sesamin exhibit an enhancing effect of α-tocopherol formation in vivo, and that the antioxidant sesaminol being more active than sesamin. Sesaminol is principally present in the refined raw salad oil though only in a small amount in the sesame seed. Recently Katsuzaki et al.[10] reported that there are three new glucosides of sesaminol in defatted sesame cake, and that antioxidative activity can be generated by treating with β-glucosidase.

4 EFFECT OF SESAME SEED ON PLASMA TOCOPHEROL LEVEL IN RATS FED HIGH ALFATOCOPHEROL DIET

Sesame seed and its lignans can improve the vitamin E status in rats fed a low α-tocopherol diet by raising the level of α-tocopherol in blood and tissues. In a series of experiments we studied the effect of sesame seed on the level of plasma tocopherol in rats fed diets containing various levels of α-tocopherol (mg/kg): (a) none, (b)10 (low), (c) 50 (normal), and (d) 250 (high) by supplementing with 20 % sesame seed containing about 50 mg/kg γ-tocopherol. Rats were divided into 8 groups, each group receiving the experimental diet for 8 weeks.

Results obtained showed that the level of α-tocopherol was increased in parallel with the dietary concentration of α-tocopherol. Furthermore, sesame seed caused a significant increase of α-tocopherol in the plasma of rats fed diets containing low, normal, and high concentrations of α-tocopherol, but it did not affect the animal group fed without α-tocopherol. On the other hand, γ-tocopherol was detected only in both (no α-tocopherol + sesame seed)- and (low α-tocopherol + sesame seed)-fed groups, but not in the normal as well as high α-tocopherol supplemented sesame seed fed animal groups. This was observed invariably despite the fact that sesame seed contains about 25 mg/100g γ-tocopherol but α-tocopherol in a negligible amount. According to recent studies using various forms of vitamin E, it is found that there is no discrimination between α- and γ-tocopherols during absorption in animal body, but following the uptake in the liver only RRR-α-tocopherol can be preferentially bound to tocopherol-binding protein and eventually secreted in conjugation with the nascent very low density lipoprotein (VLDL)(6). Therefore, it is likely that discrimination between different forms of tocopherol may occur on the hepatic tocopherol binding protein molecule. As described above, we have observed that rats fed γ-tocopherol (50mg/kg diet) showed very low level of γ-tocopherol in plasma and liver, which was elevated significantly by the addition of sesame lignans to diet containing the same amount of γ-tocopherol. These results may indicate that the binding activity of γ-tocopherol with tocopherol binding protein is weak, as previously reported by Sato et al.[7] as well as by other workers[8,11,12]. However, a possibility cannot be excluded that the presence of sesame lignans may enhance the binding activity of γ-tocopherol with the binding protein.

In this experiment, we found higher concentrations of α-tocopherol in plasma of rats fed diets containing both α-tocopherol and sesame seed in comparison with rats fed diets containing only α-tocopherol. It is of interest to note that only γ-tocopherol being detectable in either none or low α-tocopherol + sesame seed groups, but not in the normal or high α-tocopherol + sesame seed groups, despite the fact that the rats had ingested an equal amount of γ-tocopherol. These results may indicate that discrimination between α- and γ-tocopherols exists in rats fed the 20% sesame seed diets supplemented with α-tocopherol. Results also suggest that the binding activity of tocopherol isomers with tocopherol binding protein can be augmented by the sesame lignans, although discrimination between α- and γ-tocopherols will be a more likely case.

5 CONCLUSIONS

In the present experiment, we have found that sesame seed and its lignans cause a high α-tocopherol concentration in plasma and tissues in rats. Based on the results obtained, we are inclined to propose that one possible rationale for making sesame seed as a health food may exist in its enhancing effect on the vitamin E activities.

References

1. M. Namiki *Food Reviews International*, 1995 **11**, 281.
2. Y.Fukuda, M. Nagata, T. Osawa and M. Namiki *J. Am. Oil Chem. Soc.* 1986, **63**, 1027.
3. I.R.Peake and J. G. Bieri, *J. Nutr.*, 1971,**101**, 1615.
4. K.Yamashita, Y. Nohara, K. Katayama and M. Namiki *J. Nutr.*, 1992 **122**, 2440.
5. G.L. Catignani and J. G. Bieri *Biochim. Biophys. Acta* ,1977,**497**, 349.
6. H.J. Kayden and M. G. Traber *J. Lipid Res.*, 1993, **34**, 343.
7. Y. Sato, K. Hagiwara, H. Arai and K. Inoue *FEBS Lett.*, 1991, **288**, 41.
8. M.G. Traber,and H. J. Kayden *Am J. Clin. Nutr.*,1989, **49**, 517.
9. H. Yoshida, M. Yusin, J. Kuhlenkamp, A. Hirano, A. Stolz and Kaplowitz, *J. Lipid Res.*, 1992, **33**, 343.
10. H. Katsuzaki, S. Kawagishi and T. Osawa *Phytochemistry*, 1994, **35**, 773.
11. W. A. Behrens and R. Madere *Nutr. Res.*, 1983, **3**, 891.
12 W. A. Behrens and R. Madere, *J. Nutr.*, 1987, **117**, 1562.
13. K. Yamashita, Y. Iisuka, T. Imai and M. Namiki, Lipids 1995,**30**,1019

ANTIOXIDANT ACTIVITY OF SESAMIN ON NADPH-DEPENDENT LIPID PEROXIDATION IN LIVER MICROSOMES

K. Akimoto*, S. Asami*, T. Tanaka*, S. Shimizu**, M. Sugano*** and H. Yamada****

*Institute for Biomedical Research, Suntory Ltd., Shimamoto-cho, Mishima-gun, Osaka 618, Japan; **Department of Agricultural Chemistry, Kyoto University, Sakyo-ku, Kyoto 606, Japan; ***Department of Food Science and Technology, Kyushu University, Higashi-ku, Fukuoka 812-81, Japan; ****Faculty of Engineering, Toyama Prefectural University, Kosugai-machi, Imizu-gun, Toyama 939-03, Japan

1 INTRODUCTION

Sesame seed has long been known as a traditional health food or a medicinal plant;[1,2] however, the biological background of its desirable functions has not entirely been appreciated. Recently, the multiple biological functions of sesamin, the lignan exclusively occurring in sesame seeds and unroasted sesame oil, have been elucidated.[3-9] These include; (a) interference with linoleic acid metabolism, and hence, eicosanoid production, through an interference with Δ5-desaturase,[3,4] (b) hypocholesterolemic action through a simultaneous inhibition of intestinal cholesterol absorption and hepatic cholesterol synthesis,[5,6] (c) enhancement of the hepatic detoxication of chemicals and alcohol possibly through an increased microsomal activity,[7] (d) protective effect on deoxycorticosterone acetate-salt-induced hypertension and cardiovascular hypertrophy,[8] (e) protective effect on chemically-induced mammary cancer.[9] It has been known that sesamin does not have an antioxidant activity for the oxidation of edible oil and fat except the minor lignan compounds carrying a hydroxyl group, i.e. sesaminol, episesaminol and sesamolinol. But sesamin has lowered the thiobarituric acid-reactive substance (TBARS) concentration in plasma and liver lipids of rats with chemically-induced mammary cancer, indicating an antioxidant activity *in vivo*.

We have shown that sesamin is a hydroxyl radical scavenger and a potent inhibitor of NADPH-dependent microsomal lipid peroxidation, and that it is converted to a single product in this oxidative process.

2 QUENCHING OF ACTIVE OXYGEN

Active oxygen contains the superoxide radical, the hydroxyl radical, singlet oxygen and hydrogen peroxide. We examined the scavenging activities of sesamin, α-tocopherol, quercetin and catechin on the superoxide radical, hydroxyl radical and organic radical.

In the assay for the organic radical (1,1-diphenyl-2-picrylhydrazl (DPPH)) scavenging, sesamin or other chemicals as a 1.25 vol% DMSO solution, were added to 50 μM DPPH in ethanol, followed by incubation at 25°C. Their reactivity, with DPPH was determined spectophotometrically at 516 nm. As shown in Figure 1A, DPPH was quenched and bleached rapidly by, α-tocopherol, quercetin and catechin, but sesamin did not react with DPPH.

In the assay for the superoxide radical scavenging, the reaction mixture (1.0 ml), comprising 0.16 mM NADH, 6.0 μM phenazine methosalfate and 0.1 mM nitro blue tetrazolium in 0.1 mM potassium phosphate buffer, pH 7.5, was incubated at 25°C, and the formazan formed by the superoxide radical generated in the reaction mixture was measured spectrophotometrically as the increase in the absorbance at 560 nm. As shown in Figure 1B, sesamin and α-tocopherol exhibited no potential to quench the superoxide radical, but quercetin and catechin reacted with the superoxide radical, thereby inhibiting formazan formation in a dose-dependent manner.

In the assay for the hydroxyl radical scavenging, the reaction mixture (1.0 ml), comprising 0.4 mg/ml of rat liver microsomal protein in 0.1 M potassium phosphate buffer, pH 7.4, 0.3 mM NADPH, was incubated at 37°C for 20 min. Lipid peroxidation was initiated by the addition of NADPH. The TBARS concentration was determined after terminating the reaction by the addition of 2 ml of TBA reagent according to the method of Buege and Aust.[10] As shown in Figure 1C, sesamin, quercetin and catechin inhibited the NADPH-dependent microsomal lipid peroxidation in a dose-dependent manner. The IC50s were 0.4 µM for both sesamin and quercetin, and 0.35 µM for catechin. These results demonstrate that sesamin exhibits a specific scavenging activity of the hydroxyl radical.

3 PRODUCT DERIVED FROM SESAMIN ON NADPH-DEPENDENT MICROSOMAL LIPID PEROXIDATION

Isolation of the reaction product(s) derived from sesamin in the reaction was attempted. After incubating microsomes with NADPH in the presence of sesamin for two hours at 37°C, sesamin and the reaction products were extracted with chloroform. The reaction product was purified by HPLC and the purified compound was acetylated, because mass fragment peaks of the original compound were very weak, and analyzed with a mass spectrometer. As shown in Figure 2, the compound was identified as 2-(3,4-methylenedioxyphenyl)-6-(3,4-dihydroxyphenyl)-cis-dioxabicyclo[3.3.0]octane from the molecular ion peak at m/z 426, which corresponds to the molecular weight of the microsomal reaction product (m/z 342) with two added acetyl groups. Inhibitory activity of the reaction product, compared with that of sesamin in NADPH-dependent lipid peroxidation was examined, and both the product and sesamin inhibited it in almost the same dose-dependent manner.

Detection of sesamin and its metabolite in liver of sesamin-fed rats was atempted next. Eight-week-old male Sprague-Dawley rats were dosed with 500 mg/kg of sesamin, dissolved in soybean oil orally. After 0, 0.5, 1, 3, 5, 9 and 24 hours, sesamin and its metabolite in liver were extracted and analyzed by HPLC. One half-hour after administration, sesamin concentration peaked and decreased thereafter as shown in Figure 3. In contrast, metabolite concentration increased after three hours, then peaked and decreased.

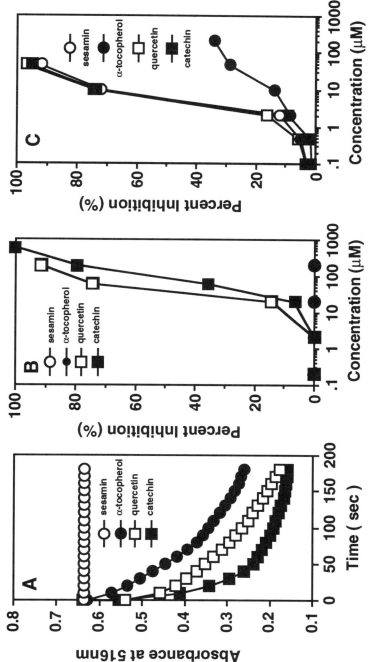

Figure 1 Effects of sesamin, α-tocopherol, quercetin and catechin on quenching of DPPH (A), the superoxide radical (B) and the hydroxyl radical (C).

244 Natural Antioxidants and Food Quality in Atherosclerosis and Cancer Prevention

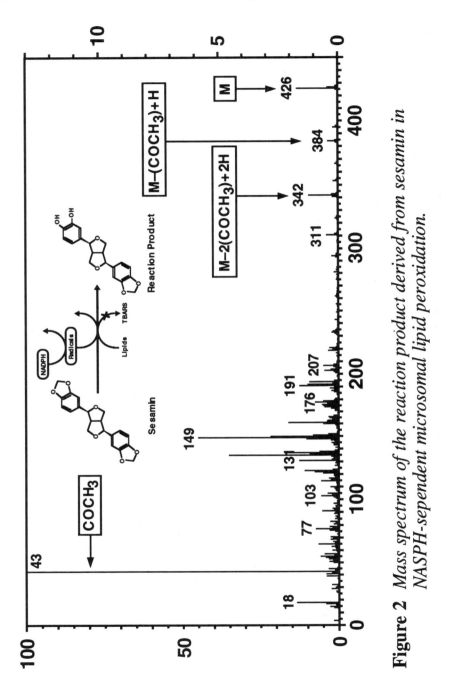

Figure 2 *Mass spectrum of the reaction product derived from sesamin in NASPH-sependent microsomal lipid peroxidation.*

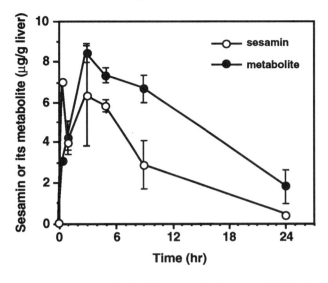

Each point represents the mean ± SE (n=3)

Figure 3 *Detections os sesamin and its metabolite in liver of sesamin-fed rats.*

4 SESAMIN AS A RADICAL SCAVENGER

Our study can be summarized by Figure 4. In the target cells or tissues, the superoxide generated by xanthine oxidase or NADH oxidase was converted into the hydroxyl radical via hydrogen peroxide. The hydroxyl radical causes oxidative cell damage. Quercetin and catechin quenched the superoxide before quenching the hydroxyl radical. But sesamin quenched the hydroxyl radical directly. This reaction product produced from sesamin can not only quench the hydroxyl radical, but also the superoxide. Sesamin is stronger than quercetin and catechin for the scavenging activity of the hydroxyl radical. The various effects of sesamin may at least partly be ascribed to its antioxidant activity.

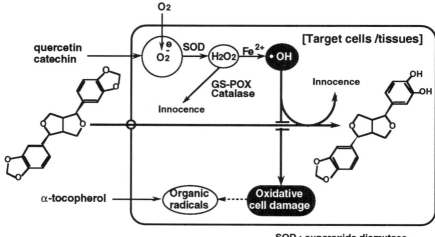

Figure 4 *Sesamin as a specific scavenger of the hydroxyl radical.*

References

1. P. Budowski and K. S. Markley, *Chem. Rev.*, 1951, **48**, 125.
2. M. Namiki and T. Kobayashi, 'Science of sesame', Asakura Shoten, Tokyo, 1989.
3. S. Shimizu, K. Akimoto, H. Kawashima, Y. Shinmen and H. Yamada, *J. Am. Oil Chem. Soc.*, 1989, **66**, 237.
4. S. Shimizu, K. Akimoto, Y. Shinmen, M. Sugano and H. Yamada, *Lipids*, 1991, **26**, 512.
5. M. Sugano, T. Inoue, K. Koba, K. Yoshida, N. Hirose, Y. Shinmen, K. Akimoto and T. Amachi, *Agric. Biol. Chem.*, 1990, **54**, 2669.
6. N. Hirose, T. Inoue, K. Nishihara, M. Sugano, K. Akimoto, S. Shimizu and H. Yamada, *J. Lipis Res.*, 1991, **32**, 629.
7. K. Akimoto, Y. Kitagawa, T. Akamatsu, N. Hirose, M. Sugano, S. Shimizu and H. Yamada, *Ann. Nutr. Metab.*, 1993, **37**, 218.
8. Y. Matsumura, S. Kita, S. Morimoto, K. Akimoto, M. Furuya, N. Oka and T. Tanaka, *Bio. Pharm. Bull.*, 1995, **18**, 1016.
9. N. Hirose, F. Doi, T. Ueda, K. Akazawa, K. Chijiiwa, M. Sugano, K. Akimoto, S. Shimizu and H. Yamada, *Anticancer Res.*, 1992, **12**, 1259.
10. J. A. Buege and S. T. D. Aust, *Method Enzymol.*, 1978, **52**, 302.

Dietary Flavonoids and Role as Natural Antioxidants in Humans

DIETARY FLAVONOIDS AND OXIDATIVE STRESS

Piergiorgio Pietta [1], Paolo Simonetti [2], Carla Roggi [3], Antonella Brusamolino [2], Nicoletta Pellegrini [2], Laura Maccarini [3] and Giulio Testolin [2]

[1] I.T.B.A., CNR, Milan, Italy
[2] Department of Food Science & Technology, University of Milan, Italy
[3] Department of Preventive, Occupational and Community Medicine - Section of Hygiene, University of Pavia, Italy

1 INTRODUCTION

The oxidative stress arising from an imbalance in the human antioxidant status (reactive oxygen species vs. defence and repair mechanisms) is responsible for the development of oxidative diseases, such as cardiovascular diseases, brain dysfunction, immune-system decline and also for aging[1,2]. Among the endogenous defences are enzymes such as superoxide dismutase, catalase and glutathione peroxidase. Besides endogenous defences, consumption of dietary antioxidants appears to be of great importance. Among these, flavonoids and related phenolic have gained increasing interest[3].

1.1 Flavonoids Occurrence and Analysis

Flavonoids are a large group of polyphenolic antioxidants that occur naturally in vegetables and fruits and beverages, such as tea, wine and beer[4,5]. The content of flavonoids in foods is still matter of debate, the information is scattered, and, generally, existing data do not always comply with nowadays analytical demands. The analysis is complicated by the fact that flavonols and flavones usually occur in plants as O-glycosides, for which no commercial standards are currently available. Therefore, in order to have quantitative results, glycosides must first be hydrolyzed into respective aglycones by acid hydrolysis[6-8]. Nevertheless, this approach may present problems related to the stability of the analytes. The analytical methods to quantify main flavonoids in selected foods and beverages need to be validated by a collaborative inter-laboratories study to make results comparable.

1.2 Flavonoid Antioxidant Properties

Flavonoids are antioxidant, since they scavenge free radicals to produce a phenoxyl radical (Figure 1).
The reduction potentials of flavonoid phenoxyl radicals are in the range 540-700 mV[9]. Thus, it is expected that the corresponding parent flavonoids efficiently inactivate various reactive oxygen species [(ROS) (OH$^\bullet$, ROO$^\bullet$, O$_2^{\bullet}$)] with higher potential (2000-950mV)[10].

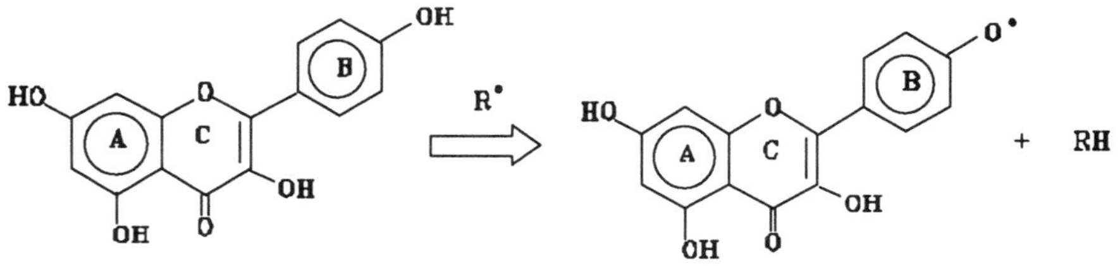

Figure 1 *Free radical (R˙) scavenges by flavonoids*

Since these ROS are produced during digestion, flavonoids or their metabolites are supposed to play their scavenging role in the digestive tract. When absorbed (as metabolites), flavonoids may shield endogenous antioxidants, like urate, vitamin E, vitamin C, and β-carotene from the aggressive ROS. Nevertheless, due to the higher potential, flavonoid phenoxyl radicals are thermodynamically able to oxidize these compounds [9,10].

1.3 Flavonoids Kinetics

Most positive evidence on the antioxidant role of flavonoids has been reached in experimental situations, without considering any synergistic effect due to the presence of other flavonoidic or non-flavonoidic components occurring in natural matrices. Moreover, preliminary studies have shown that flavonoids are only badly resorbed and quickly metabolized, and whether the metabolites themselves are responsible for the effects is a matter of study [11-13].

1.4 Flavonoids and Cardiovascular Hearth Diseases (CHD)

Several flavonoids and related phenolics have been reported to inhibit either enzymatic or non-enzymatic lipid peroxidation[14], which is an oxidative process implicated in several pathologic conditions, including atherosclerosis, hepatoxicity and inflammation. Quercetin, myricetin and phenolic acid derivatives have an activity comparable to that α-tocopherol[15]. Phenolic compounds extracted from red wine are inhibitors of LDL oxidation by macrophages which is a process involved in CHD[16]. Rutin and other flavonols can affect platelet aggregation and adhesion by their inhibitory effect on cyclo-oxygenase and lipoxygenase activity[17].

Despite this amount of information, limited evidence for the effects of flavonoids in human beings is available. In this context, our efforts have been focused on the relation between flavonoid intake and specific clinical variables related to the oxidative stress, and the results obtained in two distinct studies will be considered.

2 EXPERIMENTAL STUDIES

2.1 Dietary Flavonoid Intakes (Study 1)

Using the information provided by a food survey, the average diet of 432 subjects living in a northern Italian agricultural village (Rovescala, Italy) was established and then the food list was prepared. The food items included were purchased and, after preparation and cooking, were combined in 32 pools. The flavonoids content was determined and quantified both by HPLC/DAD and CE/DAD[18]. α-Tocopherol, β-carotene and retinol were determined and quantitated by HPLC in pools after saponification and extraction, whereas the average content of vitamin C, protein, carbohydrate and lipid in different pools were assessed using food composition tables.

Among flavonols quercetin resulted the main aglycone, whereas apigenin was the predominant flavone. Lower amounts of kaempferol and luteolin could be detected without affecting the total amount, which was then expressed as the sum of quercetin and apigenin.

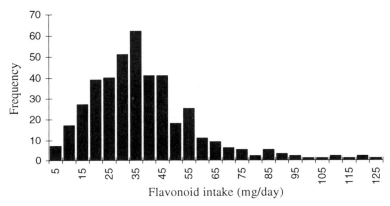

Figure 2 *Histogram of flavonoid intake of 432 Rovescala adults*

The average flavonoids intake of the Rovescala population (Figure 2) was 35.3 mg/day, with quercetin and apigenin accounting for 87.2 and 12.8% respectively.

The main sources of flavonoids in the examined population were fruits, vegetable soups and salads, whereas beverages represented a very low contribution (Table 1).

Table 1 Main sources of flavonoid in Rovescala population by tertile of intake			
Flavonoid intake, mg/day	0 - 26.0	26.1 - 38.7	> 38.7
Mean (SE), mg/day			
Fruits	6.6 (1.1)	19.5 (1.2) #	34.2 (2.9) *
Vegetable soup with legumes	4.0 (0.8)	6.9 (1.1)	12.8 (1.9) *
Salads	2.6 (0.4)	4.4 (0.5)	4.3 (0.7)
Wine	0.8 (0.2)	0.6 (0.2)	0.7 (0.2)
Vegetables	0.7 (0.2)	0.6 (0.2)	0.7 (0.1)
Total flavonoids	14.6 (1.3)	32.0 (0.8) #	52.7 (2.5) *

* $p < 0.001$ versus 0-26.0 and 26.1-38.7 mg/day flavonoid intake
\# $p < 0.001$ versus 0-26.0 mg/day flavonoid intake

As shown in table 2, besides β-carotene and vitamin C, there were no significant differences in protein, fat, carbohydrate, retinol, α-tocopherol intake from the lowest to the highest tertile of flavonoid consumption. Also the values of main clinical parameters (Total and HDL-cholesterol, APO-AI, γ-GT, ASAT, ALAT, Glycemia and blood nitrogen) were not influenced by flavonoid intake, while triglycerides diminished at increasing flavonoid intake, thus confirming the hypolipidemic activity already reported[19].

Table 2 *Dietary nutrient of partecipants by tertile of flavonoid intake*

Flavonoid intake, mg/day	0 - 26.0	26.1 - 38.7	> 38.7
Mean (SE)			
Animal proteins (g daily)	60 (5)	57 (5)	59 (5)
Vegetable proteins (g daily)	33 (3)	28 (2)	33 (2)
Saturated fat (g daily)	32 (3)	31 (3)	33 (2)
Monounsaturated fat (g daily)	47 (4)	47 (3)	51 (3)
Polyunsaturated fat (g daily)	8 (1)	8 (1)	8 (1)
Starch (g daily)	216(16)	172(18)	181(17)
Sugar (g daily)	76 (10)	100(15)	111(18)
Retinol (mg daily)	1.2 (0.3)	0.9 (0.1)	1.4 (0.2)
Ascorbic acid (mg daily)	44 (10)	118(10) #	202(10) # *
β-Carotene (mg daily)	1.6 (0.2)	2.3 (0.2) #	3.1 (0.2) # *
α-Tocopherol (mg daily)	8.3 (0.8)	8.5 (1.0)	9.1 (0.7)

\# $p < 0.05$ versus 0-26.0 mg/day flavonoid intake
* $p < 0.05$ versus 26.1-38.7 mg/day flavonoid intake

α-Tocopherol and retinol levels in plasma and RBC were within the normal range (Table 3). On the contrary, β-carotene increased significantly ($p<0.001$) in the higher tertile of flavonoid intake. This result is of particular interest, since the increase of β-carotene levels can only partly due to its dietary intake.

Table 3 *Plasma and RBC vitamins of partecipants by tertile of flavonoid intake*

Flavonoid intake, mg/day	0 -26.0	26.1 - 38.7	> 38.7
Mean (SE)			
Plasma			
α-Tocopherol, μg/ml	14.9 (1.0)	15.2 (1.0)	14.2 (1.1)
β-Carotene, μg/dl	21.1 (3.3)	19.3 (2.7)	30.6 (5.0) #
Retinol, μg/dl	58.1 (4.9)	59.2 (4.0)	53.9 (3.6)
RBC			
α-Tocopherol, μg/dl	2.24 (0.23)	2.31 (0.18)	2.26 (0.26)
β-Carotene, μg/dl	0.65 (0.18)	0.50 (0.11)	0.94 (0.30) #

\# $p < 0.001$ versus 0-26.0 and 26.1-38.7 mg/day flavonoid intake

To evaluate the effect on membrane composition, the fatty acid composition of red blood cells phosphatidylcholine (PC) was examined. Total polyunsaturated fatty acid (PUFA) concentration raised significantly ($p<0.05$) from the lowest to the highest tertile of flavonoid intake (Figure 3).

This PUFA oriented modification influences the membrane properties, mainly resulting in a reduced viscosity. When considering phosphatidylethanolamine (PE), the intake of flavonoids produced less change of fatty acid composition, probably due to the higher PUFA content of PE.

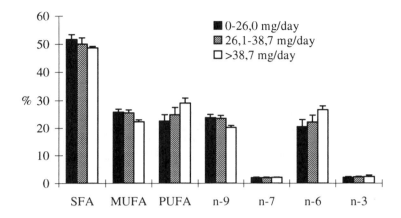

Figure 3 *Fatty acid composition of RBC phosphatidylcholine by tertile of flavonoid intake*

2.2 Tea Catechin Intake (Study 2)

Tea catechins are known to inhibit either enzymatic or non-enzymatic lipid peroxidation [14], but only limited evidence for positive effects in humans is available.

Thus fifteen healthy subjects, aged 20-40 years, from our laboratory population, were selected for a supplementation with 100 mg/die of decaffeinated green tea extracts tablet (Indena S.p.A, Milan, Italy). The examination of the subjects included a self-administered questionnaire to assess the intake of fruits and vegetables for selecting those subjects with an habitual low intake of flavonoid rich foods. Fasting venous blood samples were taken at baseline, after receiving the supplement continually for 4 weeks and after having terminated supplementation for a period of 4 weeks (wash-out).

Flavonoid and catechin content of each tablet was evaluated by a HPLC-DAD method[6]. The composition is reported in Table 4.

Table 4 *Flavonoid and catechin content of each tablet*	
Compound	mg/tablet
Epigallocatechingallate	66.7
Epicatechingallate	6.9
Epigallocatechin	1.2
Epicatechin	1.9
Catechin	0.7
Quercetin	0.15
Kaempferol	0.08
Isorhamnetin	trace

Concerning the lipid metabolism, no difference was observed for total- and HDL-cholesterol, whereas plasma triglycerides slightly decreased at the end of treatment, confirming the hypolipidemic effect previously reported in our Study 1 and by some authors[19].

Plasma β-carotene increased significantly (p<0.05) after the treatment and remained high also after the wash-out period (Table 5). Plasma α-tocopherol diminished (11%, p < 0.05) after the treatment and increased after the wash-out (24%, p < 0.05); on the contrary, RBC α-tocopherol raised continuously, and this behaviour was accompanied by a significant increase of PUFA in phospholipids membrane.

These results seem to confirm the protective effect on plasma levels of β-carotene and a sparing effect on PUFA evidenced in Study 1.

Table 5 *Plasma and RBC vitamins at the end of various treatments*

Plasma and RBC vitamins	Before receiving supplement		After receiving supplement		After terminating supplement	
Mean (SD)						
plasma retinol (μg/dl)	48.2	(11.9)	47.8	(12.2)	44.8	(10.0)
plasma α-tocopherol (μg/ml)	10.6	(1.3)	9.6	(1.6) *	11.7	(1.9) * #
plasma β-carotene (μg/dl)	32.0	(21.1)	37.9	(20.4) *	42.4	(21.3) *
plasma ascorbic acid (mg/dl)	0.94	(0.40)	0.98	(0.31)	0.72	(0.20) * #
RBC α-tocopherol (μg/g)	1.89	(0.37)	1.99	(0.26)	2.36	(0.31) * #

* $p < 0.05$ versus before supplement
\# $p < 0.05$ versus after supplement

Moreover, the intake of green tea extract resulted in a diminished ADP-induced platelet aggregation (+ 21% ADP required for the second wave) (Table 6). On the contrary, collagen-induced platelet aggregation was not affected by supplement of tea extract.

Table 6 *Platelet aggregation at the end of various treatments*

Platelet aggregation	Before receiving supplement		After receiving supplement		After terminating supplement	
Mean (SD)						
AC50 to collagen (μg/ml)	0.60	(0.06)	0.60	(0.06)	0.74	(0.25)
Minimum dose of ADP required for second wave (μM)	2.3	(1.0)	2.8	(1.1)	2.4	(1.0)

3 CONCLUSIONS

It may be concluded that all these effects, i.e. the increase of plasma and res blood cells β-carotene, the PUFA oriented modification of RBC phospholipids and the decrease of plasma triglycerides, combined with the known evidence that flavonoids inhibit free-radical oxidation of LDL, are a reasonable explication of the reduced risk of coronary heart diseases assessed in the Zutphen Elderly Study [20].

4 REFERENCES

1. O.I. Aruoma, *Fd. Chem. Toxic.* 1994, **32**, 671.
2. J. Lunec, *Biochim. Clin.* 1992, **16(2)**, 99.

3. M.J. Sanz, M.L. Ferràndiz, M.Cejudo, M.C. Terencio, B.Gil, G. Bustos, A.Ubeda, R. Gunasegaran, M.J. Alcaraz, *Xenobiotica* 1994, **24**, 689.
4. M.G.L. Hertog, P.C.H. Hollman, M.B. Katan, *J. Agric. Food Chem.* 1992, **40**, 2379
5. M.G.L. Hertog, P.C.H. Hollman, B. van de Putte, *J. Agric. Food Chem.* 1993, **41**, 1242.
6. P.G. Pietta, P.L. Mauri, C.Gardana, *HRC*, 1993, 1535.
7. P.G. Pietta, P.L. Mauri, C.Gardana, *HRC*, 1994, **17**, 616.
8. M.G.L. Hertog, P.C.H. Hollman, D.P. Venema, *J. Agric. Food Chem.* 1992, **40**,1591.
9. S.V. Jovanovic, S. Steenken, M. Tosic, B. Marjanovic, M.G. Simic, *J. Am. Chem. Soc.* 1994, **116**, 4846.
10. G.R. Buettner, *Arch. Biochem. Biophis.* 1993, **300 (2)**, 535.
11. A. Griffiths, S. Brown, M. Hackett, I.C. Shaw, *Proc. Intern. Bioflav. Symp.* 1981, 451.
12. A.M. Hackett, 'Plant flavonoid in biology and medicine: biochemical, pharmacological and structure-activity relationships', Cody V., Middleton E. and Harborne J.B., New York, 1986, p. 177.
13. P.G. Pietta, C. Gardana, P.L. Mauri, R. Maffei-Facino, M. Carini, *J. Chromatogr. B.* 1995, in press.
14. N.P. Das and A.K. Ratti, 'Plant flavonoid in biology and medicine: biochemical, pharmacological and structure-activity relationships', Cody V., Middleton E. and Harborne J.B., New York, 1986, p. 243.
15. T. Osawa, A. Ide, J.D. Su, M. Namiki, *J. Agric. Food Chem.* 1987, **35**, 808.
16. E.N. Frankel, J. Kanner, J.B.German, E. Parks, J.E. Kinsella, *Lancet* 1993, **341**, 454.
17. R.J. Gryglewski, R. Korbut, J. Robak, J. Swies, *Biochem. Pharmac.* 1987, **36(3)**, 317.
18. P.G. Pietta, P.L: Mauri, P. Simonetti, G. Testolin, *Fresenius J. Annal. Chem.* 1995, **352**, 788.
19. Z.A. Khushbaktova, V.N. Syrov, E.K Batirov, *Khim-Farm. Zh.* 1991, **25(4)**, 53.
20. M.G.L. Hertog, E.J.M. Feskens, P.C.H. Hollman, M.B. Katan, D. Kromhout, *Lancet*, 1993, **342**, 1007.

THE RELATIVE ANTIOXIDANT ACTIVITIES OF PLANT-DERIVED POLYPHENOLIC FLAVONOIDS

Nicholas J. Miller

Free Radical Research Group,
Division of Biochemistry and Molecular Biology,
UMDS, University of London,
Guy's Hospital, London SE1 9RT.

1. INTRODUCTION

The relative antioxidant activities of a range of plant-derived polyphenolic flavonoids have been assessed [1]. These substances are hydrogen-donating antioxidants by virtue of the number and arrangement of the constituent phenolic hydroxyl groups; their metal-chelating potential may also play a role in some instances. Major dietary constituents include flavon-3-ols such as quercetin (onion, tomato, berries, olive oil, red wine, tea) and kaempferol (endive, leek, radish, tea); anthocyanins such as cyanidin (blackcurrant, raspberry, strawberry) and malvidin (grapes); flavan-3-ols such as catechin and epicatechin which are abundant in tea, where they are present both in the free form and as esters of gallic acid (3,4,5-trihydroxybenzoic acid) [2]; and flavanones such as hesperetin and taxifolin (citrus fruit). Other than the flavan-3-ols in tea, these polyphenolic flavonoids are seldom found in nature in the aglycone form and are more usually present conjugated to carbohydrate residues.

Since animals cannot synthesize the flavane nucleus (2-phenyl-benzo-γ-pyrane) the sole source of them is from dietary plants, in which they are derived from the de-amination of phenylalanine and from tyrosine via para-coumaric acid. The daily intake of these substances in the American diet was estimated by Kuhnau [3] as c. 1 gm per day. There is a lot of current interest in the role of flavonoids as nutrient antioxidants and the association between high dietary intake of plant material and a low incidence of cancer and coronary heart disease. A rigorous quantitative method is presented here for characterising their antioxidant activity. The relative significance of the positions and extents of hydroxylation of their structures to the total antioxidant activity of these plant polyphenolics is discussed.

The hydroxycinnamic acids are also major phenolic constituents of the diet (tomatoes, apples, potatoes, grains). While cinnamic acid itself has no antioxidant activity, its hydroxylated derivatives are antioxidants of varying potential; for example, caffeic acid and its quinic acid ester (chlorogenic acid) have an activity in excess of that of α-tocopherol or ascorbate [4].

2. METHODS

The Trolox Equivalent Antioxidant Capacity (TEAC) can be used to compare the antioxidant activity of different compounds and to explore the antioxidant content of complex mixtures [5]. The TEAC is equal to the millimolar concentration of a Trolox solution having the antioxidant capacity equivalent to a 1.0 mM solution of the substance under investigation. The protocol is as follows. A 10 mM solution of the pure substance is prepared (in PBS or in ethanol if the compound is not sufficiently water-soluble). This solution is analysed for total antioxidant activity (TAA) using the ferrylmyoglobin/ABTS$^{\bullet+}$ assay. If it proves to have antioxidant activity, serial dilutions of the compound are analysed until an estimate is obtained of the TEAC value. Three different dilutions of the compound are then selected which produce absorbance values in the most linear region of the Trolox dose-response curve (40 - 80 % inhibition of the blank value, equivalent to an initial concentration of 1.0 - 2.0 mM Trolox). At a minimum, these three different dilutions of each stock solution are analysed in triplicate on three seperate days (n = 3, i.e. 27 determinations *in toto*). The TEAC is calculated for each dilution and the mean value of all the results derived.

3. RESULTS

TEAC values on selected polyphenols and phenols [1,2] are shown in Table 1: these values give an estimate of the relative hydrogen-donating antioxidant potentials of the pure substances against radicals generated in the aqueous phase. The enhanced values of the catechin-gallate esters epicatechin gallate (ECG) and epigallocatechin gallate (EGCG) as compared to catechin and epicatechin reflect the additional contribution from the trihydroxybenzoate, gallic acid (TEAC 3.0). Thus the epicatechin structure (TEAC 2.5), when modified to ECG by a 3-OH ester linkage to gallic acid, has an enhanced antioxidant potential (TEAC 4.9). Unsubstituted catechin and epicatechin still have TEAC values more than double those of ascorbic acid and α-tocopherol. Quercetin and catechin have the same pattern of hydroxyl group substitutions, but quercetin also has an unsaturated 2,3 bond, together with a 4-oxo function, in the C ring. These modifications increase the TEAC from 2.4 to 4.7.

If the C3-OH group is blocked, as in rutin (quercetin rutinoside) (TEAC 2.4) there is a 50% loss of antioxidant activity. Removal of the C3-OH group from the quercetin structure, as in luteolin (TEAC 2.1) produces a similar decline in activity. A greater decrease, however, is seen with taxifolin (dihydroquercetin, TEAC 1.9), where the OH substituents are maintained as in quercetin, together with the C4-oxo function, but the 2,3 bond in the C ring becomes saturated.

Re-arrangement of the pattern of hydroxyl substitution in the B ring of quercetin has a major impact on the hydrogen-donating antioxidant potential. In kaempferol, where the 3'-OH group is lost there is a decline in the TEAC from 4.7 to 1.3: kaempferol is still, however, in this respect a better antioxidant than vitamins C and E. An additional hydroxylation of the B ring, as in myricetin, also produces a decline in the TEAC (to 3.1), showing that the 3',4'-OH pattern (seen in quercetin and cyanidin) is specific to an enhanced activity. Moving the 4'-OH to the 5' position, as in morin (TEAC 2.55) also greatly reduces the antioxidant activity.

Table 1

Trolox Equivalent Antioxidant Activities of polyphenols and phenols

Compound	Free OH-substituents	TEAC	Family
epicatechin gallate		4.93±0.2 [3]	flavan-3-ol
epigallocatechin gallate		4.75±0.06 [3]	flavan-3-ol
quercetin	3,5,7,3',4'	4.7±0.1 [6]	flavon-3-ol
cyanidin	3,5,7,3',4'	4.4±0.12 [5]	anthocyanin
myricetin	3,5,7,3',4',5'	3.1±0.30 [6]	flavon-3-ol
gallic acid	3,4,5	3.0±0.05 [7]	benzoic acid
morin	2',3,4',5,7	2.6±0.02 [3]	flavon-3-ol
epicatechin	3,5,7,3',4'	2.5±0.02 [6]	flavan-3-ol
catechin	3,5,7,3',4'	2.4±0.05 [9]	flavan-3-ol
rutin	5,7,3',4'	2.42±0.06 [7]	flavon-3-ol
peonidin	3,5,7,4' (3'-OMe)	2.22±0.2 [4]	anthocyanin
luteolin	3',4',5,7	2.09±0.5 [4]	flavone
malvidin	3,5,7,4' (3',5'-di-OMe)	2.06±0.1 [4]	anthocyanin
taxifolin	3,5,7,3',4'	1.9±0.03 [6]	flavanone
naringenin	5,7,4'	1.53±0.05 [4]	flavanone
apigenin	5,7,4'	1.45±0.08 [6]	flavone
chrysin	5,7	1.43±0.07 [6]	flavone
hesperetin	3,5,7 (4'-OMe)	1.37±0.08 [3]	flavone
kaempferol	3,5,7,4'	1.34±0.08 [6]	flavonol
perlargonidin	3,5,7,4'	1.30±0.1 [6]	anthocyanin
caffeic acid	3,4	1.26±0.01 [3]	cinnamic acid
naringin	5,4'	0.24±0.02 [7]	flavanone
α-tocopherol		0.97±0.01 [3]	
ascorbate		0.99±0.04 [9]	
benzoic acid		0.00 [3]	
cinnamic acid		0.00 [3]	

[TEAC values are depicted as mean ± s.d. with the number of separate experiments in brackets]

Among the cyanidins the effect of the rearrangement of the pattern of hydroxylation of the B ring can also be demonstrated. While cyanidin (TEAC 4.4) has the same 3,5,7,3',4'-OH substitution patern as quercetin, loss of the 3'-OH group (pelargonidin) causes a decline in the TEAC to 1.3. Methoxylation of the same 3'-OH group, as in peonidin, gives a TEAC of 2.2 and the substitution of a further methoxy group in position 5' (malvidin) has a similar effect (TEAC 2.1).

While the unsubstituted benzoic and cinnamic acids have no hydrogen-donating antioxidant activity and gallic acid, with three hydroxyl substitutions, has a TEAC of 3.0, caffeic acid (3,4-dihydroxycinnamic acid) has a TEAC of 1.26. Esterification of the carboxyl group with quinic acid (producing chlorogenic acid) does not change te TEAC value, suggesting that this part of the molecule is not taking part in hydrogen donation or electron delocalisation. Chlorogenic acid is widely distrbuted in angiosperm plants and is present at relatively high concentrations in fruits such as apples [4]. The relative abundance of chlorogenic acid in fruits and vegetables makes this substance a major, if unrecognised, dietary antioxidant.

4. DISCUSSION

The proposals of Bors et al [6], based on pulse radiolysis studies, for the criteria for maximal radical scavenging activity of flavonoids are:

- the presence of the 3',4'-dihydroxy structure in the B ring,
- the presence of the 2,3-double bond in conjunction with the 4-oxo group in the C ring,
- the presence of a 5-hydroxyl group in the A ring with a 3-hydroxyl group and a 4-oxo function in the C ring.

The findings presented are in agreement with these proposals, in particular with the first two. The antioxidant activities of unesterified flavonoids against radicals generated in the aqueous phase is greatest in the cases of quercetin and cyanidin (with 3',4' dihydroxy substitutions in the B ring and conjugation between the A and B rings) which have antioxidant potentials four times that of Trolox, the vitamin E analogue. The 3',4'-di-OH effect is strongly boosted by conjugation between the A and B rings, as is demonstrated by comparing quercetin with dihydroquercetin (taxifolin). Removing the ortho-dihydroxy substitution, as in kaempferol, or the potential for electron delocalisation by removing the 2,3 double bond in C ring, as in catechin and epicatechin, decreases the antioxidant activity by more than 50%, but these structures are still more effective as hydrogen-donating antioxidants than α-tocopherol or ascorbate.

5. REFERENCES

1. Rice-Evans CA, Miller NJ., Bolwell PG, Bramley PM, Pridham JB, *Free Rad Res* 1995, 22, 375-383.
2. Salah N, Miller NJ, Paganga G, Tijburg L, Bolwell GP, Rice-Evans CA, *Arch Biochim Biophys* 1995, in press.
3. Kuhnau J, *World Reviews on Nutrition and Diet* 1976, 24, 117-120.
4. Miller NJ, Diplock AT, Rice-Evans CA, *J Agric Food Chem* 1995, 43, 1794-1801.
5. Miller NJ, Rice-Evans CA, Davies MJ, Gopinathan V, Milner A, *Clin Sci* 1993, 84, 407-412.
6. Bors W, Heller W, Michel C, Saran M (1990) In: *Methods in Enzymology*, Academic Press, New York, 1990, 186, 343-355.

REDUCED MORTALITY FROM CARDIOVASCULAR AND CEREBROVASCULAR DISEASE ASSOCIATED WITH MODERATE INTAKE OF WINE

Morten Grønbæk, Allan Deis, Thorkild I. A. Sørensen, Ulrik Becker, Peter Schnohr and Gorm Jensen

Danish Epidemiology Science Center at Institute of Preventive Medicine, Copenhagen, Denmark.

1 INTRODUCTION

The impact of alcohol intake on mortality from all causes has been described in a large number of prospective population studies from many countries. Apart from two studies, they all showed a U-shaped relation between alcohol intake and subsequent mortality. Most authors attribute the 'U' to a combination of beneficial and harmful effects of ethanol. This is based on findings from population studies of alcohol-related morbidity and cause-specific mortality, showing a decreased relative risk of coronary heart disease, and an increased risk of certain cancers and cirrhosis, with increased alcohol intake.

Two studies of the correlation between wine intake per capita in different countries and incidence of ischemic heart disease gave rise to a hypothesis of a more beneficial effect of wine than of beer and spirits. Both found an inverse relation between incidence rates of ischemic heart disease and wine consumption in different countries, but no such relation for beer consumption. This was most recently confirmed by Criqui, who validated the diagnosis, and to a certain extent excluded nutritional confounders. Ecologic studies as these three may thus give rise to a hypothesis of how different types of beverages influence the development of ischemic heart disease. But ecologic studies are not able to take us from association (correlation) to causation. Prospective population studies, capable of separating cause and effect in time, take us further in that process.

2 MATERIALS AND METHODS

2.1 Subjects

In order to examine the association between moderate intake of different types of alcoholic beverages and subsequent mortality from cardio- and cerebrovascular disease as well as from all causes, we therefore conducted a prospective population study with baseline assessment of alcohol- and tobacco consumption, income, educational level and body mass index, and 10-12 years' follow-up of cause-specific mortality. The study included 7,217 women and 5,633 men aged 30-70 years [1].

The participants filled in a self-administered questionnaire concerning various health related issues, including alcohol intake, smoking habits, school education, and household income. Weight in light clothes and height without shoes were measured, and from these the body mass index was calculated as weight divided by squared height.

2.1.1 Estimation Alcohol intake. The participants were asked in multiple choice form if they had a 'hardly ever/never', 'only monthly', 'only weekly', or 'daily' intake of beer (bottles), wine (glasses) or spirits (units). In case of a daily alcohol intake, the average number of drinks per day was reported. One bottle of beer contains 12 grams of alcohol, and this may be considered the average of the other types of drinks. In case of abstinence due to treatment (for example disulfiram) or dipsomania (n=17), this was noted, and the subject was excluded from the analysis.

During the 12 year follow-up period 1,954 (750 women and 1,204 men) died, 354 and 765 of them, respectively, from cardio- and cerebrovascular disease. In the analysis, we controlled for age, sex, smoking habits, income, educational level, and body masss index [1].

3 RESULTS AND DISCUSSION

As seen in Figure 1, subjects who drank 3-5 glasses of wine per day had a risk of dying from cardio- and cerebrovascular disease of 0.44 (0.24-0.80) as compared to those who never drank wine (relative risk set at 1.00). Those who drank 3-5 beers per day had a risk of 0.72 (0.61-0.88) as compared to those who never drank beer (Figure 2), while there was a trend towards an increased risk of death from cardio- and cerebrovascular disease among drinkers of spirits as compared to those who drank no spirits (Figure 3).

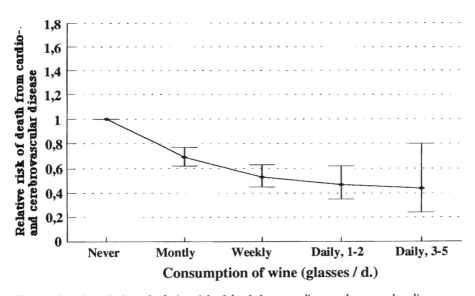

Figure 1. *Association of relative risk of death from cardio- cerebrovascular diseases with wine consumption in a Danish population sample of 7217 women and 5633 men over a 12-year period.*

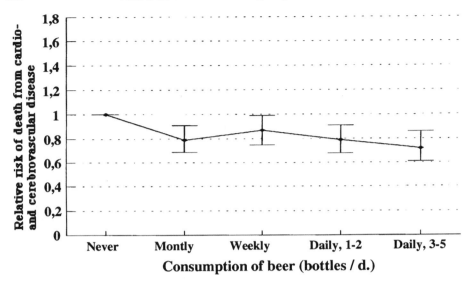

Figure 2. *Association of relative risk of death from cardio- cerebrovascular diseases with beer consumption in a Danish population sample of 7217 women and 5633 men over a 12-year period.*

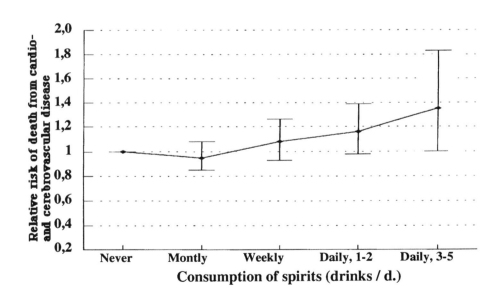

Figure 3. *Association of relative risk of death from cardio- cerebrovascular diseases with spirits consumption in a Danish population sample of 7217 women and 5633 men over a 12-year period.*

The wine - all cause mortality risk function also steadily decreased with increasing intake; from a relative risk of 1.00 for those never drinking wine through 0.51 (95% confidence limits; 0.32-0.81) among those who drank 3-5 glasses per day. In contrast, neither beer nor spirits consumption was associated with reduced risk. For spirits consumption the relative risk of dying increased from 1.00 among those who never drank to 1.34 (1.05-1.71) among those with an intake of 3-5 drinks per day. The effects of the three types of alcoholic beverages appeared to be independent of each other, and there were no significant interactions with sex, age, education, income, smoking or body mass index [1].

Wine drinking showed the same relation to risk of death from cardio- and cerebrovascular disease as to mortality from all causes.

Our finding that only wine drinking clearly reduces both risk of dying from cardio- and cerebrovascular disease, and risk of dying from other causes, may suggest that other more broadly acting factors in wine are involved [1].

References

1   Parts of these results were first published by Morten Grønbœk, Allan Deis, Thorkild I. A. Sørensen, Ulrik Becker, Peter Schnohr and Gorm Jensen, *Br. Med. J.*, 1995; **310**:1165-9.

A NEW SENSITIVE AND SPECIFIC HPLC METHOD FOR MEASURING POLYPHENOL PLASMA AND URINARY LEVELS AFTER GREEN TEA INGESTION

G. Maiani, E. Azzini, M. Salucci, A. Ghiselli, M. Serafini and A. Ferro-Luzzi.

National Institute of Nutrition, Via Ardeatina 546, 00178 Rome, Italy

1. INTRODUCTION

Polyphenols are important constituents of the human diet, widely distributed in vegetable foods (legumes, cereals, fruit) and in their products (tea, cider, wine). Because of their well established in vitro antioxidant activity, polyphenols are thought to play an important role in the prevention of chronic diseases[1]. The mechanisms through which polyphenols exert their antioxidant action are: scavenging of superoxide anions, singled oxygen[2], and lipid peroxyl radical[3]; and chelation of catalyst metal ions through ligands[4].

The consumption of polyphenol-rich foods has been associated with a lower risk for mortality from coronary heart disease in elderly men[5].

Several studies have been performed on the measurement of polyphenols in food, as well as in human body fluids such as urine and faeces[6,7]; however no reliable methods for assessing plasma polyphenol content are available as yet.

This study proposes a new sensitive, specific and reliable methodology to prepare and evaluate plasma levels of Catechin (CA), Epigallocatechin (EGC), Epigallocatechin gallate (EGCG), Epicatechin (EC), Epicatechin gallate (ECG), Caffeine and Caffeic acid in healthy subjects before and after ingestion of green tea.

2 MATERIALS AND METHODS

2.1 Blood collection

We studied four healthy women aged 20-30 years. Each subject drank 300 ml of green tea prepared by infusing 6 g of dry tea leaves in tap water at 100°C for 3 min. The amount of green was the amount capable of inducing a 41% increase of plasma antioxidant power (TRAP), as reported in a previous papern[8].

Fasting venous blood (5 ml) was collected with vacutainers containing EDTA as anticoagulant at the baseline and 30 and 50 min after tea consumption. Blood samples were centrifuged for 15 min at 3000 rpm, and the plasma stored at -40°C.

2.2 Sample preparation

1 ml of plasma was added to 1 ml of HCl (0.5M): Methanol 1:1 (V/V), vortexed for 30 sec. After incubation for 10 min at room temperature, the mixture was extracted with 2 ml of ethyl acetate. The content of the tube was vortexed for 3 min and centrifuged at 3000 rpm for 5 min. The supernatant was transferred to another test tube and the extraction repeated. The two organic layers were combined and evaporated to dryness under nitrogen flow. The residue was redissolved in 250 µl of methanol:water (3:2).

2.3 HPLC analysis

Twenty microliters of sample were injected by auto sampler (Perkin Elmer ISS 200). The polyphenols were separated by a Lichrospher 100 RP18 5 µm (25 cm x 4.6 mm) by Merck Darmstadt (Germany) column and a guard column Perisorb Supelguard LC18 2 cm by Supelco. The mobile phase consisted of two solvent systems: solvent A was demineralized water adjusted to pH 2.8 with phosphoric acid and solvent B was methanol. We used the linear linear gradients as follows:

time (min)	Flow (ml/min)	%B
1,0	1,0	13
12,6	1,0	60
0,1	0,8	80
18,0	0,8	80

Peaks were monitored with a spectrophotometric detector (Perkin-Elmer L.C. 95) set at 280 nm (AUFS 0.001, time response 2000) and stored and analysed with a personal computer (PE Nelson 1020).

The retention times of the samples were compared to those of a standard mixture for identification. A standard curve made with increasing amounts of each standard was used to read the results. The entire procedure was completed (including the stabilization time) under 50 minutes.

2.4 Standards

A stock solution containing 250 µg/ml of each catechins, 600 µg/ml caffeic acid and 100 µg/ml caffeine was obtained by dissolving pure standards in methanol, and stored at -20°C until analysis.

2.5 Standard curves:

Standard curves were obtained diluting stock solutions (methanol:water 3:2 v/v), (Catechins 0.69- 25 µg/ml, caffeine and caffeic acid 0.60-60 µg/ml) by repeated injections of a 20 µl volume.

2.6 Assay Validation

Pooled human plasma of known concentration of phenolic compounds was used. Measured amounts of these compounds were added to a reagent blank, and to pooled plasma. Samples, were treated and extracted as described above.

The reproducibility of the method was assessed by repeated analysis of a pooled human plasma spiked with standard compounds. Each determination was the result of a separate extraction as well as a separate injection. The within-day variability was assayed by measuring plasma samples in triplicate, while the day-to-day variation was established by analysing plasma duplicates for 5 days.

3 RESULTS

Table 1 shows the linearity of the detector response, the coefficients of correlation.

Table 1. *Linearity and detection limit for some phenolic compounds.*

	Range studied	Correlation Coefficient	Sensitivity
	(µg/ml)	(r)	(µg/ml)
Catechin plus Epigallocatechin	1.00 - 50	0.999	1.00
Epicatechin	1.09 - 8.75	0.999	1.09
Epicatechingallate	1.25-20.0	0.999	1.25
Caffeine plus Caffeic acid	0.69 - 60	0.998	0.69
Epigallocatechin gallate	0.90 - 25	0.999	0.90

The sensitivity of the instrumentation was verified by injecting varying amounts of the mixture of seven phenolic compounds. For all compounds the calibration curve, built by plotting the peak height against the standard concentration, was linear (r=0.999). The sensitivity (defined as the lowest measurable concentration of a compound in the sample), was estimated as the concentration which generated a peak with a height at least three-fold the baseline noise range.

Figure 1 shows the typical chromatographic patterns of a standard mixture and plasma obtained from a subject fasted for 12 hours.

Dietary Flavonoids and Role as Natural Antioxidants in Humans

Figure 1 *Comparison between a standard mixture and a plasma sample at base-line*

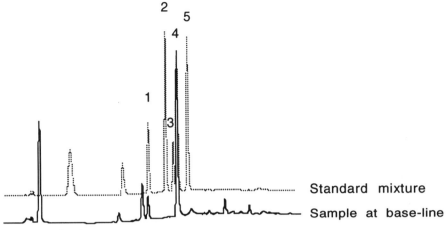

Peak identification:
1=CA plus EGC; 2=EGCG; 3=EC; 4=Caffeine plus Caffeic acid; 5=ECG

Comparing the respective retention times and absorption spectra of sample with those of pure standards, peaks 1 and 4 were identified as EGC plus catechin and caffeine plus caffeic acid, respectively. In the fasting subject there were no detectable peaks relative to EGCG, EC and ECG.

In order to confirm the retention times of each compound examined we spiked the plasma sample through standard additions (fig 2).

Figure 2 *Confirmation (by spiking) of the presumptive plasma peaks of CA plus EGC, EGCG, Caffeine plus Caffeic acid.*

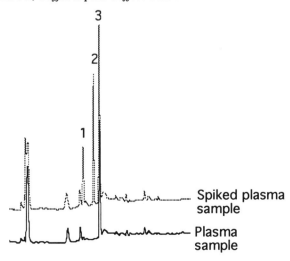

Peak identification:
1= CA plus EGC; 2= EGCG; 3= Caffeine plus Caffeic acid

Recovery and reproducibility of the assay are shown in tables 2 and 3. Recovery was checked for the compounds of interest by adding a known amount of the working standard both to a reagent blank and to pooled human plasma (Table 2).

Table 2 *Recovery (%) of added standards from reagent blank and plasma.*

	Reagent Blank %	Analytical recovery from Plasma %
Catechin plus epigallocatechin	97.8±6.6	94±12.3
Epigallocatechin gallate	99.2±7.5	100±5.9
Caffeine plus Caffeic acid	102.4± 3.3	97.1±7.8

* Mean of triplicate assays

Values obtained in the blank indicate that CA plus EGC, EGCG, and caffeine plus caffeic acid are quantitatively extracted into ethyl acetate. Analytical recovery was checked with the addition of a known amount of the working standard into the plasma pool. Each determination was the result of duplicate or triplicate analysis on different days. The within and between-day coefficient of variation was under 10% in all cases, except for catechin plus epigallocatechin.

The reproducibility of analytical procedures was tested by analysing pooled human plasma added with known amounts of working standards. Table 3 shows the good reproducibility for the phenolic compounds, and for caffeine plus caffeic acid.

Table 3 *Reproducibility of plasma sample preparation (ug/ml)*

	n°	Within run Mean ± SD	CV%	n	Between run Mean ± SD	CV%
Catechin plus epicatechin	3	13.9±0.59	4.2	6	14.28±1.8	12.0
Epigallocatechin gallate	3	6.83±0.20	3.0	6	6.92±0.43	6.1
Caffeine plus caffeic acid	3	136.0±3.0	2.0	6	133.5±8.6	6.4

n*= number of run for day

2.7 Applicability

Table 4 shows mean plasma levels of catechins and caffeine plus caffeic acid of four subjects before and after tea consumption.

Table 4 *Mean plasma levels of selected polyphenols (µg/ml) before and after green tea donsumption.*

	n°	Base - line	30 min	50 min
Catechin plus epigallocatechin	4	15.3±12.8	17.4±17.37	19.2±16.6
Epigallocatechin gallate	4	0.00	2.36±21.06	1.79±1.00
Epicatechin	4	0.00	0.00	0.00
Epicatechin gallate	4	0.00	0.00	0.00
Caffeine plus Caffeic acid	4	16.2±21.5	44.5±32.9	57.3±34.4

After overnight fasting, only catechin plus epigallocatechin, and caffeine plus caffeic acid were present in human plasma. After tea consumption we demonstrate a rapid appearance of epigallocatechin gallate (30 and 50 min) and a significant increase of caffeine plus caffeic acid (P<0.01 and P< 0.001) at 30 and 50 min. To confirm peaks identity we also examined UV spectra by a diode array detector .

Figure 3 compares the UV spectrum for EGCG obtained during the HPLC run of the plasma after green tea ingestion (A) with the spectrum of the pure standard (B).

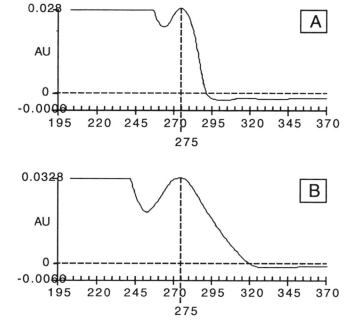

Fig 3. Spectral analyses of EGCG peak, during HPLC run, evaluated in plasma and standard solution

Spectrum A, with maxima at 275 nm, was the same as that of EGCG standard shown in B. Subsequently, plasma was analysed by a Coularray Electrochemical detector, (ESA Chelmsford, Ma. USA). These results confirmed the presence of EGCG after tea ingestion.

Figure 4 shows the UV spectrum for caffeine and caffeic acid compared with plasma spectrum.

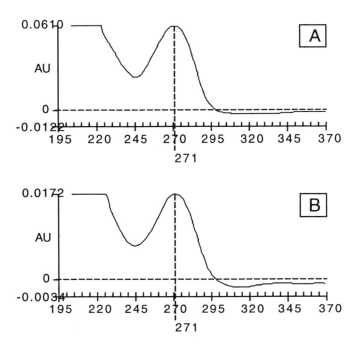

Fig 4. Spectral analyses of Caffeine peak, during HPLC run, evaluated in plasma and standard solution.

Spectrum A with maxima at 271 nm was the same of caffeine standard B. Given that caffeine and caffeic acid were coeluted by the mobile phase, it has not been possible to evaluate the individual values of these two compounds.

In order to verify the maximum absorption kinetics, one subject underwent a further blood collection at 120 minutes after tea ingestion. The results are presented in table 5 and show that no significant variation was observed in the circulating levels of catechin plus epigallocatechin, while epigallocatechin gallate undetectable levels and caffeine plus caffeic acid decreased appreciably without returning to the baseline values.

Table. 5 *Plasma levels (μg/ml) of specified polyphenols before and after green tea drinking.*

	Base - line	30 min.	50 min	120 min.
Catechin plus epigallocatechin	34.3	37.5	38.3	37.3
Epigallocatechin gallate	0.00	1.17	0.84	0.00
Epicatechin	0.00	0.00	0.00	0.00
Epicatechin gallate	0.00	0.00	0.00	0.00
Caffeine plus Caffeic acid	48.5	91.6	100.9	76.5

Polyphenols were also analysed in urine collected 6 hours after tea consumption in only one subjects. The subjects maintained the fast a part from the tea ingestion, to limit any interference from foods.

Table 6 shows the urinary levels of polyphenols.

Table 6. *Urinary levels of specified polyphenols (μg/ml) of one subject before and after green tea drinking*

	Base - line	After 6 hrs
Catechin plus epigallocatechin	185.1	309.5
Epigallocatechin gallate	0.00	8.46
Epicatechin	0.00	0.00
Epicatechin gallate	0.00	0.00

Only an appreciable amount of catechin plus epigallocatechin was detectable in the urine at base line, which was doubled 6 hours after ingestion of tea ($P<0.001$). Epigallocatechin gallate became detectable.

4 CONCLUSIONS

The method developed appears to respond to the prerequisite of research on bioactive phenolic compounds. Green tea contains catechin epigallocatechin, epigallocatechin gallate, epicatechin, epicatechingallate and caffeine. As reported by Nanjo et al[9] the major constituents of green tea are the catechins, that include (+)- catechin (1.4 %), (-)- epicatechin (5.8 %), (-)- epigallocatechin (17.6 %), (-)- epicatechingallate (12.5 %), (-)-epigallocatechin gallate (53.9 %) and others (9.8%). Caffeine has been reported to be present in green tea at 7.6% [10]. In our study the consumption of green tea produced an increase over 40 times in plasma levels of caffeine plus caffeic acid; and the peak of epigallocatechin gallate (normally absent) appeared. Our data on plasma caffeine confirm what reported in the literature[11]. Moreover, green tea consumption produces a 100 percent increase in urinary excretion of catechin plus epigallocatechin. Also, a peak of EGCG appears in urine 6 hours after green tea ingestion.

This technique provides a highly precise, sensitive, and reliable means of studying the occurrence and content of potentially important substances in biological fluids which originate from the diet.

Acknowledgements

This work was supported by the National Research Council of Italy, Special Project RAISA, Subproject 4 paper No 2367.

References

1. Hertog M.G.L., Hollman P.C.H., and van de Putte B. : Content of potentially anticarcinogenic flavonoid of tea infusion, wine and fruit juices. J. Agric Food Chem. 41: 1242-1246,1993.

2. Robak J, Gryglewski R.J. : Flavonoids are scavengers of superoxide anion. Biochem Pharmacol 37:83-88, 1988.

3. Husain S.R. Cillard J., Cillard P.: Hydroxyl radical scavenging activity of flavonoids. Phytochemistry 26:9, 2489-2491, 1987.

4. Takahama U.: Inhibition of lipoxygenase-dependent lipid peroxidation by quercetin: mechanism of anti oxidative function Phytochemistry 24: 1443-1446, 1985.

5. Hertog G.L., Feskens E.J.M., Hollman P.C.H., Katan M.B., Kromhout D. Dietary antioxidant flavonoids and risk of coronary heart disease: the Zutphen Elderly study. Lancet 342:1007-1011, 1993.

6. Hertog M.G.L., Hollman P.C.H. and Venema Dini P: Optimisation of a quantitative HPLC determination of potentially anticarcinogenic flavonoids in vegetable and fruits. J. Agr. Food Chem 40: 1591-1598, 1992.

7. Bravo L., Abia R., Estwood M.A; and Saura-Calinto F: Degradation of polyphenols (catechin and tannic acid) in the rat intestinal tract. Effect on colonic fermentation and faecal output. Br. J. of Nutr. 71-933-946, 1994.

8. Serafini M;, Ghiselli A and Ferro-Luzzi A.: In vivo antioxidant effect of green and black tea in man. Eur. J. of Clin. Nutr. 49:1995

9. Nanjo F., Honda M., Okushio K., Matsumoto N., Ishigaki F., Ishigami T. and Hara Y. : Effects of dietary Tea catechins on a-tocopherol levels, lipid peroxidation, and erythrocyte deformability in rats on high palm oil and perilla oil diets. Pharm. Soc Japan 16 (11) 1156-1159, 1993.

10. Bunker ML and McWilliams M Caffeine content of common beverages. J. Am Diet Assoc. 74:28-32, 1979.

11. Arnaud M.J.: Metabolism of caffeine and other components of coffee. Caffeine, Coffee, and Health. Ed S. Garattini pp 43-94, 1993

ANTIOXIDANT ACTIVITY OF NATURAL FOOD SUPPLEMENTATIONS: FLAVONOID RUTIN AND BIO-NORMALIZER.

James A. Osato *, Igor B. Afanas'ev ** and Ludmila G. Korkina ***

* Osato Research Foundation, Gifu 500, Japan
** Vitamin Research Institute, Moscow 117820, Russia
*** Institute or Pediatric Hematology, Moscow 117513, Russia

1 INTRODUCTION

Rutin (quercetin-3-rutinoside), a natural bioflavonoid, and Bio-normalizer (BN), a natural Japanese health food prepared by the fermentation of *Carica papaya,* are now drawing a wide interest as nontoxic therapeutic and preventive agents possessing antioxidant and free radical scavenger activities. Rutin is already widely applied for the treatment of vascular disorders (for decreasing the permeability and fragility of capillaries) while BN presumably manifests beneficial effects in the treatment of such pathologies as tumors, inflammation, allergy, immunodeficiency, etc. Since free radicals at least partly participate in the initiation of these pathologies, it seems important to study the antioxidant and free radical scavenging properties of these two natural compounds.

Earlier, we have already shown that rutin inhibited lipid peroxidation of lecitin liposomes, bovine heart microsomes,[1] and rat brain homogenates.[2] It was proposed that the inhibitory effects of rutin are explained by both the reaction with superoxide ion and other free radicals (the antioxidant effect) and the formation of an inactive iron-rutin complex (the chelating effect). The BN antioxidant activity in cell-free and cellular systems has also been studied.[3,4] In this work we enlarged the study of antioxidant properties of BN and rutin to the *ex vivo* systems, investigating the effects of BN administration to rats with lung injury induced by asbestos fibers and the effects of rutin on free radical overproduction in iron overloaded (IOL) rats.

2 MATERIALS AND METHODS

2.1 Chemicals

BN was produced by Sun-O International Inc., Gifu, Japan, by the fermentation of *Carica papaya* tropical herbal plants. Rutin, ferricytochrome c (type VIII), NADPH. lucigenin, luminol, phorbol myristate acetate (PMA), Hanks' balanced salt solution (HBSS), phosphate buffer, ADP, and bovine CuZnSOD (EC 1.15.1.1) were purchased from Sigma Chem. Co. (St. Louis, Mo.). 2-Thiobarbituric acid (TBA) was from Fluka (Buchs, Switzerland).

2.2 Animal Model of Lung Injury

Male Wistar rats of body weight ranging from 150 to 200 g were divided into 4 groups, each consisting of 24 animals. Rats of group I were given 1 ml of physiological saline intratracheally (control group). Animals of group II were injected intratracheally with 15 mg of chrysotile asbestos fibers and 10 mg of BN suspended in 1 ml of physiological saline. Animals of group III were fed daily with 50 mg of BN and standard diet during 7 days; after that, they were given 15 mg of chrysotile asbestos fibers intratracheally and fed again with 50 mg of Bio-normalizer for next 7 days. Animals of group IV were injected with 15 mg of asbestos fibers. The animals were sacrificed at 3d, 7th, 14th, and 28th days after the intratracheal injection of NaCl, asbestos fibers, or asbestos fibers with BN.

2.3 Isolation of Alveolar Cells

After rat sacrificing, the trachea was cannulated and the lungs were lavaged five times with 10 ml of HBSS at 37 °C (pH 7.4). The lavage fluid was filtered through a nylon cloth, and the filtrate was centrifuged at 300xg for 10 min. Cell content was determined in a hemocytometer and differential count was performed microscopically in the cell smears. To separate mononuclear and polymorphonuclear cells, the cell pellet was resuspended in 1 ml HBSS, layered carefully on Ficoll-Paque gradient ($\rho = 1.077$), and centrifuged at 600xg for 40 min. The upper layer consisted of mononuclear cells (macrophages and lymphocytes) and the pellet contained granulocytes (mainly neutrophils). Macrophages were identified by staining for nonspecific esterase. Macrophage and neutrophil preparations were >95% pure and 98% of the cells excluded trypan blue. The isolated cells were resuspended and stored at 4 °C in HBSS supplied with 5% fetal calf serum.

2.4 Lipid Peroxcidation of Rat Lung Homogenate

The lung tissue was dissected into small pieces, poured through the 3 mm holes for fibrotic wall separation, and homogenized thoroughly. The homogenate (10 mg) was incubated with 50 µM $FeCl_3$ and 800 µM ADP in 1 ml of 0.1 M phosphate buffer (pH 7.4) at 37 °C. The reaction was startecl by adding 40 µl of 7.5 mM NADPH in phosphate buffer. After 30 min incubation, the reaction was stopped by adding 1 ml of 15% trichloroacetic acid and 0.1 ml of 10 µM naphtol solution in ethanol. Then, 1 ml of 0.375% TBA solution was added, and the reaction mixture was boiled for 15 min. After sedimentation of precipitated proteins by centrifugation, the content of TBA reactive products was determined by measuring an absorbance at 535 nm.

2.5 Chemiluminescence Assay

The CL measurements on a Luminometer mod. 1251 (LKB, Sweden) were monitored at 37 °C and continuous mixing on a programmed IBM computer. All experiments were carried out in dublicate. Each point was a mean of 3 or 4 independent measurements. A sample of 5×10^5 cells, 400 µM luminol or lucigenin, and 0.5 ml of HBSS (pH 7.4) were placed in the 1 ml polysterene cuvette, and the CL background was measured for 5 min. After that, CL was activated by adding PMA (10 ng) or opsonized zymosan (100 µg/ml). The CL-response was measured as a mean mV signal over the 10 s intervals for a 10 min period.

2.6 Measurement of Hydroxyproline Content in Rat Lung Tissue

A sample of dried lung tissue (20 mg) was treated with 6 M hydrochloric acid (1 ml) at 100 °C for 24 hours, filtered, neutralized, and vaporized. After that, a dry residue was dissolved in 1 ml of distilled water and incubated with 1 ml of 0.03 M chloroamine B in the mixture of acetate-citrate buffer and propyl alcohol for 5 min at room temparature. Finally, the reaction mixture was treated with 1 ml of 5% p-dimethylaminebenzaldehyde at 60 °C for 20 min, and the content of hydroxyproline was determined spectrophotometrically at 558 nm.

2.7 Iron Overload in Rats

Experimental iron overload (IOL) was produced in rats by feeding diets supplemented with elemental (carbonyl) iron. Rats were divided into 4 groups: Group 1 (Control, 8 animals) was fed regular chow for 56 days; Group 2 (11 animals) was fed diets supplemented with 2.5% (wt./wt.) carbonyl iron for 42 days; Group 3 (5 animals), after feeding regular chow during the 42 days period, was daily intraperitoneally injected with 2 ml 1 mM rutin for next 10 days; Group 4 (6 animals), after feeding iron supplemented diets for 42 days, was daily interperitoneally injected with 2 ml 1 mM rutin for next 10 days. In accord with protocol, the animals of Groups 1 and 2 were killed on days 42, 52, and 56, and the animals of Groups 3 and 4 were killed on day 52.

2.8 Preparation of Liver Microsomes

The liver was perfused with 0.9% NaCl solution and homogenized with 1.12% KCl solution (1:3 v/v). The homogenate was cetrifuged at 1000xg for 20 min, then the supematant was separated and centrifuged at 105000xg for 60 min. After protein analysis by the Lowry method, microsomes were immediately used in the experiments.

2.9 Isolation of Peritoneal Macrophages

Peritoneal macrophages were prepared by peritoneal lavage with 2 ml prewarmed saline solution. The lavage fluid was filtered and centrifuged at 300xg for 10 min. Cells were resuspended, twice washed, and stored at 4 °C in HBSS. Macrophage preparations were >90% pure and >95% of the cells excluded trypan blue.

2.10 Isolation of Blood PMNs

Blood was collected from rat tail vein using heparin (5 U/ml) as an anticoagulant. Freshly obtained whole blood was separated by density gradient centrifugation with Ficoll-Hypaque. Granulocyte containing pellet was layed on dextran-metrizoate mixture for erythrocyte sedimentation. The isolated PMNs were washed twice with HBSS. The final leukocyte suspension contained more than 95% PMNs. Cell viability was maintained throughout all experiments as indicated by greater than 90% trypan blue exclusion.

2.11 Lipid Peroxidation of Rat Liver Microsomes

Microsomes (0.5 mg protein/ml) were incubated with $FeCl_3$ (50 µM) and ADP (800 µM) in 0.1 M phosphate buffer (pH 7.4) at 37 °C. The reaction was started by adding 40 µl (7.5 mM) NADPH in phosphate buffer. A total volume of incubation mixture was 1 ml. After 30 min incubation, lipid peroxidation was terminated by adding 15% trichloroacetic acid (l ml) and 10 mM ethanol solution of ß-naphthol (0.1 ml). Then, 0.375% TBA solution (1 ml) was added, and reaction mixture was heated at 100 °C for 15 min. After centrifugation of precipitated proteins, the content of TBA reactive products was determined by measuring the absorbance at 535 nm.

2.12 Measurement of the Rate of Cytochrome c Reduction by PMA–stimulated Neutrophils

PMA (100 ng/mL) was added to the incubation mixture containing leukocyte suspension (106 cells/mL), cytochrome c (50 µM), and NaN3 (2 mM) in HBSS at 37 °C, and the absorption at 550 nm was registered continuously.

2.13 Statistics

All results are from the experiments carried out in dublicate or triplicate are presented as mean ±SD. Differences were analyzed using the Student's t-test, the level of significance being set at $P < 0.05$. Each point was a mean of three independent measurements.

3 RESULTS

3.1 Effects of Bio-normalizer on Lung Inflammation and Free Radical Processes in Asbestos-injected Rats

As shown in Fig. 1, BN administration to the asbestos-injected rats substantially decreased the wef lung weight on 14th and 28th days. Furthermore, BN administration decreased the recovery of neutrophils into broncho-alveolar space and enhanced that of macrophages (data not shown). BN administration drastically decreased hydroxyproline formation in the lung tissue, which characterized the intensity of asbestos-associated fibrosis (Figure 2), and completely annihilated the asbestos-induced enhancement of lipid peroxidation in the rat lung homogenate on 28th day after the beginning of experiment (Figure 3). At the same time, the BN effect on oxygen radical production by lung phagocytes was much more complicated. BN was found to be a very effective inhibitor of hydroxyl radical formation (determined by luminol-amplified CL) without affecting substantionally superoxide production (measured by lucigenin-amplified CL). Due to that, BN administration nearly completely suppressed a sharp increase in the luminol-amplified CL/lucigenin-amplified CL ratio on 7th day in the asbestos-treated rats (Figure 4).

3.2 The Effects of Rutin on Free Radical Overproduction in Iron–overloaded Rats

Rutin administration had no effect on the content of nonheme iron in both IOL and normal liver microsomes. Similarly, rutin did not affect the rate of lipid peroxidation in microsomes of control rats but significantly (by 4 times) decreased the level of TBA reactive products in IOL microsomes (Table 1). Rutin also practically did not change the content of nonheme iron in peritoneal macrophags from both IOL and control rats but strongly inhibited spontaneous luminol- and lucigenin-amplified CL produced by IOL macrophages. Furthermore, rutin administration to normal rats did not affect the rate of cytochrome c reduction by neutrophils, but this rate became equal to zero in the case of IOL neutrophils (Table 1). The effect of rutin administration on NADPH oxidase activity was uncertain.

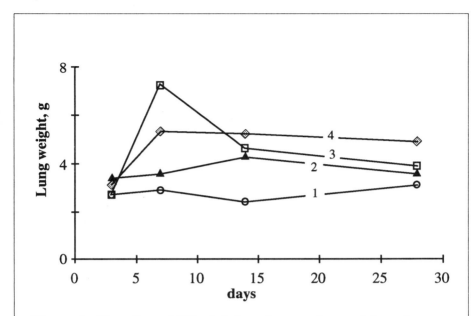

Figure 1 *The effect of BN administration to asbestos-injected rats on the wet lung weight*

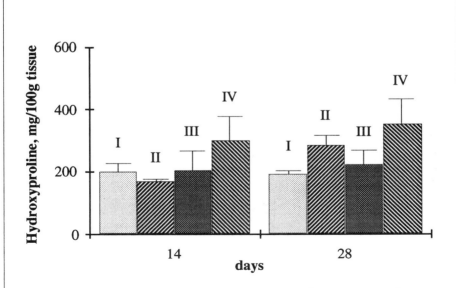

Figure 2 *The effect of BN administration to asbestos-injected rats on the hydroxyproline formation in the lung*

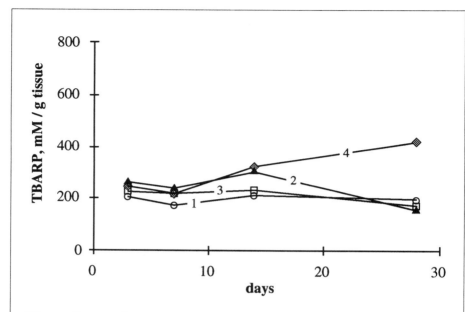

Figure 3 *Lipid peroxidation in lung homogenates of asbestos-injected rats*

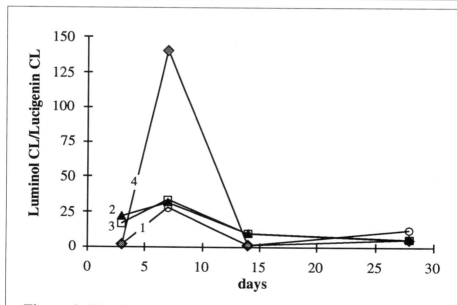

Figure 4 *The effect of BN administration on the luminol CL/lucigenin CL ratio produced by lung phagocytes*

Table 1 *The Effects of Rutin on TBA Reactive Products (TBAR) in Rat Liver Microsomes, Spontaneous Luminol- (LumCL) and Lucigenin-amplified (LucCL) Chemiluminescencein Rat Peritoneal Macrophages, and Cytochrome c Reduction by PMA-stimulated Neutrophils*

Groups of rats	TBAR nmol/mg protein	LumCL $mV*10^3$	LucCL $mV*10^3$	Rate of cyt.c. reduction nmol /min$*10^6$ cells
Group 1 (control)	3.0 ± 0.9	90 ± 4	70 ± 30	0.269 ± 0.010
Group 2 (IOL)	4.9 ± 0.1	310 ± 60	220 ± 50	0.659 ± 0.070
Group 3 (Control+rutin)	4.3 ± 2.7	170 ± 20	110 ± 40	0.624 ± 0.029
Group 4 (IOL+rutin)	1.2 ± 0.3	90 ± 20	90 ± 30	0.876 ± 0.076

IOL= Iron overloaded rats

4 DISCUSSION

Both animal models studied in this work, asbestos-induced lung injury and iron overloading, are the well-known examples of oxidative stress characterized by free radical overproduction.[5-7] Therefore, we expected that due to possessing antioxidant and free radical scavenging activities, both chosen natural substances, BN and rutin, will be able to inhibit free radical overproduction and suppress its harmful action in these animal models. Comparison of objective symptoms of lung injury in the rats with and without BN administration as well as the parameters characterized the rat organism's free radical status showed that BN indeed diminished significantly pathologic changes in the rat lungs induced by asbestos fibers instillation. Thus, the wet lung weight and the lung weight/rat weight ratio in the rats from Groups II and III approached these parameters for control animals from Group I (Figure l). BN also sharply decreased the intensity of asbestos-associated fibrosis (Figure 2). It is important that BN administration suppressed the neutrophil attachment to the inflammatory loci and enhanced macrophage recruitment and macrophage phagocytic capacity. The inhibition of neutrophil-defined stage and the stimulation of macrophage stage of inflammation can explain the normalizing effect of BN in a variety of inflammatory pathologies. BN exhibited profound inhibitory action on free radical formation in lung tissue characterized by lipid peroxidation (Figure 3). However, what is more important, BN selectively inhibited the production of the most harmful hydroxyl radicals by lung phagocytes of asbestos-injected rats (Figure 4), which are supposedly an important initiation factor of lung damage.

Our investigation of the in vivo effects of iron showed that iron overloading in rats does induce oxidative stress, which is characterized by the oxygen radical overproduction in liver microsomes, peritoneal macrophages, and blood neutrophils. This animal model permitted the study on the *in vivo effects of rutin on* oxygen radical production. The most surprising and unexpected result was a difference between its effects on normal and IOL animals. We believe that this fact demonstrates a different nature of the reactive species mediated damaging processes. Thus, rutin practically did not affect lipid peroxidation in normal microsomes but inhibited it by 75% in IOL microsomes (Table 1). Similarly, rutin was unable to inhibit spontaneous luminol- and lucigenin-amplified CL produced by macrophages or superoxide production by neutrophils from normal animals but strongly inhibited CL (by 2.5-3.5 times) produced by macrophages and completely suppressed superoxide production by neutrophils in the case of IOL rats. (It was impossible to estimate the effects of rutin administration on PMA-stimulated IOL macrophages due to a sharp decrease in the CL intensities).

Thus the ex vivo study of two natural antioxidants, Bio-normalizer and flavonoid rutin, in two different animal models of oxidative stress (asbestos-stimulated lung inflammation and iron-overloading) showed that both products are effective inhibitors of various harmful free radical-mediated processes. It was also found that these antioxidants are not only able to suppress overproduction of oxygen radicals but also selectively scavenge highly reactive hydroxyl radicals without affecting the production of physiologically important innocuous superoxide ion. These properties of BN and rutin make them prospective pharmacologic agents.

References

1. I.B.Afanas'ev, A.I.Dorozhko, A.V.Brodskii, V.A.Kostyuk, and A.I.Potapovitch, *Biochem. Pharmacol.*, 1989, **38**, 1763.

2. A.V.Kozlov, E.A.Ostrachovich, and I.B.Afanas'ev, *Biochem. Pharmacol.*, 1994, **47**, 795.

3. L.A.Santiago, J.A.Osato, M.Hiramatsu, R. Edamatsu, and A.Mori, *Free Rad.Biol.Med.*, 1991, **11**, 379.

4. J.A.Osato, L.G.Korkina, L.A.Santiago, and I.B.Afanas'ev, *Nutrition*, 1995, **12**, 0000.

5. L.G.Korkina, A.D.Durnev, T.B.Suslova, Z.P.Cheremisina, N.Daugel-Dauge, and I.B.Afanas'ev, *Mutat.Res.* 1992, **265**, 245.

6. T.Takeuchi, K.Morimoto, H.Kosaka, and T.Shiga, *Biochem.Biophys.Res.Commun.*, 1993, **194**, 57.

7. L.M.Fletcher, F.D.Roberts, M.G.Itving, and L.W.Powell, *Gastroenterology*, 1989, **97**, 1011.

ANTIOXIDANT ACTIVITY OF SILIBININ AGAINST HAEM-PROTEIN DEPENDENT LIPID PEROXIDATION AND DNA DAMAGE INDUCED BY BLEOMYCIN-FE(III)

L. Mira, M. Silva and C. F. Manso

Centro de Metabolismo e Endocrinologia da Universidade de Lisboa
Faculdade de Medicina .
1600 Lisboa. PORTUGAL

1 INTRODUCTION

Silibinin dihemisuccinate (SDH) is a water-soluble form of silibinin, the main structural isomer constituent of the flavonoid mixture termed silymarin.[1] Flavonoids are phenolic compounds of plant origin with hepatoprotective effects which have been partially attributed to their potential abilities to scavenge oxygen free radicals and to chelate metal ions.[2-4]

In a previous paper, we have shown that SDH is a scavenger of biologically relevant reactive species and it has some degree of metal binding capability.[5-6]

In this work, in order to characterize better antioxidant properties of SDH, we have studied its ability to inhibit lipid peroxidation induced by myoglobin-H_2O_2. A potential pro-oxidant activity of SDH was also examined by studying the ability to aggravate oxidative damage on DNA in the presence of bleomycin-Fe(III) complex.

Myoglobin can react with H_2O_2 to give ferryl myoglobin species, which are capable of catalysing lipid peroxidation.[7] Ferryl myoglobin may appear after reperfusion of the ischemic heart contributing to free radical damage of myocytes.[8] In addition, as a result of violent exercise, muscle can be injured releasing myoglobin into the circulation, where it could interact with any H_2O_2 available to cause oxidative damage.[9]

One way of testing for pro-oxidant activity of a compound is to study its ability to accelerate DNA damage induced by the bleomycin-Fe(III) complex. This complex is inert, but the addition of a compound capable of reducing Fe(III) to Fe(II), produces rapid DNA damage under aerobic conditions.[10]

2 MATERIALS AND METHODS

2.1 Chemicals

All reagents were of the highest quality available from Sigma Chemical Co. (St. Louis, MO, U.S.A.) or from Merck (Darmstadt, Germany). Ferric chloride was from Fluka (Buchs, Switzerland). SDH, disodium salt, was a gift from Madaus AG (Köln, Germany).

2.2 Myoglobin-Dependent Lipid Peroxidation

Commercial equine heart myoglobin was reduced with 10-fold molar excess of solid sodium dithionite under aerobic conditions and then purified by gel filtration on a Sephadex G-25 column.[11] The total haem concentration was calculated through the molar absorptivity coefficient of myoglobin at 525 nm. Arachidonic acid and microsomal peroxidation induced by myoglobin-H_2O_2 were carried out as described by Evans et al.[9]. Each assay contained, in a final volume of 1 ml, 25 mM NaH_2PO_4/Na_2HPO_4 buffer (pH 7.4), 0.4 mM arachidonic acid or 0.25 mg of microsomal protein, 50 µM myoglobin, SDH and 0.5 mM H_2O_2. The reaction mixtures were incubated for 10 min (20 min for microsomes) at 37° and peroxidation measured by the formation of thiobarbituric acid (TBA) reactive substances.[12] The TBA reactivity was assayed in the presence of butylated hydroxy-toluene (BHT) in the final concentration of 0.009% (w/v).[6] In control experiments we verified that SDH did not interfere with the TBA test; when SDH was added just before TBA, the absorbance values obtained were the same as those when it was omitted.

2.3 Spectral Analysis

Myoglobin-H_2O_2 mixtures without and in the presence of 1 mM SDH were spectrophotometrically analyzed by scanning 500-700 nm at 2 min intervals on a Pye Unicam UV2-100 spectrophotometer.

2.4 DNA Damage

The bleomycin assay, carried out as described by Jedding et al.[13] with some modifications, was used to study the ability of SDH, instead of ascorbic acid, to damage DNA in the presence of bleomycin-Fe(III) complex. Each assay contained in a final volume of 1 ml, 0.2 mg DNA, 0.05 mg bleomycin, 5 µM $FeCl_3$, 5 mM $MgCl_2$, 30 mM KH_2PO_4-KOH buffer (pH 7.0), SDH and 150 µM ascorbate (when added). The reaction mixtures were incubated for 30 min at 37°. The reaction was stopped adding 0.1 ml of 0.1 mM EDTA. DNA damage was evaluated by the formation of TBA reactive substances in the presence of BHT. This assay was also used for assessing the ability of SDH to inhibit DNA lesion through the bleomycin-Fe(III)/ascorbic acid system.

3 RESULTS AND DISCUSSION

Hydrogen peroxide results from spontaneous or enzymatic dismutation of superoxide anion radical ($O_2^{\cdot-}$) and it also can be formed directly by some oxidases.[14] The oxidation of oximyoglobin ($Mb^{II}O_2$) by excess H_2O_2 has been shown to yield ferryl myoglobin (Mb^{IV}) and a transient free radical form of ferryl myoglobin ($^{\cdot}Mb^{IV}$), which are very reactive species.[15] In addition, the excess of H_2O_2 can cause haem breakdown to release iron ions.[16] Both ferryl formation and release of iron are able to promote free

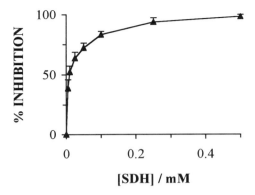

Figure 1 *Effect of silibinin dihemisuccinate (SDH) on myoglobin-H_2O_2 dependent microsomal lipid peroxidation.* Means ± S.D.; n = 4-5.

radical reactions on biological molecules such as lipids and proteins.[15]

As expected, a mixture of myoglobin and H_2O_2 stimulated peroxidation[7,9] of arachidonic acid and rat liver microsomes. SDH, at concentrations up to 250 µM, inhibited strongly the microsomal lipid peroxidation (Figure 1). Higher concentrations of SDH (up to 1 mM) were needed to inhibit in the same extent the peroxidation of arachidonic acid (Figure 2).

In order to evaluate if lipid peroxidation inhibition by SDH was due to its ability to scavenge ferryl species, a spectrophotometric analysis of ferryl myoglobin without SDH and in the presence of SDH was done. In Figure 3, it can be observed the spectra of ferryl myoglobin, produced after the addition of H_2O_2 to oximyoglobin, which are stable for at least 20 min. When SDH was added the characteristic peaks of ferryl myoglobin at around 550 and 580 nm progressively disappear and simultaneosly there is an increase in the shoulder at 630 nm, characteristic of metmyoglobin. The loss

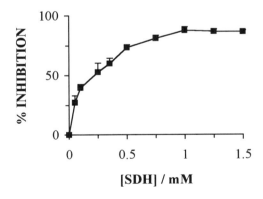

Figure 2 *Effect of SDH on myoglobin-H_2O_2 dependent arachidonic acid peroxidation.* Means ± S.D.; n = 4-5.

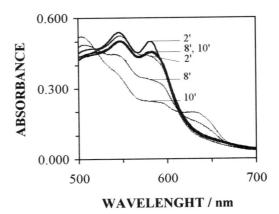

Figure 3 *Spectrum modifications of myoglobin-H_2O_2 mixture as a function of time.* Each assay contained 25 mM NaH_2PO_4/Na_2HPO_4 buffer (pH 7.4), 50 µM myoglobin and 0.5 mM H_2O_2; without SDH (——) and in the presence of 1 mM SDH (-----).

of ferryl myoglobin spectrum indicates that SDH is a "quencher" of ferryl species. Probably, silibinin is oxidized by the ferryl protein into a radical which seems unable to initiate lipid peroxidation.

Pro-oxidant properties have been attributed to flavonoids.[10] However, in the bleomycin assay, SDH, tested at concentrations up to 3 mM, had no pro-oxidant activity. On the contrary, SDH inhibited the pro-oxidant effect of ascorbic acid on bleomycin-dependent DNA damage (Figure 4). This effect must be due to its capacity of binding Fe(III) without reducing it to Fe(II).

As it was previously verified[6] in the deoxyribose assay, SDH did not also stimulate $HO^{\bullet}$ generation in reaction mixtures containing Fe(III)-EDTA and H_2O_2.

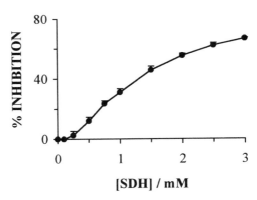

Figure 4 *Effect of SDH on DNA damage induced by bleomycin-Fe(III)/ascorbate.* Means ± S.D.; n = 4.

The results of this work highlight the idea that silibinin possess useful antioxidant properties, but unlike other flavonoids (quercetin and myricetin) it does not present pro-oxidant effects in the deoxyribose and bleomycin assays.

4 REFERENCES

1. E. Bosisio, C. Benelli and O. Pirola, *Pharmacol. Res.*, 1992, **25**, 147.
2. J. Robak and R. J. Gryglewski, *Biochem. Pharmacol.*, 1988, **37**, 837.
3. A. Mora, M. Payá, J. L. Ríos and M. J. Alcaraz, *Biochem. Pharmacol.*, 1990, **40**, 793.
4. M. Mourelle, L. Favari and J. L. Amezcua, *J. Appl. Toxicol.*, 1988, **8**, 351.
5. M. L. Mira, M. M. Silva and C. F. Manso, *Life Chem. Reports*, 1994, **12**, 55.
6. L. Mira, M. Silva and C. F. Manso, *Biochem. Pharmacol.*, 1994, **48**, 753.
7. Kanner J. and Harel S., *Arch. Biochem. Biophys.*, 1985, **237**, 314.
8. C. E. Cooper, 'Free Radical Damage and its Control', C. A. Rice-Evans and R. H. Burdon, Eds., Elsevier, Amsterdam, 1994, Vol. 28, Chapter 3, p. 67.
9. P. J. Evans, R. Cecchini and B. Halliwell, *Biochem. Pharmacol.*, 1992, **44**, 981.
10. M. J. Laughton, B. Halliwell, P. J. Evans, and J. R. S. Hoult, *Biochem. Pharmacol.*, 1989, **38**, 2859.
11. A. Puppo and B. Halliwell, *Biochem J.*, 1988, **249**, 185.
12. L. Mira, L. Maia, L. Barreira and C. F. Manso, *Arch. Biochem. Biophys.*, 1995, **318**, 53.
13. I. Jedding, P. J. Evans, D. Akanmu, D. Dexter, J. D. Spencer, O. I. Aruoma, P. Jenner and B. Halliwell, *Biochem. Pharmacol.*, 1995, **49**, 359.
14. D. C. Borg, 'Oxygen Free Radicals in Tissue Damage', M. Tarr and F. Samson, Eds., Birkhäuser, Boston, 1993, Chapter 2, p.12.
15. D. Galaris, G. Buffinton, P. Hochstein and E. Cadenas, 'Membrane Lipid Oxidation', C. Vigo-Pelfrey, Ed., CRC Press, Boca Raton, 1990, Vol. I, Chapter12, p. 269.
16. A. Puppo and B. Halliwell, *Free Rad. Res. Comms.*, 1988, **4**, 415.

INFLUENCE OF A GARLIC EXTRACT ON OXIDATIVE STRESS PARAMETERS IN HUMANS

T.G. Scherat[1, 4*], W.G. Siems[2], R. Brenke[3], H. Behrends[1], M. Jakstadt[1], E. Conradi[1] and T. Grune[1]

[1] Clinics of Physical Therapy and Rehabilitation, Medical Faculty (Charité), Humboldt University Berlin, Schumannstr.20/21, D-10098 Berlin
[2] Herzog-Julius Hospital for Rheumatology and Orthopaedics, D-38655 Bad Harzburg
[3] Krankenhaus Simbach, D-84359 Simbach
[4] Humaine Clinics Bad Saarow, Department for Internal Medicine, Pieskower Str. 33, D-15526 Bad Saarow
* address for correspondence

1 INTRODUCTION

Garlic (Allium sativum Linné) has been used for medicinal purposes throughout the last centuries. Since Lehman studied the antibacterial properties of garlic, lipid lowering effects, inhibition of platelet aggregation, influence on drug metabolism and anticarcinogenic actions have been reported (1-5). Kourounakis (6) found that Allium sativum presented antioxidant activity and allicin, an ingredient of garlic was a good hydroxyl radical scavenger in vitro. Previously also Phelps & Harris (7), Popov et al. (8), Lewin & Popov (9) and Heinle & Betz (10) reported antioxidant effects of aqueous garlic extract. Robak and Gryglewski (11) described antioxidant properties of flavonoids, which are in a smaller amount ingredients of garlic too.

Taking notice of the proved antioxidant activities of nutrient or plant preparations like garlic the influence of a nine week lasting daily oral application of a standardized garlic preparation on malondialdehyde (MDA) concentration in human blood plasma and on levels of reduced (GSH) and oxidized glutathione (GSSG) in circulating human erythrocytes was investigated in a therapy trial in humans.

2 MATERIAL AND METHODS

25 healthy volunteers (13 female, 12 male, mean age 35.2 ± 15.1 years) were treated for 9 weeks with daily doses of 900 mg dried garlic powder equivalent to Sapec®/ Kwai® (Fa. Lichtwer Berlin) containing 1.3 % allicin with an allicin release of 0.6 %.

For the investigations two age groups were separated out of the one described above: 16 patients younger than 30 years (7 male, 9 female, mean age 24.6 ± 2.1 years) and 9 patients older than 40 years (4 female, 5 male, mean age 53.9 ± 7.7 years).

For the measurement of glutathione and MDA (thiobarbituric acid reactive substances) venous blood samples were token before starting this study, after 3, 6 and 9 weeks. Samples were obtained early in the morning after a 12 hour fast. EDTA anticoagulated blood was used for determination of glutathione, blood samples for MDA measurement were anticoagulated by sodium citrate.

2.1 MDA analysis

MDA plasma concentrations were determined by HPLC separation of MDA thiobarbituric acid adducts based to the method of Wong et al. (12).

2.2 Glutathione status

Erythrocytic GSH levels were estimated photometrically similar to the method described by Beutler et al. (13). GSSG was measured by the fluorimetric method of Hissin and Hilf (14). GSH and GSSG determination was calibrated with commercial reagents by Boehringer Mannheim Ltd.

2.3 Statistics

After determination of mean value and standard deviation significance ($p \leq 0.05$) was tested with Wilcoxon and Wilcox' test.

3 RESULTS

A significant decrease of MDA levels is found after 9 weeks treatment with garlic by 63% of the initial value (Table 1). The initial values of MDA-levels in the group of older patients are significantly higher compared to the other group, so a decrease of the MDA concentration by 50 % in younger and by 76 % in older patients was found (also Table 1). At the end of the treatment period no difference between the age groups concerning MDA levels is found.

Table 1 *MDA concentrations [μmol/l plasma] before and during nine weeks treatment period with Allium sativum (* significance $p \leq 0.05$; ** significance $p \leq 0.01$)*

	garlic - all patients	garlic < 30 years	garlic > 40 years
before treatment	0.90 ± 0.51	0.69 ± 0.39	1.29 ± 0.48
after 3 week	0.66 ± 0.40	0.65 ± 0.35	0.68 ± 0.49*
after 6 weeks	0.51 ± 0.30*	0.51 ± 0.28	0.50 ± 0.34*
after 9 weeks	0.33 ± 0.17**	0.34 ± 0.14*	0.30 ± 0.22**

Table 2 *GSH concentrations [μmol/l erythrocytes] before and during nine weeks treatment with Allium sativum (* significance $p \leq 0.05$)*

	garlic - all patients	garlic < 30 years	garlic > 40 years
before treatment	1.66 ± 0.84	1.63 ± 0.92	1.72 ± 0.72
after 3 week	2.05 ± 0.89	1.79 ± 0.91	2.49 ± 0.69*
after 6 weeks	2.13 ± 0.60	1.95 ± 0.60	2.47 ± 0.46*
after 9 weeks	2.32 ± 1.22*	2.31 ± 1.28*	2.34 ± 1.19

Table 3 *Glutathione ratio (=100 * 2 GSSG / [GSH + 2 * GSSG]) [%] before and during nine weeks treatment with Allium sativum (* significance p ≤ 0.05)*

	garlic - all patients	garlic < 30 years	garlic > 40 years
before treatment	6.7 ± 6.2	7.4 ± 7.6	5.6 ± 2.3
after 3 week	4.9 ± 3.4	5.5 ± 4.0	3.6 ± 0.8
after 6 weeks	4.1 ± 1.9*	4.4 ± 2.0*	3.6 ± 1.6*
after 9 weeks	4.1 ± 3.0*	3.8 ± 2.9*	4.5 ± 3.3

In the red blood cells a significant increase of GSH (Table 2) accompanied by no change in the GSSG concentration is found. The glutathione ratio which is the ratio of GSSG thiol groups related to total glutathione thiol groups shows a similar course as MDA and has a significantly decreasing trend after 9 weeks treatment with garlic (Table 3). Concerning the whole glutathione status there is no difference between the two age groups.

4 DISCUSSION

The results suggest that the investigated standardized garlic preparation has good antioxidant activity in humans in vivo. These findings agree with other recent in vitro and in vivo investigations describing antioxidant effects of garlic extracts (6-10).

In this therapy trial an effective decrease of MDA plasma levels was registered after treatment with garlic. The true mechanism of this effect remains unclear, but a few can be discussed now. On the one hand it's possible that Allium sativum acts as a free radical scavenger itself like shown in vitro (7, 10).

As for allicin (the so called active principle of garlic [15]), an interesting structural analogy with dimethylsulfoxide (DMSO), a well known radical scavenger is to be seen, especially because the free electrons of the -SO-group in DMSO are responsible for its radical scavenging function (Figure 1 and 2).

$$CH_2=CH-CH_2-\underset{\underset{\overline{\underline{O}}:}{|}}{S}-\underline{\overline{S}}-CH_2-CH=CH_2 \qquad CH_3-\underset{\underset{\overline{\underline{O}}:}{|}}{S}-CH_3$$

Figure 1 *Allicin*   **Figure 2** *DMSO*

On the other hand it has to be considered if selenium as a garlic ingredient (16) causes the decrease of the oxidant burden via increase of selenium dependent enzyme activity.

Garlic also can act as a modulator of several enzymes, e.g. as an inhibitor of 3-hydroxy-3-methylglutaryl-CoA-reductase (4), acetyl-CoA-synthetase (17) and cytochrome P-450 IIE1 (18,19) and as an activator of glutathione-S-transferase (3, 20, 21) glutathione-peroxidase and -reductase (3) and cytochrome P-450 IIB1 (19). Especially via influencing the glutathione related enzymes and the cytochrom P-450 enzyme systems changes in the pro- and antioxidant balance might be possible.

Furthermore it can be supposed that the sulfur containing ingredients of garlic [a summary is given by Dorant et al. (22)] act as GSH precursors and regenerators, as enzyme inducers of GSH synthesis and regeneration, or as antioxidants themselves, as shown by Dansette et al. (23) for several sulfur containing compounds. It may also suggested that ingredients of garlic are cystein donors and thus inducing GSH synthesis. The two last hypotheses have to be discused especially with regard to the fact that GSH levels increased by about 40 % during the nine week therapy period with garlic.

It was found that older people had higher initial values of MDA levels. Because MDA is a measure of the amount of free radicals an organism is faced with this result seems to underline the fact that aging is accompanied with increasing levels of free radicals. After nine weeks of treatment the MDA concentrations in older people show no difference to the MDA levels of the group consisting of younger people. It may be suggested that a treatment or an additional supply with a garlic preparation is especially beneficial for older people. Because free radicals are involved in several chronic and degenerative diseases therapeutical usable antioxidants like garlic could play a more important role in disease prevention and probably also treatment.

It can be presumed that the investigated standardized garlic extract seems to be an effective natural occuring antioxidant. The mechanisms of the action of Allium sativum seem to be manyfold, not at last because of the great variety of ingredients and metabolites.

Further investigations have to be done to evaluate the effectiveness of Allium sativum, especially to get better informations on dose dependent effects and to investigate the ways of the antioxidant action of garlic and its ingredients.

6 REFERENCES

1. T. Brosche and D. Platt, *Fortschr. Med.*, 1990, **108**, 703
2. C. Legnani, M. Frascaro, G. Guazzaloca, S. Ludovici, G. Cesarano and S. Coccheri, *Arzneimittelforschung*, 1993, **43**, 119
3. A.K. Maurya and S.V. Singh, *Cancer Lett.*, 1991, **57**, 121
4. R.V. Om Kumar, A. Banerji, C.K. Ramakrishna Kurup and T. Ramasarma, *Biochim. Biophys. Acta*, 1991, **1078**, 219
5. J.G. Dausch and D.W. Nixon, *Prev. Med.*, 1990, **19**, 346
6. P.N. Kourounakis and E.A. Rekka, *Res. Commun. Chem. Pathol. Pharmacol.*, 1991, **74**, 249
7. S. Phelps and W.S. Harris, *Lipids*, 1993, **28**, 475
8. I. Popov, A. Blumstein and G. Lewin, *Arzneimittelforschung*, 1994, **44**, 602
9. G. Lewin and I. Popov, *Arzneimittelforschung*, 1994, **44**, 604
10. H. Heinle and E. Betz, *Arzneimittelforschung*, 1994, **44**, 614
11. J. Robak and R.J. Gryglewski, *Biochem. Pharmacol.*, 1988, **37**, 837
12. S.H.Y. Wong, J.A. Knight, S.M. Hopfer, O. Zaharia, C.N. Leach Jr. and F.W. Sundermann Jr., *Clin. Chem.*, 1987, **33**, 214
13. E. Beutler, O. Duron and B.M. Kelly, *J. Lab. Clin. Med.*, 1963, **61**, 882
14. H.J. Hissin and R. Hilf, *Anal. Biochem.*, 1976, **74**, 214
15. E.D. Wills, *Biochem. J.*, 1956, **63**, 514
16. C. Ip, D.J. Lisk and G.S. Stoewsand, *Nutr. Cancer*, 1992, **17**, 279
17. M. Focke, A. Feld and H.K. Lichtenthaler, *FEBS Letters*, 1990, **261**, 106
18. J.F. Brady, H. Ishizaki, J.M. Fukuto, M.C. Lin, A. Fadel, J.M. Gapac and C.S. Yang, *Chem. Res. Toxicol.*, 1991, **4**, 642
19. J.F. Brady, D. Li, H. Ishizaki and C.S. Yang, *Cancer Res.*, 1988, **48**, 5937
20. V.L. Sparnins, G. Barany and L.W. Wattenberg, *Carcinogenesis*, 1988, **9**, 131
21. V.A. Gudi and S.V. Singh, *Biochem. Pharmacol.*, 1991, **42**, 1261
22. E. Dorant, *Br. J. Cancer*, 1993, **67**, 424
23. P.M. Dansette, 'Antioxidants in therapy and preventive medicine', I. Emerit (ed.), Plenum Press, New York, 1990, p. 209

PROOXIDANT ACTIVITIES OF FLAVONOLS: A STRUCTURE ACTIVITY STUDY.

J. Gaspar[1,2], I.Duarte Silva[1], A. Laires[1,3], A. Rodrigues[1,2], S. Costa[1] and J. Rueff[1].

[1]Dept. Genetics, Faculty of Medical Sciences UNL, Rua da Junqueira 96, 1300 Lisboa, Portugal. [2]Universidade Lusófona de Humanidades e Tecnologias, Largo do Leão 9, 1000 Lisboa, Portugal. [3]Faculty of Sciences and Technology UNL, P-2825 Monte da Caparica, Portugal.

1 INTRODUCTION

The flavonoids comprise a group of more than 2000 derivatives of flavone with variable degrees of ring hydroxylation and O-glycosidic substitution of these hydroxyls. These compounds are ubiquitous in vegetal products and occur essentially as glycosides in order to prevent toxic effects in the cells where they are produced.[1] Flavonol glycosides are non genotoxic but they can be hydrolysed, by the action of bacterial populations present in gut flora giving rise to the free aglicones, which can behave as genotoxicants.[2-3] The average human intake of flavonoids is estimated at about 1 g per day including about 50 mg of quercetin,[4] the major flavonol in vegetable products.

Flavonols (3-hydroxyflavones) have been the subject of various studies concerning their ability to behave as anticarcinogens and as antioxidants but some studies have shown that some of these compounds could also behave as prooxidants[5] and mutagens in several short term mutagenicity tests.[4,6]

There is a large scarcity of data concerning the carcinogenicity of flavonoids. The results obtained concerning the potential carcinogenic activity of quercetin are not conclusive. An incidence of intestinal, bladder and kidney tumours was reported[7-9] but these results were not confirmed by other groups.[10-11]

The mutagenic and prooxidant activities of flavonoids seem to be strongly dependent on their structure, and the knowledge of the structural features associated with these activities could allow a fast screening of flavonoids without prooxidant and genotoxic activities, suitable to use as an antioxidant. We have carried out studies with 3 flavonols structurally related (galangin, kaempferol and myricetin) (Figure 1) in order to establish the structural requirements associated with their prooxidant activity.

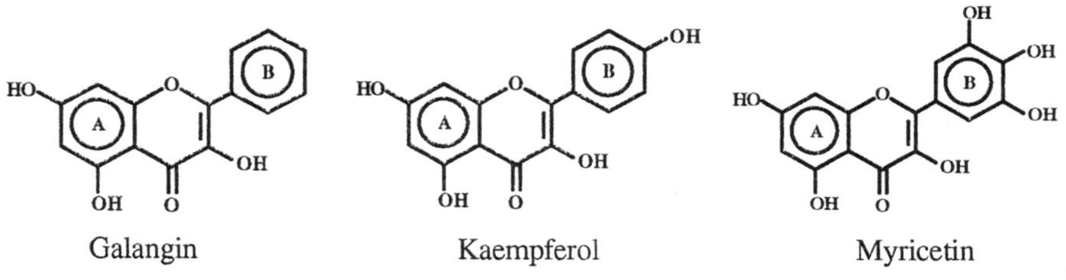

Galangin   Kaempferol   Myricetin

FIGURE 1. Chemical structure of the three flavonols studied.

We have assessed the induction of plasmidic DNA strand breaks and the degradation of 2-deoxyribose in the presence and in the absence of Fe^{3+}/EDTA in order to estimate the prooxidant properties of the different molecules studied and also the induction of chromosomal aberrations in V79 cells.

2 MATERIALS AND METHODS

2.1 Chromosomal Aberration Assay   V79 Chinese Hamster cells were cultured in Ham´s F-10 medium (Sigma) supplemented with 10 % new born calf serum (Sigma), penicillin (50 IU/ ml), streptomycin (50 µg/ml) and amphotericin B (25µg/ml) (Biological Inc., Israel) and incubated at 37 °C, under an atmosphere of 5% CO_2 in 50 cm^2 culture flasks with 5 ml of medium.   20 h cultures (approx. 1×10^6 cells) were washed with Ham´s F-10 medium reconstituted in phosphate buffer 0.01M pH 7.4 and grown in this medium for 2 h in the presence of the test chemical (dissolved in DMSO, final concentration 1%). After the treatment cells were washed with culture medium and grown for another 15 h. Colchicin was added, at a final concentration of 6 µg/ml, and cells were grown for a further 3h.   When metabolic activation was required 500 µl of S9 Mix were added to 4.5 ml of medium.   Cells were then harvested by trypsinization.   After 3 min hypotonic treatment with KCl 0.56% (w/v), at 37 °C, cells were fixed with methanol/ acetic acid (3:1) and slides were prepared and stained with Giemsa (4% in phosphate buffer 0.01M pH 6.8) for 10 min. Two independent experiments were carried out with each test compound and 100 metaphases were scored for each dose level treatment group in each experiment. Scoring of the different types of aberrations followed the criteria described by Rueff et al.[12]

2.2 Detection of Hydroxyl Radicals   Hydroxyl radicals were measured by the deoxyribose assay according to Laughton et al.[5] by incubating for 2 h at 37 °C, 1.2 ml of a reaction mixture composed of potassium phosphate buffer pH 8.0 (10 mM), the flavonols studied in concentrations up to 27 µM, deoxyribose (2.8 mM) and $FeCl_3$ (20 µM) and EDTA (100 µM). Hydrogen peroxide was used as a positive control (1.42 mM) for hydroxyl radical generation.   Deoxyribose degradation by hydroxyl radicals was measured by the TBA method using 1 ml of tricloroacetic acid (2.8 %) and 1 ml of TBA (1 %) in 0.05 M NaOH. The mixture was incubated at 100 °C for 15 min, cooled and the absorbance measured at 532 nm. For each assay six independent experiments were performed.   Negative controls (iron plus EDTA and no flavonol or flavonol without iron and EDTA) were performed and the controls without flavonol were subtracted in each experiment.

2.3 Assessment of direct damage to DNA in vitro   *Escherichia coli* strain DH1 containing plasmid pUC18 was grown in LB medium in the presence of ampicillin (60 µg/ml). Plasmid DNA was extracted and purified using a Quiagen Kit-pack (Diagen, Düsseldorf, Germany). The plasmid was ressuspended in H_2O and the DNA concentration was estimated spectrophotometrically according to Sambrook et al.[13]   Plasmidic DNA preparation typically contained 85 % covalently-closed circular supercoiled plasmidic DNA.   DNA breakage was assessed by the decrease of supercoiled plasmidic DNA. The reaction mixture, in a final volume of 15 µl, contained (final concentrations): plasmidic

DNA (2 ng/ml), the flavonols studied up to a concentration of 23 µM, $FeCl_3$ (6.7 µM), EDTA (33 µM) and potassium phosphate buffer pH=8.0 (9.3 mM). The reaction mixture was incubated for 2 h at 37 °C with gentle shaking and the reaction mixture was stopped by the addition of 15 µl of electrophoresis loading buffer [dextran blue (0.0085 % w/v), Ficoll (0.05 % w/v), dextran sulphate (0.0017 % w/v) and TAE (Tris-acetate-EDTA; 10 x 20 % w/v))]. Electrophoretic separation (3.5 V/cm) and quantification of the different DNA plasmidic forms was performed in a Gene Scanner System 362 (Applied Biosystems) with Genescan 672 software using 1 % agarose gels in TAE buffer (pH 8)[13] stained with ethidium bromide (50 µg/l). For calculation of the percentage of supercoiled DNA, a correction factor of 1.4 was applied to account for the relatively lower fluorescence of supercoiled DNA compared to with the other forms of plasmidic DNA.[14]

2.4 Rat liver enzymes Preparation of S9 and S9 Mix was carried out as described by Maron and Ames.[15]

3 RESULTS AND DISCUSSION

Of the molecules studied, myricetin strongly induced plasmidic DNA strand breakage and deoxyribose degradation, in the presence of Fe^{3+}/EDTA, at pH 8. Kaempferol, in the presence of Fe^{3+}/EDTA, also induces plasmidic DNA strand breakage, but fails to promote deoxyribose degradation. Galangin did not show any activity in what concerns the induction of plasmidic DNA strand breakage and deoxyribose degradation (Figure 2).

The induction of plasmidic DNA strand breakage and deoxyribose degradation is completely inhibited in the presence of catalase and DMSO when quercetin was used as a model molecule, suggesting that the prooxidant activities of these molecules are due to the production of hydroxyl radicals.[16] It has been previously reported that quercetin at alkaline pH conditions could give rise to superoxide anion.[17] In fact the superoxide anion could give rise to hydrogen peroxide which in the presence of Fe^{3+}/EDTA can lead to the formation of hydroxyl radical which could be the ultimate species responsible for the prooxidant activities observed with this flavonol.[16]

Quercetin[16] and myricetin strongly induce chromosomal aberrations in V79 cells in the absence of S9, and this effect was also observed although at a lesser extent for kaempferol (Table 1 and Figure 3). This effect is not observed for galangin (Table 1 and Figure 3). The induction of chromosomal aberrations in V79 cells by myricetin is decreased in the presence of S9, suggesting that the antioxidant enzymes present in S9 (e.g. catalase, SOD) could prevent the formation of radicalar species involved in the genotoxicity of this molecule as it was observed for the flavonol quercetin.[17] The induction of chromosomal aberrations in V79 cells by kaempferol is increased by the presence of S9. S9 is able to metabolise kaempferol giving rise to quercetin that could be involved in the increase of genotoxicity of kaempferol in the presence of S9 (data not shown).

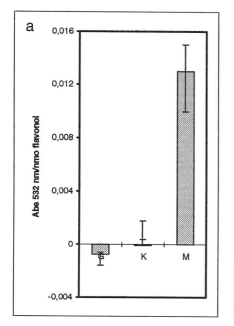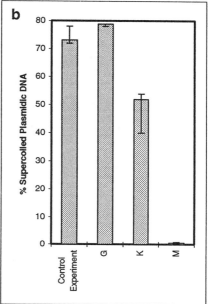

Figure 2. a) Production of TBA reactive products, arising from deoxyribose degradation at pH=8 and **b)** induction of plasmidic DNA strand breakage by flavonols (23 µM) at pH=8 and in the presence of Fe^{3+}/EDTA. Control experiments were carried out in the absence of flavonols. (G - Galangin, K - Kaempferol, M - Myricetin).

Table 1 Induction of chromosomal aberrations by flavonoids in Chinese hamster lung cells (V79-MZ) with and without rat liver S9 mix.

GALANGIN (V79-MZ)

Dose µg/ml	Ctg	Ctb	Int	Chg	Chb	Dic	Oth	>10	TAIG	TAEG	%ACIG	%ACEG
0 exp1	0	0	0	0	0	0	0	0	0	0	0	0
exp2	0	1	0	0	0	0	0	0	1	1	1	1
5 exp1	3	1	0	0	0	1	0	0	5	2	5	2
exp2	1	1	0	0	0	0	0	0	2	1	2	1
15 exp1	1	0	0	0	0	0	0	0	1	0	1	0
exp2	0	1	0	0	0	0	0	0	1	1	1	1
30 exp1	1	0	0	0	0	0	0	0	1	0	1	0
exp2	1	2	1	0	0	0	0	0	4	3	4	3
MMC 1µg/ml												

Table 1 (Cont.) Induction of chromosomal aberrations by flavonoids in Chinese hamster lung cells (V79-MZ) with and without rat liver S9 mix.

GALANGIN (V79-MZ + S9)

Dose µg/ml	Ctg	Ctb	Int	Chg	Chb	Dic	Ot	>10	TAIG	TAEG	%ACIG	%ACEG
0 exp1	1	0	0	1	0	0	0	0	2	0	2	0
exp2	5	0	0	0	0	0	0	0	5	0	5	0
5 exp1	1	0	0	0	0	0	0	0	1	0	1	0
exp2	1	1	1	0	0	1	0	0	4	3	4	3
15 exp1	5	3	2	0	0	0	0	0	10	5	8	4
exp2	2	0	0	0	0	0	1	0	3	1	3	1
30 exp1	3	0	0	0	0	0	0	0	3	0	2	0
exp2	2	0	1	0	0	0	0	0	3	1	3	1
CP	18	12	3	3	0	3	1	0	40	19	31	19
6 µg/ml	5	3	4	0	1	5	0	5	18	13	19	14

KAEMPFEROL (V79-MZ)

Dose µg/ml	Ctg	Ctb	Int	Chg	Chb	Dic	Oth	>10	TAIG	TAEG	%ACIG	%ACEG
0 exp1	3	1	0	1	0	0	0	0	5	1	5	1
exp2	3	1	0	0	0	0	0	0	4	1	4	1
5 exp1	2	0	0	0	0	0	0	0	2	0	2	0
exp2	2	1	0	0	0	0	0	0	3	1	3	1
15 exp1	15	3	1	1	0	1	0	0	21	5	18	5
exp2	18	4	1	1	0	1	0	0	25	6	21	4
30 exp1	15	8	9	2	0	0	0	0	34	17	25	11
exp2	19	7	3	1	0	1	1	0	32	12	22	10
40 exp1	22	13	12	1	0	0	0	0	48	25	30	15
exp2	13	7	5	1	0	0	1	0	27	13	13	13
MMC 1 µg/ml	5	7	8	1	0	2	0	2	23	17	20	15

KAEMPFEROL (V79-MZ + S9)

Dose µg/ml	Ctg	Ctb	Int	Chg	Chb	Dic	Oth	>10	TAIG	TAEG	%ACIG	%ACEG
0 exp1	6	0	0	0	0	0	0	0	6	0	5	0
exp2	5	0	0	0	0	0	0	0	5	0	5	0
5 exp1	5	1	0	0	0	0	0	0	6	1	6	1
exp2	3	0	0	0	0	1	0	0	4	1	4	1
15 exp1	9	13	0	2	0	0	0	0	24	13	16	10
exp2	5	9	0	1	0	0	0	0	15	9	15	9
30 exp1	21	8	3	2	2	7	0	0	43	20	34	16
exp2	5	6	3	0	1	8	0	0	23	18	20	17
40 exp1	21	13	4	1	2	2	1	0	44	22	36	21
exp2	19	15	4	1	1	3	0	0	43	23	34	20
CP	13	16	5	2	0	4	0	0	40	25	27	19
6 µg/ml	6	8	18	0	1	4	1	1	38	32	22	20

Table 1 (Cont.) Induction of chromosomal aberrations by flavonoids in Chinese hamster lung cells (V79-MZ) with and without rat liver S9 mix.

MYRICETIN (V79-MZ)

Dose µg/ml	Ctg	Ctb	Int	Chg	Chb	Dic	Oth	>10	TAIG	TAEG	%ACIG	%ACEG
0 exp1	2	0	0	0	0	0	0	0	2	0	2	0
exp2	2	0	0	0	0	0	0	0	2	0	2	0
5 exp1	5	0	0	1	1	0	0	0	7	1	7	1
exp2	2	0	0	0	0	0	0	0	2	0	2	0
10 exp1	2	2	0	0	0	1	0	6	5	3	11	9
exp2	1	1	3	0	2	1	0	8	8	7	13	12
15 exp1	1	2	3	1	2	0	0	9	9	7	16	15
exp2	4	1	0	1	1	2	1	10	10	5	19	15
20 exp1	4	5	0	0	0	1	0	2	10	6	11	8
exp2	0	2	1	1	1	3	0	5	8	7	13	12
MMC	4	9	0	8	1	3	4	2	29	17	24	16
1 µg/ml	4	8	0	6	4	5	0	0	27	23	17	15

MYRICETIN (V79-MZ + S9)

Dose µg/ml	Ctg	Ctb	Int	Chg	Chb	Dic	Oth	>10	TAIG	TAEG	%ACIG	%ACEG
0 exp1	0	0	0	0	0	0	0	0	0	0	0	0
exp2	1	0	0	0	0	0	0	0	1	0	1	0
5 exp1	1	0	0	1	0	0	0	0	2	0	2	0
exp2	0	0	1	0	0	0	0	0	1	1	1	1
10 exp1	2	1	2	0	0	1	0	0	6	4	5	3
exp2	0	0	0	0	0	0	0	0	0	0	0	0
15 exp1	2	0	3	2	0	1	0	2	8	4	8	4
exp2	1	0	1	0	1	0	1	0	4	3	4	3
20 exp1	1	0	0	0	0	0	0	0	1	0	1	0
exp2	3	1	0	0	0	0	0	1	4	1	4	2
CP	4	4	12	2	0	5	1	1	28	22	23	18
6 µg/ml	5	5	4	0	1	5	1	0	21	16	20	15

Ctg. - Chromatid gap, Ctb. - chromatid break, Int-interchange, Chg - chromosome gap, Chb - chromosome break, Dic-dicentrics, Oth-others, >10 - multi aberrant cells, TAIG - total of aberrations including gaps, TAEG - total of aberrations excluding gaps, % ACIG - % aberrant cells including gaps; %ACEG - % aberrant cells excluding gaps.

Considering the similarities between the results in the different systems used (chromosomal aberrations in V79 cells, induction of strand breakage in plasmidic DNA and deoxyribose degradation) we can not rule out the involvement of reactive oxygen species in the induction of chromosomal aberration in V79 cells by some flavonols (e.g. quercetin and myricetin).

The results obtained concerning the potential prooxidant properties of galangin, kaempferol and myricetin show that the number of OH groups in the B ring is strongly related with these properties, increasing the prooxidant activities as a consequence of the increase of OH groups. Galangin, a molecule without OH groups in the B ring, did not show any genotoxic and prooxidant activity. However myricetin with 3 OH groups on the B ring showed a high prooxidant activity when compared with kaempferol (1 OH group in the B ring).

It has been suggested that the genotoxic activities of flavonols could also be related with the presence of adjacent hydroxyl groups, and without these structural features flavonols could well prove to be innocuous and yet retain antimutagenic activity,[18] which could be also associated with antioxidant activity.

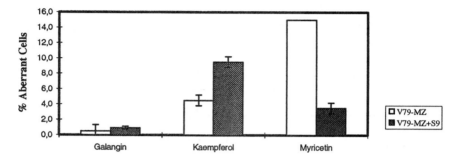

Figure 3- Induction of chromosomal aberrations in V79 cells by flavonoids (15 µg/ml), in the presence and in the absence of rat liver S9 mix.

The increase of reactive oxygen species in the intestine, as a consequence of exposure to prooxidants, has been pointed out as a risk factor in intestine carcinogenesis.[19] However considering that the human exposure to flavonoids occurs essentially by food products we can not rule out that the simultaneous intake of prooxidant flavonols with other antioxidant molecules present in vegetables, could lead to a decrease of the potential risk associated with the exposure to these molecules.

The application of these methodologies to other compounds structurally related could be an important tool to estimate the prooxidant activities of this important family of molecules, leading to a more accurate screening of potential antioxidant flavonoids.

ACKNOWLEDGEMENTS

This work is supported by JNICT (Programs PFMRH and PRAXIS XXI) and by the European Commission.

REFERENCES

1. J.B. Harborne and C.A. Williams 'Flavone and flavonol glycosides'. in: J.B. Harborne and T.J. Marby (eds), The flavonoids: Advances in research. Chapman and Hall Ltd, London, 1982, p. 261
2. J.P. Brown, *Mutat. Res.*, 1980, **75**, 243
3. A. Laires, P. Pacheco and J. Rueff, *Fd Chem. Toxic.*, 1989, **27**, 437
4. J.P. Brown and P.S. Dietrich, *Mutat. Res.*, 1979, **66**, 223
5. M.J. Laughton, B. Halliwell, P.J. Evans and J.R. Hoult, *Biochem. Pharmacol.*, 1989, **38**, 2859
6. J. Rueff, A. Laires, H. Borba, T. Chaveca, M.I. Gomes and M. Halpern, *Mutagenesis*, 1986, **1**, 179
7. A.M. Pamucku, S. Yalçiner, J.F. Hatcher, and J.T. Bryan, *Cancer Res.*, 1980, **40**, 3468
8. H. Erturk, Nunoya T., J.F. Hatcher, A.M. Pamuku and J.T. Bryan, *Proc. Am. Assoc. Cancer Res.* 1983, **24**, 53
9. J.K. Dunnick and J.R. Hailey, *Fundam. Appl. Toxicol.*, 1992, **19**, 423
10. I. Hirono, I. Ueno, S. Hosaka, H. Takanishi, T. Matsushima, T. Sugimura and S. Natori, *Cancer Lett.*, 1981, **13**,15
11. M. Hirose, S. Fukushima, T. Sakata, M. Unui and N. Ito, *Cancer Lett.*, 1983, **21**,23
12. J. Rueff, A. Brás, L. Cristovão, J. Mexia, M. Sá da Costa and V. Pires, *Mutat. Res.*,1993, **289**, 197
13. J. Sambrook, E.F. Fritsch and T Maniatis 'Molecular Cloning: A Laboratory Manual'. Cold Spring Harbor Laboratory Press, Cold Spring Harbor, NY, 1989,
14. B. Epe, J. Hegler and D. Wild, *Carcinogenesis,* 1989, **10**, 2019
15. D.M. Maron and B. Ames, *Mutat. Res.*,1993, **113**, 173
16. J. Gaspar, A. Rodrigues, A. Laires, F. Silva, S. Costa, M.J. Monteiro, C. Monteiro and J. Rueff, *Mutagenesis,* 1994, **9**, 445
17. J. Rueff, A. Laires, A. Brás, H. Borba, T. Chaveca, J. Gaspar, A. Rodrigues, L. Cristovão and M. Monteiro,. in M.W. Lambert and J. Laval (eds) 'DNA Repair Mechanisms and Their Biological Implications in Mammalian Cells'. Plenum Press, New York, 1989, p. 171
18. P.E. Hartman and D.M. Shankel, *Environ. Mol. Mutagenesis*, 1990, **15**, 145
19. C. F. Babbs, *Free Radical Biology & Medicine*, 1990, **8**, 191

HPLC METHOD FOR SCREENING OF FLAVONOIDS AND PHENOLIC ACIDS IN BERRIES: PHENOLIC PROFILES OF STRAWBERRY AND BLACK CURRANT

S. Häkkinen[1,2], H. Mykkänen[2], S. Kärenlampi[3], M. Heinonen[4] and R. Törrönen[1,2]

Departments of [1]Physiology, [2]Clinical Nutrition and [3]Biochemistry and Biotechnology, University of Kuopio, P.O. Box 1627, FIN-70211 Kuopio, Finland
[4]Department of Applied Chemistry and Microbiology, University of Helsinki, P.O. Box 27, FIN-00014 University of Helsinki, Finland

1 INTRODUCTION

Flavonoids and phenolic acids, the natural phenolic compounds present in foods of plant origin, are suggested to have health-promoting effects[1]. For example, the flavonoids quercetin, kaempferol and myricetin and the phenolic acids ellagic, gallic, caffeic and ferulic acids have been reported to possess antioxidative and anticarcinogenic properties[1,2].

The present work is part of a larger project on flavonoid and phenolic acid content of Finnish berries, their bioavailability and effects on carcinogen metabolism, conducted at the University of Kuopio. A method for qualitative screening of the phenolic profile would be useful in the characterization of the phenolics in berries.

Reversed-phase high performance liquid chromatography (RP-HPLC) is a well-known method for the separation of complex mixtures of phenolic compounds in plant extracts and wines[3,4]. However, most of the previous studies on phenolics in berries are based on TLC methods[5,6], and usually only a single phenolic compound or a limited group of phenolics has been analyzed[7,8].

In this paper, an RP-HPLC method for the screening of flavonoids and phenolic acids present in berries is described and preliminary data on the phenolic profiles of strawberry and black currant are reported.

2 MATERIALS AND METHODS

Freeze-dried samples of berries (1.0 g) were extracted and hydrolyzed in a solution of 1.2 M HCl in 50 % aqueous methanol (v/v) by a method modified from that of Hertog et al.[8] After refluxing at 85 °C for 2 h with constant swirling, the cooled extract was sonicated and filtered prior to injection (20 µl) into the HPLC system.

The effect of the sample preparation procedure on phenolic compounds was determined using a mixture of pure standards without berries.

An ODS-Hypersil C18 column (3 µm, 100 mm length, 4 mm i.d.) was used with a ternary HPLC system in which the pH of the eluting solvent was changed in a gradient of increasing hydrophobicity[4]. The solvents were: Solvent A = 50 mM dihydrogen ammonium phosphate (pH 2.6); Solvent B = 20 % of solvent A and 80 % acetonitrile; Solvent C = 0.2 M orthophosphoric acid (pH 1.5). The total run time was 75 min and flow rate 0.5 ml/min.

Diode array detection was used for the identification of the compounds. The UV spectra and retention times of the peaks were compared with those of the standards. Ellagic acid was detected at 260 nm, gallic acid and flavan-3-ols at 280 nm, cinnamic acids at 320 nm and flavonols at 360 nm. The peak areas were measured at the absorbance maxima for each compound class.

3 RESULTS

The method developed facilitated the separation and identification of 10 phenolic compounds (5 flavonoids and 5 phenolic acids) in the berries (Figure 1 and Table 1). While the recovery of gallic acid, flavonol aglycone and ellagic acid standards was 59-120 %, that of cinnamic acids and flavan-3-ols was only 18-32 % (Table 1).

The phenolic profiles for strawberry and black currant were different: in strawberry the major peak represented ellagic acid, in black currant p-coumaric acid (Table 2).

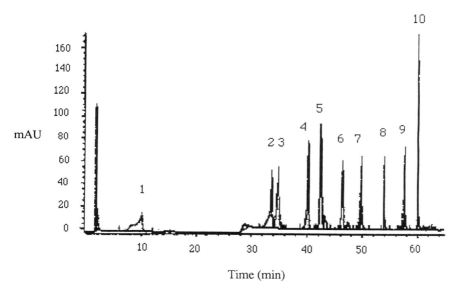

Figure 1 *HPLC chromatogram of the standards monitored at 280 nm. See Table 1 for peak identifications.*

Table 1 *Recoveries and Retention Times of the Standards*

Peak Number	Standard	Recovery[1] %	Retention Time[2] min
1	Gallic acid	59	9.7-11.8
2	(+)-Catechin	32	33.6-34.7
3	Caffeic acid	25	34.9-36.0
4	(-)-Epicatechin	18	39.5-41.5
5	p-Coumaric acid	28	42.1-44.2
6	Ferulic acid	23	45.6-48.5
7	Ellagic acid	120	48.3-50.5
8	Myricetin	95	54.0-54.9
9	Quercetin	94	57.7-59.1
10	Kaempferol	120	60.2-61.1

[1]Recovery was determined using pure standards (without berries). The values are means of 5 determinations.

[2]The values are ranges from 15 determinations.

Table 2 *Relative Contents (% of total peak areas) of Phenolics present in Black Currant and Strawberry*

Group of Phenolics	Strawberry	Black Currant
Ellagic acid	35-40	not detected
Gallic acid	3-8	4
Hydroxycinnamic acids (p-coumaric, caffeic and ferulic acids)	24-32	56-60
Flavonol aglycones (quercetin, kaempferol and myricetin)	5-7	25-30
Flavan-3-ols [(+)-catechin and (-)epicatechin]	24-29	10-15

Recoveries of standards have been taken into account.

4 DISCUSSION

Using a simple acid hydrolysis for sample preparation and a ternary HPLC system with an increasing gradient of hydrophobicity it was possible to separate and identify 10 phenolic compounds in freeze-dried berry samples.

The method still has quantitative limitations since the recovery was low for flavan-3-ols and cinnamic acids. However, the recoveries of the standards are based on analyses of a standard mixture without any sample. Recoveries of the standard compounds will be tested in berry samples and the extraction and hydrolysis method will be developed in order to improve the recovery of flavan-3-ols and cinnamic acids.

These qualitative profiles for strawberry and black currant agree with some previous, although quite variable, data for phenolics in these berries[5-7].

References

1. E. Middleton Jr and C. Kandaswami C 'The Flavonoids. Advances in Research since 1986', J. B. Harborne, Chapman & Hall, London, 1994, p. 619.
2. M. Strube, L. O. Dragsted and J. C. Larsen, 'Naturally Occuring Antitumourigens. I Plant Phenols', The Nordic Council of Ministers, 1993.
3. A. Rommel and R. E. Wrolstad, *J. Agr. Food Chem.*, 1993, **41**, 1237.
4. R. M. Lamuela-Raventos and A. L. Waterhouse, *Am. J. Enol. Viticult.*, 1994, **45**, 1.
5. H. Stöhr and K. Herrmann, *Z. Lebensm. Unters. Fors.*, 1975, **159**, 31.
6. H. Stöhr and K. Herrmann, *Z. Lebensm. Unters. Fors.*, 1975, **159**, 341.
7. E. M. Daniel, A. S. Krupnick, Y.-H. Heur, J. A. Blinzler, R. W. Nims and G. D. Stoner, *J. Food Comp. Anal.*, 1989, **2**, 338.
8. M. G. L. Hertog, P. C. H. Hollman and D. P. Venema, *J. Agr. Food Chem.*, 1992, **40**, 1591.

Dietary Fibers and Inositol Phosphates in Foods and Nutrition

DIETARY FIBER AND ITS APPLICATIONS: DEFINITIONS, ANALYTICAL METHODS AND THEIR APPLICATIONS.

L. Prosky[1] and S.C. Lee[2]

[1]U.S. Food and Drug Administration, Center for Food Safety and Applied Nutrition
Office of Special Nutritionals, Division of Programs and Enforcement Policy
200 C Street, S.W., Washington, DC 20204
[2]Kellogg Company, Science and Technology Center, 235 Porter Street, Battle Creek, MI 49016-3423

1. INTRODUCTION

It was originally envisioned in the U.S. that a method for dietary fibre would be necessary for the nutrition label to be meaningful and still meets the requirements set forth in the Nutrition Labeling and Education Act proposal first put out for comment in 1990 and the final rule published in the Federal Register in January 1993[1]. Under the auspices of the AOAC, method 985.29[2,3], the enzymatic-gravimetric, method was collaboratively studied by many scientists from approximately 20 countries. The method was found to be simple, rugged, accurate and precise such that the method was approved and given official final action status by the AOAC in 1986. In a series of papers, the same method was extended to the determination of insoluble dietary fibre (IDF) [991.42][4] and soluble dietary fibre (SDF) [993.19][5]. Methods 985.29, 991.42 and 993.19 all use the same basic procedure with phosphate buffer. Another method mentioned by the U.S. Food and Drug Administration and the U.S. Department of Agriculture in the Federal Register as one of the two methods to be used for labeling of foods for TDF, IDF, and SDF is 991.43[6]. This method is similar to the first-named method, 985.29, using the same three enzymes (heat stable -amylase, protease, and amyloglucosidase) and similar incubation conditions but substituting 2-(N-morpholino)ethanesulfonic acid-tris(hydroxymethyl)aminomethane, [MES-TRIS], buffer for phosphate buffer. The results using the MES-TRIS buffer for determination of fibre are similar to those obtained using the phosphate buffer. Several other methods, both gravimetric types and those producing results that agree with TDF values obtained using 985.29 have been adopted by the AOAC. These include an enzymatic-gravimetric type, 992.167[7], which sums insoluble and soluble fiber fractions, a nonenzymatic type for foods with little or no starch (993.21)[8] and a method recently accepted for official action by the AOAC for the determination of TDF based on assays for the components of TDF, neutral sugars, uronic acids residues and Klason lignin (994.13)[9]. The results of those of values obtained for TDF compare favorably with those obtained by the enzymatic-gravimetric method 985.29 and the modifications introduced in 991.42 and 993.19.

2. MATERIAL AND METHODS

The food samples analyzed in these studies were (1) corn bran, (2) iceberg lettuce, (3) oats, (4) potatoes, (5) raisins, (6) rice, (7) rye bread, (8) soy isolate, (9) wheat bran, (10) whole wheat flour, (11) white wheat flour, (12) nonvegetarian mixed diet, (13) lacto-ovo vegetarian mixed diet, (14) Fabulous fiber, a mixture of malto dextrin, whey, psyllium hulls, guar gum, pectins, vitamins and minerals, (15) a high fibre cereal from Farma Foods, (16) butter beans, (17) French beans, (18) kidney beans, (19) Brussels sprouts, (20) cabbage, (21) carrots, (22) chick peas, (23) okra, (24) onions, (25) parsley, (26) turnips, (27) apples, (28) apricots, (29) Calimyrna figs, (30) peaches, (31) prunes, (32) barley, (33) rye flour, (34) soy bran, (35) wheat germ, (36) green beans, (37) Mission figs and (38) sugar beet fibre. Many of these samples were used in several collaborative studies for determining TDF, IDF and SDF. The samples were all dried by lyophilization and ground to a uniform size of 350 um in a Microjet 10 centrifugal mill (Quartz Technology, Inc., Westbury, NY). Samples which resisted drying because of their high sugar content were desugared (extracted) with 10 volumes of 85% methanol, three times each).

The reagents to be used in the determination of TDF, IDF and SDF are phosphate buffer, 0.08M, pH 6.0; NaOH, 0.275 N; HCl, 0.325N; ethanol 78 and 95%; Celite C-211, acid washed; and the three enzymes, Termamyl, protease and amyloglucosidase.

The analytical scheme for the separation and determination of TDF, IDF, and SDF is presented in Figure 1.

(1) 1 g sample
V
(2) Gelatinization and Termamyl incubation
pH 6.0, 30 min., 100°C
V
(3) Protease incubation, pH 7.5, 30 min., 60°C
V
(4) Amyloglucosidase incubation, pH 4.5, 30 min., 60°C
V
(5) Precipitation with 4 volumes ethanol
V
(6) Filtration
V
(7) Washing with ethanol and acetone
V
(8) Drying (TDF)
V
(9) Correction for undigested protein and ash

For IDF and SDF

start with (4) above
(4) Amyloglucosidase incubation, pH 4.5, 30 min., 60°C
V
Filtration
(through fritted crucible containing Celite)
V   V
Residue   Filtrate
(wash with ethanol   (add 4 vol. of ethanol)
and acetone)
V   V
(IDF)   Filtration
V
Residue
(SDF)

Figure 1 : *AOAC scheme for determining TDF, IDF, and SDF.*

3. RESULTS AND DISCUSSION

The measures of precision for determining TDF (Table 1) were obtained in a collaborative study conducted in 1984[3] and accepted as official final action by the AOAC. The two samples with high RSD_R's were rice and soy isolate, which indeed may be unimportant for nutrition labeling because of their low dietary fibre content (1.04 and 1.42%, TDF respectively). All other foods had a TDF content of 2.78% or more and gave good results. The RSD_R's varied from 1.56 for corn bran to 9.80% for white wheat flour.

Table 1 : *Measures of precision for determining TDF*

Product	Average TDF (%)	Repeatability RSD_r (%)	Reproducibility RSD_R (%)	Laboratories
Corn bran	86.86	0.56	1.56	9
Oats	11.03	5.30	5.30	8
Potatoes	7.25	5.66	7.49	8
Rice	1.04	45.62	63.71	8
Rye bread	6.58	3.94	5.29	9
Soy isolate	1.42	66.25	66.25	7
Wheat Bran	42.65	2.33	2.66	9
White wheat flour	2.78	5.55	9.80	8
Whole wheat flour	12.57	3.67	5.92	9

In a second collaborative study[4] both IDF and SDF were determined. Only the IDF values (Table 2) are reported because they were accepted official final action by the AOAC International. The RSD_R's of 22 foods varied from 3.68 for soy bran to 19.44% for prunes, half the samples had an $RSD_R<10\%$.

A third collaborative study was conducted because it was felt that the Celite was the major cause of high RSD_R's in the determination of SDF in previous study. Indeed the RSD_R values improved significantly and the SDF method was accepted by the AOAC International[5].

The AOAC method for total dietary fiber was compared to the Englyst nonstarch polysaccharide (NSP) method[11,12] (Englyst et al. 1988; Englyst and Hudson 1987) in a recent study coordinated by the European Community (EC) Bureau of Reference[13] (Hollman et al. 1993). Wheat flour, rye flour, and haricot beans (Figures 2, 3 and 4) were analyzed in 10 laboratories by the AOAC method and in 5 laboratories by the Englyst GLC procedure. The AOAC method[3] yielded a high level of agreement. The smaller intra-and inter-laboratory repeatability and reproducibility coefficients of variation obtained with this method led to certified values being based on the AOAC method[14,15,16]. The set of results obtained by the Englyst procedures[11,12] exhibited considerably larger scatter and could not be certified at this time. It was interesting to note that the average TDF results for the samples were similar to the values obtained for the NSP method, and not as different as had previously been believed. In general the range of estimates for TDF and NSP in different laboratories using the same method may be wider than the differences in the mean values obtained with both methods.

Table 2: *Measures of precision for determining IDF*

Product	Average IDF (%)	Repeatability RSD$_r$ (%)	Reproducibility RSD$_R$ (%)	Laboratories
Apples	55.57	0.92	4.55	4
Apricots	44.92	0.86	8.22	5
Barley	4.30	9.92	14.33	12
Brussels sprouts	30.23	2.27	7.89	15
Butter beans	17.36	2.34	11.31	10
Cabbage	21.60	4.00	7.79	9
Calimyna figs	43.07	5.59	18.40	5
Carrots	32.29	5.38	11.39	12
Chick peas	16.69	10.38	16.80	12
French beans	25.64	3.23	5.87	10
Kidney beans	16.33	4.53	6.39	13
Mission figs	33.61	2.76	12.09	6
Okra	24.15	6.43	13.57	14
Onions	13.32	6.51	11.79	12
Parsley	34.39	3.56	13.64	12
Peaches	39.53	2.17	6.16	6
Prunes	46.18	6.11	19.44	6
Raisins	49.18	5.51	19.30	8
Rye flour	11.81	4.87	8.62	15
Soy bran	65.24	1.40	3.68	13
Turnips	21.38	6.60	16.61	12
Wheat germ	15.67	4.54	6.13	9

4. CONCLUSIONS

The AOAC method for total dietary fiber in foods is within the commonly accepted definition of dietary fiber as the remnants of plants components resistant to hydrolysis by the alimentary enzymes of humans. The procedure for its determination is rather simple and the more experience the laboratory has in performing the TDF analysis the lower the RSD$_R$. For nutrition label purposes, the TDF information is more than sufficient for giving concise, accurate information about the TDF content as well as the physiological benefit of that amount of TDF.

Acknowledgements

The Figures 2-4 are reproduced with the permission of the FPM Group, Oak Park, Illinois 60304 and appeared in the Conference and Exposition on Food Product Development Marketing & Technology, held in Secaucus, NJ November 19-20, 1991.

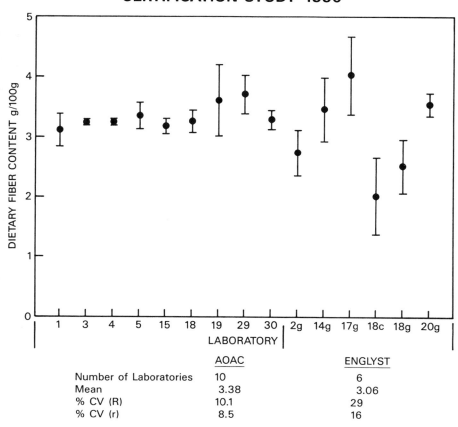

Figure 2 *Total Dietary Fiber of Wheat Flour by the AOAC and Englyst Procedures.*

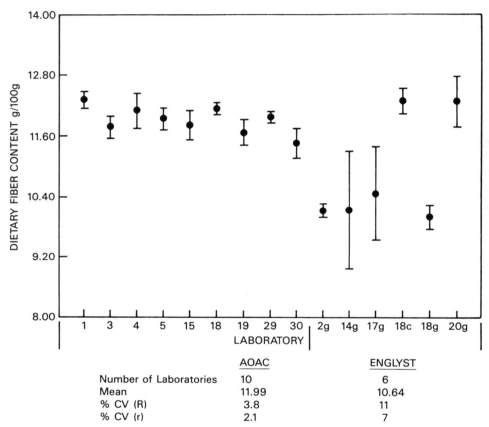

Figure 3. Total Dietary Fiber of Haricot Beans by the AOAC and Englyst Procedures.

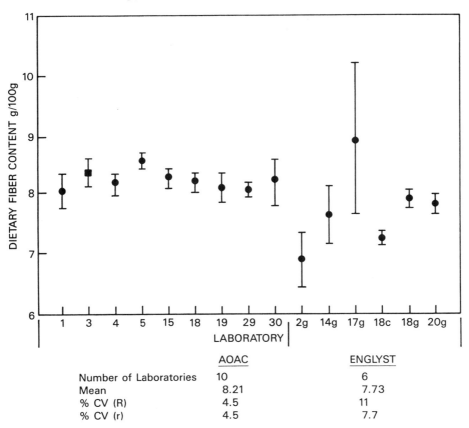

Figure 4. Total Dietary Fiber of Rye Flour by the AOAC and Englyst Procedures

References

1. Federal Register, 1993, Par IV, Department of Health and Human Services, Vol. 58, No. 3, p. 2066.
2. Prosky, L., Asp, N.-G., Furda, I., DeVries, J.W., Schweizer, T.F. and Harlan B.F., J. Assoc. Off. Anal. Chem., 1984, **67,** 1044.
3. Prosky, L., Asp, N.-G., Furda, I., DeVries, J.W., Schweizer, T.F. and Harland, B.F., J. Assoc. Off. Anal. Chem., 1985, **68,** 677.
4. Prosky, L., Asp, N.-G., Schweizer, T.F., DeVries, J.W., and Furda, I., J. AOAC Int., 1992, **68**, 360.
5. Prosky, L., Asp, N.-G., Schweizer, T.F., DeVries, J.W., Furda, and Lee, S.C., J. AOAC Int., 1994, **77**, 690.
6. Lee, S.C., Prosky, L. and DeVries, J.W., J. Assoc. Off. Anal. 1992, **75**, 395.
7. Mongeau, R. and Brassard, R., J. AOAC Int., 1993, **76**, 923.
8. Li, B.W. and Cardozo, M.S., J. AOAC Int., 1994, **77**, 687.
9. Theander, O., Aman, P., Westerlund, E., Andersson, R., and Pettersson, D., J. AOAC Int., 1995, in the press.
10. Theander, O., Aman, P., Westerlund, E., Andersson, R., and Pettersson, D., J. AOAC Int., 1994, **77**, 687.
11. Englyst, H.N. and Cummings, J.H., J. Assoc. Off. Anal. Chem., 1987, 71, 808.
12. Englyst, H.N. and Hudson, G.J., Fd. Chem., 1987, **24**, 63.
13. Hollman, P.C.H., Boenke, A. and Wagstaffe, P.J., Fresenius J. Anal. Chem., 1993, **345**, 179.
14. Prosky, L., Asp, N.-G., Schweizer, T.F., DeVries, J.W. and Furda, I., J. Assoc. Off. Anal. Chem., 1988, **71**, 1017.
15. Prosky, L. and Lee, S.C., in Cereal science and technology impact on a changing Africa, eds. Taylor, J.R.N., Randall, P.G. and Viljoen, J.H., CSIR, Pretoria, South Africa, 1993, p. 203.
16. Prosky, L. and DeVries, J.W., Controlling dietary fiber in food products, Van Nostrand Reinhold, New York, 1992.

DIETARY FIBRE INTAKES AND TRENDS IN EUROPEAN COUNTRIES

Y. Mälkki

Cerefi Ltd
Lyökkiniemi 24
FIN-02160 Espoo

1 INTRODUCTION

For 30 years ago Burkitt and Trowell introduced the concept of dietary fibre and hypothetized that is is one of the reasons of the so-called Western civilization diseases. This has resulted in an increasing amount of studies on dietary fibre, the number of publications being now more than 500 annually, and also dietary recommendations on the intake of fibre. The purpose of the present paper is to review, how the present intakes in European countries match with the recommendations. The statistical background of this paper is mainly a survey[1] recently made in the framework of the European COST 92 Action "Metabolic and physiological aspects of dietary fibre in food".

2 TARGET LEVELS

Earlier recommendations of the level of total dietary fibre intakes were presented by individual scientists or by scientific bodies. These recommendations have later been reviewed and adopted by public administration. The official recommendations of the intake vary between 18 g non-starch polysaccharides/day (United Kingdom) and 3.0 to 3.5 g dietary fibre/MJ $\approx$ 30-40 g/day (Scandinavian countries). Recommendations exist also for some subgroups of population: thus Anderson and Akanji[2] recommend for diabetics an intake of 40 g/day or 3.6-6.0 g/ MJ. The American Health Foundation recommends that children above 3 years of age should eat their age plus five grams of dietary fibre daily[3]. Recommendations on the quality of fibre are less quantitative. The most common concept is, that the share of soluble fibre should be 30 to 60% of the total dietary fibre intake.

3 ESTIMATION OF INTAKE

The most common method to estimate dietary fibre intakes on national level is to rely on food balance sheets, which are made in each country using standardized principles. The greatest advantage of this method is the commensurability of results from various countries. Although non-food use of the commodities is taken into account in the principle in this statistics, some of it is evidently remaining, and furthermore, the figures are not corrected for food waste. Household surveys seem to give figures more close to the real consumption, but are not corrected for waste. This correction is made in food basket studies. Individual food intake studies based on questionnaires often overestimate the fibre intake. The same is more or less true for 24 hours recall studies. The most reliable methods are weighing records and analyzing of duplicate portions, but the high costs bound limit their use to small group studies. When converting intakes of food items to fibre, differences arise from the analytical methods used in the study, or from the food tables used for calculations.

Following examples can elucidate the differences given by various methods. In Norway, household budget survey gave 5 g/d higher dietary fibre intakes than the national food balance sheets[1]. In an Italian study[1], the error of the questionnaire method, as compared to 7-day weighing method, was estimated to be 7.9 g/day/ person. Daily intake of resistant starch, included in the figures using the AOAC analytical method for dietary fibre, but not included in the non-starch polysaccharide determination, is estimated to be in Sweden 4 to 5 g[4], but can be in extreme cases up to 40 g/d. Amount of lignin, which is included as dietary fibre in the AOAC method but not in the non-starch polysaccharide method, is in the Finnish diet in the average 3.3 g[1].

4 NATIONAL AVERAGE INTAKES

Dietary fibre intakes according to the COST 92 survey are presented in Table 1.

Table 1. *Average Daily Intakes of Dietary Fibre (Adapted from Ref. 1)*

Country	Fibre g/day/pers	Calcul. Basis	Anal.Method	Year
Austria	23.1	Balance sheets	AOAC	90-91
Belgium	19-23	24 h recall	Dutch tables	89
Croatia	19.0	Household survey	NSP	90
Denmark	21.4	Balance sheets	AOAC	85-87
	18.4	Balance sheets	NSP	85-87
Finland	23.3	Balance sheets	NSP + lignin	90
France	15.9	Household survey	Southgate	89
Germany (FRG)	23.5	Food basket	Tables	85-86
(GDR)	21.1	Food survey	AOAC	88
Italy	25.0	Household survey	Tables	90
Norway	23	Balance sheets	AOAC,Southg	90
	18	Household survey	AOAC,Southg	90
Spain	15.1	Household survey	NSP	91
Sweden	21.0	Balance sheets	Tables	88
Switzerland	21.8	Various,corr.f. losses	Tables	87
United Kingdom	12.1	Balance sheets	NSP	89

The figures indicate that in general the intakes are at the lower part of the range of recommendations, and lower than recommendations in the United Kingdom, Spain and France. When differences due to methodology are taken into account, differences between the countries are not great. The largest intakes are in the Nordic countries, German-speaking countries and Italy.

The three main sources of dietary fibre are cereals, vegetables including potatoes, and fruits (Figure 1). Other important sources are pulses and nuts, which contribute markedly especially in the Mediterranian countries.

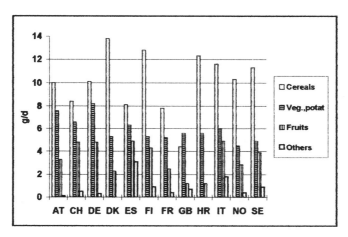

Figure 1.   *Sources of dietary fibre (Adapted from data of reference 1).*

The highest intakes of total dietary fibre are in countries where also intake of fibre from cereals is high. Exceptions are Croatia, having a high intake of fibre from cereals, but the total fibre intake is lower than average, and Switzerland, where vegetables and fruits form a great part of fibre intake. Intakes of fibre from vegetables vary between 4 and 8 g/day, and are highest in Germany, Austria and Switzerland.

In most European countries the intake levels were 30 to 50 g/d at the beginning of the century, and again high during the years immediately after the second World War. With increasing standard of living there has been a continuous decrease in fibre intake, which levelled in most countries between 1970-1980. The decline still continues in the United Kingdom and Spain. After 1970-1980, the intake levels have been rather stable. The only remarkable elevation has been reported from Italy indicatig an increased intake from fruits and vegetables.

Intake from cereals has been declining until 1970-1980, partly due to a decrease of total consumption of cereals, partly to a shift to more refined flours. Within the group vegetables, intake of fibre from potatoes has been declining remarkably. This has been compensated by an increase of intake from green and root vegetables. Annual variations of fibre intake from fruits are great. In several countries intakes of fibre from fruits now show a declining tendency after a long period of elevation.

Statistical figures of the share of soluble fibre exist from only a few countries (Table 2).

Table 2. *Intake of Soluble Dietary Fibre (Adapted from Data of Reference 1)*

Country	Soluble Fibre		Year
	g/d	%	
Finland	5.3	23.6	1990
Italy	8.9	36.0	1980-84
Spain	8.7	57.6	1991
United Kingdom	5.3	44.9	1990

5 REGIONAL AND INDIVIDUAL STUDIES

Within countries differences in intake exist notably due to ethnic and/or life-style differences. Thus in Belgium in 1989 the average intake of fibre was in the north 21.9 g/d, and in the south 20.0 g/d, respectively. This is mainly due to a higher total consumption of bread and in particular wholemeal bread in the north. In Italy, fibre intake was 1980-84 in the north 18.1 g/d, whereas in the center and south it was 22.1 g/d.

Individual and daily differences in fibre intake are surprisingly great. In two recent Austrian studies[5,6], the range of individual daily fibre intakes varied from 7.5 to 52.1 g/d, and the standard deviations between individuals on the same period were from 3.9 to 13.5 g/d. In a third Austrian study[7], the broadest range of individual daily intake within one week was from 9.2 to 27.6 g/d.

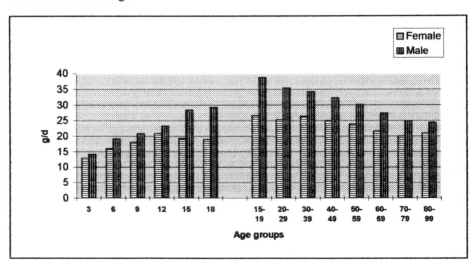

Figure 2. *Dietary fibre intake at different ages. Data for children and adolescents from 1980[8], for adults from 1967-1976[9].*

Variations due to age and sex are illustrated in Figure 2, adapted from results of two different Finnish studies. The difference in the level between these studies derives partly from the methodology, partly from the time of collecting the experimental material. The intakes follow closely the energy needs of each age group. Fibre densities calculated from these figures vary for children and adolescents between 2.2 to 2.6 g/MJ, and for adults, in the earlier study, between 2.3 (highest age group) to 2.9 g/MJ (group 15 to 19 years). Differences in fibre density are smaller than in the total fibre intake, but in contrast to the total intake, female persons from 9 years onwards have 0.05 to 0.2 g/MJ higher fibre density than males in their diet. The small variations in the fibre density reflect evidently the role of common meals of the families in the level of fibre intake.

Similar differences between the sexes are reported from most other countries. In Belgian studies the intake of fibre by women was 3 g/d less than that of men[1.] In Germany 1988, the intake of fibre by women was 17.3 g/day ≈ 1.84 g/MJ, and by men 20.7 g/day ≈1.53 g/MJ. Contrary to these figures, in United Kingdom a higher total non-starch polysaccharide intake (12.5 g/day) by women than by men (11.2 g/day, respectively) was found.

6 POSSIBILITIES TO IMPROVE FIBRE INTAKES AND THEIR EFEFCTS ON HEALTH

Recommendations to increase intake of food items rich in dietary fibre have had a positive, but meager effect. This is evidently due to the natural resistance to change dietary and consumption habits bound to the culture. When these are altered, especially an increase in the consumption of meat, fish and dairy products reflects in reduction of the use of ceral products and potatoes. Reduced need of energy reflects often in reductions in staple foods. A positive tendency in the development has been the increase in the consumption of vegetables, which is probably more often a result of the need to diminish the total energy intake than a conscious increase of fibre containing food.

Due to the reduced need of energy, addition of new items in the daily menu is possible only by replacing some items usually consumed. Another, probably more effective way, is to create changes within the food items commonly consumed. An obvious approach is to replace a part of white flour with wholemeal flour, or to add cereals rich in fibre, such as rye, oats or barley in the bakery products. Several of the countries which have a high average fibre intake, *e.g.* Austria, Denmark and Finland, also have a high consumption of rye flour. Another commodity the consumption of which can easily be increased is potatoes. The diminution of their consumption is only very recent, and higher figures can be returned by increasing the variety in the supply of industrially prepared potato products, as it has recently happened in Norway. An increase of product varieties available and their consumer appeal is also a key for a new increase in the consumption of cereal products, which most effectively would elevate the total dietary fibre intake.

Increasing the total dietary fibre intake should, however, not be a value of its own. It is known that different types of fibre have different physiological effects, and that these depend not only of the composition and content of fibre, but also of the physical state and functional properties. Some of the effects can also be achieved by other substances such as resistant starch and oligosaccharides. The content and nature of physiologically active non-fibre components accompanying fibre are different in various fibre-containing food sources. With new information on the specific effects of different kinds of fibre, also more specific recommendations should be elaborated. It is already recognized, that the health-promoting effects in cardiovascular diseases, diabetes and colon cancer are effected by soluble fibre components, and even not by all of these. Recording intake of total dietary fibre will probably in a near future remain as a general indication of a diet rich in health-promoting factors, including both fibre components and accompanying minor components. For health-promoting effects, more specific information of the content of effective components and their physical properties should be given.

References

1. J.H. Cummings and W. Frølich, 'Dietary fibre intakes in Europe'. Commission of European Communities, Luxembourg 1993. 89 pp.
2. J.W. Anderson and A.O. Akanji, In: 'CRC Handbook of Dietary Fiber in Nutrition', 2nd Edition, G.A.Spiller, Editor. CRC Press, Boca Raton, FL., U.S.A., 1993, p. 462
3. A.E. Sloan, *Food Technology* 1995, **49** (3), 32
4. H. Andersson, I. Bosaeus, L. Ellegård, A.M. Langkilde and T. Schweizer, In: 'Dietary Fibre and Fermentation in the Colon', Y. Mälkki and J.H. Cummings, Editors. Commission of European Communities, Luxembourg 1995 (in press)
5. H. Pintscher, Physiologische Wirkung von Erbsenkleie. Diplomarbeit, 1989. (cf. ref. 1, p. 25)
6. C. Holler, M. Just and K. Irsigler, Physiologische Wirkung von veredelter Erbsenkleie. Forschungsprojekt Forschungsförderungsfonds für die gewerbliche Wirtschaft, 1991 Nr. 6/559/2607. (cf. ref. 1, p. 24)
7. Anon., Forschungsinstitut der Ernährungswirtschaft: Untersuchungen von 200 Proben verzehrsfertiger Nahrung auf Quecksilber. In: 'Beitrage Umweltschutz, Lebensmittelangelegenheiten, Veterinärverwaltung'. Forschungsberichte, Bundesministerium für Gesundheit un Umweltschutz, Vienna, Austria, 1, 1982. (cf. ref. 1, p. 26)
8. S. Kimppa, L. Räsänen and P. Varo, Ravintokuitu lasten ja nuorten ruokavaliossa (Dietary fibre in the diet of children and young persons). *Kotitalous* 1987, **51** (5-6), 6. (cf. ref. 1, p. 45)
9. R. Korpela, Valio Ltd, Helsinki, Finland. Unpublished. (cf. ref. 1, p.44).

DIETARY FIBRE AND PHYTIC ACID: POTENTIAL HEALTH IMPLICATIONS OF THEIR INCREASED INTAKE

Brittmarie Sandström

Research Department of Human Nutrition/Center for Advanced Food Research, The Royal Veterinary and Agricultural University, Frederiksberg, Denmark

1 INTRODUCTION

The scientific interest in potential positive health implications of a high fibre intake started more than 20 years ago with the observations of Dennis Burkitt during his work in Africa. He noticed a dramatic difference in incidence of almost all diseases between African populations and populations in UK or USA[1]. He also observed that there was a direct relation between stool size of the population in a community and the size of its hospitals. In a text supplied with a set of slides produced by Dennis Burkitt he writes:"doubling the size of stools in a community might allow a great reduction in the size of hospitals, while still meeting the medical requirements of the community". The way to increase stool size is to increase the dietary fibre intake[2]. The question is, however, would an increased fibre intake, resulting in increasing stool weight, also have an impact on health? A lot of research has been undertaken to try to prove or disapprove Dennis Burkitt's hypothesis regarding the positive health effects of dietary fibre.

2 RECOMMENDED DIETARY FIBRE INTAKE

Some countries have today official recommendations for dietary fibre intake. The British COMA report[3] suggest an intake of 18 g of non-starch polysaccharides per day and the Nordic recommendations[4] are 3 g dietary fiber/MJ i.e. approximately 25-30 g/day. Present intakes are however much lower, typical 10-12 g NSP and 20 g dietary fiber and it is clear that to reach the recommendations substantial dietary changes are needed.

The scientific basis for these recommendations is entirely the effect on bowel disease. A recent meta-analysis has shown that the risk for bowel cancer and diverticulosis is significantly reduced at stool weight above 100 g per day and that this is achieved with a fiber intake of approximate 12 g non-starch polysaccharides[2]. However, for many in the general population dietary fibre has a "health image" that goes far beyond its effect on GI-function. Positive effects have been claimed e.g. on coronary heart disease, obesity, blood pressure, other cancer forms, etc. On the other hand, an increased dietary fibre intake could have negative effects through impairment of the utilization and availability of nutrients, especially mineral availability.

3 PHYSICO-CHEMICAL CHARACTERISTICS AND PHYSIOLOGICAL EFFECTS OF DIETARY FIBER

As dietary fibre per definition is not degraded in the small intestine, and only to some extent is metabolised in the colon, the putative physiological effects are related to physico-chemical characteristics in the bowel. Water solubility, water binding, ability for gelling, ion exchange capacity etc. could affect the absorption and the metabolism of other nutrients and might be the explanation for any beneficial effect (table 1). In the mouth, stomach and upper small intestine dietary fiber could affect satiety and the rate of absorption of nutrients (table 2). The morphology and motility of the intestine, including the cell proliferation of the large colon, are also assumed to be affected by dietary fibre intake. Considering these potential effects on the environment in the intestine, it is not surprising that health effects have been proposed.

Table 1 *Physico-chemical effects of dietary fiber*

Properties	Characteristics	Effect
Water-holding		Volume
Solubility	Viscosity	Motility, gastric emptying
	Gelling	Micelle formation
Binding ability	Ion exchange	Cation binding
	Adsorption of organic molecules	Bile acid/cholesterol metabolism
Degradability	Fermentability	Bacterial growth, SCFA production

Table 2 *Gastrointestinal effects of dietary fiber*

Sites	Effect
Mouth	Food intake
	Satiety
Stomach	Gastric emptying
Small intestine	Absorption of nutrients
	Mobility
	Intestinal morphology
Large intestine	Fecal weight
	Stool frequency
	Transit time
	Bacterial metabolism
	Cell proliferation

4 DIETARY FIBRE AND CHRONIC DISEASES

4.1 Coronary Heart Disease

Ancel Keys, one of the pioneers of epidemiology in coronary heart disease research, probably best known for the 7-countries study, suggested already in the 1960'ies that fibre-rich foods might have a positive influence on blood lipids[5]. His original view was that only fat was important for blood lipids, but in one of his experimental dietary intervention studies, he observed that at the same fat intake and fat quality an Italian type of diet had a much more pronounced effect on blood lipids than an American diet. He attributed this difference to a negative effect of the sugar content of the American diet; however, it could as well have been a positive effect of the fruits and vegetables in the Italian diet.

A large number of studies have investigated the effect of fibre isolates or concentrates on blood lipids[6]. These studies have rather consistently shown that high doses of viscous gelling fibers like guar gum and pectin have a decreasing effect on cholesterol levels. Oat bran has in some studies shown effect, however, rather high doses seem to be needed. The effect of natural fibre rich food sources is much less pronounced. Only when combined with a reduction in fat intake, effects similar to that of the isolated fibers have be observed.

We have evaluated the effect on blood lipids of a high fiber diet in randomised cross-over studies in healthy subjects[7]. Two different diets were served, a low-fibre diet providing 15-20 g of fibre was compared to a high fibre diet with only the fibre sources changed and all other components identical. A similar cross-over study was performed adding 100 g of oat bran to a low-fiber diet. The main results from both studies were that the effects of an increased fibre intake was depending on the fat quality of the diet. With a fat quality typical for northern European countries i.e. a high intake of saturated fatty acids, a substantial increased intake of fruits, vegetables and whole-grain cereals had no effect on blood lipids. Only if the fat quality also was improved by reducing the saturated fatty acids and increasing the polyunsaturates, there was a significant effect.

However, fibre rich foods could have effects on coronary disease in other ways not related to changes in serum cholesterol levels. Two recent secondary intervention studies suggest that an increased intake of fruits and vegetables reduce the mortality in myocardial infarct patients[8, 9]. Despite rather small differences in serum cholesterol levels, dramatic differences were seen in mortality in the intervention groups.

4.2 Obesity

A number of intervention studies suggest that is difficult to become obese on high fiber diets. Spontaneous weight losses at *ad libitum* intake of high fibre, low-fat diets have been observed in several studies[10]. Whether this effect can be ascribed to the dietary fibre or to other dietary changes is still unknown.

4.3 Cancer

There is strong and consistent evidence that a high fruit and vegetable intake reduce cancer risk[11] (table 3). Although some studies have indicated an independent effect of dietary fiber, the putative preventive effect is most likely not due to the fibre *per se*, but to components associated with fibre-rich foods, e.g. antioxidants and flavonoids.

5 DIETARY FIBER AND MINERAL UTILIZATION

Many fibre rich sources, especially cereal fibre sources, have a high content of phytic acid, a strong antagonist of mineral and trace element absorption. Single meal studies have shown that phytate has a strong negative effect on zinc and iron absorption[12]. Does this also apply to total diets with realistic fiber intakes? To test this we have followed iron status in a group of 30 subjects over 8 months after a dietary change to high fibre diet[13]. Compared to habitual intake they increased their intake of whole-grain cereals and vegetables. This resulted in an increase in fibre intake from 23 to 35 g per 10 MJ, and a relatively high phytate intake, 1.3 mmol. Their habitual intake of phytate was estimated to approximately 0.5 mmol per day. However, the subjects also increased their vitamin C intake, which is a strong enhancer of iron absorption. Surprisingly, this type of diet, which had a positive effect on many risk markers for coronary heart disease, gave reductions in serum ferritin levels, ie. decreased iron stores in the majority of the subjects. These results indicate that certain types of high fibre diets with a high phytate content might induce a risk for impairment of mineral and trace element status in the population.

Table 3 *Epidemiological studies of fruit and vegetable intake and cancer prevention, adapted from Block et al[11]*

Cancer site	Number of studies	Significantly reduced risk	Significantly increased risk
All sites	170	132	6
Lung	25	24	0
Esophagus	16	15	0
Stomach	19	17	1
Colorectal	35	20	3
Breast	14	8	0
Prostate	14	4	2

Table 4 *Potential health implications of an increased intake of fiber rich foods*

Target	Effect/role
Bowel disease	Protective
CHD	Facilitates fat reduction
	Potential positive effect of fiber associated components
Cancer	Protective effect of fruits and vegetables
Obesity	Facilitates fat and energy reduction
Mineral status	Fruits/vegetables: mainly positive effects
	Cereals/legumes: risk for negative effects

References

1. D.P. Burkitt and H.C. Trowell (eds.). 'Refined carbohydrate foods and disease; some implications of dietary fibre', Academic Press, London, 1975.
2. J.H. Cummings, S.A. Bingham, K.W. Heaton and M.A. Eastwood. *Gastroenterol.*, 1992, **103**, 1783.
3. Dietary Reference Values for Food Energy and Nutrients for the United Kingdom. Department of Health. HMSO, London, 1991.
4. Nordic Committee on Foods, Nordic Council of Ministers. 'Nordic Nutrition Recommendations', Nordic Council of Ministers, Copenhagen, 1989, 2nd edn.
5. A. Keys, J.T. Anderson and F. Grande. *J. Nutr.* 1960, **70**, 257.
6. J.W. Anderson, D.A. Deakins, T.L. Floore, B.M. Smith and S.E. Whitis. *Cr. Rev. Food Sci. Nutr.*, 1990, **29**, 95.
7. B. Sandström, A.-M. Sørensen, and T. Leth. In: 'Mechanisms of action of dietary fibre on lipid and cholesterol metabolism', D. Lairon. (ed.), Commission of the European Communities, Luxembourg, 1993, p. 153-158.
8. R.B. Singh, S.S. Rastogi, R. Verma, B. Laxmi, R. Singh, S. Ghosh, and M.A. Niaz. *Br. Med. J.*, 1992, **304**, 1015.
9. M. de Lorgeril, S. Renaud, N. Mamelle, P. Salen, J.-L. Martin, I. Monjaud, J. Guidollet, P. Touboul, and J. Delaye. *Lancet*, 1994, **343**, 1454.
10. A. Astrup and A. Raben. *Eur. J. Clin. Nutr.*, 1992, **46**, 611.
11. G. Block, B. Patterson and A. Subar. *Nutr. Cancer*, 1992, **18**, 1.
12. L. Rossander, A.-S. Sandberg, and B. Sandström. In: 'Dietary fibre - a component of food', T.F. Schweizer et al. (eds.), Springer-Verlag, London, 1992, p. 197-216.
13. B. Sandström. In: 'Nutritional, Chemical and Food Processing Implications of Nutrient Availability, U. Schlemmer (ed.), Bioavailability '93, Ettlingen, 1993, p. 159-163.

ANALYSIS AND CONSUMPTION OF DIETARY FIBER IN GERMANY

E. Rabe

Federal Center for Cereal, Potato and Lipid Research in Detmold and Münster, Institute for Milling and Baking Technology, D-32756 DETMOLD

1. METHODS OF DIETARY FIBER DETERMINATION

In Germany, the determination of dietary fiber is carried out according to the AOAC-procedure (phosphate-buffer; offical method 00.00-18 since December 1988). The International Association for Cereal Science and Technology has also accepted the method (ICC-Draft-Standard No. 156). At the moment, a ringtest for the German official methods for the analysis of food, evaluating the statistical values for the MES-TRIS-buffer-method (with separate determination of soluble and insoluble fiber) is under way. One of the results is, that the sum of soluble and insoluble fiber is 90-105 % (mean 96 %) of that for total dietary fiber. The EC is performing a certification study with both AOAC methods, the ENGLYST and the UPPSALA method; at the moment there are no results. In general nowadays it is confirmed, that resistant starch as well as lignin is part of the dietary fiber complex. We believe that the AOAC method is best for routine work and for nutritional labelling.

Although standardized, there is the possibility to use a long stem funnel with ash-free paper filter (125 mm diameter, Schleicher & Schüll 589[1], Schwarzband) instead of glass crucibles and by this, to make the method a little cheaper. These typ of funnels were used a lot in former times for gravimetric determinations. They are characterizied by a long tube (diameter 7 cm; 15 cm long, i.d. 3 mm); by this the filtration is aided by capillary action. It has several preferences:
- the ash determination can be done in a usual porcelain or gold/platinum crucible (the sintered glass crucibles often break because of the temperature differences in the muffle furnace)
- during protein determination (Kjeldahl method), there is no danger in bumping, caused by the Celite
- no vacuum source is needed
- the filter paper is much cheaper than the crucibles
- the filtation time is the same and
- the results are the same
- no cleaning of crucibles

Table 1: *Dietary fiber content in flours*

flour	dietary fiber, %		
	soluble	insoluble	total
wheat flour type 405 *)	2.0	1.2	3.2
wheat flour type 550	2.2	1.3	3.5
wheat flour type 1050	2.1	3.1	5.2
wheat flour type 1700	2.3	6.9	9.2
whole meal wheat flour	2.3	7.7	10.0
rye flour type 815	2.6	3.9	6.5
rye flour type 997	3.0	3.9	6.9
rye flour type 1150	3.3	4.4	7.7
rye flour type 1800	3.3	8.7	12.0
whole meal rye flour	3.3	10.2	13.5

*) mean ash content is 0.405 g/100 g flour d.b.

2. CONTENT OF DIETARY FIBER IN GRAIN, BREAD AND BREAKFAST CEREALS

Among the different types of flour, there is a good correlation between the dietary fiber content and the ash content (table 1). The German bread is based on wheat and rye and mixtures therefrom. Also, flour, meal or whole kernels (alone or together) are used. Additionally, the incorporation of several ingredients of grain and plant or animal origin (like oat, linseed, onion, butter, milk, ham) into the recipe is possible. This results in a great variety of breads and rolls. The different ingredients and their varying proportions lead to a great range of the dietary fiber contents in breads. Small bakery products differ by size only, not by composition.

The main ingredients of müsli are cereals (mostly oat flakes), oilseeds and (dried) fruit; most popular products at the moment are chocolate and fruit müsli (with about 30 % fruit). In the müsli-bars the ingedients are mixed together with a binder, mostly

Table 2: *Dietary fiber content in typical german breads*

bread	dietary fiber, %		
	soluble	insoluble	total
toast bread	1.8	2.0	3.8
wheat bread	2.0	2.1	4.2
wholemeal wheat bread	2.0	4.9	6.9
mixed wheat rye bread	2.7	2.1	4.8
mixed rye wheat bread	2.2	3.8	6.0
rye bread	2.4	4.4	6.8
wholemeal rye bread	2.6	6.3	8.9
crispbread	3.7	9.8	13.5

Table 3: *Dietary fiber content in breakfast cereals*

breakfast cereal	soluble	dietary fiber, % insoluble	total
cornflakes	1.2	2.8	4.0
müsli	2.6	11.7	14.3 *)
müsli bars	3.0	4.2	7.2 *)
oat flakes	4.5	5.0	9.5
oat bran	8.2	10.4	18.6
wheat bran	3.6	45.7	49.3
rice, white, cooked	0.3	0.2	0.5
rice, wholemeal, cooked	0.5	0.5	1.0
pasta, cooked	1.1	0.4	1.5
pasta, whole meal, cooked	3.7	0.7	4.4

*) differences, depending on the composition

on a sugar basis. The different ingredients and their varying proportions lead to a great range of the dietary fiber contents in cereal based food (table 3). The consumption of breakfast cereals such as müsli or müsli bars is small, but still increasing.

In cereals the insoluble dietary fiber is the dominant fraction. Soluble hemicelluloses are found mostly in the endosperm, insoluble hemicelluloses and cellulose in the outer parts of the kernel. Therefore, with increasing extraction, mainly the content of insoluble dietary fiber is raising, while the amount of soluble fiber is nearly constant. The composition of dietary fiber depends of the kind of plant (table 4), its age, and the conditions during growth. An increase of the dietary fiber intake only by increasing the amount of fruit and vegetable is not possible because of their low fiber content and high water content, resp. (table 5). Fruit and vegetable dietary fibers contain more uronic acid and galactose. Most of the soluble fiber is degraded in the gut by microbes, yielding mostly short chain fatty acids. The dietary fiber from cereals that is more effective on the stool weight and transit time, on the other hand, contains more

Table 4: *Composition of dietary fiber from different material*

	total dietary fiber (g/100g)	cellulose	% of total dietary fiber hemicellulose	pectin
rye, wholemeal	15.7	12.1	79.0	6.1
wheat, wholemeal	11.5	71.0	15.8	5.3
wheat bran	50.5	67.4	23.6	8.5
soy bran	64.6	32.7	61.0	0
white cabbage	1.9	32.0	28.9	2.4
apple	1.9	47.1	23.1	0

Table 5: *Dietary fiber content in fruit and vegetables*

fruit or vegetable	soluble	dietary fiber, % insoluble	total
grapefruit	0.3	0.3	0.6
cucumber	0.1	0.8	0.9
tomato	0.5	0.8	1.3
pineapple	0.5	0.9	1.4
onion	0.7	0.7	1.4
grapes	0.4	1.2	1.6
peach	0.9	0.8	1.7
chinese leaves	0.2	1.5	1.7
spinach	0.5	1.3	1.8
mushroom (Champignon)	0.4	1.5	1.9
potato	1.3	0.6	1.9
strawberry	0.8	1.2	2.0
green peppers	0.3	1.7	2.0
apple	1.2	1.1	2.3
red cabbage	0.8	1.7	2.5
carrot	1.4	1.5	2.9
broccoli	1.3	1.7	3.0
brussel sprouts	1.1	3.3	4.4
raspberry	1.0	3.7	4.7

hemicelluloses, composed of arabinose and xylose. Independent from their composition, the total dietary fiber in Germany is considered to supply no energy.

3. CONSUMPTION OF DIETARY FIBER

In the seventies, the consumption of bread in Germany was very low, in the meantime it has reached 81,3 kg/head/year, wich is 223 g/head/day. Bread makes up 65,8 kg (81 %), and rolls 15,5 kg (19 %). Most of the bread (50 %) is made from mixtures of wheat and rye. Wholemeal bread constituts 22 %. The part of the dark, coarse breads and wholemeal breads has risen to 20 % of the total consumption. The consumption of breakfast cereals such as müsli, cornflakes or müsli bars is small, but an increasing segment. At the moment the consumption is 1.4 kg/head/year. A consumption of 30 g dietary fiber/day/person is recommended with half of it coming from cereal products. Because of their high water content, fruits and vegetables have less dietary fiber than white bread. Among usual consumed products, wheat and rye whole meal breads contain the highest amounts.

The advertising of dietary fiber in Germany is regulated (recommendation, no food law) in a way, that the value of the daily consumption is taken as the base: a daily intake of the product should provide at least 10 % of the 30 g dietary fiber/day/ person. If advertising claims, that a food contains dietary fiber, it has to provide at least 10 %, if advertising claims, that it is rich in dietary fiber, it has to provide at least 20 % of the daily recommended intake, that means 3 g or 6 g dietary fiber per serving (amount of it has to be named), respectively. In this way, the different dietary fiber content (fresh product) of food such as bread, vegetables, yoghurt, müsli, or bran was eliminated. Since the daily consumption of bread and rolls is about 220 g, all bread could be advertised as "fiber containing". However, bread and fine baked goods can only be advertised, if they contain 3 g or 6 g dietary fiber per 100 g. This was done, to avoid, that products, produced from low extraction flour would be advertised as "fiber containing" and to reserve this attibute for wholemeal products, which contain >3 g (e.g. wholemeal bread). These values of 3 and 6 g, resp. are used for enriched bread, too. Most fine baked goods (table 6, without whole meal), wheat- and toast breads do not contain 3 g dietary fiber and therefore, cannot be advertised as "fiber containing".

Most of the 25 million tons of bread and rolls per year, produced in the EC, is consumed in Germany, followed by the scandinavian countries, Belgium and Italy. Whole-meal products are consumed mainly in the northern countries. In the seventies the consumption of breakfast cereals in Germany was very low (69 kg/year/head in 1965), in the meantime it has reached 81,3 kg/year/head, 65,8 kg (81 %) originating from bread and 15,5 kg from rolls. Most of the bread (50 %) is made from mixtures of wheat and rye; 22 % is wholemeal bread. The part of the dark, coarse breads and wholemeal breads has increased to 20 %. In 1984 the dietary fiber intake from bread and baked goods and breakfast cereals was 10.2 g/day for men and 8.3 g/day for women. Potatoes, vegetables, fruit, and other products provided 3.0 g/day for men and 2.3 g/day for women. This resulted in an total intake of 21.4 g dietary fiber/day for men and 18.4 g dietary fiber/day for women. The consumption of breakfast cereals as müsli today is 1.4 kg/year/head.

Dietary fiber shows a good swelling capability, water binding capacity, and the ability to bind metal ions and other components. These attributes are of special significance for the physiological effects of dietary fiber. Apart from that, the kind of dietary fiber, the particle size, and the amount consumed, are of influence on the physiological effects. By binding water and being consumed by bacteria itself, this results in an increased growth of microorganisms in the gut, and a raise of the stool weight. The motion of the gut is enhanced by the abundance and the fermentation products and shortens the transit time of the stool. Bread from coarse cut whole meal is more effective than bread from whole meal flour. Pectin, which is rich in dietary fiber from fruit and vegetables, results in an increase in the excretion of bile acids, thereby lowering the cholesterol content in serum. A slowing-down of the reception of carbohydrates resulting in a smaller availability of bloodglucose is still being discussed; most effective in this is dietary fiber from fruit or vegatable. A decrease in availability of minerals (e.g. by ion exchange) is not detrimental considering the usual consumption of bread, because the layers rich in dietary fiber are also rich in minerals. Mineral losses are more than compensated. Another advantage of fiber-enriched bread is that it has less fat than normal bread.

Table 6: *Dietary fiber content of fine baked goods*

fine baked good	dietary fiber, %		
	soluble	insoluble	total
Madeira cake / pound cake	0.3	0.3	0.6
Stollen	0.3	1.0	1.3
butter cookie	0.6	0.8	1.4
strawberry tart / gateaux	1.2	0.8	2.0
Danish pastry	1.1	1.0	2.1
plum cake	1.1	3.8	4.9
rusk / zwieback	1.6	3.6	5.2

4. DIETARY FIBER - ENRICHMENT

One of the possibilities to enrich the dietary fiber content of cereal based products, is to use whole meal products instead of flour with a low ash content. The dietary fiber content of fine baked goods is especially low (table 6). Today there are even many fine baked goods available, that are made from whole meal flour. Because of the amounts of fat, sugar, and eggs used in the recipe, these whole-meal-fine-baked-goods contain clearly not so many dietary fiber than whole meal bread. Nevertheless, even these fine baked goods can raise the dietary fiber intake. The best way, to enhance the consumption significantly is to enrich breads or fine baked goods with products, high in dietary fiber. Todays industry offers a great variety of these products, high in soluble or insoluble dietray fiber. One of the most commonly used are brans from different kernels.

When incorporating dietary fiber from different sources in baked goods, the baker has to adjust the amount of water added (with addition of 10 % bran the amount of water is to be raised 3-5 %). Since the dough quality changes, too, the fermentation conditions have to be controlled carefully. In general, the fermentation tolerance is reduced. Depending on the type of bread, there may be a reduction of bread volume. Therefore breads with big volumes should not be used for dietary fiber enrichment. If more than 5 to 10 parts of bran per 100 parts of flour are used, a mixture of different brans shows best results. A good example is a mixture from wheat bran and soy bran. Especially suited are with wheat breads, toast breads, mixed weat rye breads, and rolls made from wheat. Highest dietary fiber contents can be obtained with rye breads, made from meal.

DIETARY FIBER, ß-GLUCAN AND INOSITOL PHOSPHATE CONTENTS OF SOME BREAKFAST CEREALS

S.P. Plaami and J.T. Kumpulainen

Agricultural Research Centre of Finland, Food Research Institute, Laboratory of Food Chemistry
FIN-31600 Jokioinen, Finland

1. INTRODUCTION

Insoluble dietary fiber (IDF) has been shown to prevent diverticular disease of the colon and possibly colon cancer [1]. Soluble dietary fiber (SDF), such as ß-glucan in oats and barley, has important therapeutic effects in diabetes [2] and hypercholesterolemia [3]. Especially the incorporation of oat products into the diet has been shown to cause a modest reduction in blood cholesterol level [4]. Oats are usually consumed in the form of breakfast cereals. Moreover it has been shown that cooking increases the solubilities of ß-glucans and hence the viscosities of rolled oats [5,6] rye, wheat and barley flakes [5]. Cooking may therefore enhance the cholesterol lowering effect of cereal flakes by increasing the viscosity in the gastrointestinal tract which is one of the mechanisms by which soluble fiber affects plasma cholesterol concentrations [7]. Phytic acid (IP6), which occurs with DF in cereals, has also been suggested to be useful in the therapy of cardiovascular diseases [8,9,10] and cancer [11]. Phytic acid, however, is notorious for its ability to limit the bioavailability of cationic essential mineral elements [12].

This paper will present the contents of IDF and SDF, ß-glucan [13] and myoinositol hexa-, penta-, tetra- and triphosphates (IP6-3) found in breakfast cereals consumed in Finland. The results will be discussed in terms of their effects on mineral bioavailability.

2. MATERIALS AND METHODS

The sample collection and pretreatment has been presented in detail earlier [13,14]. The samples collected included all breakfast cereals sold in Finland.

SDF and IDF contents were determined by the enzymatic-gravimetric method of Asp et al. (1983) [15] and ß-glucans by the enzymatic method of McCleary and Glennie-Holmes [16]. Contents of IP3-6 were quantified by an HPLC-method [17,18].

Accuracy of the analysis was tested by reference materials (RMs). The ARC/CL Wheat flour RM [16] was used in the DF analysis. ARC/CL Wheat flour and ARC/CL Potato powder RMs were used for analytical quality control of inositol phosphates determination [20,14]. The reference materials used for ß-glucan were barley flours supplied by Biocon Ltd [14].

3. RESULTS AND DISCUSSION

Results of the quality control programme are presented in Tables 1 and 2.

Table 1 *Analytical accuracy of DF and ß-glucan determinations as tested by reference materials.*

Reference material	n*	Value obtained DF content (% of d.w.)	Certified value
ARC/CL Wheat flour	3	4.0±0.3	4.08±0.27
		ß-glucan content (% of d.w.)	
Barley flour	11	3.4±0.08	3.2 ± 0.2
Barley flour	7	4.68±0.17	4.4 ± 0.2

n* = number of duplicate determinations

Table 2 *Reproducibility of inositol phosphate (IP6-3) determinations as tested by the ARC/CL Wheat flour and Potato powder reference materials.*

Reference material	n^a	IP6	IP5	IP4	IP3
			$\mu mol/g d.w.$		
ARC/CL Wheat flour RM	6	5.29±0.67	0.83±0.03	0.13±0.02	traces
ARC/CL Potato powder	8	1.72±0.17	0.47±0.06	traces	traces

a = number of analyses

Contents of TDF, ß-glucans, and inositol phosphates are presented in Figures 1, 2 and 3.

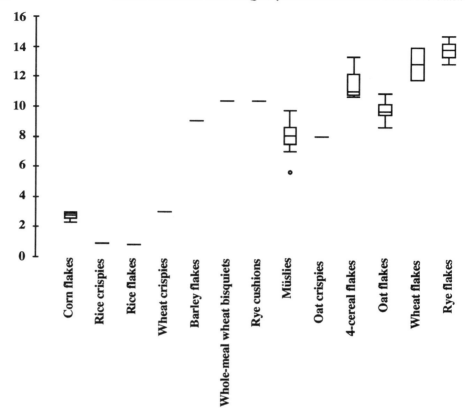

Figure 1 Total Dietary Fiber(*TDF*) *contents of breakfast cereals (% of fresh weight). Boxplots represent medians, range and 25 and 75 per centiles. Line represents the mean value of 1 brand only.*

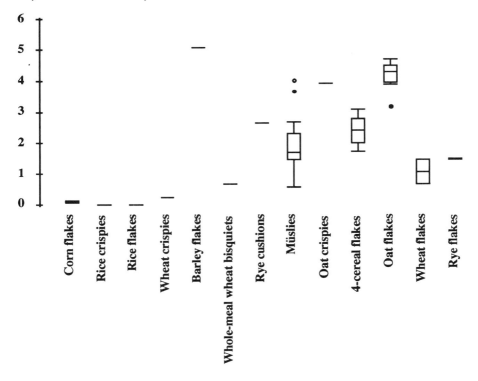

Figure 2 *β-glucan contents of breakfast cereals (% of fresh weight). Boxplots represent medians, range and 25 and 75 percentiles. Line represents the mean value of 1 brand only.*

A significant proportion of the total DF in oat, rye, barley, and wheat products and müslies was soluble. ß-glucan contents of oat cereals were also high. However, the highest ß-glucan contents were encountered in barley flakes. It has previously been shown that the solubilities of barley ß-glucans are higher than those of oat flakes [5]. Since the beneficial effects of soluble fiber are largely due to its viscosity effect [7, 21], cooked barley likely has the same beneficial effects on cholesterol metabolism as oats [4].

In general, phytic acid and DF contents were parallel in the breakfast cereals studied (Figure 3).

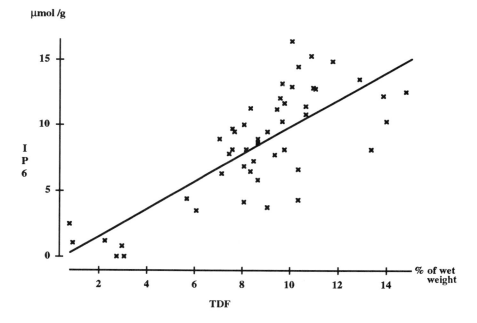

Figure 3 *Correlation between phytic acid (IP6) and Total Dietary Fiber (TDF) contents of breakfast cereals (Pearson Product-Moment Correlation 0.77).*

Heterogeneity of the ingredients and effects of processing caused inconsistencies. Contents of DF and phytic acid were highest in wheat, oat and rye flakes, slightly lower in müslies and scanty in cornflakes and rice cereal products. Effects of processing were detectable in puffed and baked cereal products. DF contents in these products were at the same level as those in flakes made of the same cereals, but lower in phytic acid contents and higher in inositols, with a lower number of phosphate groups.

Phytate has been reported to decrease the bioavailability of Ca, Zn, Mg and Fe but does not exert an influence on Se, Cu and Mn utilization in humans [22]. However, the chemical environment of the meal in which phytate is present is highly significant when considering its effects on mineral absorption. Partly digested ingredients of a meal in the stomach can form soluble complexes with minerals and thus prevent the formation of insoluble mineral-phytate complexes [23,24]. Such ligands are histidine, citric acid [26] and lactic acid [27] and they can be found in proteins and peptides [25] as well. In addition, ascorbic acid can counteract the inhibition of phytic acid in iron absorption [28]. The formation of insoluble complexes between phytic acid and minerals also depends on the mineral element concentrations. For example, Ca supplementation has been found to increase Mg absorption in the pig since phytic acid favors complex formation with Ca before Mg [29].

Many factors affect the absorption of mineral elements by phytic acid. Phytic acid contents may be rather high in breakfast cereals. On the other hand, the mineral contents of such products are also high and, when consumed as a part of a mixed meal, the decrease in the bioavailability of mineral elements due to phytic acid is compensated.

References
1. B. Reddy, C. Sharma, L. Mathews, A. Engle, K. Laakso, K. Choi, P. Puska, R. Korpela, Mutat. Res., 1985, **152**, 97-105.
2. D. Jenkins, C. Rainey -MacDonald, A. Jenkins, and G. Benn, 'CRC Handbook of Dietary Fiber in Human Nutrition', G.A.Spiller, Ed., CRC Press, Boca Raton, FL., 1986, pp. 327-344.
3. Anderson 'CRC Handbook of Dietary Fiber in Human Nutrition', G.A.Spiller, Ed., CRC Press, Boca Raton, FL., 1986, pp.349-359
4. C. Ripsin, J.K. Keenan, D.R. Jacobs, P.J. Elmer, R.R. Welch, L. Van Horne, K. Liu, W.H. Turnbull, F.W. Thye, M.K. Kestin, M. Hegsted, D.M. Davidson, L.D. Dugan, W. Demark-Wahnefried, S. Beling, J. Amer. Med. Assoc., 1992, **267**, 3317-3325.
5. S.Plaami, submitted to J. Food and Agric.
6. S.H. Yiu, J. Weisz and P.J. Wood, Cereal Chem., 1991, **68**, 372-375.
7. D.D. Gallaher, C.A. Hassel, and K. Lee, J. Nutr., 1993, **123**, 1732-1738.
8. J.M. McCord, *Circulation*, 1991, **83**, 1112-1114.
9. J.L. Sullivan, *Am. Heart J.*, 1980, **117**, 1177-1188.
10. K.L. Empson, T.P. Labuza, and E. Graf, 1991, **56**, 560-
11. I. Vucenic, K. Sakamoto, M. Bansal, and A. Shamsuddin, Cancer Letters, 1993, **75**, 95-102.
12. A-S Sandberg and U. Svanberg, J Food Sci., 1991, **56**, 1330-1333.
13. S. Plaami and J. Kumpulainen, *J. Food Comp. and Anal.*, 1993, **6**, 307-315.
14. S. Plaami and J. Kumpulainen, *J. Food Comp. and Anal.*, 1995, **8**, 324-335.
15. N.-G. Asp, C.-G. Johansson, H.A. Hallmer, and M. Siljeström, *J.Agric. Food Chem.*, 1983, **31**, 476.
16. B.V. McCleary and M. Glennie-Holmes, J. Inst. Brew., 1985, **91**, 285-295.
17. A.-S. Sandberg and R. Ahderinne, *J. Food Sci.*, 1986, **51**, 547-550.
18. A.-S. Sandberg, H. Andersson, N.G. Carlsson, and B. Sandström, *J. Nutr.*, 1987, **117**, 2061-2065.
19. S. Plaami and J. Kumpulainen, *Fresenius J. Anal. Chem.*, 1991, **339**, 886-888.
20. S. Plaami and J. Kumpulainen, *J. Assoc. Off. Anal. Chem.*, 1991, **74**, 32-36.
21. I.Wang, R.K. Newman, W.C. Newman and P.J. Hofer, *J. Nutr.*, 1992, **122**, 2297-2297.
22. E.R. Morris, 'Phytic acid. Chemistry & applications', E. Graf (Ed.), Minneapolis, USA, Pilatus Press, pp. 57-76.
23. B.Sandström, *Proceedings of the Nutrition Society*, 1988, **47**, 161-167.
24. Y. Kim, C.E. Carpenter, and A.W.Mahoney, *J. Nutr.*, 1993, **123**, 940-946.
25. B. Sandström, B.Arvidsson, Å. Cederblad, and E. Björn-Rasmussen, *Am. J. Clin. Nutr.*, 1980, **33**, 739-745.
26. D.B. Lyon, *Am. J. Clin. Nutr.*, 1984, **39**, 190-195.
27. A.-S. Sandberg, *Adv. Exp. Med. Biol.*, 1991, **289**, 499-508.
28. L. Hallberg, M. Brune, and L. Rossander, *Am. J. Clin. Nutr.*, 1989, **49**, 140-144.
29. A.-S. Sandberg, T. Larsen, and B. Sandström, *Diet. J. Nutr.*, 1993, **123**, 559-566.

BIOAVAILABILITY OF ESSENTIAL TRACE ELEMENTS WITH SPECIAL REFERENCE TO DIETARY FIBRE AND PHYTIC ACID

Brittmarie Sandström

Research Department of Human Nutrition/Center for Advanced Food Research, The Royal Veterinary and Agricultural University, Frederiksberg, Denmark

1 INTRODUCTION

Carefully performed dietary intake studies, e.g. using the duplicate portion technique, can provide us with data for dietary intake of trace elements. However, in order to relate dietary intake to dietary requirements of trace elements, the total trace element content of a diet is only of limited use unless the absorption and utilization of the elements, the so called bioavailability, can be taken into account.

1.1 Factors Influencing Mineral Availability

A number of physiological factors, e.g. age, body demand for trace elements, determine the degree of mineral absorption. In addition, the dietary composition has a large impact on availability. The chemical form of the element, the presence of antagonist or promoters for absorption, competition between trace elements and the physical availability of the element in the gut are the main factors of importance for the availability and absorption.

An example of the impact of the availability on the recommended intakes of zinc is given in table 1[1]. A pregnant woman in the third trimester has a high physiological requirement of zinc. If the diet has a high availability, she could cover her needs with a dietary intake of 5 mg/d, especially if she is adapted to this low intake level. At a low dietary availability, approximately five-fold as much zinc is needed to cover her physiological requirement.

Table 1 *Lower limits (mg/d) of safe ranges of population mean intakes of zinc for selected groups (FAO/WHO/IAEA 1993)*[1]

Availability Requirement	High		Moderate		Low	
	Basal[1]	Normative[2]	Basal	Normative	Basal	Normative
Child, 1-3	2.1	3.3	3.4	5.5	7.9	11.0
Females	2.5	4.0	4.0	6.5	9.4	13.1
Males	3.6	5.6	5.7	9.4	13.4	18.7
Pregnant 3rd trimester	5.1	8.0	8.0	13.3	18.7	26.7

[1] Assumes that adaptive reductions in endogenous losses have taken place in response to a low intake.
[2] Allows some safety margin.

Dietary trace element availability especially of iron and zinc is to a large extent related to the dietary fibre and the phytate content of the diet (table 2). The highest availability is assigned to refined animal protein based diets and a very low availability is assumed from fibre rich diets, especially those with a high phytate content. A phytate zinc molar ratio of 15 has been suggested as the critical level for a poor availability of zinc. In many unrefined stable foods e.g. soy beans, whole wheat, sorghum the molar ratio exceeds 20.

The physiological impact of a low availability is demonstrated by documented cases of zinc deficiency in man. Despite a high total zinc intake impaired growth and development has been observed in teenagers in Egypt and Iran subsisting on diets with a high fiber and phytate content and a low animal protein intake[2, 3].

Table 2 *Provisional criteria for categorizing diets according to the potential availability of zinc (adapted from[1])*

- *High availability:*
refined animal protein based diets;
semisynthetic animal-protein based formula diets

- *Moderate availability:*
mixed animal protein based diets;
lacto-ovo or vegan diets not based on unrefined cereals;
phytate/zinc ratio 5-15 or <10 if high calcium content

- *Low availability:*
diets based on unrefined, unfermented cereal grains with negligible amounts of animal protein;
phytate/zinc >15;
more than 50% of energy from unrefined cereals or legumes

2 DIETARY FIBRE AND TRACE ELEMENT AVAILABILITY

Originally it was suggested that a low availability from unrefined cereal based diets was due to a high fiber content. From a pure chemical view dietary fibre, especially isolated dietary fibre, has a rather high cationic binding capacity, suggesting that fibre *per se* could have an effect on mineral availability. If tested *in vitro*, dried samples of fibre rich foods, have a mineral binding capacity, which is close to or in the same range as can be seen for commercial ionic exchange resins[4] (table 3).

Table 3 *Ion exchange capacity of selected foods[3]*

Food	Ion exchange capacity meq/g dry weight
Apple	1.9
Cabbage	2.4
Orange	2.4
Carrot	3.1
Commercial ionic exchange resins	2-3

However, relatively large doses of isolated of fibre components, cellulose, pectin, guar gum, and other sources have not shown any negative effect on iron and zinc absorption in humans[5].

2.1 Fiber Associated Components and Trace Element Availability

Some fibre rich foods contain components which act as antagonistic ligands for mineral and trace element absorption. In practice, probably the most important antagonistic ligand for trace elements is phytic acid, myoinositol hexa-phosphate. It is present in many unrefined cereals and legumes. The inositol ring, with its six phosphate groups, is in its natural state in foods bound to potassium and magnesium. It dissociates in the stomach and can potentially bind to other minerals and trace elements.

Single meal studies in human subjects have shown that iron availability is strongly influenced by phytate[5]. Already at rather low phytate concentrations (molar ratio 1:1), the effect on iron absorption is quite dramatic, reducing absorption to very low levels.

The effect of phytate on zinc absorption is less dramatic. Depending on the total composition of the diet the level at which phytic acid has significant negative effects varies, but molar ratios of approximately 6-10 seems to be critical for zinc[5]. Addition of phytic acid in the same amounts as found in whole meal bread to a white bread reduces absorption to the same low level and the same is true, if phytate is added to a cow's milk formula[6, 7]. On the other hand, if the phytate content of whole-grain cereals is reduced by fermentation or other techniques, trace element absorption is improved[5, 6].

A reduction in phytate content is necessary if one wants to improve iron availability, as there does not seem to be any way to adapt to a high phytate diet. Comparing iron absorption in vegetarians used to a high fibre and phytate intake over many years and control subjects eating a typical Western mixed diet, showed no difference in absorptive ability[8].

Regarding other trace elements there are very few data available on the effect of fibre or phytate on absorption and utilization. Copper absorption has been measured with a stable isotope technique in a small number of subjects. Contrary to what was seen for zinc in the same study, copper absorption did not seem to be influenced by addition of rather large amounts of phytate[9].

Some fibre rich foods also contain other potential complexing ligands for trace elements, especially some of the polyphenols. It seems that a certain configuration of the components is required to reduce absorption. Two adjacent hydroxyl groups seems to be necessary for mineral binding[10]. At high doses, which fortunately only are found in a few food products (aubergine, tea, spinach and some herb spices), the negative effect of some of these phenolic components on iron absorption is as dramatic as of the phytic acid.

2.2 Components Facilitating Trace Element Absorption

Fortunately, fibre rich foods also contain ligands facilitating mineral availability. Low molecular weight organic acids like ascorbate, citrate, lactate or amino acids help to keep the minerals in a soluble form and might also prevent them from complexing with other agents. In practice it meals that ascorbic acid, that facilitates iron absorption, can counteract the strong effect of phytate on iron absorption and at high doses restore a reasonable availability[11].

3 PHYSICAL AVAILABILITY OF TRACE ELEMENTS

In fibre rich foods the physical availability of trace elements is also of importance for the degree of absorption. Disrupting the cell walls by food processes will probably lead to a higher availability, but could also release some of the complexing agents and the net effect is not always easy to predict. High temperature processes could have the opposite effect by creating poorly digestible compounds. Producing corn flakes from the corngrits, for example reduces zinc absorption by approximately 30%[12].

4 CONCLUSIONS

From what we know today there does not seem to be any negative effect on trace element availability of fibre *per se*. Instead the positive and negative effects observed at a high intake of fiber-rich foods can be ascribed to the presence of components associated with the fibers or simultaneously present in fiber rich foods. Phytic acid has a negative effect on iron and zinc absorption, but seems not to influence the absorption from other trace elements. Some of the food polyphenols have a strong negative effect on iron availability. Low molecular weight organic acids can counteract some of the negative effects of phytate and has especially a positive effect on iron availability. The overall impact of these positive and negative things depends on the total composition of the diet, the total trace element content, the degree of processing of the foods etc. Unfortunately, we cannot today, not even for iron, where the most thorough knowledge of availability is present, predict the exact availability from the diet. However, some rough estimates can be done, based primarily on the amount of phytic acid in diet.

References

1. WHO/FAO/IAEA Expert Consultation. Trace Elements in Human Nutrition and Health. WHO, Geneva, in press.
2. H.H. Sandstead, A.S. Prasad, A.R. Schulert, Z. Farid, A. Miale Jr, S. Bassily and W.J. Darby. *Am. J. Clin. Nutr.,* 1967, **20**, 422.
3. J.A. Halsted, H.A. Ronaghy, P. Abadi, M. Haghshenass, G.H. Amirhakemi, R.M. Barakat and J.G. Reinhold. *Am. J. Med.,* 1972, **53**, 277.
4. AA. McConnel, M.A. Eastwood and W.D. Mitchell. *J. Sci. Food Agric.,* 1974, **25**, 1457.
5. L. Rossander-Hultén, A.-S. Sandberg and B. Sandström. In: 'Dietary fibre - a component of food - nutritional function in health and disease', T. Schweizer and C. Edwards (eds.), Springer-Verlag, London, 1992, p. 195-216.
6. B. Nävert, B. Sandström, and Å. Cederblad. *J. Nutr.* 1985, **53**, 47.
7. B. Lönnerdal, Å. Cederblad, L. Davidsson and B. Sandström. *Am. J. Clin. Nutr.,* 1984, **40**, 1064.
8. M. Brune, L. Rossander and L. Hallberg. *Am. J. Clin. Nutr.,* 1989, **49**, 542.
9. J.R. Turnlund, J.C. King, B. Gong, W.R. Keyes and M.C. Michel. *Am. J. Clin. Nutr.* 1985, **42**, 18.
10. M. Brune, L. Rossander and L. Hallberg. *Eur. J. Clin. Nutr.* 1989, **43**, 547.
11. L. Hallberg, M. Brune and L. Rossander. *Am. J. Clin. Nutr.* 1989, **49**, 140.
12. G.I. Lykken, J. Mahalko, P.E. Johnson, D. Milne, H.H. Sandstead, W.J. Garzia, F.R. Dintzis and G.E. Inglett. *J. Nutr.* 1986, **116**, 795.

DIETARY FIBER AND INOSITOL PHOSPHATE CONTENTS OF FINNISH BREADS

S.P. Plaami, J.T. Kumpulainen and R. L. Tahvonen

Agricultural Research Centre of Finland, Food Research Institute, Laboratory of Food Chemistry.
FIN-31600 Jokioinen, Finland

1. INTRODUCTION

Dietary fiber (DF) has been demonstrated to have an important role in human nutrition. Increased DF consumption, specifically in the natural form, has thus been recommended by Finnish public health authorities [1]. Insoluble fiber may have a role in the prevention of large bowel cancer [2,3] while soluble fiber has been found to produce hypocholesterolemic effects [4,5]. Phytic acid (myoinositol hexaphosphate, IP6) is one of the substances associated with DF [6]. When cereal grains are processed into foods for human consumption, high temperatures cause part of the phytate to break down into lower inositol phosphates [7]. In breadmaking phytate reduction by the endogenous phytases of cereals and the added yeast can also be significant [8,9].

Phytic acid has detrimental effects on mineral element absorption [10]. On the other hand, however, phytic acid may play a beneficial antioxidative role in the etiology of cardiovascular diseases by limiting intestinal absorption of free iron and other transitory elements [11,12,13]. Furthermore, phytic acid has been shown to decrease the incidence of colonic cancer [14]. It has also been suggested, that it can exert its antineoplastic effect by reducing cell division via the lower inositol phosphates [15,16].

Cereal products are the principal source of dietary fiber [17,18] and phytates [19] in the Finnish diet. Due to the nutrient composition and low fat content of bread, increased consumption of bread has been advocated in Finland.

This paper will present the results on contents of DF [17] and inositol hexa-, penta-, tetra-, and tri-phosphates (IP6-IP3) [19] in Finnish breads. The results have been extrapolated to estimate the effect of increased DF intake from higher bread consumption on phytic acid intake.

2. METHODS

Sample collection and pretreatment methods have been presented in detail earlier [14,16]. The present samples represented all bread types available in Finland. DF determinations were performed on fresh samples and inositolphosphate determinations on pooled freeze-dried samples of the same bread types obtained from different bakeries.

Soluble and insoluble DF contents were determined by the enzymatic-gravimetric method of Asp et al. (1983) [20] and an HPLC method was used to quantify IP3-6 contents [21,22].

Analytical accuracy was tested by reference materials (RMs). The ARC/CL Wheat flour RM [23] and BCR Rye flour RM 381 were used in the DF analysis and the ARC/CL Wheat flour RM was the reference material for inositolphosphates, too [24,16].

3. RESULTS

Results of the analytical quality control program are presented in Tables 1 and 2.

Contents of DF and inositol tri-, tetra-, penta- and hexaphosphates in different bread groups are presented in Figures 1 and 2.

Highest total DF contents were encountered in crispbreads and sourdough rye breads. Breads made from a combination of wheat, potato, and / or oat also had relatively high total DF contents. Coffee breads and wheat bread (French bread) had the same total DF contents. On the other hand, wheat breads with varying ash contents, or made from a combination of wheat and rye, had clearly higher total DF contents. The whole meal wheat bread group included breads sold as whole meal or graham bread, although the actual whole meal contents varied from 20 to 100%. Hence the DF contents of that group were not as high as might have been expected, being clearly lower than those in the rye bread group, for instance.

Total IP6-3 contents found in the present samples closely paralleled total DF values (Figure 3). Highest values were detected in crispbread, Finnish sourdough bread and bread made of wheat and oat. In contrast, no correlation existed between DF and inositol hexaphosphate contents (Figure 4), suggesting decomposition of IP6 during processing.

Table 1 *Analytical accuracy for dietary fiber (DF) determinations of breads as tested by reference materials.*

Reference material	n	DF content (Mean ± SD) (% of dry wt.)			Certified value
		INSOLUBLE DF	SOLUBLE DF	TOTAL DF	TOTAL DF
ARC/CL Wheat flour[23]	16	2.5±0.25	1.5±0.26	4.0±0.32	4.08±0.27
BCR Rye flour RM 381	9	4.7±0.27	3.7±0.27	8.4±0.41	8.2±0.2

Table 2 *Reproducibility of inositol phosphate determinations as tested by the ARC/CL wheat flour reference material.*

Reference material	n^a	IP6	IP5	IP4	IP3
			μmol/g dry wt..		
ARC/CL Wheat flour RM	14	5.93±0.54	1.20±0.04	0.38±0.06	n.d.[b]

a = number of analysis = number of days
b = not detected
IP3= inositol triphosphate; IP4= inositol tertaphosphate; IP5= inositol pentaphosphate;
IP6= inositol hexaphosphate;

Per capita phytic acid intake from cereal products has been roughly estimated at 370 mg / d in Finland[19]. Low intake is mainly due to the high decompositon of IP6 during baking. In particular, it was detected that in Finnish dark sourdough rye bread (the main single source of DF in the Finnish diet), almost all the IP6 had hydrolysed into inositols with a lower number of phosphate groups. In addition into lowering the phytic acid content, sourdough breadmaking may also improve the nutritional features of starch [25].

Finnish crispbreads were devided into two groups based on IP6 contents. In group one, the phytic acid contents were detected in imported crispbreads. In the other group most of IP6 had hydrolysed into lower inositols due to endogenous phytase activity during the souring process. DF contents of crispbread samples varied slightly and no corresponding significant differences were detected.

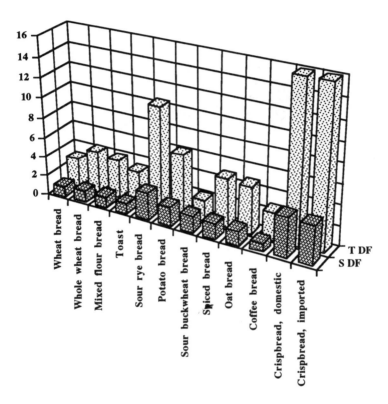

Figure 1 *Contents of total (TDF) and soluble (SDF) dietary fiber in Finnish breads (% of wet weight).*

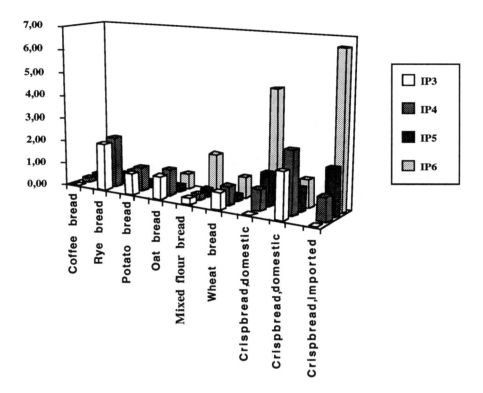

Figure 2 *Contents of inositol phosphates (IP3-IP6) in Finnish breads (µmol / g fresh weight).*

In conclusion, increased DF consumption in the form of bread (especially sourdough bread) in Finland does not appear to involve a risk due to an increased phytic acid intake. Increased bread consumption would further raise the intake of essential elements, as high mineral element contents are found in those Finnish breads with high DF levels[26]. Thus no correlation was found between IP6 and Zn and Mg contents in the Finnish bread samples studied (Figures 5 and 6), although the Pearson Product-Moment correlations between TDF and Zn and Mg contents were 0.87 and 0.89 correspondingly. Zn has been reported to be more adversely affected by phytate[27,28,29]. Phytate may also inhibit Mg-absorption[30]. Iron absorption may be similarly affected by phytic acid [31,32]. At the time of the present sample collection, wheat flour was fortified with iron in Finland and thus no significant differences were encountered in the high iron contents of the samples.

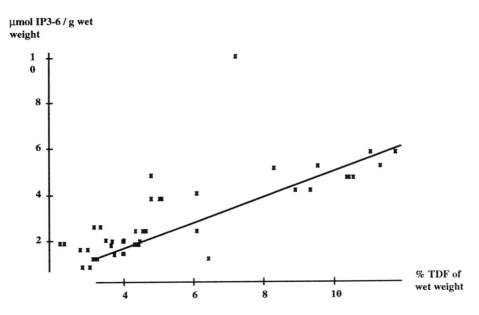

Figure 3 *Correlation between total inositol phosphate (IP3-6) and total dietary fiber (TDF) contents of Finnish breads (Pearson Product-Moment correlation 0.77).*

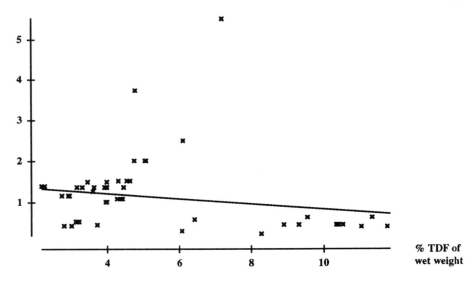

Figure 4 *Correlation between phytic acid (IP6) and total dietary fiber (TDF) contents of Finnish breads (Pearson Product-Moment correlation -0.197).*

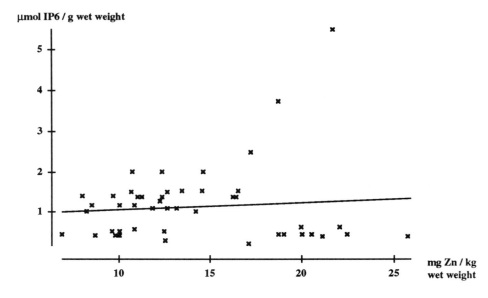

Figure 5 *Correlation between phytic acid (IP6) and total Zn contents of Finnish breads (Pearson Product-Moment correlation 0.089).*

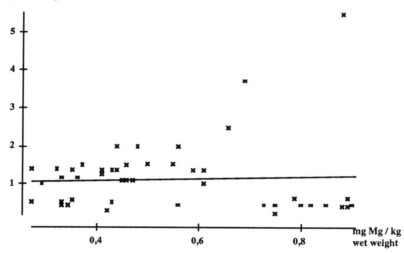

Figure 6 *Correlation between phytic acid (IP6) and total Mg contents of Finnish breads (Pearson Product-Moment correlation 0.064).*

References
1. Anonymous, ´ Dietary Guidelines and Their Scientific Principles´, The Government Printing Centre, Helsinki, 1987, Vol. **3**, pp. 44-45.
2. A. Hague, A.M. Manning, K.A. Hanlon, L.I. Huschtscha, D. Hart and C. Paraskeva. *Int. J. Cancer*, 1993, **55**, 498-505.
3. A.M. Berggren, I. Björck, M. Nyman, *J. Sci. Food Agric.*, 1993, **63**, 397-406.
4. H. Matheson, I.Colon and J. Story, *J. Nutr.*, 1995, **125**, 454-458.
5. S. Glore, D. Van Treeck, A. Knehans, M. Guild, *J. American Dietetic Assoc.*, 1994, **94**, 425-436.
6. E.R. Morris and A.D. Hill, *J. Food Comp. and Anal.*, 1995, **8**, 3.
7. N.R. Reddy, M.D. Pierson, M.D. Sathe and D.K. Salukhe, ´Phytates in cereals and legumes´, *Adv. Food Res.*, 1989, **28**, 39-56.
8. J.M. McKenzie-Parnell and N.T. Davies, *Food Chem.*, 1986, **22**, 181-192.
9. M. Larsson and A.-S. Sandberg, *J. Cereal Sci.*, 1991, **14**, 141-149.
10. A.-S. Sandberg, N.G. Carlsson, and U.Svanberg, ´Bioavailability 88 - Chemical and biological aspects of nutrient availability´, 1988, Henry Ling Ltd. Great Britain, ISBN 0-85186-856-8, p. 158-160.
11. J.M. McCord, *Circulation*, 1991, **83**, 1112-1114.
12. J.L. Sullivan, *Am. Heart J.*, 1980, **117**, 1177-1188.
13. K.L. Empson, T.P. Labuza, and E. Graf, 1991, **56**, 560-
14. E. Graf and J. Eaton, *Nutrition and cancer*, 1993, **19**, 11-19.
15. V. Tomazik, D.Fabian, and A.M.Shamshuddin, *Cancer Letters*, 1992, **65**, 9-13.
16. A.M. Shamsuddin and A.Ullah, *Carcinogenesis*, 1995, **10**, 625-626.
17. S. Plaami and J. Kumpulainen, *J. Food Comp. and Anal.*, 1994, **7**, 134-143.

18. P. Varo and P. Ekholm, Dietary fat and fibre and the health of the Finnish population, 1992, Turku: Publications of the Social Insurance Institution, Finland, ML:115, 178 pp.
19. S. Plaami and J. Kumpulainen, *J. Food Comp. and Anal.*, 1995, **8**, 324-335.
20. N.-G. Asp, C.-G. Johansson, H.A. Hallmer, and M. Siljeström, *J.Agric. Food Chem.*, 1983, **31**, 476.
21. A.-S. Sandberg and R. Ahderinne, *J. Food Sci.*, 1986, **51**, 547-550.
22. A.-S. Sandberg, H. Andersson, N.G. Carlsson, and B. Sandström, *J. Nutr.*, 1987, **117**, 2061-2065.
23. S. Plaami and J. Kumpulainen, *Fresenius J. Anal. Chem.*, 1991, **339**, 886-888.
24. S. Plaami and J. Kumpulainen, *J. Assoc. Off. Anal. Chem.*, 1991, **74**, 32-36.
25. H. Liljeberg, C.Lönnerdahl, and I. Björck, *J. Nutr.*, 1995, **125**, 1503-1511.
26. R.Tahvonen and J. Kumpulainen, *J.Food Comp. and Anal.*, 1994, **7**, 83-93.
27. J.R. Turnlund, J.C. King, W.R. Keyes, M.C. Maynard, *Am. J. Clin. Nutr.*, 1984, **40**, 1071-1077.
28. B.Nävert, Sandström, B, and Å. Cederblad, *British J. Nutr.*, 1985, **53**, 47-53.
29. J.R.Zhou, E.J. Fordyce, V. Raboy, D.B. Dickinson, M-S Wong, R.A. Burns, and J.W. Erdman Jr., *J. Nutr.*, 1992, **122**, 2466-2473.
30. D.B. Lyon, Am. J. Clin. Nutr., 1984, **39**, 190-195.
31. A.-S. Sandberg, N.-G. Carlsson and U. Svanberg, *J. Food Sci.*, 1989, **54**, 159-186.
32. A.-S. Sandberg and U. Svanberg, *J. Food Sci.*, 1991, **56**, 1330-1333.

Natural Anticarcinogenic Compounds in Diets and Cancer Prevention

LIGNANS AND ISOFLAVONOIDS: EPIDEMIOLOGY AND A POSSIBLE ROLE IN PREVENTION OF CANCER

H. ADLERCREUTZ

Department of Clinical Chemistry,
University of Helsinki, Meilahti Hospital,
FIN-00290 Helsinki, Finland

1 INTRODUCTION

In Asian countries such as Japan and China the incidence of prostate, breast and colon cancer is low compared to that in the Western world, particularly compared to USA. In Finland the incidence of these cancers is lower than in USA, but clearly higher than in Japan, but the rate is steadily increasing both in Japan and in Finland. Differences in fat consumption, originally being very low in Japan, explain some of the differences in incidence between Japan, the West-European countries and USA, but they do not explain the incidence differences between Finland and USA because of the similar animal fat consumption in these countries. It seems obvious that in addition to causative factors, protective factors in the diet may play an essential role.

2 LIGNANS AND ISOFLAVONOIDS, OCCURRENCE AND METABOLISM

The detection and identification in human urine and other biological fluids of the two groups of hormone-like compounds, the lignans and isoflavonoids,[1-5] and the observation that their excretion showed a positive correlation with fiber intake[6] and a negative correlation with the incidence of the above-mentioned cancers, led us to formulate a hypothesis according to which these compounds, originating in the diet, may play a protective role with regard to many Western diseases.[4,5,7-11] The isoflavonoids, occurring in high amounts in soybeans and soy products, and the lignans, derived mainly from whole grain bread, various seeds, fruits, berries and vegetables, have been shown to influence not only sex hormone metabolism and biological activity but also intracellular enzymes, protein synthesis, growth factor action, malignant cell proliferation and angiogenesis. Most of these effects suggest that they may inhibit cancer growth but may also prevent cancer initiation.

The pioneers in the phytoestrogen research worked in the veterinary field. They observed great fertility problems,[12] particularly in sheep but also in cattle, as a result of consumption of forage plants, mainly clover and alfalfa, containing high amounts of isoflavonoid phytoestrogens and coumestrol.[13-15] The infertility in sheep is brought about by many factors among other by interference with spermatozoa in the genital tract, abnormal transport of ova, and interference with implantation and decreased sensitivity of the hypothalamic-hypophyseal system to feedback regulation by estradiol. Such problems have, however, not been reported in human vegetarian subjects or Japanese consuming high amounts of phytoestrogen-containing foods. Lindner in his review in 1976[13] in fact stated that "there is no published evidence that herbal estrogens reaching the human from any of these sources is of pathogenic significance". Until 1981-84 it seems that nobody had suggested that the isoflavonoid phytoestrogens could be beneficial for human health. In opposite many relatively recent publications suggest that some environmental estrogens, including the phytoestrogens, could have negative effects in man.[16,17]

Following detection of the two mammalian lignans, enterolactone and enterodiol,[2,3] and of the isoflavonoid equol many other metabolites of these diphenols were identified.[4,5,18-22]. The main sources of lignans are various seeds like linseed (secoisolariciresinol), sesame seed (matairesinol), various grains (matairesinol and secoisolariciresinol) and whole soybeans. Isoflavonoids occur mainly in soybeans and various soy products (tofu, soy milk, miso) (genistein and daidzein, free or as glycosides), except soy sauce, and to a lesser extent in other legumes. Recently we found that tea contains both secoisolariciresinol and matairesinol in relatively high amounts.

The precursors of the isoflavonoids found in the human organism occur in plants as glycosides and the gut bacteria are responsible for the hydrolysis of these compounds and further conversion to many biologically active metabolites.[19,22] The lignan precursors (matairesinol and secoisolariciresinol) in grain seem to occur in the aleurone and fiber layer. The aleurone layer contains 1-3 cell layers and is very tightly bound to the outer fiber layer and lignans are present only in minute amounts in refined meal products. After removal of two methyl and two hydroxyl groups the plant lignans matairesinol and secoisolariciresinol are converted to the two main mammalian lignans, enterolactone and enterodiol, respectively. Enterodiol is then partly oxidized to enterolactone.[19]

3 ANTICANCER EFFECTS OF LIGNANS AND ISOFLAVONOIDS

In addition to numerous other biological activities many of these compounds are antiproliferative and anticarcinogenic. This has been demonstrated in many *in vitro* cell culture and *in vivo* animal studies.(review in[20]) Some lignans and isoflavonoids bind to the nuclear type II estrogen binding sites[23] and may in this way exert an antiproliferative effect regarding cells stimulated by estradiol.[24] Nuclear type II estrogen binding sites occur in breast, prostate, and other cancer cells.[25-27] The isoflavonoid genistein, the most interesting of these compounds because of its high concentration in plasma and urine of Japanese subjects,[28,29] is a tyrosine-kinase, angiogenesis, and topoisomerase II inhibitor and stimulates differentiation in many types of cells including leukemia cells.(ref. in[11,20,30]) It antagonizes the effect of epidermal growth factor and other growth factors. Genistein inhibits proliferation of numerous different types of malignant cells in culture, but relatively high concentrations are needed. Many cells are stimulated at lower concentrations and the possible cancer protective role of genistein *in vivo* in human subjects is still uncertain.

In cell cultures, in the presence of estradiol, physiological concentrations of enterolactone antagonize estradiol with regard to breast cancer cell growth.[23] In addition, enterolactone is a moderate aromatase inhibitor entering the cells and inhibiting aromatase both in choriocarcinoma cells as well as in preadipocytes.[31-33] Of great interest is the observation that genistein and coumestrol inhibit the 17ß-hydroxysteroid dehydrogenase type I.[34] However, adding genistein to the breast cancer cell culture did not inhibit the proliferative effect of estrone. A combination of aromatase inhibition by lignans and flavones with inhibition of the conversion of estrone to estradiol in malignant cells could theoretically result in prolongation of the promotional phase of estrogen dependent cancer.

All the compounds are antioxidative, particularly genistein. Consequently these compounds may contribute to the prevention of other chronic diseases like coronary heart disease and may in the gut prevent oxidation of cocarcinogens to carcinogens (see below).

4 EPIDEMIOLOGY

The observation of very low urinary excretion of lignans and equol in postmenopausal breast cancer patients[4] compared to vegetarians and very high excretion in chimpanzees, highly resistant to the induction of breast cancer by various toxic compounds[35] led us to suggest that these compounds may be protective with regard to hormone-dependent cancer and colon cancer.[4,5,8,36] In human subjects the highest isoflavonoid values are observed in Japanese men and women, in macrobiotics and vegans followed by lactovegetarians.[20] Omnivorous subjects have usually very low values.[4,20,37] Oriental immigrants to Hawaii have already within a few months low isoflavonoid values in urine.[20] In Finland lignan excretion tends to be the highest in North Karelian subjects with less cancer risk, less in subjects living in Helsinki and the lowest values are found in breast cancer patients and some omnivorous subjects.[37] The lignan excretion is higher in Finland compared to USA, but in Japan we found the lowest values in urine.[28] However, in plasma, the biologically active sulfate+free fraction was similar or even higher in some Japanese compared to Finnish men.[29] This was obviously due to less glucuronidation in the Japanese subjects, because their plasma concentration of lignan glucuronides was very low. Thus the Japanese have relatively high concentrations of the sulfate + free fraction of both plasma isoflavonoids and lignans.

5 PHYTOESTROGENS AND PROSTATE CANCER

Despite of the same incidence of latent and small or non-infiltrative prostate carcinomas as in the Western countries[38] the mortality of Japanese men in prostate cancer is very low. Decreased prostate cancer risk has been found in men of Japanese ancestry in Hawaii[39] who consume rice and tofu, a soybean product containing isoflavonoids in great quantities.[28] That diet may be an important factor in the promotional stage of the disease is shown in two studies[40,41] suggesting that environmental factors later in life can have a substantial impact on the likelihood of developing clinically detectable prostate cancer. Despite of the high fat intake the prostate cancer incidence in Finland and some European countries like France is much lower than in USA, but much higher than in Japan. Because prostate cancer is hormone-dependent we have postulated, that the diet in countries with low or relatively low cancer risk may contain higher amounts of cancer-protective compounds affecting hormone metabolism or action.[11,28] Among such compounds lignans and isoflavonoids seem of particular interest because their close structural relationship with estrogens. In addition, various other types of flavonoids may be involved.

In vitro studies with prostate cancer cells in culture have demonstrated that genistein and daidzein inhibit cell proliferation.[42-45] However, *in vivo* in rats genistein did not affect the growth of subcutaneously implanted MAT-LyLu prostate cancer cells.[45] Recently we observed that intake of soy or rye bran inhibited the growth of implanted prostate cancer in rats (Landström, M., Zhang, J., Åhman, P., Bergh, A., Damber, J.-E., Hallmans, G., Wähälä, K., Mazur, W., and Adlercreutz, H., to be published). The lignan enterolactone and its plant precursor matairesinol inhibit prostate cancer cell growth *in vitro*[43] and the rye bran contains matairesinol and another precursor for enterolactone called secoisolariciresinol. It was interesting to observe that in the rye bran fed rats the amount of enterolactone excreted in urine exceeded by far the amount of matairesinol and secoisolariciresinol in the rye bran consumed by the rats. This points to other mammalian lignan precursors in rye bran than the two ones measured.

Using the developmentally estrogenized mouse model, it has been shown that the development of dysplastic changes in prostate was delayed by soy feeding. When animals were given soy-free diet from fertilization onwards, most of them had dysplastic lesions at the age of 9 months, but in a group given soy diet the number of animals showing prostatic dysplasia was significantly lower. At the age of 12 months, the difference between the two groups had diminished and was not anymore statistically significant.[46,47] Morphologically these dysplastic lesions are similar to prostatic intraepithelial neoplasia (PIN) in the human prostate. Although no progression to carcinomas with invasion to surrounding tissues or metastasis can be demonstrated, the tissue changes, observed in developmentally estrogenized mice, suggest an increased potential for benign and malignant growth. Of interest is also that soy intake prevents prostatitis in rats.[48]

6 PHYTOESTROGENS AND BREAST CANCER

The early observation that postmenopausal breast cancer patients, living in Boston MA (USA) and followed for one year, had a very low grain fiber intake combined with a very low excretion of lignans in urine suggested that the lignans may play a role as protective compounds in breast cancer.[4] As already mentioned the lignans or genistein have been shown to inhibit estrogen-stimulated proliferation of human MCF-7 breast cancer cells in culture[23,49,50] and they are moderate inhibitors of the aromatase enzyme[32,33] and bind (like daidzein) to the type II nuclear estrogen-binding site.[23] Furthermore, they seem to stimulate SHBG synthesis[23] which leads to reduced circulating levels of free estradiol and testosterone. All these biological activities may reduce breast cancer risk. Even more data, both *in vitro* and *in vivo*, is available suggesting a protective effect of isoflavonoids with regard to breast cancer (see lit. in[20,51]). Feeding of soy to experimental animals reduces breast cancer risk and treatment with genistein postpartum in rats reduces the development of mammary tumors when dimethylbenz(a)anthracene was administered after genistein treatment.[52] A precocious maturation of undifferentiated terminal end buds to more differentiated lobules may account for neonatal genistein treatment protecting against chemically induced mammary cancer. Epidemiological evidence obtained in Singapore indicates that soy intake is associated with lower breast cancer risk in women,[53] but another epidemiological study in Shanghai and China did not support the view that soy intake protects against breast cancer.[54]

7 LIGNANS, ISOFLAVONOIDS AND COLON CANCER

It has been suggested that the lignans may be protective also with regard to colon cancer.[3,8] Recently, we observed a high lignan excretion in subjects with a low risk of colon cancer (see discussion[55]). Lignan excretion is also high in Finnish subjects living in areas with lower colon cancer risk.[37] Epidemiological evidence obtained in Japan[56] points to lower colon cancer incidence in areas with high tofu consumption. This is now being further investigated. Both breast and colon carcinogenesis is reduced in rats fed flax seed containing high amount of soy secoisolariciresinol.[57,58] Due to their phenolic structure, lignans and flavonoids have antioxidative properties[8,59-61] and may prevent conversion of procarcinogens to carcinogens or eliminate free radicals in the gut reducing colon cancer risk.

8 CONCLUSIONS

Plant food that contains isoflavonoids and lignans may play a role in the prevention of several types of cancer and particularly the so-called Western cancers. The concentrations in plasma of these compounds, particularly the lignans, may easily reach biologically active levels without toxic effects. By inhibiting the effect of growth factors and angiogenesis, genistein may be a general inhibitor of cancer growth. To call soy isoflavonoids the natural equivalent to the breast cancer antiestrogen drugs is not indicated on the basis of our present knowledge about the rather different mechanisms of action of these compounds. Genistein stimulates differentiation of many types of leukemic cells and by its property to modulate drug transport, genistein may prove to be a good addition to established cancer therapy. Lignans on the other hand seem in the light of recent research to be at least as potent as isoflavonoids in preventing initiation and promotion of cancer and may perhaps be used also in connection with cancer therapy. The described biological effects might be used as a preventive strategy for other Western diseases not discussed in this connection, such as cardiovascular diseases and osteoporosis, due to the antioxidative and estrogenic effects. In the light of present knowledge, it should be kept in mind that it is to be preferred to consume original food, or food modified only slightly, instead of consuming isolated or synthetic compounds. This is particularly true for the lignans. The evidence obtained is not yet sufficient for any specific dietary recommendations (e.g. kind of food, amount per day) and further work is needed to establish the role of these natural anticancer compounds in human health and disease. Intake of soy, whole-grain crisp or soft rye bread, other whole-grain bread, fruits, berries and vegetables is in any case a good choice.

Acknowledgements
The work carried out in the Department of Clinical Chemistry, University of Helsinki was supported by grants from NIH (grant no. 1 R01 CA56289-01) and by the Comprehensive 10-year Strategy for Cancer Control, Ministry of Health and Welfare, Japan.

References

1. K. D. R. Setchell and H. Adlercreutz, J. Steroid Biochem., 1979, 11, xv.
2. S. R. Stitch, J. K. Toumba, M. B. Groen, C. W. Funke, J. Leemhuis, J. Vink and G. F. Woods, Nature, 1980, 287, 738.
3. K. D. R. Setchell, A. M. Lawson, F. L. Mitchell, H. Adlercreutz, D. N. Kirk and M. Axelson, Nature, 1980, 287, 740.
4. H. Adlercreutz, T. Fotsis, R. Heikkinen, J. T. Dwyer, M. Woods, B. R. Goldin and S. L. Gorbach, Lancet, 1982, 2, 1295.
5. C. Bannwart, T. Fotsis, R. Heikkinen and H. Adlercreutz, Clin. Chim. Acta, 1984, 136, 165.
6. H. Adlercreutz, T. Fotsis, R. Heikkinen, J. T. Dwyer, B. R. Goldin, S. L. Gorbach, A. M. Lawson and K. D. R. Setchell, Medical Biology, 1981, 59, 259.
7. K. D. R. Setchell, A. M. Lawson, S. P. Borriello, R. Harkness, H. Gordon, D. M. L. Morgan, N. Kirk, H. Adlercreutz, L. C. Anderson and M. Axelson, Lancet, 1981, 2, 4.
8. H. Adlercreutz, Gastroenterology, 1984, 86, 761.
9. C. Bannwart, H. Adlercreutz, T. Fotsis, K. Wähälä, T. Hase and G. Brunow, Finn. Chem. Lett.,1984, 120.
10. H. Adlercreutz, "Progress in Diet and Nutrition" (Frontiers of Gastrointestinal

Research 14), C. Horwitz and P. Rozen (eds.), S. Karger, Basel, 1988, p.165.
11. H. Adlercreutz, Scand. J. Clin. Lab. Invest., 1990, 50 (Suppl 201), 3.
12. H. W. Bennets, E. J. Underwood and F. L. Shier, Aust. Vet. J. 1946, 22, 2.
13. H. Lindner, Environ. Quality Safety, 1976, 5, 151.
14. D. A. Shutt, Endeavour 1976, 35, 110.
15. K. Verdeal and D. S. Ryan, J. Food Protect., 1979, 42, 577.
16. B. Register, B., M. A. Bethel, N. Thompson, D. Walmer, P. Blohm, L. Ayyash, and C. Hughes, Proc. Soc. Exp. Biol. Med. 1995, 208, 72.
17. R. Stone, Science, 1994, 265, 308.
18. M. Axelson, D. N. Kirk, R. D. Farrant, G. Cooley, A. M. Lawson and K. D. R. Setchell, Biochem. J. 1982, 201, 353.
19. K. D. R. Setchell and H. Adlercreutz, "Role of the Gut Flora in Toxicity and Cancer", I. Rowland (ed.), Academic Press, London, 1988, p.315.
20. C. H. T. Adlercreutz, B. R. Goldin, S. L. Gorbach, K. A. V. Höckerstedt, S. Watanabe, E. K. Hämäläinen, M. H. Markkanen, T. Mäkelä, K. T. Wähälä, T. A. Hase and T. Fotsis, J. Nutr. 1995, 125, 757S.
21. G. E. Kelly, C. Nelson, M. A. Warin, G. E. Joannou and A. Y. Reeder, Clin. Chim. Acta, 1993, 223, 9.
22. G. E. Joannou, G. E. Kelly, A. Y. Reeder, M. Waring and C. Nelson, J. Steroid Biochem. Molec. Biol. 1995, 54, 167.
23. H. Adlercreutz, Y. Mousavi, J. Clark, K. Höckerstedt, E. Hämäläinen, K. Wähälä, T. Mäkelä and T. Hase, J. Steroid Biochem. Molec. Biol., 1992, 41, 331.
24. B. M. Markaverich, R. R. Roberts, M. Alejandro, G. A. Johnson, B. S. Middleditch and J. H. Clark, J. Steroid Biochem. 1988, 30, 71.
25. B. M. Markaverich, M. Varma, C. L. Densmore, A. A. Tiller, T. H. Schauweker, T.H. and R. R. Gregory, Int. J. Oncol., 1994, 4, 1291.
26. B. M. Markaverich, T. H. Schauweker, R. R. Gregory, M. Varma, F. S. Kittrell, D. Medina and R. S. Varma, Cancer Res. 1992, 52, 2482.
27. L. M. Larocca, M. Giustacchini, N. Maggiano, F. O. Ranelletti, M. Piantelli, E. Alcini and A. Capelli, J. Urol. 1994, 152, 1029.
28. H. Adlercreutz, H. Honjo, A. Higashi, T. Fotsis, E. Hämäläinen, T. Hasegawa and H. Okada, Am. J. Clin. Nutr., 1991, 54, 1093.
29. H. Adlercreutz, H. Markkanen and S. Watanabe, Lancet, 1993, 342, 1209.
30. T. Fotsis, M. Pepper, H. Adlercreutz, G. Fleischmann, T. Hase, K. Montesano and L. Schweigerer, Proc. Natl. Acad. Sci. USA, 1993, 90, 2690.
31. H. Adlercreutz, C. Bannwart, K. Wähälä, T. Mäkelä, G. Brunow, T. Hase, P. J. Arosemena, J. T. Jr. Kellis and L. E. Vickery, J. Steroid Biochem. Molec. Biol., 1993, 44, 147.
32. D. R. Campbell and M. S. Kurzer, J. Steroid Biochem. Mol. Biol., 1993, 46, 381.
33. C. F. Wang, T. Mäkelä, T. Hase, H. Adlercreutz and M. S. Kurzer, J. Steroid Biochem. Mol. Biol., 1994, 50, 205.
34. S. Mäkelä, M. Poutanen, J. Lehtimäki, M. L. Kostian, R. Santti and R. Vihko, Proc. Soc. Exp. Biol., 1995, 208, 51.
35. P. I. Musey, H. Adlercreutz, K. G. Gould, D. C. Collins, T. Fotsis, C. Bannwart, T. Mäkelä, K. Wähälä, G. Brunow and T. Hase, Life Sci. 1995, 57, 655.
36. H. Adlercreutz, P. I. Musey, T. Fotsis, C. Bannwart, K. Wähälä, T. Mäkelä, G. Brunow and T. Hase, Clin. Chim. Acta, 1986, 158, 147.
37. H. Adlercreutz, T. Fotsis, C. Bannwart, K. Wähälä, T. Mäkelä, G. Brunow and T. Hase, J. steroid Biochem. 1986, 25, 791.
38. R. Yatani, I. Chigusa, K. Akazaki, G. N. Stemmerman, R. A. Welsh and P. Correa, Int. J. Cancer, 1982, 29, 611.
39. R. K. Severson, A. M. Y. Nomura, J. S. GROVE and G. N. Stemmerman, Cancer

40. H. Shimizu, R. K. Ross, L. Bernstein, R. Yatani, B. E. Henderson and T, M. Mack, Br. J. Cancer, 1991, 63, 963.
41. L. Le Marchand, L. N. Kolonel, L. R. Wilkens, B. C. Myers and T. Hirohata, Epidemiology, 1994, 5, 276.
42. G. Peterson. and S. Barnes, Prostate, 1993, 22, 335.
43. H. Adlercreutz, S. Mäkelä, L. Pylkkänen, R. Santti, J. Kinzel, M. Van Reijsen, H. Markkanen, E.-L. Kämäräinen, S. Watanabe, T. Fotsis, K. Wähälä, T. Mäkelä and T. Hase, Proc. Am. Assoc. Cancer Res. 1995, 36, 687.
44. E. Kyle, R. C. Bergan and L. Neckers, Proc. Am. Assoc. Cancer Res. 1995, 36, 338 (abst. 2310).
45. H. R. Naik, J. E. Lehr and K. J. Pienta, Anticancer Res. 1994, 14, 2617.
46. S. Mäkelä, L. Pylkkänen, R. Santti and H. Adlercreutz, "Euro. Food Tox. III", Proceedings of the Interdisciplinary Conference on Effects of Food on the Immune and Hormonal Systems, Institute of Toxicology, Swiss Federal Institute of Technology & University of Zürich, CH-8603 Schwerzenbach, Switzerland, 1991, p.135.
47. S. I. Mäkelä, L. H. Pylkkänen, R. Santti and H. Adlercreutz, J. Nutr., 1995, 125, 437.
48. O. P. Sharma, H. Adlercreutz, J. D. Strandberg, B. R. Zirkin, D. S. Coffey and L. L. Ewing, J. Steroid Biochem. Mol. Biol., 1992, 43, 557.
49. G. Peterson and S. Barnes, Biochem. Biophys. Res. Commun., 1991, 179, 661.
50. M. C. Pagliacci, M. Smacchia, G. Migliorati, F. Grignani, C. Riccardi and I. Nicoletti, Eur. J. Cancer, 1994, 30A, 1675.
51. H. Adlercreutz, Environ. Health Perspect., (in press).
52. C. A. Lamartiniere, J. Moore, M. Holland and S. Barnes, Proc. Soc. Exp. Biol. Med., 1995, 208, 120.
53. H. P. Lee, L. Gourley, S.W. Duffy, J. Estéve, J. Lee and N.E. Day, Lancet, 1991, 337, 1197.
54. J.-M. Yuan, Q.-S. Wang, R. K. Ross, B. E. Henderson and M. C. Yu, Br. J. Cancer, 1995, 71, 1353.
55. J. T. Korpela, R. Korpela and H. Adlercreutz, Gastroenterology, 1992, 103, 1246.
56. S. Watanabe and S. Koessel, J. Epidemiol., 1993, 3, 47.
57. M. Serraino and L. U. Thompson, Nutr. Cancer, 1992, 17, 153.
58. M. Serraino and L. U. Thompson, Cancer Lett, 1992, 63, 159.
59. M. Naim, B. Gestetner, A. Bondi and Y. Birk, J. Agric. Food Chem., 1976, 24, 1174.
60. H. Lu and G. T. Liu, Planta Medica, 1992, 58, 311.
61. H. C. Wei, L. H. Wei, K. Frenkel, R. Bowen and S. Barnes, Nutr. Cancer, 1993, 20, 1.

ANTICANCER EFFECTS OF FLAXSEED LIGNANS

L.U. Thompson, L. Orcheson, S. Rickard, M. Jenab, M. Serraino, M. Seidl, and F. Cheung

Department of Nutritional Sciences
University of Toronto
Toronto, Ontario
Canada M5S 1A8

1 INTRODUCTION

Lignans are a class of compounds which have been suggested to have anticancer effects.[1] They are diphenolic in nature and contain a dibenzylbutane skeleton.[2] Lignans, present in higher plant species,[3] are converted to the mammalian lignans - enterodiol (ED) and enterolactone (EL) - by colonic bacterial flora.[4] The major precursor of ED is secoisolariciresinol diglycoside (SD), while that of EL is matairesinol.[5] EL may also be produced through the oxidation of ED (Figure 1). Once formed, mammalian lignans undergo enterohepatic circulation and some are excreted in the urine.[2] Urinary mammalian lignan excretion correlates directly with the amount of lignan precursors present in foods.

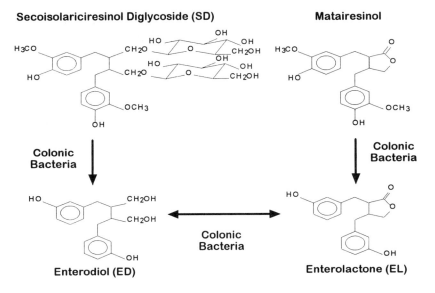

Figure 1 *Formation of mammalian lignans from their plant precursors*

Several observations indicate that lignans may protect against cancer. For example, breast cancer patients and individuals at high risk for breast and colon cancer have been shown to excrete significantly lower levels of mammalian lignans compared to vegetarians or those at lower risk of developing these diseases.[6] Also, lignans are similar in structure to estrogens, isoflavonoid phytoestrogens, and tamoxifen which have weak estrogenic/antiestrogenic properties (Figure 2).[7] Moreover, some lignans have been shown to have antimitotic, antiestrogenic, antiviral, antibacterial, and antifungal properties.[7]

Flaxseed contains 75 to 800 times more lignans than over 60 other plant foods[8] making it the richest known source of lignans (Table 1). Since flaxseed is a food that may influence cancer development, we designed a series of short- and long-term experiments to test the hypothesis that flaxseed has a protective effect against both mammary and colon tumorigenesis. This paper summarizes the results of such studies.

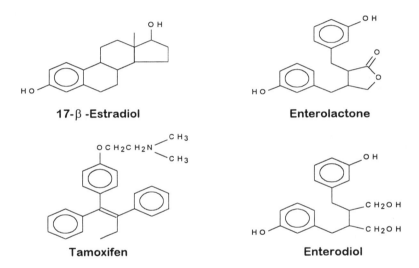

Figure 2 *Chemical structure of mammalian lignans, estradiol and tamoxifen*

Table 1 *Mammalian Lignan Production from Various Foods (µg/100 g, as is basis)*

Food Group	Number of Foods	Range	Mean
Defatted Flax	1	67,541	67,541
Flax	1	52,679	52,679
Other oilseeds	4	161-1,130	638
Legumes	7	201-1,287	562
Cereal brans	5	181-651	486
Cereals	9	115-924	359
Vegetables	27	21-407	144
Fruits	7	35-181	84

2 MAMMARY TUMORIGENESIS

2.1 Short-Term Study

In our first study we examined the ability of flaxseed to reduce early markers of mammary cancer risk (cell proliferation and nuclear aberrations) in female rats.[9] Rats were fed either a basal high fat diet (BD) or BD supplemented with 5 or 10% full fat (F) or defatted (DF) flaxseed for 4 weeks. Epithelial cell proliferation, measured by tritiated thymidine labeling index and mitotic index, was determined in the individual structures of the mammary gland in one sub-group of animals. Nuclear aberrations per 100 epithelial cells was determined in the other sub-group after treatment with the mammary carcinogen dimethylbenzanthracene (DMBA).

Significant findings include: (1) a reduction in the mitotic index in the terminal end bud (TEB) of the mammary glands of rats fed 5 and 10%F by 52 and 55%, respectively (Figure 3), (2) a 39% reduction in the labeling index in the TEB of rats fed 5%F, and (3) a 59-66% reduction in nuclear aberrations of the TEB in the F and DF groups.

Excretion of urinary lignans significantly increased with flaxseed intake and a significant negative relationship was observed between urinary lignan excretion and the nuclear aberration in the total structures of the mammary gland (TEB, terminal ducts, and alveolar buds; Figure 4).

Overall, the results suggest that the protective effect of flaxseed may be related to its lignans and that a high level of supplementation may not be necessary for reduction of cancer risk since increasing the level of F to 10% produced no further reduction in early risk markers. However, the cancer-protective effect of other components of flaxseed, such as α-linolenic acid (ALA), remains unclear.

2.2 Long-Term Studies

2.2.1 Effect of Flaxseed on Tumor Initiation and Promotion. A long-term study with tumor development as the endpoint was conducted in order to confirm the results of the short-term study.[10] Weanling rats were fed either the BD or the BD supplemented with 5%F, the level seen to be most effective in the previous study.[9] At 50 days of age,

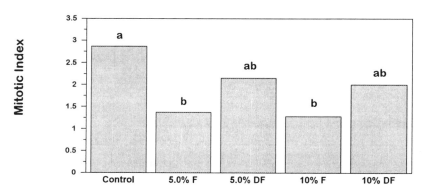

Figure 3 *Mammary epithelial cell proliferation in teminal end buds.* Bars with different letters are significantly different ($p<0.05$)

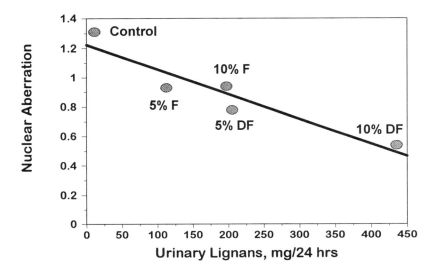

Figure 4 *Relationship between nuclear aberration and urinary lignan excretion in rats fed ground flaxseed (F) or defatted flaxseed meal (DF).*[9]

the rats were treated with DMBA. One week after carcinogen administration, one-half of the BD group was fed 5%F (to determine the effect of flaxseed on the promotion stage of carcinogenesis) and one-half of the 5%F group was fed the BD only (to determine the effect of flaxseed on the initiation stage of carcinogenesis).

Tumor incidence was reduced 21-32% by feeding 5%F at either the initiation or promotion stage of tumorigenesis. The number of tumors developing in each group was lowest for the rats fed 5%F at the initiation stage only followed by those fed 5%F throughout the study. In contrast, the tumor size was smallest for the group fed 5%F at the promotion stage only (Figure 5).

These results show that fewer tumors are initiated in the presence of flaxseed and those that are initiated grow to a lesser extent. These results are in agreement with the short-term study.[9] The effects may be due to the lignan component of flaxseed but a differing mechanism of action at the various stages of tumorigenesis and the role of the flaxseed oil as indicated by the incorporation of ALA into the tumors and mammary glands of the flaxseed-fed rats[10] may have contributed to the complexity of the results.

2.2.2 Effect of SD on Tumor Promotion. In order to compare the anticancer effect of mammalian lignans with that of flaxseed, it was necessary to isolate and purify SD from flaxseed[11] and then test its tumor-inhibiting ability. In our next experiment, purified SD was fed to rats at a daily dose of 1.5 mg (approximate amount taken by rats fed 5% F) starting one week post-DMBA for 20 weeks.[12] Although there were no significant differences in tumor incidence or tumor volume, the group fed SD had 47% fewer tumors ($p<0.05$) than the control group fed the BD. This is the first *in vivo* study to show conclusive evidence that providing a precursor of mammalian lignans to rats has a protective effect against tumorigenesis.

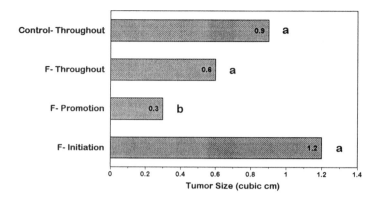

Figure 5. *Size of Tumors in Rats Fed 5% Flaxseed (F) Either Throughout or Only at The Initiation or Promotion Stage of Tumorigenesis.* Bars with different letters are significantly different (p<0.05).

2.2.3 *Effect of Flaxseed and Its SD and Oil Fractions at the Later Stage of Tumorigenesis.* A fourth study was designed to test whether flaxseed and its mammalian lignan precursor, SD, can reduce the growth of tumors at a later stage of tumorigenesis and to determine what role, if any, the oil component of flaxseed may play.[13] Female rats were fed the BD until 13 weeks post-DMBA at which time a majority of the rats had palpable tumors of about 1-2 cm in diameter. The rats were divided into 5 treatment groups having similar numbers of established tumors, mean tumor burden, tumor diameter, and body weight. They received the BD or BD supplemented with either 2.5%F, 5%F, 1.82% flaxseed oil (the amount equivalent to that in 5%F), or 1.5 mg SD.

After 7 weeks of treatment, the volume of the established tumors in all the treatment groups was significantly smaller (over 50%) compared to the control group. The volume of new tumors which appeared during treatment was significantly smaller for the SD-fed rats and approached significance in the case of the 2.5%F group. The number of new tumors was also lowest in the SD group which agreed with an earlier experiment that showed a reduction in the number of tumors with SD.[12]

A significant negative correlation was observed when the volume of established tumors at the end of treatment was related to urinary lignan excretion in the control, SD, 2.5%F, and 5%F groups but not when the flaxseed oil group was included. These results indicate that the tumor-inhibiting effect of flaxseed is largely due to its lignan component. A reduction in established tumor growth produced by flaxseed oil may be due to another mechanism.

3 COLON TUMORIGENESIS

3.1 Short-Term Study

Mammalian lignans are produced in the colon and, as a result, this is the most likely site for a cancer-protective effect of lignans. In a short-term study, male rats were treated with the carcinogen azoxymethane (AOM) and were fed either BD or BD supplemented with either 5 or 10% F or DF for 4 weeks.[14] In all treatment groups, there was a significant reduction in the number of aberrant crypts (AC; 41-53%) and aberrant crypt foci (ACF; 48-57%) which indicates that flaxseed has a protective effect against colon cancer since AC and ACF have been suggested to be early markers of colon cancer[15] (Figure 6).

Urinary lignan excretion was significantly increased in the F and DF groups indicating that the lignans may be responsible for the effects seen. However, the relationship between reduction in risk markers and the level of F or DF intake or lignan excretion was not linear. This indicates that there may be a limit to the protective ability of lignans, as was seen in the mammary tumorigenesis studies,[9-13] or that there may be interactive effects with other flaxseed constituents.

3.2 Long-Term Study

To fully establish the role of mammalian lignans in colon tumorigenesis, a long-term study utilizing purified SD was conducted. Carcinogen-treated rats were fed for 100 days either the BD or the BD supplemented with either 2.5 or 5%F, 2.5 or 5%DF, or 1.5 mg SD.[16] The effect of the oil could be differentiated from the effect of the lignans since the 2.5 and 5%F diets contained an amount of lignan precursors equivalent to the 2.5 and 5%DF diets, respectively. As in the mammary studies,[11-13] 1.5 mg SD was estimated to be equivalent to the daily intake provided by the 5%F diet.

The results of this study showed a consistent reduction in the number of AC and ACF in the treatment groups[16] which is in agreement with our short-term study.[14] Also, the number of AC per ACF and the size of the ACF were significantly lower in all of the treatment groups compared to the control. The SD group had effects similar to the DF and F diet groups and a significant negative correlation was seen between urinary lignan excretion and the number of AC per ACF indicating that the effect of flaxseed is primarily due to SD. In contrast to our mammary tumorigenesis studies,[9-13] it appears as though the ALA in flaxseed, at up to the 5% level, did not influence AC morphology since the results of the F and DF groups did not differ significantly.

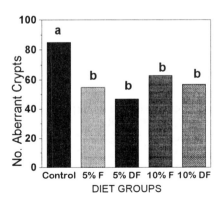 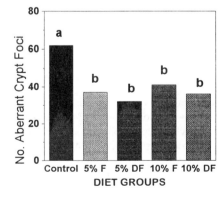

Figure 6. *Number of Aberrant Crypts and Aberrant Crypt Foci in Male Rats Fed Ground Flaxseed (F) or Defatted Flaxseed Meal (DF).* Bars with different letters are significantly different ($p<0.05$).

4 MECHANISM OF LIGNAN ACTION

The mechanism by which lignans from flaxseed influence tumorigenesis has not been fully established. Given the structural similarity of lignans to estrogenic compounds, lignans may influence mammary tumorigenesis by acting as a weak estrogen/antiestrogen[17-26] as have been demonstrated *in vitro* (Table 2). Other potential mechanisms may be involved in non-hormone dependent cancers, such as colon cancer, including inhibition of angiogenesis[27] and lipid peroxidation.[28] Paradoxically, cecal beta-glucuronidase activity, a colon cancer risk marker, had a significant positive relationship with AC morphology,[16] suggesting the complexity of the lignan action in the colon.

An antiestrogenic effect of flaxseed and its mammalian lignan precursor has been demonstrated in our laboratory.[29] Female rats were fed either BD or BD supplemented with SD (0.75, 1.5, and 3.0 mg/day) or flaxseed (2.5, 5, and 10%) for 4 weeks. The results showed a dose-dependent cessation or lengthening of the estrous cycle which approached the effect produced by tamoxifen (0.2 mg/day).[29] A similar lengthening of the luteal phase of the menstrual cycle in premenopausal women ingesting flaxseed (10 g/day)[30] has been observed suggesting that flaxseed has the ability to influence sex hormone metabolism *in vivo*.

The results of our studies show that lignans and flaxseed likely operate via the interaction of many of these mechanisms depending on the dose, the timing of administration, and the type of tumor.

5 FLAXSEED LIGNAN CONCENTRATION

Lignans are very rich in flaxseed but their concentration can vary widely. When we tested ten flaxseed varieties grown at the same Canadian location in the same year, the total lignans significantly differed and ranged from 0.955 to 2.700 µmol/g. Significant differences were also observed when the same varieties were grown at different locations. Likewise, the year of harvest significantly affected the lignan concentration. Because of these variables, each batch of flaxseed should be tested for lignans prior to use in any tumor experimentation.

Table 2 *Biological Effects of Lignans and Phytoestrogens*

Biological Effects	Reference
Stimulate growth of estrogen-dependent breast cancer cells (MCF-7, T47D)	17,18
Inhibit estradiol-stimulated growth of breast cancer cells (MCF-7, ZR-75-1)	17,19
Bind to nuclear type II binding site; compete with estradiol	20,21
Inhibit estrogen-stimulated RNA synthesis in rat uterus	22
Induce expression of estrogen responsive protein pS2 in MCF-7 cells	23
Bind to rat and human α-fetoprotein; compete with estrogen for binding site	24
Stimulate sex hormone binding globulin synthesis by human liver cancer cells	25
Inhibit aromatase enzyme	26
Inhibit angiogenesis	27
Antioxidative	28

6 CONCLUSION

Results to date indicate that flaxseed has a protective effect against mammary and colon cancer which is in part due to the mammalian lignans produced from their precursor SD. The effect appears to result from the interaction of several mechanisms including the influence on sex hormone metabolism. While flaxseed is a rich source of lignans, one has to consider the large variability in their content due to the variety, growing location, and year of harvest.

7 ACKNOWLEDGEMENT

The assistance of H. Fong and L. Luyengi in the preparation of SD as well as financial assistance from the National Sciences and Engineering Research Council, Health Canada, Flax Council, and Flax Growers of Western Canada are gratefully acknowledged.

8 REFERENCES

1. H. Adlercreutz, *Gastroenterol.*, 1984, **86**, 761.
2. K. D. R. Setchell, In: 'Flaxseed in Human Nutrition', S. C. Cunnane and L. U. Thompson (eds.), AOCS Press, Champaign, Ill, 1995, Chapter 5, p. 82.
3. J. L. Hartwell, *Cancer Treat Rep.*, 1976, **60**, 1031.
4. K. D. R. Setchell, A. M. Lawson, E. Conway, N. F. Taylor, D. N. Kirk, G. Cooley, R. D. Farrant, S. Wynn, and M. Axelson, *Biochem J.*, 1981, **197**, 447.
5. S. P. Borriello, K. D. R. Setchell, M. Axelson, and A. M. Lawson, *J. Appl. Bacteriol.*, 1985, **58**, 37.
6. H. Adlercreutz, T. Fotsis, C. Bannwart, K. Wahala, T. Makela, G. Brunow, and T. Hase, *J. Steroid Biochem.*, 1986, **25**, 791.
7. K. D. R. Setchell and H. Adlercreutz, In: 'Role of the Gut Flora in Toxicity and Cancer,' I.R. Rowland (ed.), Academic Press, London, 1988, pp. 315.
8. L. U. Thompson, P. Robb, M. Serraino, and F. Cheung, *Nutr. Cancer*, 1991, **16**, 43.
9. M. Serraino and L. U. Thompson, *Cancer Lett.*, 1991, **60**,135.
10. M. Serraino and L. U. Thompson, *Nutr. Cancer*, 1992, **17**, 153.
11. S. Rickard, L. Orcheson, M. Seidl, F. Cheung, H. Fong, L. Luyengi, and L. U. Thompson, *Fed. Am. Soc. Exp. Biol. J.*, 1993, **7**, A62.
12. L. U. Thompson, L. Orcheson, S. Rickard, and M. Seidl. *Adv. Exp. Med. Biol.*,1994, **364**, 150.
13. L. U. Thompson, S. Rickard, and L. Orcheson, M. Seidl. *Proc. Am. Assoc. Cancer Res.*, 1995, **36**, 114.
14. M. Serraino and L. U. Thompson, *Cancer Lett.*, 1992, **63**, 159.
15. M. C. Archer, W. R. Bruce, C. C. Chan, D. E. Corpet, A. Medline, L. Roncucci, D. Stamp, and X. M. Zhang, *Environ. Health Persp.*, 1992, **98**, 195.
16. M. Jenab and L. U. Thompson, *Proc. Am. Assoc. Cancer Res.*, 1995, **36**, 114.
17. Y. Mousavi and H. Adlercreutz, *J Steroid Biochem. Molec. Biol.*, 1992, **41**, 615.
18. W. V. Welshons, C. S. Murphy, R. Koch, G. Calaf, and V. C. Jordan, *Cancer Res. Treat.*, 1987, **10**, 169.

19. T. Hirano, K. Fukuoka, K. Oka, T. Naito, K. Hosaka, H. Mitsuhashi, and Y. Matsumoto, *Cancer Invest.*, 1990, **8**, 595.
20. B. M. Markaverich, R. R. Roberts, M. A. Alejandro, G. A. Johnson, B. S. Middleditch, and J. H. Clark, *J. Steroid Biochem.*, 1988, **30**, 71.
21. H. Adlercreutz, Y. Mousavi, J. Clark, K. Hockerstedt, E. Hamalainen, K. Wahala, T. Makela, and T. Hase, *J. Steroid Biochem. Molec. Biol.*, 1992, **41**, 331.
22. A. P. Waters and J. T. Knowler, *J. Reprod. Fert.*, 1982, **66**, 379.
23. N. Sathyamoorthy, T. T. Y. Wang, and J. M. Phang, *Cancer Res.*, 1994, **54**, 957.
24. B. Garreau, G. Vallette, H. Adlercreutz, K. Wahala, T. Makela, C. Benassayag, and E. A. Nunez, *Biochem. Biophys. Acta*, 1991, **1094**, 339.
25. H. Adlercreutz, K. Hockerstedt, C. Bannwart, S. Boigu, E. Hamalainen, T. Fotsis, and A. Ollus, *J. Steroid Biochem.*, 1987, **27**, 1135.
26. H. Adlercreutz, C. Bannwart, K. Wahala, T. Makela, G. Brunow, T. Hase, P. J. Arosemena, J. T. Kellis, Jr., and L. E. Vickery, *J. Steroid Biochem. Molec. Biol.*, 1993, **44**, 147.
27. T. Fotsis, M. Pepper, H. Adlercreutz, G. Fleischman, T. Hase, R. Montesano, and L. Schweigerer, *Proc. Natl. Acad. Sci.*, 1993, **90**, 2690.
28. R. Amarowicz, U. Wanasundara, J. Wanasundara, and F. Shahidi, *J. Food Lipids*, 1993, **1**, 111.
29. L. Orcheson, S. Rickard, M. Seidl, F. Cheung, and L. U. Thompson, *Fed. Am. Soc. Exp. Biol. J.*, 1993, **7**, A291.
30. W. R. Phipps, M. C. Martin, J. W. Lampe, J. L. Slavin, and M. S. Kurzer, *J. Clin. Endocrinol. Metab.*, 1993, **77**, 1215.

COMPOUNDS IN PLANTS INDUCING DETOXIFYING AND ANTIOXIDATIVE ENZYMES.

Lars O. Dragsted

Institute of Toxicology
Danish National Food Agency
DK-2900 Soborg, Denmark

1 INTRODUCTION

Several classes of natural compounds commonly found in edible plants are known to interact with xenobiotic metabolising enzymes and antioxidative enzymes and to modify their activities. These include the ubiquitous plant phenols (1), e.g. flavonoids and phenolic acids, degradation products of *Brassica* glucosinolates (2), e.g. isothiocyanates and indolecarbinols, the allylic sulphides of onions and garlic (3), and many others. Interaction with xenobiotic activating and detoxifying enzymes include direct inhibition or enhancement without *de novo* protein synthesis as well as enzyme induction. Interactions with phase I and phase II enzymes activities in the mammalian xenobiotic biotransformation forms part of the cancer-preventive actions of many low molecular weight plant constituents (4). Several phase II enzymes, including quinone reductase (QR) and glutathione transferases (GST), represent important groups of inducible enzymes which protect against many carcinogens (5). Some glutathion transferases, particularly GST type 8-8, also catalyse the degradation of certain membrane oxidation products and form part of the Se independent glutathione peroxidases (6). There is very limited information regarding the influence of cancer-preventive plant constituents on other antioxidative enzymes, e.g. superoxide dismutases (SOD), catalase (CAT) and Se-dependent glutathion peroxidase (Gpx). In the present paper, the doses necessary to elicit en effect on these various enzymes are briefly reviewed in light of the actual concentrations of the active compounds in the diet.

2 MODULATION OF THE ACTIVITIES OF XENOBIOTIC METABOLISING ENZYMES

Conney and coworkers (7) observed a strong inhibition of the microsomal oxygenases involved in zoxazolamine-metabolism by polyhydroxylated flavonoids *in vitro*, whereas the relatively unpolar polymethoxylated flavonoids found mainly in citrus fruits had the opposite effect, doubling the enzyme activities within minutes both *in vitro* and *in vivo*. Interestingly, with 5,6-benzoflavone induced microsomes, the effect of these unpolar flavonoids reversed. This has recently been corroborated by Obermeier et al. (8), who found that polar as well as unpolar flavonoids are strong inhibitors of the CYPIA family of monooxygenases. This inhibitory effect of flavonoids on carcinogen activating enzymes may explain the strong effect of topical quercetin or myricetin towards covalent binding of the carcinogen benzo[*a*]pyrene to mouse lung and epidermal DNA (9).

The gallic acid dimer, ellagic acid, is found in raspberries, strawberries and walnuts, but is mainly bound in hydrolysable tannins (1). It is not known whether these tannins are hydrolysed in the gastrointestinal tract of the human. Based on Danish consumption data on fresh berries, jams and similar products (10), it is estimated that 3-4 mg of free ellagic acid and four or five times more of the polymer may be consumed in average per individual per day. Very low concentrations of ellagic acid were observed to decrease the metabolic activation of carcinogens *in vitro* (11). *In vivo* experiments in animals indicate that halving the activities can be achieved by as little as 200 µg/kg bw/day of ellagic acid (12), an amount which is probably consumed by individuals with a high daily intake of the mentioned products. No data exist regarding the effects of ellagic acid on enzyme activities in humans, however.

3 INDUCTION OF PHASE I ENZYMES BY PLANT CONSTITUENTS

The induction of the various cytochrome P-450 monooxygenases (CYP) proceeds by several mechanisms. The three families CYPI, CYPII and CYPIII of these monooxygenases are involved in the biotransformation of xenobiotics as well as of endogenous compounds. The CYPIA and to some extent the CYPIIA and CYPIIIA subfamilies are induced by unpolar, planar molecules which interact with the *Ah*-receptor. The enzymes are involved in the activation of several important carcinogens in our environment, including polycyclic aromatic hydrocarbons and heterocyclic amines. The enzymes are also involved in metabolism of estradiol into less estrogenic compounds thereby decreasing the growth stimulus to breast cancer cells. The unpolar flavonoid model compounds, flavone and flavonol, as well as the polymethoxyflavones from citrus peel, tangeretin and nobiletin, are able to induce this class of enzymes (13-14), but about two orders of magnitude higher doses are neded compared to what we could get in the diet. As shown by Wattenberg (15) only flavone among these flavonoids can induce the metabolism of benzo[*a*]pyrene at much lower doses in rats. Indole-3-carbinol, which is formed by the action of myrosinase on the glucosinolates found in cruciferous vegetables is a much stronger inducer of CYPIA monooxygenases. Experiments in humans with doses that only exceed what we could get in the diet by a factor of 2-5 have produced a significant induction (16). Diets with a high content of brussels sprouts or broccoli are also able to induce these enzymes in experimental animals as well as in humans (17-19). It is not known to what extent these inductive effects of dietary crucifers are involved in their anticarcinogenic effects in animals (4) or possibly in humans (20).

CYPIIB and CYPIIE are induced through other, presently unknown receptor systems. No structural determinants are known for the induction of CYPIIB, and the classical inducer used experimentally, phenobarbital, is known to be able to increase the toxicity of some xenobiotics while decreasing the toxicity of others. D-limonene which is very abundant in citrus oils has been shown to be an inducer of this enzyme subfamily, but the doses necessary to produce an effect are extremely high, about 1-5% in the diet (21). This is several orders of magnitude above the calculated dose of 0.1-1 mg/kg/day that the Danish population could get in their diets based on consumption of citrus products (10) and assuming 100 ppm D-limonene in them all. CYPIIE is induced by ethanol (22), and is involved in the oxidation of a range of organic solvents, including benzene, as well as in the activation of nitrosamines. CYPIIE is suppressed 4-5 times by feeding with ellagic acid (23) or with some isothiocyanates (24). The latter compounds belong to another group of glucosinolate degradation products from mustards, garden cress and crucifers (25). An evaluation of Danish intake rates indicate that the mean intake is not higher than 0.1 mg/kg/day, and possibly 10 times higher in those who ingest the largest proportion of precursors (26). This is two to three orders of magnitude less than what is needed to suppress CYPIIE activities (27).

Isothiocyanates also suppress CYPIA and CYPIIIA, whereas they may induce CYPIIB (28, 29). Several other compounds have effects on more than one of these subfamilies. Indoles have been shown to induce CYPIIIA in addition to CYPIA (30-33), whereas tangeretin at high doses induces CYPIA as well as CYPIIB activities (13, 34). On the other hand, grapefruit juice was found to suppress the activity of CYPIA2 when given at a dose of 4-8 g/kg/day to rats. This is interesting since similar doses have been shown to inhibit CYPIA2 dependent metabolism of caffeine in humans (35). High doses (2.5% in the drinking water) of aqueous green tea extract, rich in catechins and their gallate esters, have been shown to increase the activities of CYPIA and CYPIIB as well as the fatty acid omega-hydroxylase, CYPIVA, whereas CYPIIIA activity was depressed (36).

4 PHASE II ENZYME INDUCTION BY PLANT CONSTITUENTS

The phase II enzymes of xenobiotic metabolism include enzymes which add water to activated compounds, e.g. epoxide hydrolase (EH) and QR, as well as transferases of for example sulphate, glucuronic acid and glutathione. These enzymes are generally regarded as detoxifying enzymes, but examples exist for a role of several phase II activities in the activation of carcinogens.

Classical examples are the involvement of (EH) in the activation of Benzo[a]pyrene, the involvement of sulphotransferases and acetyl transferases in the ultimate activation of aromatic amines, and the activation of haloalkanes by GST. Most of these enzymes are inducible by compounds which interact with the Ah-receptor but other induction mechanisms exist. Tangeretin as well as indole-3-carbinol and its acid degradation products have been found to induce these enzymes (30, 31, 34). Induction was observed at dose levels where CYPIA activities are also induced (30, 31, 32, 37), indicating that induction was a consequence of interaction with the Ah-receptor. In the case of indoles this may be of significance for consumers of high levels of crucifers. In the special case of flavonol, induction of phase II enzymes commenced at doses two orders of magnitude less than necessary to induce CYPIA activities, indicating that other induction mechanisms may also be involved (34). D-limonene is another good example of a substance that induces GST at doses at least an order of magnitude less than those necessary for CYP induction (21), in this case CYPIIB and CYPIIC (38).

Other receptors are known to lead to the induction of selected phase II activities without affecting Ah. Importantly the work of Talalay and coworkers (5, 39, 40) has shown that all Michael acceptors are inducers of QR through the electrophile responsive element in the promoter region of its gene and also that many antioxidants induce this activity through the antioxidant responsive element. A large number of antioxidants exist among low molecular weight plant constituents, and several of them have been shown to induce QR *in vitro* (41) or *in vivo*, often together with UDP-glucuronosyl transferases (UDPGT), GST, EH and also glutathione reductase (GR). Polar flavonoids significantly induced UDPGT and GST at doses where no interaction with CYP enzymes have been observed so far (13, 42). Dosing with lower levels comparable to those found in the human diets, i.e. 0.1-1 mg/kg/day (43) have not been reported for this group of plant phenolics, but ellagic acid was observed to induce GST at the same exceedingly low dosage level where it suppressed CYPIA (12), indicating that plant phenols at dietary relevant levels may be important inducers of phase II activities. This is supported by the observation that a low dose of 0.2% green tea in the drinking water was able to induce GST and QR in mice (44).

Isothiocyanates are also very potent inducers of GST and QR and iberine, an isothiocyanate formed from broccoli glucosinolates, induced QR in rodent intestine by a factor of eight at doses of only about 10 mg/kg/day for one week (45). Still this dose exceeds human intake levels by at least one order of magnitude (25, 26). The anticarcinogenic effects of isothiocyanates seem to follow closely their ability to induce phase II enzymes in experimental animals (26) and this group of substances may therefore not be involved in the cancer preventive effects of diets rich in cruciferous vegetables. This is supported by data from feeding studies where various diets supplemented with cabbage, brussels sprouts or broccholi were given to rodents (46, 47) or to humans (19, 48). Whenever GST is induced in these studies, CYPIA activity is increased at a similar dose level, indicating that interaction with the *Ah*-receptor is responsible for the induction of phase II enzymes by constituents of these vegetables, possibly indoles. Alfalfa has a similar pattern of induction in rodents whereas soya beans and onions are inducers by a different receptor mechanism since they induce EH and GST without affecting CYPIA (49).

5 MODULATION OF ANTIOXIDATIVE ENZYMES

The induction of GST, which is also participating in peroxidase activities in the plasma membrane has been discussed above. Relatively few reports on modulation of other antioxidative enzyme activities exist in the literature, and most effects were found at relatively high doses compared to what humans might get in the diet. Products from *Allium* species like onions and garlic contain high levels of allylic sulphides, which are potent inducers of both GST and Gpx (3). On the other hand, indole-3-carbinol decreased the activity of Gpx and also of SOD while inducing GST and GR (31). Soybean meal at very high doses increased CAT activity in rats. The inducing constituents have not been identified (50). Green tea in the drinking water was shown to induce Gpx, GR and CAT in mice at low dose levels, indicating that plant phenolics may be able to induce these activities even at dietary relevant levels (44).

6 CONCLUSIONS

The evidence that natural compounds in plants are able to modulate the activities of enzymes involved in activation and detoxication of carcinogens has been briefly reviewed with a focus on the relation between active doses and dietary exposures. Two groups of compounds with different mechanisms are expected to modulate enzyme activities at dietary relevant doses. Firstly, the indole-glucosinolates or other products in cruciferous vegetables seem to be able to induce both phase I and phase II enzymes, presumably by interaction with the *Ah*-receptor, at high dietary levels. The effects of long-term medium dietary levels of indoles in humans have not been tested, and it is therefore possible that induction of only phase II activities by a different mechanism could be achieved at lower doses. Secondly, some plant phenols at low doses seem to induce phase II enzymes only, while suppressing some phase I activities in experimental animals. The efficiency of phase II enzyme induction in humans has not been ascertained. However, effects of grapefruit juice on CYPIA2 in humans indicate that phenolic or other constituents in this juice have a suppressing effect on the enzymes involved in the activation of heterocyclic amine carcinogens (35, 51). More studies on specific groups of vegetables given in the diet to experimental animals and to humans may lead to the identification of the constituents which are of significant importance for modulating these enzymes, even at the low concentrations, at which they are found in the human diet.

References

1. M. Strube and L. O. Dragsted, 'Naturally occuring antitumourigens I: plant phenols', Nordic Council of Ministers, Seminar reports, Copenhagen, 1993, **605**, p. 111.
2. C. A. Bradfield and L. F. Bjeldanes, In: Nutritional and Toxicological Consequences of Food Proccessing, M. Friedman (Ed.), Plenum Press, New York, 1991, **13**, p. 153.
3. V. A. Gudi and S. V. Singh, Biochem. Pharmacol., 1991, **42**, 1261.
4. L. O. Dragsted, M. Strube and J. C. Larsen, Pharmacol. Toxicol., 1993, **72 suppl.**, 116.
5. P. Talalay, M. J. De Long and H. J. Prochaska, Proc. Natl. Acad. Sci. USA, 1988, **85**, 8261.
6. B. Mannervik, Chem. Scripta, 1987, **27A**, 121.
7. J. M. Lasker, M.-T. Huang and A. H. Conney, J. Pharmacol. Exp. Ther., 1984, **224**, 162.
8. M. T. Obermeier, R. E. White and C. S. Yang, Xenobiotica, 1995, **25**, 575.
9. M. Das, W. A. Khan, P. Asokan, D. R. Bickers and H. Mukhtar, Cancer Res., 1987, **47**, 760.
10. P. H. Larsen and J. H. Jensen, 'A Danish Dietary Survey', National Food Agency of Denmark Publications, 1985, **108**, 1.
11. R. W. Teel, R. Dixit and G. D. Stoner, Carcinogenesis, 1985, **6**, 391.
12. M. Das, D. R. Bickers and H. Mukhtar, Carcinogenesis, 1985, **6**, 1409.
13. M.-H. Siess, M. Guillermic, A. M. Le Bon and M. Suschetet, Xenobiotica, 1989, **19**, 1379.
14. M.-H. Siess, A. M. Le Bon and M. Suschetet, J. Toxicol. Env. Health, 1992, **35**, 141.
15. L. W. Wattenberg, M. A. Page and Leong J. L., Cancer Res., 1968, **28**, 934.
16. J. J. Michnovicz and H. L. Bradlow, Nutr. Cancer, 1991, **16**, 59.
17. J. J. P. Bogaards, B. van Ommen, H. E. Falke, M. I. Willems and P. J. van Bladeren, Fd. Chem. Toxic., 1990, **28**, 81.
18. J. J. P. Bogaards, H. Verhagen, M. I. Willems, G. van Poppel and P. J. van Bladeren, Carcinogenesis, 1994, **15**, 1073.
19. K. Vistisen, H. E. Poulsen and S. Loft, Carcinogenesis, 1992, **13**, 1561.
20. K. A. Steinmetz and J. D. Potter, Cancer Causes and Control, 1991, **2**, 325.
21. J. A. Elegbede, T. M. Maltzman, C. E. Elson and M. N. Gould, Carcinogenesis, 1993, **14**, 1221.
22. D. W. Nebert, D. R. Nelson and R. Feyereisen, Xenobiotica, 1989, **19**, 1149.

23. D. H. Barch and C. C. Fox, Cancer Lett., 1988, **44**, 39.
24. H. Ishizaki, J. F. Brady, S. M. Ning and C. S. Yang, Xenobiotica, 1990, **20**, 255.
25. K. Sones, R. K. Heaney and G. R. Fenwick, J. Sci. Food Agric., 1984, **35**, 712.
26. M. Strube and L. O. Dragsted, 'Naturally Occurring Antitumourigens II: Organic Isothiocyanates', Nordic Council of Ministers, Copenhagen, 1994.
27. H. Ishizaki, J. F. Brady, S. M. Ning and C. S. Yang, Carcinogenesis, 1990, **20**, 255.
28. Z. Guo, T. J. Smith, E. Wang, N. Sadrieh, Q. Ma, P. E. Thomas and C. S. Yang, Carcinogenesis, 1992, **13**, 2205.
29. Z. Guo, T. J. Smith, E. Wang, K. I. Eklind, F. L. Chung and C. S. Yang, Carcinogenesis, 1993, **14**, 1167.
30. H. G. Shertzer and M. Sainsbury, Food Chem. Toxicol., 1991, **29**, 237.
31. H. G. Shertzer and M. Sainsbury, Food Chem. Toxicol., 1991, **29**, 391.
32. O. Vang, M. B. Jensen and H. Autrup, Carcinogenesis, 1990, **11**, 1259.
33. D. M. Stresser, G. S. Bailey and D. E. Williams, Drug Metab. Dispos. 1994, **22**, 383.
34. M.-H. Siess, A. M. Le Bon and M. Suschetet, J. Toxicol Environ. Health., 1992, **35**, 141.
35. U. Fuhr and K. Klittisch, Naunun Schmiedebergs Arch. Pharmacol., 1991, **345 (suppl. 1)**, R7.
36. A. Bu-Abbas, M. N. Clifford, C. Ioannides and R. Walker, Fd. Chem. Toxic., 1995, **33**, 27.
37. L. W. Wattenberg, M. A. Page and J. L. Leong, Cancer Res., 1968, **28**, 934.
38. T. H. Malzman, M. Christou, M. N. Gould and C. R. Jefcoate, Carcinogenesis, 1991, **12**, 2081.
39. Benson, A. M., M. J. Hunkeler and P. Talalay, Proc. Natl. Acad. Sci. USA, 1980, **77**, 5216.
40. T. Prestera, W. D. Holtzclaw, Y. Zhang and P. Talalay, Proc. Natl. Acad. Sci. USA, 1993, **90**, 2965.
41. H. J. Prochaska, A. B. Santamaria and P. Talalay, Proc. Natl. Acad. Sci. USA, 1992, **89**, 2394.
42. M.-H. Siess, C. Brouard, M.-F. Vernevaut, M. Suschetet, In ' V. Cody, E. Middleton, J. B. Harborne and A. Beretz (Eds.) Plant Flavonoids in Biology and Medicine, Alan R. Liss Inc., New York', 1988, 147.
43. M. H. J. Hertog, Ph. D. Thesis, University of Wageningen, 1994.
44. S. G. Khan, S. K. Katiyar, R. Agarwal and H. Mukhtar, Cancer Res., 1992, **52**, 4050.
45. A. M. Kore, E. H. Jeffrey and M. A. Wallig, Fd. Chem. Toxic., 1993, **31**, 723.
46. R. McDanell, A. E. M. McLean, A. B. Hanley, R. K. Heaney, and G. R. Fenwick, Fd. Chem. Toxic., 1989, **27**, 289.

47. V. L. Sparnins, P. L. Venegas and L. W. Wattenberg, J. Natl. Cancer Inst., 1982, **68**, 493.
48. H. M. Wortelboer, C. A. DeKruif, A. A. J. van Iersel, J. Noordhoek, B. J. Blaauboer, P. J. van Bladeren and H. E. Falke, Fd. Chem. Toxic., 1992, **30**, 17.
49. C. A. Bradfield, Y. Chang and L. F. Bjeldanes, Fd. Chem. Toxic., 1985, **23**, 899.
50. T. E. Webb, P. C. Stromberg, I. Abou, R. W. Curley Jr., and M. Moeschberger, Nutr. Cancer, 1992, **18**, 215.
51. H. Wallin, A. Mikalsen, P. F. Guengerich, M. Ingelman-Sundberg, K. E. Solberg, O. J. Rossland and J. Alexander, Carcinogenesis, 1990, **11**, 489.

A CROSS-SECTIONAL STUDY ON RELATIONSHIP BETWEEN SERUM LEVELS OF CAROTENOIDS AND PERIPHERAL DISTRIBUTION OF LYMPHOCYTE SUBSETS IN JAPANESE ADULTS

Y. ITO[1], R. SASAKI[2], Y. NIIYA[3]

1: School of Health Sciences, Fujita Health University, Toyoake-shi, Aichi , 470-11, Japan
2: Department of Public Health, Aichi Medical University, Nagakute-cho, Aichi, 480-11, Japan
3: Daido Hospital, Nagoya, 457, Japan

1 INTRODUCTION

It was suggested that the cancer prevention associated with carotenoids might be related to its ability of antioxidant and immune enhancement[1,2]. There were many reports that carotenoids such as β-carotene reduced the lipid peroxidation in human [3,4]. In contrast, it was shown in the immunological studies on cancer prevention that β-carotene administration to human increased the peripheral blood distribution of lymphocyte subsets such as Tcells and NK cells [5,6]. However, it was recently reported the different results which β-carotene administration to human did not affect with the peripheral distribution of T cell and NK cell subsets[7].

In this study, the relationship between serum carotenoid levels and peripheral blood distributions of lymphocyte subsets in the healthy adults was investigated by the cross-sectional study.

2 SUBJECTS AND METHODS

The subjects enrolled were 134 healthy male adults aged from 20-59y and selected randomly from Japanese workers in a steel factory, as shown in TABLE 1.

Serum levels of carotenoid such as β-carotene, cryptoxanthin and zeaxanthin/lutein were determined by a HPLC method[8]. The peripheral distributions of lymphocyte subsets were estimated by a fluorocytometry using specific antibodies for various cell markers as follows; T cell markers: CD3 (T3), CD4 (T4), CD45 RA (2H4), CD8 (T8) and CD11b (MO1), NK cell markers: CD56 (NKH1), CD57 (Leu7) and CD16 (Leu11), and B cell marker: CD20 (B1).

In the statistical analyses, the variables of serum carotenoid levels and blood counts of lymphocyte subsets were converted to logarithmic values. The analyses of t-test and multiple regression of the statistical package of StatView were performed.

TABLE 1 Characteristics of the subjects

No. of subjects		134	(100.0)
Age	20 - 29	25	(18.9)
	30 - 39	32	(23.9)
	40 - 49	52	(38.8)
	50 - 59	25	(18.7)
Smoking habit			
	never smoker	39	(29.1)
	ex-smoker	16	(11.9)
	current smoker	79	(59.0)
Alcohol drinking habit			
	non drinker	16	(11.9)
	irregular drinker	26	(19.4)
	regular drinker	92	(68.7)

3 RESULTS AND DISCUSSION

2.1 Serum levels of carotenoids and peripheral blood distributions of lymphocyte subsets

Serum geometric means of β-carotene, cryptoxanthin and zeaxanthin /lutein in the subject were 0.66, 0.382 and 0.987 μ mol/l, respectively. As these studies were conducted in February, serum β-carotene levels were lower in comparing to the other results of Japanese populations, while serum cryptoxanthin levels were higher[9].

The mean and geometric mean values of percentages and counts of lymphocyte subsets in peripheral blood were shown in TABLE 2. These results were agreed with the other reports[10,11].

TABLE 2   Distribution of peripheral lymphocyte subsets in the subjects

Subset	Percentages (%)	Counts (counts/mm³)	
T3	69.1 (8.6)	1759	[1204 - 2660]
T4	46.5 (8.7)	1170	[756 - 1780]
Th	30.3 (6.9)	756	[476 - 1221]
Ti	16.2 (6.3)	381	[192 - 699]
T8	31.8 (7.3)	795	[509 - 1311]
Tc	21.5 (6.6)	524	[315 - 909]
Ts	10.4 (4.9)	241	[122 - 469]
NKH1	19.9 (8.2)	468	[216 - 872]
Leu7	14.4 (6.1)	337	[148 - 710]
Leu11	8.8 (5.3)	192	[76 - 415]
B1	15.8 (5.1)	385	[215 - 688]

Percentages : mean value (S.D.). Counts : geometric mean value [10% - 90 %] .

2.2 Relationship between serum carotenoid levels and peripheral blood distribution of lymphocyte subsets

Serum carotenoid levels were influenced by age, consumption of cigarettes smoked and alcohol drunk, and serum values of cholesterol, BUN and γ-GTP activity. So, the regression analyses were conducted using the age, alcohol consumption, serum levels of cholesterol, BUN and γ-GTP activity as independent variables (TABLE3). On peripheral blood counts of lymphocyte subsets, serum β-carotene levels were inversely associated with peripheral blood counts of T cell subsets and Leu 7 subset. However, there were no significant relations between peripheral blood counts of lymphocyte subsets and serum β-carotene levels in the adults without smoking habit.

Serum zeaxanthin/lutein levels were not significant related with T cell subsets, while those were significantly and positively associated with Leu11 subset. Further, the trend for peripheral blood counts of Leu11 subset was demonstrated to be much in non-smokers with high zeaxanthin /lutein levels. In addition, although serum β-carotene and cryptoxanthin levels were not significantly related with NK cell subsets, non-smokers with high β-carotene levels tended to be high counts of Leu11 subsets.

Preipheral blood counts of T cells such as T3 and T4, and B cell subsets were positively associated with consumption of cigarette smoking, while those of NK cell subsets were inversely related with cigarette smoking. In addition, serum levels of β-carotene and cryptoxanthin were inversely and strongly related with cigarette smoking comparing to serum zeaxanthin/lutein levels[12]. These facts reflect to inverse association between serum levels of β-carotene or cryptoxanthin and peripheral blood counts of T cell and B cell subsets. However, NK cell subsets, especially Leu 11 subset with activated NK cell were positively and significantly associated with serum zeaxanthin/lutein levels, although they were inversely associated with cigarette smoking.

TABLE 3 Relationship between peripheral blood counts of lymphocyte subsets and serum carotenoid levels in the subjects

Dependent variable (counts/mm³)	Serum carotenoid level (μ mol/l)					
	β-Carotene		Zeaxanthin/lutein		Cryptoxanthin	
	All	Non-SM	All	Non-SM	All	Non-SM
Lymphocytes	-0.259**	-0.146	-0.117	-0.047	-0.301**	-0.155
T3	-0.303***	-0.197	-0.164	-0.147	-0.277**	-0.102
T4	-0.287**	-0.247$	-0.150	-0.208	-0.290**	-0.161
Th	-0.280**	-0.230	-0.156$	-0.210	-0.368***	-0.199
Ti	-0.146	-0.115	-0.051	-0.061	-0.075	-0.040
T8	-0.225**	-0.186	-0.024	-0.022	-0.320**	-0.190
Ts	-0.207 *	-0.241	0.025	0.080	-0.257**	-0.237
Tc	-0.196 *	-0.127	-0.096	-0.143	-0.270**	-0.090
NKH1	-0.156	-0.222	0.072	0.192	-0.220*	-0.174
Leu7	-0.206 *	-0.161	0.006	0.144	-0.162	-0.122
Leu11	-0.042	-0.012	0.200 *	0.251	-0.167$	-0.181
B1	-0.258 **	-0.131	-0.217*	-0.261	-0.217*	-0.170
Number	134	55	134	55	134	55

Data were represented as standardized regression coefficients from calculation of multiple regression analyses using the age, consumption of alcohol drunk, and serum values of total cholesterol, BUN and γ-GTP activity as the adjusted variables. Serum values of carotenoids and peripheral blood counts of lymphocyte subsets were converted the logarithmic values.
All: all subjects, Non-SM: non-smoker; $ < 0.10, * : p < 0.05, ** : p < 0.01, *** : p < 0.001.

4 CONCLUSION

Relationship between serum levels of carotenoids and distributions of lymphocyte subsets in peripheral blood of the healthy 134 male adults, aged from 20-59y was investigated.
It was demonstrated that serum β-carotene levels were negatively associated with the peripheral blood counts of T3 and T4 of T cell subsets, while there was no significant relation between serum β-carotene levels and the peripheral counts of T cell subsets (T3, T4 and T8), NK cell subsets (NKH1, Leu 11 and Leu 7) and B cell subset (B1) in the non-smokers. In contrast, serum zeaxanthin/lutein levels were positively and significantly associated with the peripheral blood percentages and counts of activated NK cell subsets (Leu11 and NKH1) in both all subjects and the non-smokers.
So, it seemed that the healthy adults with high serum zeaxanthin /lutein levels could keep to be present high distribution of activated NK cells in peripheral blood.

REFERENCES

1. R. Peto, R. Doll J.D. Burckley and M.B. Sporn, Nature 1981, 290, 201.
2. A. Bendich and J.A. Olsen FASEB J., 1989, 3, 1927.
3. G.W. Burton and K.U. Ingold, Science, 1984, 224, 569.
4. Y. Ito, R. Shinohara, R. Sasaki, K. Yagyu, N. Mizutani, K. Okamoto, S. Suzuki and K. Aoki, Vitamins (Japan), 1994, 68, 569.
5. M. Alexander, H.H. Newmark and R.G. Miller, Immunol. Letters, 1985, 9, 221.
6. R.H. Prabhala, H.S. Garewal, M.J. Hicks, R.E. Sampliner and R.R. Watson, Cancer, 1991, 67, 1556.
7. P.A. Daudu, D.S. Kelley, P.C. Taylor, B.J. Burri and M.M. Wu, Am. J. Clin. Nutr., 1994, 60, 969.
8. Y. Ito, J. Ochiai, R. Sasaki, S. Suzuki, Y. Kushuhara, Y. Morimitsu, M. Otani and K. Aoki, 1990, 194, 131.
9. Y. Ito, R. Sasaki, S. Suzuki, T. Yasui, H. Hishida, M. Otani and K. Aoki, Vitamins (Japan) 1994, 68, 351.
10. G. van Poppel, S. Spanhaak, and T. Ockhuizen, Am.J.Clin. Nutr., 1993, 57, 402.
11. T. Murata, H. Tamai, T. Morinobu, M. Manago, H. Takenaka, K. Hayashi and M.Mino, Am. J. Clin. Nutr., 1994, 60, 597.
12. Y. Ito, R. Sasaki, S. Suzuki and K. Aoki, Int. J. Epidemiol., 1991, 20, 615.

ANTICARCINOGENIC PROPERTIES OF LYCOPENE

Yoav Sharoni and Joseph Levy

Department of Clinical Biochemistry
Faculty of Health Sciences
Ben-Gurion University of the Negev and Soroka Medical Center
Beer Sheva 84105, Israel.

1 SUMMARY

Consumption of carotenoids (plant pigments) has frequently been inversely correlated with cancer incidence. We describe here the anticarcinogenic properties of lycopene, the major tomato carotenoid, and compare it to those of α- and β-carotene. We conclude that lycopene is the more potent inhibitor of endometrial, lung and mammary cancer cell growth. Lycopene inhibited the growth of these cancer cells in a dose-dependent manner - half maximal inhibition was at ~2 µM. This effect was detected after 24 hours of incubation. In contrast, the effects of α- and β-carotene were evident only after the second day of incubation and the cells escaped inhibition at a later time. In contrast to cancer cells, human fibroblasts were less sensitive to inhibition by lycopene and these cells also gradually escaped inhibition. Lycopene also inhibited IGFs stimulated cancer cell growth. Preliminary results suggest that lycopene intervenes in the signal transduction mechanism of IGF-I. When mammary cancer cells were stimulated by IGF-I to grow faster, a lower lycopene concentration (0.4 µM) was inhibitory. However, this concentration was ineffective in slow-growing cells. This may imply that there is a differential effect of lycopene on transformed cells as compared to normal cells. Lycopene effects were also tested in the DMBA-induced rat mammary tumor model for human hormone dependent mammary cancer. Lycopene-treated rats developed fewer and significantly smaller mammary tumors when compared to the control or β-carotene treated rats. These results should stimulate additional experiments that will elucidate the mechanism of lycopene action and its effects on tumor growth in vivo.

2 INTRODUCTION

Since mammals cannot synthesize carotenoids, including lycopene, these pigments are obtained mainly from vegetables and fruits. In Western countries lycopene is the major carotenoid in human plasma; its concentration (~0.8 µM) is more than twice that of β-carotene or the sum of lutein and zeaxantin together.

Over the past several years there has been a growing interest in the ability of carotenoid pigments to act as cancer-preventive agents. The main emphasis has been on β-carotene, and this research has yielded hundreds of reports. However, two recent intervention studies which tested the protective effect of β-carotene against cancer have revealed negative findings.[1,2] In the Finnish (ATBC) study[1] an unexpectedly higher incidence of lung cancer was observed among men who received β-carotene when compared with subjects who did not receive this carotene. In the other study[2] the efficacy of β-carotene in preventing colorectal adenoma (a precursor of invasive carcinoma) was tested in randomly assigned patients. From the results it is clear that there was no evidence that β-carotene reduced the incidence of adenomas.

These results seem to suggest that factors other than β-carotene which are present in diets rich with vegetables and fruits act as cancer-preventive agents. It is quite possible that lycopene is one of these factors; however only a few studies have stressed its protective role. Several epidemiological studies of patients suffering from various malignancies have suggested a cancer-preventative role for lycopene. A study of serum lycopene in cervical intra-epithelial neoplasia patients[3] has revealed a strong inverse relationship: When fasting venous blood samples were assayed for serum carotenoids, the results indicated that serum lycopene and dietary intake (tomatoes) of lycopene were inversely correlated with cancer incidence. No such results were obtained for the other tested carotenoids. A low level of serum lycopene was also observed in patients who subsequently developed bladder- and pancreatic-cancers.[4,5] Serum was collected from 25,802 persons in Washington County, MD and kept frozen at -70°C for more than 12 years. The results showed that serum levels of lycopene and selenium were lower than the matched controls.

We present new evidence for the anticarcinogenic role of lycopene which emerged from experiments with in vitro and in vivo models. These results are backed by studies which investigate the biochemical mechanisms that may explain the action of lycopene.

3 MATERIALS AND METHODS

3.1 Carotenoids and Peptides. Synthetic β-carotene and lycopene were produced and generously donated by Hoffmann-La Roche (Basel, Switzerland). Natural α and β-carotene and lycopene extracted from fruits or vegetables were purchased from Sigma (MO, USA). Various lycopene preparations extracted from tomatoes were donated by Makhteshim (Beer Sheva, Israel). Tetrahydrofuran (THF) containing 0.025% butylated hydroxytoluene (BHT) was purchased from Aldrich (WI, USA).

3.2 Cells. The following cells were employed for the experiments: MCF-7 human mammary cancer cells, Ishikawa endometrial cancer cells originating from a well-differentiated tumor. NCI-H226 squamous cell lung cancer cells, and NCI-H345 small cell lung cancer cells. Foreskin primary human fibroblasts in the 15th to 23rd passage.

3.3 Carotenoid Solutions. Carotenoids were dissolved in THF containing 0.05% BHT as an antioxidant, at a concentration of 2 mM. These stock solutions were added to the cell culture medium under vigorous stirring and the medium was filtered (Millex-HV, 0.45 μm, Millipore). The final THF concentration of 0.5% did not have any significant effect.

3.4 Cell Growth. For growth studies, cells were seeded into 96 multiwell plates (5,000-10,000 cells per well) in medium containing 3% fetal calf serum. One day later the medium was changed to one containing the solubilized carotenoid which was replaced daily.

3.5 Cell Proliferation Assay by [³H]thymidine Incorporation. After incubation, [³H]thymidine (1.25 µC/well) was added. After 2.5 hr the incorporation was stopped by addition of unlabeled thymidine (0.5 µmole). The medium was discarded and the cells were trypsinized and collected on a glass-fiber filter with a cell harvester. Radioactivity was determined by scintillation counting (Packard Downers Grove, Illinois) or radioactive image analyzer (BAS 1000, Fuji, Japan).

4 RESULTS AND CONCLUSIONS

4.1 in vitro studies

Lycopene was found to be a potent inhibitor of the growth of mammary (Figure 1), endometrial and lung cancer cells.[6] The inhibition was dose-dependent - half maximal inhibition was at 2 µM. Lycopene was observed to be a more effective inhibitor than α- and β-carotene. For example, at least a ten-fold higher concentration of α- and β-carotene were needed for growth inhibition of MCF-7 mammary cancer cells. Furthermore, inhibition of cell growth by lycopene was detected after 24 hours of incubation. In contrast, α- and β-carotene effects were evident only after the second or third day of incubation.

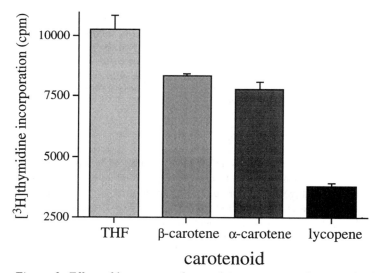

Figure 1: *Effect of lycopene and α- and β-carotene on the growth of MCF-7 mammary cancer cells grown for two days in the presence of 0.05% THF or 3µM carotenoids solubilized in THF. Cell proliferation was evaluated by the [³H]thymidine incorporation method. The results are the mean ± SEM of three different experiments each done in ten replicates.*

Carotenoid preparations from the following sources were used in the experiments: Sigma: lycopene, α-carotene and β-carotene. Makhteshim: lycopene - 5% tomato oleoresin, crystalline (about 95%) purified from the tomato oleoresin. Hoffman La-Roche: all-trans synthetic lycopene from two batches. All preparations of lycopene tested were effective in inhibiting mammary cancer cell growth. However the various preparations differed in potency. The reason for this difference is not clear but it suggests that the inhibitory effect is an intrinsic property of lycopene or its metabolites, and does not result from an unrelated contaminant.

In order to confirm the inhibitory effect of lycopene on cell growth, we compared measurements of [^{3}H]thymidine incorporation with direct cell counting by a Coulter counter. While in both analytical methods an inhibitory effect of lycopene on MCF-7 cell growth was observed, a decrease in [^{3}H]thymidine incorporation was detected after the first day of lycopene application, whereas a reduction of cell number was evident only after the second day. These results indicate that suppression of cell growth by lycopene is secondary to its inhibition of DNA synthesis.

One of the reasons for the uncontrolled growth of many cancer cells is the augmented secretion of autocrine or paracrine growth factors. Recently we showed that IGFs are potent autocrine mitogens for breast- and endometrial-cancer cells.[7,8] In order to understand the mechanism by which the growth of mammary cancer cells is inhibited by lycopene, we investigated if it can interfere with IGF-I induced growth stimulation. MCF-7 cells were incubated for one to three days with IGF-I (30 nM), and increasing concentrations of lycopene. In order to reduce the interference of serum-derived growth factors these experiments were performed in the presence of 0.1% charcoal stripped serum.

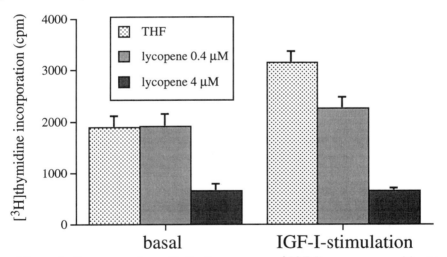

Figure 2: Faster growing cells (in the presence of IGF-I) are more sensitive to lycopene treatment. For experimental details see legend to figure 1. MCF-7 mammary cancer cells were incubated for two days without or with IGF-I (30 nM) and two concentrations of lycopene, as depicted in the figure. The results are the mean±SEM of three different experiments each done in ten replicates.

IGF-I significantly stimulated [^{3}H]thymidine incorporation, whereas lycopene at high concentration (4 μM) inhibited both basal and IGF-induced cell proliferation (Figure 2). It is impressive that lycopene is more effective in inhibiting fast growing cells than those growing slowly. When 0.4 μM lycopene was applied, [^{3}H]thymidine incorporation was inhibited only in cells growing in the presence of IGF-I. The same lycopene concentration was ineffective in cells growing slowly in the absence of IGF-I. The inhibition of IGF-I induced cell growth was associated with a decrease in IGF-induced tyrosine phosphorylation of cellular substrates, which suggests that lycopene intervenes in the signal transduction mechanism of the IGF-I receptor.

Lycopene also inhibited the growth of human skin fibroblasts,[6,9] although these cells were less sensitive to inhibition. After the first day of incubation, [^{3}H]thymidine incorporation in cancer cells was inhibited by about 75% and in fibroblasts by 50%. Moreover, fibroblasts gradually escape from the inhibitory effect of lycopene. On the third day of incubation [^{3}H]thymidine incorporation in fibroblasts was inhibited only marginally, while inhibition in endometrial cancer cells remained in the order of the first day.

Lycopene inhibition of cancer cell growth has been reported by others. Wang et al[10] have shown that lycopene is an effective growth inhibitor of glioma cells which are cancer cells having a neuronal origin. Countryman et al.[11] have recently reported that lycopene inhibits the growth of HL-60 cells, a well known model for promyelocytic leukemia. Wang's group proceeded with their studies on an in vivo model of glioma cells transplanted in rats, and demonstrated that also in this system lycopene is an effective inhibitor.[12]

4.2 in vivo studies

We extended our in vitro studies to an in vivo model of carcinogenesis aimed at evaluating the role of lycopene and β-carotene on the induction and progression of tumors in rats. The DMBA-induced rat mammary tumor is an excellent model for studying hormone-dependent human breast cancer, as the tumor growth rate is easily manipulated by estrogens[13] and other hormones.[13-15]

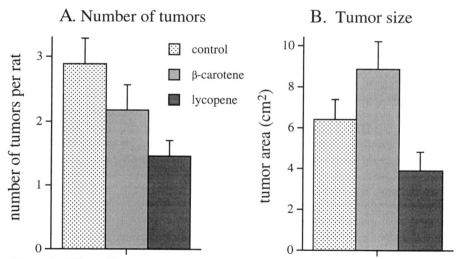

Figure 3: Effect of lycopene and β-carotene on the induction and progression of DMBA induced rat mammary tumors. The rats were treated as described in the text. The number of tumors per rat (A) and tumor area (B) is shown at the end of the follow-up period (four month since DMBA administration). Tumor size was measured with a caliper. Values are the mean ± SEM.

Mammary tumors were induced in female Sprague-Dawley rats, as described previously.[13-15] Rats were fed ad libidum with Purina chow which consists of 50% corn and 50% synthetic ingredients. This provides a diet with a very low carotenoid content. Treatment with the various carotenoids (10 mg/kg by i.p. injection) was begun two weeks prior to DMBA-tumor induction. The experiment was repeated three times and the total number of rats in each group for all three experiments was about forty.

The number and size of the tumors at the end of a four month-follow-up period is seen in Figure 3. Lycopene reduced the number of tumors while β-carotene had no statistically significant effect (Figure 3A). Lycopene also had a marginal effect on the area of the tumors (Figure 3B). Surprisingly, the tumor area in the β-carotene treated group was larger than that of the control and lycopene-treated groups. The difference in tumor size between the lycopene and β-carotene treated groups was statistically significant.

4.3 Mechanism of action

Several mechanistic studies were performed for the purpose of explaining how carotenoids attenuate the induction and proliferation of tumors:

4.3.1 Anti-oxidant Activity. Reactive oxygen species arising in tissues can damage DNA, proteins, carbohydrates and lipids. Lycopene exhibits the highest physical quenching rate constant with singlet oxygen.[16] In a recent study[17] it was found that the ability of β-carotene to quench NOO· radicals is two to three-fold lower than that of lycopene. NOO· radicals are present in tobacco smoke and may cause cancer by reacting with various cell components, especially by damaging DNA and by causing oncogenic mutations. The antioxidant activities of lycopene are probably also important in the prevention of cardiovascular diseases.

4.3.2 Effect on the Expression of a Gene Encoding Connexin43 - A Gap Junction Protein. The observation that both β-carotene and canthaxanthin can inhibit malignant transformation caused by either methylcholanthrene - (MCA) or X-radiation in C3H/10T1/2 cells has been recently reviewed.[18] This study has been extended to other carotenoids. In addition to β-carotene and canthaxanhin, α-carotene and lycopene were also effective in inhibiting MCA-induced malignant transformation.[19] In these studies the effective dose (0.3 μM) of carotenoids, including lycopene, were lower than those found in previous studies of cell proliferation (10 μM). In a recent review, Wolf[20] interpreted the low concentration needed for carotenoid action to mean that it acts by catalytic or genomic (receptor) mechanisms, rather than as an antioxidant which may require higher concentrations. For example, in a liposome system, the antioxidant effect of β-carotene occurred at about 20 μM.[21] Support for the genomic mechanism is found in recent reports by Bertram and colegues[19,22] which showed that carotenoids increase gap junctional communication between cells and induce the synthesis of connexin43, a component of the gap junction structure. This effect was independent of provitamin-A or the antioxidant properties of the carotenoids. Loss of gap junctional communication may be important for malignant transformation, and its restoration may reverse the malignant process.[23]

These studies were recently extended by Kennedy and Krinsky[24] who reported that natural, but not synthetic, β-carotene is effective in inhibiting X-ray and phorbol ester-induced transformation in several cell lines including C3H/10T1/2 cells. The explanation for this observation is not clear.

4.3.3 Intervention in Autocrine Growth Factors Action. The importance of the IGF autocrine system in the growth regulation of endometrial and mammary cancer has been established in recent reports.[7,8] The observation that lycopene is a potent inhibitor of IGF-induced growth in addition to basal growth may point to a mode of action of this carotenoid.[6] Our initial observation that lycopene impairs IGF-I-induced tyrosine phosphorylation indicates that the signal transduction mechanism of this and possibly other growth factors is modified. Because of the complexity of the IGF system, which includes two different growth factors that may interact with three receptors and six different binding proteins, the effect of lycopene on all these components should be studied. Such information may help in understanding the anticancer properties of lycopene.

5 CONCLUSION

Although lycopene has attracted less attention than β-carotene as a potential anti-cancer agent, it has now begun to gain increasing importance. In order to be established as an anti-carcinogenic agent in humans, the above mentioned studies should be expanded and this should promote intervention studies with lycopene in various populations.

6 REFERENCES

1. O. P. Heinonen, J. K. Huttunen, D. Albanes, *et al.*, Effect of vitamin E and beta carotene on the incidence of lung cancer and other cancers in male smokers. *New Engl. J. Med.* 1994, **330**, 1029.
2. E. R. Greenberg, J. A. Baron, T. D. Tosteson, *et al.*, Clinical trial of antioxidant vitamins to prevent colorectal adenoma. *New Engl. J. Med.* 1994, **331**, 141.
3. J. VanEenwyk, F. G. Davis and P. E. Bowen, Dietary and serum carotenoids and cervical intraepithelial neoplasia. *Int. J. Cancer* 1991, **48**, 34.
4. K. J. Helzlsouer, G. W. Comstock and J. S. Morris, Selenium, lycopene, alpha-tocopherol, beta-carotene, retinol, and subsequent bladder cancer. *Cancer Res* 1989, **49**, 6144.
5. P. G. Burney, G. W. Comstock and J. S. Morris, Serologic precursors of cancer: serum micronutrients and the subsequent risk of pancreatic cancer. *Am. J. Clin. Nutr.* 1989, **49**, 895.
6. J. Levy, E. Bosin, B. Feldman, Y. Giat, A. Miinster, M. Danilenko and Y. Sharoni, Lycopene is a more potent inhibitor of human cancer cell proliferation than either α-carotene or β-carotene. *Nutr. & Cancer* 1995, In Press.
7. D. Kleinman, C. T. Roberts Jr., D. LeRoith, A. V. Schally, J. Levy and Y. Sharoni, Growth regulation of endometrial cancer cells by Insulin-like Growth Factors and by the Luteinizing Hormone-Releasing Hormone antagonist SB-75. *Regulatory Peptides* 1993, **48**, 91.
8. E. Hershkovitz, M. Marbach, E. Bosin, *et al.*, Luteinizing Hormone-Releasing Hormone antagonists interfere with autocrine and paracrine growth stimulation of MCF-7 mammary cancer cells by Insulin-like Growth Factors. *J. Clin. Endo. Metab.* 1993, **77**, 963.

9. Y. Sharoni and J. Levy, Lycopene, the major tomato carotenoid, inhibits endometrial and lung cancer cell growth. in "Proceedings of the XVI international cancer congress" (eds. R. S. Rao, M. G. Deo and L. D. Sanghvi) Monduzzi Editore, Bologna, 1994, **1** 641.
10. C.-J. Wang and J.-K. Lin, Inhibitory effects of carotenoids and retinoids on the in vitro growth of rat C-6 glioma cells. *Proc. Natl. Sci. Counc. B. ROC* 1989, **13**, 176.
11. C. Countryman, D. Bankson, S. Collins, B. Man and W. Lin, Lycopene inhibits the growth of the HL-60 promyelocytic leukemia cell line. *Clin. Chem.* 1991, **37**, 1056.
12. C. J. Wang, M. Y. Chou and J. K. Lin, Inhibition of growth and development of the transplantable C-6 glioma cells inoculated in rats by retinoids and carotenoids. *Cancer Letters* 1989, **48**, 135.
13. J. Levy, Y. Liel, B. Feldman, L. Aflallo and S. M. Glick, Peroxidase activity in mammary tumors: Effect of tamoxifen. *Eur. J. Cancer. Clin. Oncol.* 1981, **17**, 1023.
14. Y. Sharoni, D. Radian and J. Levy, Membranal tyrosine protein kinase activity (but not cAMP dependent protein kinase activity) is associated with growth of rat mammary tumors. *FEBS Lett.* 1985, **189**, 133.
15. M. L. Johnson, J. Levy and J. M. Rosen, Isolation and characterization of casein producing and non-producing cell populations from 7,12 dimethyl- benz(a)anthracene-induced rat mammary carcinomas. *Cancer Res.* 1983, **43**, 2199.
16. P. Di Mascio, S. Kaiser and H. Sies, Lycopene as the most efficient biological carotenoid singlet oxygen quencher. *Arch. Biochem. Biophys.* 1989, **274**, 532.
17. F. Bohm, J. H. Tinkler and T. G. Truscott, Carotenoids protect against cell membrane damage by the nitrogen dioxide radical. *Nature Medicine* 1995, **1**, 98.
18. J. S. Bertram, A. Pung, M. Churley, T. d. Kappock, L. R. Wilkins and R. V. Cooney, Diverse carotenoids protect against chemically induced neoplastic transformation. *Carcinogenesis* 1991, **12**, 671.
19. L.-X. Zhang, R. V. Cooney and J. Bertram, Carotenoids enhance gap junctional communication and inhibit lipid peroxidation in C3H/10T1/2 cells: relationship to their cancer chemopreventive action. *Carcinogenesis* 1991, **12**, 2109.
20. G. Wolf, Retinoids and carotenoids as inhibitors of carcinogenesis and inducers of cell-cell communication. *Nutr. Rev.* 1992, **50**, 270.
21. T. A. Kennedy and D. C. Liebler, Peroxyl radical scavenging by beta-carotene in lipid bilayers - effect of oxygen partial pressure. *J. Biol. Chem.* 1992, **267**, 4658.
22. L. X. Zhang, R. V. Cooney and J. S. Bertram, Carotenoids up-regulate connexin-43 gene expression independent of their provitamin-A or antioxidant properties. *Cancer Res.* 1992, **52**, 5707.
23. A. Hotz-Wagenblatt and D. Shalloway, Gap junctional communication and neoplastic transformation. *Crit. Rev. Oncogenesis* 1993, **4**, 541.
24. A. R. Kennedy and N. I. Krinsky, Effects of retinoids, beta-carotene, and canthaxanthin on UV- and x-ray-induced transformation of C3H10T1/2 cells in vitro. *Nutr & Cancer* 1994, **22**, 219.

FLAVONOIDS AND EXTRACTS OF STRAWBERRY AND BLACK CURRANT ARE INHIBITORS OF THE CARCINOGEN-ACTIVATING ENZYME CYP1A1 IN VITRO

L. Kansanen[1,2], H. Mykkänen[2] and R. Törrönen[1,2]

[1]Department of Physiology and [2]Department of Clinical Nutrition
University of Kuopio
P.O. Box 1627, FIN-70211 Kuopio, Finland

1 INTRODUCTION

Flavonoids and phenolic acids are widely distributed in most vegetables, fruits and berries and thus commonly present in our foods. In plants flavonoids occur as glycosides which are hydrolyzed to aglycones in the intestine. However, the metabolism and bioavailibility of these phenolic compounds are still poorly known in man[1].

Some flavonoids and phenolic acids have been shown to have anticarcinogenic effects in vitro and in vivo[2]. One possible mechanism is the modification of the activities of enzymes responsible for the activation or detoxification of carcinogens. CYP1A1, one of the isozymes of cytochrome P450, has an important role in the metabolic activation of carcinogens (e.g. polycyclic aromatic hydrocarbons) in mammalian tissues[3].

The aim of this study was to investigate the effects of some model flavonoids and phenolic acids, as well as extracts of strawberry and black currant, on CYP1A1 in vitro. This work is part of a larger project on flavonoid and phenolic acid content of Finnish berries, their bioavailability and effects on carcinogen metabolism, conducted at the University of Kuopio.

2 MATERIALS AND METHODS

The model flavonoids studied were the flavonols quercetin and its glycoside rutin, myricetin and kaempferol, and the flavanone naringenin and its glycoside naringin. The phenolic acids studied were ellagic, gallic, ferulic, caffeic, chlorogenic and p-coumaric acids. Each model compound was dissolved in DMSO.

Freeze-dried strawberry and black currant (1 g without seeds) were extracted with 50 % aqueous methanol and hydrolyzed with 1.2 M HCl, as described by Häkkinen et al.[4] The solvents were evaporated and the residues corresponding to 200 mg (dry weight) of berries were dissolved in 2 ml of DMSO.

The effects of the model compounds and berry extracts on CYP1A1 were studied using the subclone Hepa-1c1c7[5] of the mouse hepatoma cell line Hepa-1. To increase the CYP1A1 level, the cells were pretreated with TCDD, a potent inducer of CYP1A1. The enzymatic activity of CYP1A1 was detected as AHH (aryl hydrocarbon hydroxylase, i.e. hydroxylation of benzo(a)pyrene to its fluorescent 3-hydroxy metabolite), as described earlier[6].

The model compounds (final concentration from 5 to 500 μM) or berry extracts (corresponding to 0.25 to 2.5 mg dry weight of berries) were added to the AHH reaction mixture with 200 μl of the cell suspension. The final concentration of DMSO did not exceed 2.5 %. DMSO alone served as a solvent control. AHH activities detected in the presence of the test compounds or berry extracts were compared to that of the control (AHH activity produced by the TCDD-induced cells in the absence of the test compounds or extracts).

3 RESULTS

The flavonoid aglycones produced a concentration dependent inhibition of AHH activity (Figure 1). The approximate IC50 values (inhibitor concentration which produces a 50 % inhibition) for quercetin, myricetin, kaempferol and naringenin were 50, 75, 25 and 25 µM, respectively. At the concentration range studied, the phenolic acids and flavonoid glycosides had no effect on AHH activity.

Also the extracts of strawberry and black currant were effective inhibitors of CYP1A1. The extracts corresponding to about 1 mg (dry weight) of berry caused a 50 % inhibition of AHH activity (Figure 2).

4 DISCUSSION

The results of this in vitro study suggest that flavonoid aglycones and extracts of strawberry and black currant are effective inhibitors of CYP1A1, whereas flavonoid glycosides and phenolic acids are not.

Flavonoids have been reported to possess structure-activity relationships as inhibitors of cytochrome P450 enzymes.[7-11] The present results agree with the previous studies[7,10] indicating that increasing the number of hydroxyl groups on the B ring from 1 (naringenin and kaempferol) to 2 (quercetin) or 3 (myricetin) decreases the inhibitory effect of flavonols. The IC50 value of myricetin was higher than that of quercetin or kaempferol in this study.

The flavanone naringenin with a saturated 2C-3C bond was as potent an inhibitor of CYP1A1 as kaempferol with a 2C-3C double bond. This is in contrast to the results of Chae et al. suggesting that naringenin is a less active inhibitor of benzo(a)pyrene metabolism than kaempferol in rats[8] and in hamster embryo cells[9]. Also Siess et al.[11] have reported that flavanones are less potent inhibitors of EROD activity in man and in the rat than the corresponding flavonols.

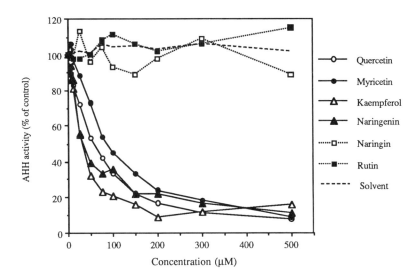

Figure 1 *Effects of flavonoids and their glycosides on CYP1A1-dependent AHH activity. The values are means of two independent experiments.*

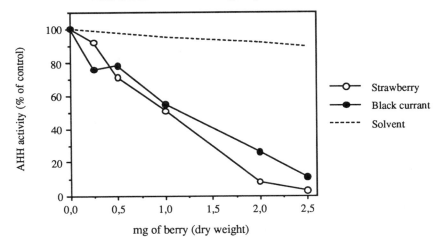

Figure 2 *Effects of extracts of strawberry and black currant on CYP1A1-dependent AHH activity. The values are means of two independent experiments.*

Similarly to the flavonoid aglycones, the extracts of strawberry and black currant inhibited the CYP1A1-dependent AHH activity. Although the qualitative phenolic profiles of these berries are different[4], their CYP1A1 inhibition properties seem rather similar. The flavonols (quercetin, myricetin and kaempferol) shown to be present in strawberry and black currant[4] may contribute to the inhibitory effect of these berry extracts.

These in vitro results suggest that flavonols as well as strawberry and black currant containing these flavonols are able to inhibit the activation of carcinogens by CYP1A1.

References

1. E. Middleton Jr and C. Kandaswami, 'The Flavonoids. Advances in Research since 1986', J. B. Harborne, Chapman & Hall, London, 1994, p. 619.
2. M. Strube, L. O. Dragsted and J. C. Larsen, 'Naturally Occuring Antitumourigens. I Plant Phenols', The Nordic Council of Ministers, 1993.
3. A. J. Paine, *Hum. Exp. Toxicol.*, 1995, **14**, 1.
4. S. Häkkinen, H. Mykkänen, S. Kärenlampi, M. Heinonen and R. Törrönen, in this issue.
5. O. Hankinson, *P. Natl Acad. Sci. USA*, 1979, **76**, 373.
6. P. Kopponen, E. Mannila and S. Kärenlampi, *Chemosphere*, 1992, **24**, 201.
7. M.-H. Siess, A. Pennec and E. Gaydou, *Eur. J. Drug Metab. Ph.*, 1989, **14**, 235.
8. Y.-H. Chae, C. B. Marcus, D. K. Ho, J. M. Cassady and W. M. Baird, *Cancer Lett.*, 1991, **60**, 15.
9. Y.-H. Chae, D. K. Ho, J. M. Cassady, V. M. Cook, C. B. Marcus and W. M. Baird, *Chem.-Biol. Interact.*, 1992, **82**, 181.
10. Y. Li, E. Wang, C. J. Patten, L. Chen and C. S. Yang, *Drug Metab. Dispos.*, 1994, **22**, 566.
11. M.-H. Siess, J. Leclerc, M.-C. Canivenc-Lavier, P. Rat and M. Suschetet, *Toxicol. Appl. Pharm.*, 1995, **130**, 73.

COUMARIN 7-HYDROXYLATION IS NOT ALTERED BY LONG-TERM ADHERENCE TO A STRICT UNCOOKED VEGAN DIET

A-L. Rauma[1], A. Rautio[2], O. Pelkonen[2], M. Pasanen[2], R. Törrönen[3] and H. Mykkänen[1]

[1]Department of Clinical Nutrition and [3]Department of Physiology, University of Kuopio, P.O. Box 1627, FIN-70211 Kuopio, Finland, and [2]Department of Pharmacology and Toxicology, University of Oulu, FIN-90220 Oulu, Finland

1 INTRODUCTION

Diet and various food processing methods, such as frying and grilling, are recognized among the factors modulating the biotransformation reactions. Essential macro- and micronutrients and nonnutrient components of diet have been shown to induce, inhibit and stimulate specific P450 isozymes.[1-6]

Many edible plants contain coumarins[7-9], bioactive substances that are currently of interest due to their suggested beneficial health effects.[10-11] The predominant pathway in man for the metabolism of coumarin is 7-hydroxylation by the P450 isozyme CYP2A6.[10,12] 7-Hydroxycoumarin (7-OHC) is excreted in the urine and its urinary excretion as a function of time has been shown to be a reliable estimate of the CYP2A6 activity in humans.[13]

By using the in vivo coumarin test developed by Rautio et al[13], we wanted to examine whether a long-term adherence to an uncooked strict vegan diet ("living food diet") will influence the activity of CYP2A6. The "living food diet" is composed of plant foods (vegetables, fruits, nuts, legumes) which are eaten raw or after fermentation or sprouting. Consequently, these vegans receive different amounts of essential macro- and micronutrients and nonnutrients as compared to omnivores. It was also hoped that the present study could clarify the large interindividual variation in coumarin hydroxylase activity observed in earlier studies on humans.[13,14]

2 SUBJECTS AND METHODS

The subjects were 20 female and one male long-term adherents of a strict uncooked vegan diet ("living food diet") who had consumed this diet for 5.2 years (range 0.7-14). Each vegan subject was matched for sex, age and social status with an omnivorous control subject. Mean values for age and height were similar in the groups, but the vegans had significantly lower body mass index. Five of the vegans had originally started this diet due to an illness (1 x multiple sclerosis, 1 x allergy, 3 x rheumatoid disease), and three control subjects suffered from allergy and one from lactose intolerance. At the time of the study, none of the participants used any regular medication. All participants were nonsmokers. Informed consent was obtained from each subject before the study began. The study was accepted by the Ethical Committee of Kuopio University and Central Hospital.

Food intake was estimated using 5-day dietary records, and nutrient intakes were calculated by the Nutrica computer program (Social Insurance Institution, Finland) using the Finnish nutrient database supplemented with the analytical data on nutrient content of vegetable foods commonly used by the Finnish vegans.[15]

"In vivo coumarin test" was performed on a weekend morning of the subject's own choice. The test kit included written instructions, a glass beaker, a graduated cylinder, 4 small plastic sample tubes, and a coumarin capsule (Venalot^R, Schaper & Brummer, Ringelheim, FRG) containing 5 mg coumarin and 25 mg rutosides. Two hours after ingestion of the coumarin capsule, the subjects were allowed to eat breakfast. All vegans were advised to have a similar breakfast, containing buckwheat porridge, 2 dl rejuvelac (a fermented wheat drink) and 2 dl carrot juice, and the controls were advised to avoid fruits, berries and other vegetables during the morning of the coumarin test. Urine was voided before the administration of coumarin and at 2, 4 and 6 h thereafter, and 7-OHC in the urine samples was determined as described by Rautio et al.[13]

The SPSS/PC computer program and the Wilcoxon test (paired comparisons of the food intake data), the Mann-Whitney test (coumarin index), and the Spearman rank correlation were used in the statistical analysis.

3 RESULTS

The vegans consumed more root crops, pulses and nuts, and fruits and berries, and less cereals than the omnivorous controls. Of the vegetables and fruits, the vegans consumed significantly more *Cruciferous* vegetables and significantly less citrus fruits than the omnivores. These differences in dietary patterns resulted in differences in intakes of essential nutrients. The vegans received significantly less protein, vitamin D, riboflavin, niacin, vitamin B_{12}, calcium, iodine, and selenium, and significantly more fiber, beta-carotene, vitamin E, vitamin C, thiamin, magnesium and iron than the omnivorous controls. As expected, the 24-h urinary excretion of sodium by the vegans was significantly lower than that of the controls, indicating a good compliance to the no-salt vegan diet.

The baseline urine samples had no detectable amounts of 7-OHC, and the total excretion of 7-OHC after administration of 5 mg coumarin varied considerably. The total urinary excretion of 7-OHC in the vegan group was 2.91 mg (SD 0.86) and in the control group 3.20 mg (SD 0.80). The percentage excretion ranged from 23 to 85 and from 39 to 92 in the vegan and control groups, respectively (Figure 1). No significant differences between the groups were observed. Similarly, there were no significant differences in the rate of excretion of 7-OHC as indicated by the coumarin index (7-OHC excretion during the first 2 h expressed as percentage of the total excretion), and no significant correlations of the coumarin index with intakes of foods were found, other than that of the intake of wheatgrass juice by the vegans ($r=-0.60$, $p<0.01$).

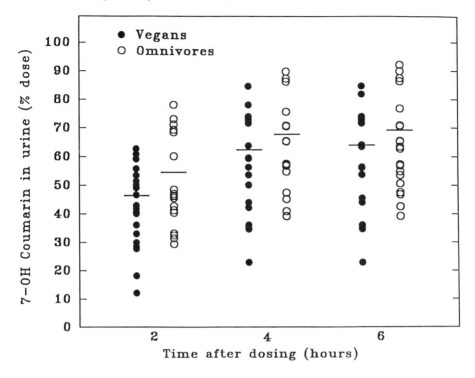

Figure 1 *Urinary excretion of 7-hydroxycoumarin as a function of time in the long-term adherents of a strict uncooked vegan diet ("living food diet") and in the omnivorous controls after oral administration of 5 mg coumarin*

4 DISCUSSION

Despite the very different dietary patterns the rate of coumarin biotransformation did not differ between the vegan and control subjects of the present study. The consumption of many foods (especially *Cruciferous* vegetables, nuts and pulses), and consequently, the intakes of essential nutrients known to influence the activities of P450 enzymes were significantly different.[6] The fact that the foods in the "living food" regimen are eaten uncooked lessens the intakes of foreign compounds (PAH and heterocyclic amines in cooked foods) known to influence the activities of P450 enzymes, while the intakes of other substances (e.g. phenolic compounds and hydrolyzation products of glucosinolates) may be increased by the traditional food preparation methods, such as sprouting and fermentation.[4,16]

The rate of urinary excretion of 7-OHC in the present study was in accordance with that obtained in earlier studies on Finnish and Turkish volunteers.[13,14] The similarities of the mean levels and of the interindividual variability in the extent and rate of 7-OHC excretion in the vegan and control groups indicate that the dietary habits may not account for the large variation observed in earlier studies[13,14] in coumarin hydroxylase activity.

The present study shows that long-term adherence to the "living food diet", a strict vegan diet, does not significantly influence the rate and extent of coumarin biotransformation. Since the "living food diet" consumed by the vegans in the present study differs from the typical Western omnivorous diet in many factors affecting biotransformation processes (i.e. food choice, food preparation methods, nutrient supply), its effects on the other P450 isozymes need to be investigated.

References

1. W. R. Bidlack, R. C. Brown and C. Mohan, Fed. Proc., 1986, **45**, 142.
2. J. N. Hathcock, Pharmacol. Therapeut., 1988, **38**, 53.
3. C. S. Yang and J.-S. H. Yoo, Pharmacol. Therapeut., 1988, **38**, 53.
4. K. E. Anderson and A. Kappas, Annu. Rev. Nutr., 1991, **11**, 141.
5. C. S. Yang, J. F. Brady and J.-Y. Hong, FASEB J., 1992, **6**, 737.
6. F. P. Guengerich, Am. J. Clin. Nutr., 1995, **61**, 651S.
7. C. E. Searle, Chemical Carcinogens. 722 p. Monograph 173. American Chemical Society, Washington DC, 1976.
8. G. W. Goodwin and E. I. Mercer, Introduction to Plant Chemistry p.534-536. Pergamon Press, Oxford, 1983.
9. M. Tinel, J. Belghiti, V. Descatoire, G. Amouyal, P. Letteron, J. Geneve, D. Larrey and D. Pessayre, Biochem. Pharmacol., 1987, **36**, 951.
10. D. Egan, R. O'Kennedy, E. Moran, D. Cox, E. Prosser and R. D. Thornes, Drug Metab. Rev., 1990, **22**, 503.
11. K. A. Steinmetz and J. D. Potter, Cancer Caus. Control, 1991, **2**, 427.
12. A. J. Cohen, Food Cosmet. Toxicol., 1979, **17**, 277.
13. A. Rautio, H. Kraul, A. Kojo, E. Salmela and O. Pelkonen, Pharmacogenetics, 1992, **2**, 227.
14. M. Iscan, H. Rostami, M. Iscan, T. Guray, O. Pelkonen and A. Rautio, Eur. J. Clin. Pharmacol., 1994, **47**, 315.
15. M. Rastas, R. Seppänen, L. R. Knuts, R.-L. Karvetti and P. Varo, Nutrient composition of foods. The Social Insurance Institution, Helsinki, 1993.
16. K. E. Anderson, E. J. Pantuck, A. M. Conney and A. Kappas, Fed. Proc., 1985, **44**, 130.

EFFECT OF CHLOROPHYLLIN AND OTHER NATURAL ANTIOXIDANTS UPON THE RESPIRATORY BURST OF HUMAN PHAGOCYTES FROM SMOKERS AND NON-SMOKERS.

L. Benítez and M. A. Wens

Oncological Research Unit, Oncology Hospital,
National Medical Center, IMSS.
Apartado Postal 107-014, 06700 Mexico, D.F. Mexico.

1 INTRODUCTION

Considerable information has been rapidly accumulating indicating the prevalent role of micronutrients, such as vitamins, and other natural compounds, as antioxidants, hormone regulators and with other noncoenzymatic functions[1]. Much of this evidence comes from to epidemiological studies showing consistently an inverse relationship between the risk of some types of cancer and consumption of fresh vegetables and fruits[2,3]. Although the antioxidant and antimutagenic effects of a number of vitamins (i.e., E, C, D and A) have been clearly determined, and its possible use as chemopreventive agents has reached clinical trials[4,5], very little information exists on the ubiquitous molecule of green plants, chlorophyll.

The chlorophyll-related chromophore chlorophyllin has been used for treatment of some minor human illnesses with no reported toxic effects[6-8]. In addition a number of reports have appeared demonstrating the antimutagenic activity of chlorophyllin in various biological models, showing that this molecule prevents carcinogen-DNA binding in vivo and complex formation with mutagens and procarcinogens[9]. It seems possible that part of the antimutagenic activity of this compound is exerted by its capacity to quench reactive oxygen species (ROS), but very little information exists in this regard.

The generation of ROS in stimulated phagocytes known as "respiratory burst" (RB) has been used as a suitable in vivo model to study the antioxidant properties of several molecules, mainly different vitamins. It has been demonstrated that smokers have a higher RB in alveolar macrophages and circulating phagocytes and that oral administration of vitamin E, L-selenium methionine ascorbic acid, retinoic acids and beta carotenes in smokers decreases significantly the RB in PMNs[10-13]. Therefore, we used this model to assess the antioxidant capacity of chlorophyllin in human circulating polymorphonuclear leukocytes (PMNL). The general objective of this experiments was to conduct a comparative study of the antioxidant activity of chlorophyllin with two well known natural antioxidants, ß-carotene and ascorbic acid in stimulated PMNL of smokers and non-smokers in intact human phagocytes.

2 MATERIAL AND METHODS

Blood was obtained from 12 asymptomatic cigarette smokers and 12 age-matched control subjects. Smokers had smoked at least 1 pack of cigarettes per day for 2 to 10 years. A modified microtechnique for studying chemiluminescence response in circulating PMNLs was used[14]. Briefly PMNs were separated from heparinized blood samples in one step using a density gradient and centrifuged at 400 g for 35 minutes (Polimorphprep^{TM}, Nycomed Pharma). Cell concentration was adjusted to 3×10^6 PMNs/ml and incubated for 20 minutes with increasing antioxidant concentrations. Beta carotene and chlorophyllin were added from 5 to 100 μmol and ascorbic acid from 250 to 2500 mmol final concentration (Chlorophyllin, ascorbic acid and ß-carotene obtained from Sigma Chemical Co). As phagocytic particle, opsonized Zymosan was used (Zymosan from Sigma Chemical Co). Luminol was added at a concentration of 10^{-4}M and measurements were made in a luminometer (BioOrbit 1251). The peak of the curve measured in mV was used as the value of maximum chemiluminescence.

3 RESULTS

Chemiluminescence of stimulated PMNLs from smokers showed a consistently higher value than that obtained in non smokers, thus indicating a higher production of ROS in the former group as compared to the latter ($p < 0.01$).
In both groups, the inhibitory effect of the antioxidant molecules used was similar and all showed a typical dose-response curve. Chlorophyllin however, was the most active since at 10 μM concentration there was almost a 30% inhibition of the RB and at 50 μM the RB was inhibited to approximately 80% (Figure 1-a). ß-carotene showed a similar dose-response curve at the same concentrations (Figure 1-b), while ascorbic acid was much less active, needing concentrations one thousand times higher at the mM range (Figure 1-c).

4 DISCUSSION

A number of studies have demonstrated the antimutagenic effect of chlorophyllin in different short-term tests such as the Ames test, micronucleus and micro-screen phage induction[15-17]. Other investigations have shown its anticarcinogenic and chemopreventive action in experimental carcinogenesis[9,18]. It has also been demonstrated that chlorophyllin is a potent inhibitor of mutagenicity of tobacco products[19].
The mechanisms by which chlorophyllin exerts its chemopreventive and antimutagenic activity is still unclear, but there are indications that it forms complexes with carcinogenic and mutagenic compounds and that it might involve accelerated degradation of the ultimate carcinogen[20-22]. Another possible mechanism to explain the antimutagenic anticarcinogenic action is the inhibition of free radicals, mainly those derived from oxygen.
A variety of natural antimutagenic compounds, such as ß-carotene, retinols, and Vitamin C act as potent antioxidants and thus prevent the oxidative damage to different cell components, particularly DNA. It is generally agreed that this is the main mechanism of chemopreventive action from these compounds in fixed and

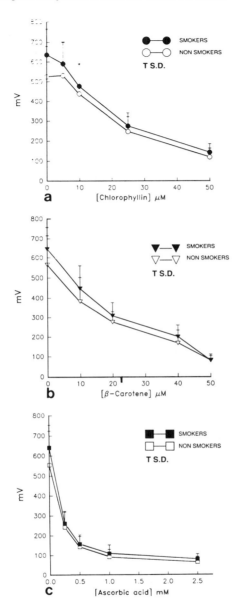

Figure 1 *Dose-response curves of chemiluminescence inhibition (measured in mV) of PMNL respiratory burst by three different molecules at increasing concentrations* **a)** *chlorophyllin;* **b)** *ß-carotene;* **c)** *ascorbic acid.*

circulating phagocytes of smokers[23-25]. These cells are capable of generating a variety of oxygen free radicals, mainly oxygen superoxide upon stimulation, and this mechanism in turn can produce the endogenous damage of DNA and other cell components involved in the pathogenesis of a number of diseases. Our results demonstrate that chlorophyllin has a potent antioxidant activity in stimulated human PMNL both from smokers and non smokers. This activity is comparable and even better than that of ß-carotene, a well known oxygen quencher, and much better than ascorbic acid, also a well studied antioxidant in plasma. The dose response curve obtained in our experiments demonstrates its pharmacological action at very low concentrations in the micromolar range.

In view of these findings it is reasonable to assume that chlorophyllin anticarcinogenic properties are due in part to its antioxidant activity. Studies in our laboratory to be published elsewhere show that this compound inhibits also the powerful oxidants present in cigarette smoke[26,27] that are responsible for the production of pulmonary emphysema and lung cancer in smokers. Since chlorophyll is found in most vegetables containing other natural antioxidants a synergistic action of these compounds might be possible.

Other studies would be needed to see if this compound is present in diets rich in green vegetables known to be protective for the development of some cancers.

References

1. H. Sies, W. Stahl and A. R. Sundquist, Ann N Y Acad Sci., 1992, **669**, 7.
2. R. G. Ziegler and A. F. Subar, . In: A. Bendich and C. E. Butterworth, Jr., "Micronutrients in Health and in Disease Prevention", Marcel Dekker, Inc., New York, N Y, 1991, Chapter 5, p. 97.
3. G. Block, Ann N Y Acad Sci., 1992, **669**, 280.
4. D. B. Menzel, Ann N Y Acad Sci, 1992, **669**, 141.
5. H. Garewal and G. J. Shamdas. In: A. Bendich and C. E. Butterworth, Jr., "Micronutrients in Health and in Disease Prevention", Marcel Dekker, Inc., New York, N Y, 1991, Chapter 6, p. 127.
6. R. W. Young and J. S. Bergei, Use of chlorophyllin in the care of geriatric patients. J. Am. Geriat. Soc., 1980, **28**, 46.
7. R. Tawashi, M. Cousineau and G. Denis, Urol. Res., 1982, **10**, 173.
8. N. A. Krasnikova, Byul. Eksp. Biol. Med., 1973, **76**, 99.
9. R. H. Dashwood, Carcinogenesis. 1992, **13**, 113.
10. J. Clausen, Biol. Trace Element Res., 1991, **31**, 281.
11. P. W. Ludwig, B. A. Schwartz, J. R. Hoidal and D. E. Niewoehner, Am. Rev. Respir. Dis., 1985, **131**, 828.
12. J. R. Hoidal, R. B. Fox, P. A. LeMarbe, R. Perri and J. E. Repine, Am. Rev. Respir. Dis., 1981, **123,** 85.
13. R. Anderson, A. J. Theron and G. J. Ras, Am. Rev. Respir. Dis., 1987, **135**, 1027.
14. R. J. Selvaraj, A. J. Sbarra, G. B. Thomas, C. L. Cetrulo and G. W. Mitchell, Jr., J. Reticuloendothel. Soc., 1982, **31**, 3.

15. K. K. Park, Y. J. Surh, B. C. Stewart and J. A. Miller, Proc Annu Meet. Am. Assoc. Cancer Res., 1994, **35,** A828.
16. K. Pupatwibul, Diss. Abstr. Int., 1993, **53**, 3285.
17. L. Romert, M. Curvall, D. Jenssen, Mutagenesis, 1992, **7**, 349.
18. Z. L. Wu, J. K. Chen, T. Ong, H. E. Brockman and W. Z. Whong, Teratog. Carcinog. Mutagen., 1994, **14**, 75.
19. L. Romert, T. Jansson, M. Curvall and D. Jenssen, Mutat. Res., 1994, **322**, 97.
20. S. Arimoto, S. Fukuoka, C. Itome, H. Nakano, H. Rai and H. Hayatsu, Mutat. Res., 1993, **287**, 293.
21. N. Tachino, D. Guo, W. M. Dashwood, S. Yamane, R. Larsen and R. Dashwood. Mutat. Res., 1994, **308**, 191.
22. D. Guo and R. Dashwood, Proc. Annu. Meet. Am. Assoc. Cancer Res., 1993, **34**, A737.
23. P. W. Ludwig and J. R. Hoidal, Am. Rev. Respir. Dis., 1982, **126**, 977.
24. T. Nakayama, M. Kaneko, M. Kodama and Ch. Nagata, Nature, 1985, **314**, 462.
25. A. Theron and R. Anderson, Am. Rev. Respir. Dis., 1985, **132**, 1049.
26. W. A. Pryor, B. J. Hales, P. I. Premovic and D. F. Church, Science, 1983, **220**, 425.
27. D. F. Church and W. A. Pryor, Environ. Health Perspect., 1985, **64**, 111.

Contaminants in Foods and Diets

GEMS/FOOD-EURO: A WHO/FAO/UNEP APPROACH TOWARDS BETTER QUALITY FOODS IN EUROPE

Peter Weigert

World Health Organization
Regional Office for Europe
European Centre for Environment and Health
Via Vincenzo Bona, 67
I-00156 Rome, Italy

1 INTRODUCTION

National authorities have the responsibility and obligation to ensure the supply of wholesome food for their population; this includes assuring that potentially toxic chemicals are not present in food and drinking-water at levels that may adversely affect the health of consumers, as well as food trade and food safety. The monitoring and control of food is not only essential for consumer protection but, at the same time, can be used to assess public health risks arising from the presence of potentially toxic chemicals in food.

To assist Member States of the WHO European Region to develop, strengthen and improve programmes for the effective and efficient monitoring and assessment of dietary exposure to potentially toxic chemicals, the WHO Regional Office for Europe of the World Health Organization organised consultations to recommend the most appropriate approach in the development of a European programme. Following this advice, the Regional Office established the WHO European Programme on Monitoring and Assessment of Dietary Exposure to Potentially Hazardous Substances (GEMS/Food-EURO) in November 1992 through its European Centre for Environment and Health, Rome Division. GEMS/Food-EURO is the special European part of the former global Joint UNEP/FAO/WHO Food Contamination Monitoring Programme, better known as GEMS/Food; GEMS stands for GLOBAL ENVIRONMENTAL MONITORING SYSTEM. This programme contains components for water and air in addition to food.

2 ORGANIZATION AND STRUCTURE OF GEMS/FOOD-EURO

The GEMS/Food-EURO programme has been established as a WHO-network in cooperation with all Member States of the WHO European region through their nominated National Contact Points. The overall responsibility for the management of the GEMS/Food-EURO programme lies with the Food Safety unit of the European Centre for Environment and Health, Rome Division, of the Regional Office for Europe of the World Health Organization. The first meeting of the National Contact Points, 18 - 20 November 1992 in Berlin, Germany, endorsed the establishment of a Steering Committee for the periodic review of the progress of the programme. A Coordinating Centre was established to assist WHO/EURO in coordination of component activities, in particular with regard to technical cooperation. Technical Sub-Committees for Analytical Quality Assurance, for data management, for assessment and evaluation and for veterinary drug residues were set up to plan, manage and evaluate activities of selected components.

3 SPECTRUM OF CONTAMINANTS

The GEMS/Food-EURO programme reflects, furthermore, the specific priorities and needs of the European Region. Therefore, the programme collates data not only on contaminants such as heavy metals, pesticides, industrial chemicals and mycotoxins, as the global programme does, but also on nitrates and nitrites, radionuclides and veterinary drug residues which have or may have importance for the European Member States. In order not to overload the programme, microbiological agents and bacterial toxins etc. in food are excepted.

4 PROGRAMME COMPONENTS

The GEMS/Food-EURO programme comprises the following components:
- activities related to sampling plans and sampling procedures
- activities related to sample preparation and methods of analyses
- Analytical Quality Assurance
- data management (data collection, data handling)
- assessment and evaluation
- reporting and publication
- technical cooperation

A look at this list shows that all such components are needed if reliable and comparable data and information are to be obtained - and only data of such quality are worth being dealt with in an international environment, particularly in the view of the potential consequences for consumers health and for trade which could arise if erroneous data are used for legislation or similar activities.

5 REPRESENTATIVITY OF DATA

The components related to sampling and sample handling try to assure the representativity of the information given for a certain country or region. Other components contribute to this aim as well, particularly those activities related to the assurance of analytical quality.

6 ANALYTICAL QUALITY ASSURANCE

From the very beginning of the GEMS/Food-EURO programme Analytical Quality Assurance (AQA) has been one of the most important components and one which up to now also best demonstrates the success of the programme: Thanks to the very active GEMS/Food-EURO - Technical Subcommittee on Analytical Quality Assurance in the first 12 months (1993/1994) five collaborative studies, or better to say proficiency tests, were carried out comprising heavy metals, pesticides, aflatoxins, patulin and nitrates. The materials for the trials were donated by Member States. The "ISO/IUPAC/AOAC International harmonised protocol for proficiency testing of analytical laboratories" was adopted for use in GEMS/Food-EURO. In those five testing rounds 139 laboratories from 23 European countries participated. The results of those five proficiency tests were partially satisfactory, but partially quite unsatisfactory. The latter is especially true for the proficiency test for nitrates; perhaps the reason for the unsatisfactory results was that it was the first proficiency test ever carried out in such a huge international programme.
    Bearing all this in mind, it was therefore decided not to rely only on discussion within the participating laboratories, but to organise a workshop to identify problem areas, to discuss weak and/or critical points within the whole analytical procedure and, if possible, to provide assistance to such laboratories in need of it. The aim was to reach an overall improvement

through the workshop. Furthermore, the workshop was expected to contribute to a sound basis for further actions in AQA measures in Europe. In the workshop but also through other means the participating institutions were made aware of the importance of Analytical Quality Assurance with its components, i.e. the internal, the external quality assurance and Good Laboratory Praxis (GLP).

We feel that this approach is one of the best and most cost-effective ways to improve the analytical quality of data produced in Europe; without direct follow-up through discussions of the problems no real improvement can be achieved as past experience in nearly all programmes demonstrates. The improvement of analytical quality not only contributes to a more reliable data base for GEMS/Food-EURO but is particularly important for the overall improvement of the situation in the Member States and therefore has a much broader impact.

At present additional tests are underway for radionuclides, nitrates and pesticides; we hope that this part of the programme will also continue in the future, although we face major financial problems at present; therefore we have to look for contributions from Member States to assist the programme in the future. I would like to invite everybody to join the programme as we on our part try to bridge with other ongoing international programmes, for example with the FAO European Cooperative Research Network on Trace Elements, Natural Antioxidants And Contaminants.

7 DATA MANAGEMENT

The improvement of data management also contributes to the representativity of the collated data on dietary studies or analytical results. To assist at a practical level, GEMS/Food-EURO provided all interested National Contact Points free of charge with a computer package which can run on any Personal Computer for the harmonised data transfer to the Coordinating Centre; some additional computer tools such as coding lists for foods and contaminants, data transmission tools and so on were also provided by the programme to the participating institutions. This programme package improves the data quality remarkably by avoiding many transmission errors through various plausibility checks during the data entry process, but also by ensuring that all necessary information is submitted which is needed for a meaningful interpretation and evaluation of the analytical results.

8 EVALUATION OF EXPOSURE

Reliable evaluation of analytical results of measurements of substances in foods and diets is a prerequisite for the realistic estimation of food contamination and exposure evaluation. Within GEMS/Food-EURO measures are undertaken to assure the reliability and comparability of data handled in this programme and to provide decision-makers in the Member States and the scientific community with sound information and a reliable evaluation of these data.

I would like to present another initiative of GEMS/Food-EURO which will help to achieve this aim: During the collation of contamination data on various food items and diets, there are generally some analytical results reported as non-detectable (ND) or non-quantifiable (NQ). Such qualitative results may be combined with quantified (numerical) data usually expressed as statistics, such as median or 90th percentile, to assess the exposure of populations to potentially hazardous substances.

Manipulation of data sets by various methods can heavily influence the result of such a calculation, especially if a large number of results are ND or NQ. To demonstrate the effects, some hypothetical calculations for four pesticides using a theoretical diet were performed. Table 1 gives an overview of how calculation methods can influence the results. With regard to the specific Acceptable Daily Intake (ADI), enormous but unrealistically high percentages of the ADI can be obtained, in spite of the fact that most analytical values were ND.

In order to find a satisfactory solution for this problem we organised two international workshops in 1994 and 1995 to discuss current methods for evaluating data on low level contamination of food. In particular the workshops addressed the problem of combining quantified and non-quantified analytical data. The workshops reached a number of conclusions that are going to be published in a few weeks time to propose procedures on how to calculate medians or means under various conditions. We are quite sure that this activity will improve the evaluation of exposure data remarkably and provide, from the scientific point of view, a much better basis for further actions if any are needed.

Table 1 *Influence of Calculation Methods on Exposure Description*

Contaminant	Percentage of ADI if		
	ND = 0	ND = 50 % LOD	ND = LOD
Aldrin + Dieldrin	39.2	78.5	157.1
Chlordane	9.1	18.3	36.6
Endrin	28.5	57.1	114.2
Heptachlor	22.1	66.4	132.8

ADI = Acceptable Daily Intake: mg/kg Body Weight per day

9 CONCLUSION

GEMS/Food-EURO was established as a cooperative effort of the National Contact Points of 26 Member States of the WHO European region in order to improve the overall health situation in Europe. Its components contribute not only to its own growth and value but even more to an improvement of the food safety situation in the participating Member States. The programme assists in minimising the dietary exposure of Europeans to potentially hazardous substances; I therefore think everybody will agree with me when I make the statement that better quality of food in Europe will be achieved by minimising contaminants and residues and, this is exactly the overall aim of GEMS/Food-EURO.

RISKS OF DETRIMENTAL HEALTH EFFECTS BY DIETARY INTAKES OF HEAVY METALS IN VARIOUS POPULATIONS.

G. F. Nordberg

Department of Environmental Medicine
Umeå University, S-90187 Umeå. Sweden

1 INTRODUCTION

Human dietary intake of toxic metals and their compounds occur as a result of natural occurrence and/or their use in various industrial applications. Since several of the metals are essential to human and animal nutrition, a basic level of intake of some metallic compounds is required in order to maintain health. When there is excessive exposure, various adverse effects may occur, ranging from clinical manifestations of unambiguous poisoning or death to chronic effects of low level exposures sometimes of a subtle nature, but in other situations, e.g. when cancer has been induced, of serious concern for health and survival. Since the industrial use of metals is constantly expanding and recycling occurs only to a limited extent at present, exposures of concern to health may affect a growing number of individuals in the human population. It is thus of considerable importance to be able to assess the risks to human health that may be associated with such exposures in order to prevent adverse influence on health. Basic principles for risk assessment were presented a long time ago by the ICOH Scientific Committee on the Toxicology of Metals [1,2]. Considerations relating to the carcinogenicity [3], reproductive toxicity [4] and immunotoxicity [5] of metals and their compounds as well as detailed considerations on specific metals have later been dealt with during workshops organized by the SCTM Arsenic: [6], Cadmium: [7,8], Mercury: [9] and these evaluations in combination with those made within WHO, particularly by IARC and in the International Programme on Chemical Safety form the basis for the present presentation.

2 RISK ASSESSMENT BASED ON DOSE-RESPONSE AND CRITICAL EFFECT CONCEPTS

While excessive exposures give rise to severe effects and even death of the individual, the effects that are of particular concern in relation to preventive measures in the public health area, are those effects that occur at relatively low exposures. Such effects have been termed Critical effects [1] since these effects and their dose-response relationships can be regarded as a critical piece of information when discussion preventive measures such as e.g. exposure limits. The concept of critically in this context is primarily an operational definition of importance in public health. It may be extended however, to an organ and cellular level and thus permit evidence from experimental toxicology research to be used in a systematic way to provide information of relevance for risk assessment of importance for public health. The selection of the critical effect among the various effects that may be elicited in humans by exposure to metallic compounds thus is a very central part in the risk assessments based on these concepts.

A spectrum of dose response curves can be assumed to exist for human exposure to a metallic compound. Effects of increasing severity occur when the dose or exposure is increased. At high doses severe effects occur such as life threatening clinical disease and death. The LD50 dose can be derived from this curve. At lower doses, effects representing mild clinical diseases, for example certain types of anaemia may occur. Obviously, anaemla is a clinical effect which should be prevented in environmentally exposed populations. If a mild clinical effect is chosen as a criterion for standard setting, it is evident that only a very small proportion of an exposed group can be accepted to have such an effect as a result of exposure. It is difficult to obtain reliable information on the dose response curve at its lower end, because of statistical difficulties. Nevertheless, for effects of several metals, clinical effects must be used as the critical effect in the absence of more detailed information. In some instances, however, precursors (or biomarkers) of clinical effects, in this example a biochemical change in blood indicating dysfunction of blood formation may be classified as the critical effect. Since the indicator effect occurs at lower doses it is sufficient that it is prevented at a higher prevalence in the population in order for the clinical effect to be prevented. When critical effects of the kind that implies a functional impairment of cells and tissues of a subclinical nature can be identified, they often are particularly useful as a basis for preventive action. These considerations are valid for threshold type effects (deterministic effects) i.e. effects with S-shaped dose response curves and the definitions of critical effects in this case can be carried to an organ an cellular level. The critical organ being the organ in which the critical effect appears. The critical concentration in that organ is the lowest concentration that triggers the effect, as discussed in detail by the Task Group on Metal Toxicity [1], the Task Group on Metal Interactions [2] and Nordberg and Strangert [10]. For effects of the kind discussed, i.e. those with S-shaped dose-response curves, (i.e. deterministic effects), a threshold can be identified in the affected organ (critical organ) and also when considering the total dose taken up by a human being. There is also a practical threshold on the population level. A critical concentration of the metal can be identified at a cellular and organ level both in individuals and in populations.

While dose response curves of the deterministic, or S-shaped type are often found, it is considered that for some effects, so called stochastic effects, dose response curves have a different shape. The shape of dose response curves is frequently the subject of considerable discussion and controversy since, it is extremely difficult in epidemiological studies to determine a low dose response.

For toxic effects which are of a non threshold (or stochastic) nature, the dose-response curve has a linear component that extends all the way down to zero dose. An example of such a non threshold or stochastic effect is the induction of cancer, which is considered to be dependent on a process of random interaction between the chemical compound and DNA. It has however been increasingly recognized lately, that the linear extrapolation down to zero dose is not valid in many instances, since DNA-repair functions and other protective biochemical mechanisms that exist in most cells can eliminate low level effects on DNA at low doses. Risks therefore usually are lower than what is predicted by the linear extrapolation [11]. In the interest of safety it is, however, still considered justified to assume a linear component of the dose-response curve extending all the way down to zero dose for stochastic effects such as cancer. The distinction between damage mechanisms of the threshold (deterministic) and non-threshold (stochastic) type has been discussed by Nordberg and Strangert 1985 [12].

For a stochastic effect with a dose-response curve that has a linear component in the low dose range, this response will be dominating over deterministic effects in this low dose range. The serious nature of cancer means that this effect must be prevented at a very low response level, since even a low incidence of such a disease will be of concern. For this reason, it is likely that for compounds that have been identified as carcinogenic, the carcinogenic effect will be the critical effect which dominates the risk assessment at low doses.

From the considerations given, it is obvious, that a first step in an assessment of risks of detrimental health effects from exposure to metals should include considerations of what effect could occur as a result of exposure and in particular to try to identify the critical effect. Subsequently, a quantitative estimation of dose response relationships for the critical effect is of crucial importance.

As pointed out in the foregoing text, the shape of the dose-response curve for the critical effect is of crucial importance. This shape of the dose-response curve is a result of variability in deposition of the metallic compound in the critical organ on the one hand and variability in sensitivity of that organ to damage, on the other hand, cf [10,13]. As discussed, the nature of toxicity is important as is also the quality of the overall database used for assessment. In the following, two examples i.e. oral exposure to cadmium and methylmercury respectively, via the general environment, will be briefly reviewed in order to illustrate these points.

3 RISK ASSESSMENT- DETERMINISTIC EFFECTS

In the following text two examples of risk assessments for threshold type or deterministic effects will be described, i.e. those pertaining to exposures to cadmium and methylmercury. For both of these substances there is widespread human exposure in many countries.

3.1 Cadmium

The involvement of cadmium as an etiological factor for the development of the so called itai-itai- disease is now well established [14]. This disease appeared after the second world war in Toyama prefecture in Japan. Bone disease (osteomalacia and osteoporosis) in combination with renal disease were characteristic features. It was recognized already in the 1960s by the Japanese Government that cadmium in rice was an important causative agent. While bone disease developed only in persons with long term excessive exposure to cadmium and who belonged to segments of the population with particular demand for calcium, like women who had given birth to many children, renal disease, mainly of a tubular type, occurred also in other segments of the population and in persons with less exposure to cadmium. It could thus be concluded that the early effects of cadmium in this type of exposure were those occurring in the kidney. This also agreed with previous experience from the occupational setting, where renal dysfunction has been shown to be the earliest and thus the critical effect of cadmium on exposed workers [15,16]. Cadmium has also been shown to cause cancer in experimental animals and in humans after exposure to airborne cadmium in the occupational setting [8,17] and it is presently discussed to what extent the carcinogenic effect should be regarded as a critical effect when exposures are via the oral route.

3.2 Methylmercury

Although there are considerable species differences in the disposition and metabolism of methylmercury, there is widespread distribution of this compound in the mammalian organism, with considerable uptake in sensitive tissues, such as the brain and the fetus.

During the epidemic of methylmercury poisoning in Iraq in 1972, [18] it was possible to define the distribution of biological half-times in the human population. It was demonstrated that the mean biological half-time is approximately 64 days and there is a subpopulation with a longer half-time, between 110-120 days [19].

During the Iraqi epidemic, it was also possible to define the distribution of sensitivity among individuals by relating observed clinical symptoms with measurements of hair and blood mercury levels [20]. At an estimated body burden of about 300 mg, some lethal cases of poisoning occur. At lower body burdens, effects such as deafness, dysarthria, ataxia and parestesia occur. The most sensitive effect in adult persons is paresthesia. No biochemical precursor for this effect is known and paresthesia as a manifestation of neurotoxicity of methylmercury must therefore be taken as the critical effect for adults based on this set of information.

The data collected during the kaqi epidemic of methylmercury poisoning was obtained during exposure periods of weeks, but can be used to calculate long term intakes giving rise to various levels of response [10]. By combining the information about distribution of biological half-life and the distribution in sensitivity, the dose response curve in terms of a relationship between long-term exposure to methylmercury and occurrence of effects, particularly the critical effect, can be calculated. The statistical uncertainty in the risk estimate can also be estimated by the use of mathematical models based on different assumptions. Long term intake (for a 50 kg individual) of 0.1 mg per day (corresponding to a hair level of approximately 30 µg/g) would correspond to a risk level of 0.5% or 2.5% depending on mathematical model chosen for extrapolation. This is the situation for adults and the data is shown in order to demonstrate how data on sensitivity and variation in metabolic parameters can be used in order to calculate dose-response relationship for the critical effect and the corresponding statistical uncertainty of the estimated response level.

It is well known that, on a population level, the critical effect for methylmercury is not paresthesia in adults, but rather neurotoxicity during the development of the fetus, exposed via the mother during pregnancy. Information has been gathered from the exposures in Iraq [18, 21] on the risk for development of mental retardation in children in relation to the concentration of methylmercury in their mother's hair during pregnancy.

Different bases for interpreting the data in terms of drawing dose-response curves and defining uncertainty intervals have been demonstrated in the WHO report. Increasing risks of CNS-signs appear when the maternal hair concentration increases above 10 mg/kg (corresponding approximately to a daily intake of 0.03 mg in a mother of 50 kg body weight). It is obvious that the neurotoxic effects in the fetus, causing developmental impairment, should be considered as the critical effect of methylmercury in humans when whole populations including pregnant females are exposed, since these effects occur at considerably lower exposure levels than those giving rise to effects in adults.

4 REPRODUCTIVE AND DEVELOPMENTAL TOXICITY

Data on reproductive and developmental toxicity of metallic compounds which is of possible significance in public health has been compiled based on considerations during the Scientific Committee on the Toxicology of Metals symposium published by Clarkson et al 1983 [22] and updated by Clarkson et al 1985 [4] and also considering some more recent data [23].

Neurotoxic effects of methylmercury on the fetal brain which should be considered as a critical organ, with mental retardation being the critical effect, is one of the best established examples of this type of effect, constituting the critical effect of whole population exposure. However, also for lead neurobehavioural deficiencies and a decrease of IQ is considered to be the critical effect from whole population exposure based on effects on the infant brain. Whether lead can also give rise to male or female reproductive hazards at exposure levels of a similar magnitude as the lowest ones giving rise to effects on the fetal and infant brain is still a matter of discussion. It is known that arsenic can give effects on the developing brain based on the data from the Morinaga poisoning cases in Japan, but also some more recent data indicate that this may be an important and possible critical effect of arsenic while, however, effects in terms of carcinogenicity are usually considered to be the critical effects of inorganic arsenic exposure when there is long-term exposure to whole human populations. There is some discussion still concerning whether effects of vascular nature on the placenta can occur in cadmium exposures, but recent data has not been convincing and makes it less likely that this effect should be regarded as a critical effect.

5 EFFECTS RELATED TO THE IMMUNE SYSTEM

A category of effects which is of potential great importance as critical effects are immune mediated effects and immunotoxicity. However, it has not been the rule in evaluations of occupational exposures, to take into account hypersensitivity (e g TLV-ACGIH). The same can be said for most of the evaluations of dietary exposures. Based on a joint workshop between the Scientific Committee on the Toxicology of Metals and the CEC, [5] it was recognized that metals in specific chemical forms may give rise to a number of immunerelated effects. These include hypersensitivity reactions of the immediate type I from chromium, cobalt, nickel and platinum; hypersensitivity involving cytotoxicity (type II) from gold, immune complex mediated hypersensitivity (type III) from mercury or gold and the common, cell mediated or delayed type-IV hypersensitivity from beryllium, chromium, cobalt, nickel and zirconium. Autoimmune reactions can be induced by gold or mercury, enhancement of immunity by selenium and zinc and suppression of immunity can occur as a result of exposure to cadmium, cobalt, lead, mercury or organotin. In addition some effect of non-specific immunity were identified. Because of the complicated dose-response relationships and because it has been considered impossible to reduce exposures below the levels triggering hypersensitivity reactions in sensitized individuals, immune-related effects and in particular hypersensitivity reactions have not usually been recognized as critical effects. However, it has still been considered important from a practical point of view to try to reduce the occurrence of hypersensitivity of the skin from e.g. hexavalent chromium in the cement industry and such programmes have been implemented recently. This successful experience shows that it is possible in some cases to define practical thresholds for hypersensitivity and to bring down exposures below levels that cause sensitization [5].

6 CARCINOGENIC EFFECTS

The importance of considering carcinogenicity as a critical effect in risk assessments of metallic compounds has already been alluded to. The serious nature of cancer means that this effect must be prevented at a very low response level and, as aleady mentioned, this means that for metallic compounds which are considered as carcinogenic, the carcinogenic effect will often be considered as the critical effect. Consequently, carcinogenicity has attracted considerable attention in evaluations by the Scientific Committee on the Toxicology of Metals and workshops on carcinogenic effects have been organized [3, 24].

When considering the chronology of development of knowledge on carcinogenicity of metallic compounds it is evident that case reports and epidemiological evidence were the initial observations for the metallic compounds which were first recognized as carcinogens, such as trivalent arsenic, chromium and nickel, while carcinogenicity in animals were confirmed at a later stage. For the more recently recognized metallic carcinogens, such as beryllium and cadmium, the course of events have been the opposite one and for the two candidate carcinogens methylmercury and antimony trioxide, sufficient animal data are considered to be present according to IARC. According to IARC principles, the latter substances should be considered as possible human carcinogens and measures be taken to reduce exposures as much as possible. However, until these substances have been declared human carcinogens, the possible carcinogenic effect will probably not be considered as the critical effect.

The carcinogenic effects are considered as the critical effects in most instances when a metallic compound has been declared a human carcinogen. For trivalent arsenic, for hexavalent chromium and for nickel the carcinogenic effect is usually considered to be the critical effect and these substances are generally regulated as carcinogens in accordance with the fact that they are all considered as carcinogenic to humans based on evaluations by IARC. Also beryllium has recently [17] been classified as a human carcinogen. The reason why it has not been stated clearly that the carcinogenicity of Be is the critical effect is a suspicion that berylliosis, an effect which is most probably an immune-related one might occur in a low incidence at extremely low exposures, possibly even lower than those giving rise to cancer. For cadmium it has until recently been customary to use the quantitative data on nephrotoxicity as a basis for preventive action and this effect has been considered as the critical effect. However, according to a recent evaluation by IARC [17], cadmium has now been recognized as a human carcinogen and this effect should be considered as a first hand choice when selecting the critical effect. However, while quantitative dose-response data are relatively good for the nephrotoxic effects of cadmium, there is only very limited quantitative dose-response data for the carcinogenic effects of cadmium (related to the risk of developing lung cancer in occupational inhalation exposures). It is thus difficult at present to use a quantitative estimate of the carcinogenic risks from cadmium exposure as a basis for regulation or estimation of health based guidelines of exposure limits and the most reasonable way to deal with this situation may be to use an additional safety factor on risk estimates for other effects in order to prevent carcinogenicity, particularly in inhalation exposures. The carcinogenicity will most probably be considered as a critical effect in the future and cadmium be regulated as a carcinogen [8]. For lead, methylmercury and antimony trioxide the available evidence still is not at a level so as to warrant the carcinogenicity to be used as a critical effect and these substances are in most instances not yet regulated as carcinogens.

It is obvious that identification of the critical effect in long-term low-level exposure for various metallic compounds requires a comprehensive set of information from epidemiological as well as experimental studies. The distinction between stochastic and deterministic effects and related differences in shapes of dose-response curves are of great importance. A reasonably complete set of information that allows a scientifically based quantitative risk assessment is only available for a limited number of metallic compounds. It is reassuring to see that these concepts and principles, developed within the framework of the Scientific Committee on the Toxicology of Metals have been used in risk assessments by the international scientific community and it is hoped that they will prove to be helpful in providing a framework for data collection and risk assessment also in the future.

References

1. G. Nordberg, Ed., 'Effects of dose-response relationships of toxic metals', Elsevier Publishing Co, Amsterdam, 1976.
2. G. F. Nordberg, *Environ. Health Perspect.*, 1978, **25**, 3.
3. S. Belman, G. Nordberg, *Environ. Health Perspect.*, 1981, **40**,1.
4. T. W. Clarkson, G. F. Nordberg, P. R. Sager, *Scand. J .Work Environ. Health*, 1985, **11**, 145.
5. A. D. Dayan, et al., Eds., 'Immunotoxicity of Metals and Immunotoxicology'. Proceedings of an International Workshop, Plenum Press, New York and London, 1990.
6. B. A. Fowler, Ed., Proceedings of the International Conference on Environmental Arsenic, in Fort Lauderdale, Florida, USA, October 5-8,1976. *Environ. Health Perspect.*, 1977, 19.
7. B. A. Fowler, Ed., Proceedings of an International Conference on Environmental Cadmium, in Bethesda, Maryland, USA, June 7-9,1978. *Environ. Health Perspect.*, 1979, 28.
8. G. F. Nordberg, L. Alessio and R. F. M. Herber, Eds., 'Cadmium in the Human Environment: Toxicity and Carcinogenicity. International Agency for Research on Cancer, Lyon, 1992. IARC Scientific Publications No 118.
9. T. Suzuki, T. Clarkson and N. Imura, Advances in Mercury Toxicology. Seminar University of Tokyo, August 1-3, 1990,
10. G. F. Nordberg and P. Strangert, *Environ. Health Perspect.*, 1978, **22**, 97.
11. T. Abelson, *Science,* 1994, **26**, 1507.
12. G. F. Nordberg and P. Strangert, in 'Methods for Estimating Risk of Chemical Injury: Human and Non-Human Biota and Ecosystems', V. B. Vouk, G. C. Butler, D. G. Hoel and D. B. Peakall, Eds. J Wiley Publ and SCOPE, Chichester, England, 1985 pp. 477.
13. WHO, 'Assessing Human Health Risks of Chemicals: Derivation of Guidance Values for Health-based Exposure Limits', Environmental Health Criteria 170, World Health Organization, Geneva, 1994.
14. WHO, 'Cadmium', Environmental Health Criteria 134, World Health Organization, Geneva, 1992.
15. L. Friberg, C. G. Elinder, T. Kjellström and G. F. Nordberg, 'Cadmium and Health: A Toxicological and Epidemiological Appraisal. Vol I, Exposure Dose and Metabolism', CRC Press, Boca Raton, FL, 1985.
16. L. Friberg, C.-G. Elinder, T. Kjellström and G. F. Nordberg, 'Cadmium and Health: A Toxicological and Epidemiological Appraisal. Vol II, Effects and Responses', CRC Press, Boca Raton FL, 1986.
17. IARC, 'Beryllium, cadmium, mercury and exposure in the glass manufacturing

industry', IARC Monographs on the evaluation of carcinogenic risks to humans International Agency for Research on Cancer, Lyon, 1993, vol. 58.
18. WHO, 'Methylmercury', Environmental Health Criteria 101, World Health Organization, Geneva, 1990.
19. F. Bakir, et al., *Science,* 1973, **181**, 230.
20. H. Al-Shahristani and K. M. Shihab, *Arch. Environ. Health,* 1974, **27**, 342.
21. D. O. Marsh, et al., *Arch.Neurol.,* 1987, **44**, 1017.
22. T. W. Clarkson, G. F. Nordberg and P. R. Sager, Eds., 'Reproductive and developmental toxicity of metals'. Plenum Press, New York, 1983.
23. M. Berlin, et al., in 'Cadmium in the Human Environment: Toxicity and Carcinogenicity', G. F. Nordberg, R. F. M. Herber and L. Alessio, Eds. International Agency for Research on Cancer, Lyon, 1992. IARC Scientific Publications No 118, p. 257.
24. G. F. Nordberg and S. Skerfving, Eds., 'Biological Monitoring, Carcinogenicity and Risk Assessment of Trace Elements ', *Scand. J. Work Environ. Health,* 1993, **19**, Supplement 1.

TOXICITY AND INTAKES OF VARIOUS FORMS OF HEAVY METALS

M. Nordberg* and G.F. Nordberg**

*   Institute of Environmental Medicine, Karolinska Institutet, S-171 77 Stockholm
** Department of Environmental Medicine, Umeå University, S-901 87 Umeå, Sweden

1 INTRODUCTION

Human beings are exposed to several metals. Many metals are regarded as essential and important for life. Also essential metals may however cause adverse health effects under certain circumstances. This occurs when, for example exposure is excessive, exposure route is not the physiological one or the chemical form-species is toxic to the human being. Toxicity of metals is dependent on speciation of the toxic metals[1,2]. Different factors such as intake of protein, iron status and interaction with other metals influence metabolism and toxicity of metals[3,4]. This review presents toxicity and intake of metals and, when appropriate, emphasizes the importance of various forms of metals. Various species of metals have been shown in various foodstuff for metals such as cadmium, mercury and arsenic, sometimes recognized in the form of metal binding proteins. Changes in mobility of metals caused by acidification of soil and lakes change human exposure via the food chain and drinking water[5]. Increased bioavailability is expected for mercury as methylmercury, cadmium, lead and aluminum. On the other hand certain trace elements eg selenium become less available in the ecological system.

2 METALS IN FOOD

Concentration of metals changes in fish and crops with pH in lakes and soils. For mercury a cycle of mercury in ecological system is described. Various species[6,7] that is inorganic mercury as vapor, inorganic mercury and various organic compounds of mercury may be emitted in nature and finally all these end up as methylmercury through bioconversion mechanisms of microorganisms. An example of another metal that exist in various species depending on pH of the media is aluminum[8] for which the chemical species-forms varies with pH. Free Al-ions exist chemically in increased concentration below pH 5 and at 2.5 almost only free Al^{3+} exists. Concentration of metals in food and drinking water varies with metal and foodstuff. Cadmium exist on µg/kg to mg/kg basis in beef meat, kidney, seafood, wheat grains and rice. However cadmium exists in liver and kidney as metallothionein and in other foodstuff as various metal binding proteins of molecular weight from 7-24 KDa[9]. The proteins have for example been identified in various oysters by gel chromatography[2] to exist as different low molecular weight metal binding proteins one of which is regarded as metallothionein and another one is a non-MT-protein.

Mercury in fish from oceans is reported[6] to occur at concentration levels of 0.003-0.25 mg Hg/kg wet weight for non-predators and for predators corresponding figures are 0.004-1.8 mg Hg/kg wet weight. In the same document data on the relationship between consumption-concentration of methylmercury and intake is given. However regulation of exposure to methylmercury via fish has so far been a national decision. In Sweden it is regarded of specific concern to advise certain population groups to restrict their intake of methylmercury. In the general population mercury vapor exposure has not been regarded as a problem. However it has been shown[10] that mercury vapor in intraoral air is related to dental amalgam fillings. For a majority an uptake of mercury occurs as a result of exposure to methylmercury and mercury vapor simultaneously[6,7]. Arsenic is present in drinking water as III and V valent inorganic arsenic that are regarded as highly toxic. In fish and seafood organic arsenic compounds have been identified as arsenobetaine and arsenocholine. These species[8] are much less toxic than the inorganic forms. Concentration of lead in foodstuff depends on if canned or uncanned[8]. Lead in cans originate from the solder and are restricted in many countries. Food stuff obtained as vegetables or from animals close to traffic has shown high content of lead before restrictions on lead in petrol.

3 INTAKE OF METALS

Content and intake of metals in foodstuff and drinking water is restricted by national authorities. In the Scandinavian countries intake of cadmium is around 10-25 µg/day. This could be compared to the intake reported from noncontaminated areas of Japan of 65 µg/day. Lead is in the same range as for cadmium. Selenium intake is regarded as low if lower than 0.03 mg/day[5]. As mentioned various species of metals occur in food. Since the metabolism and thus the metabolic models for various species varies, tolerance limits vary accordingly. International organizations and national regulatory agencies issue recommendations concerning dietary intake. Metabolic models are developed from the knowledge of factors such as metabolism, absorption, distribution, excretion, biotransformation and biological half-time for specific species of metal.

In the metabolic model for cadmium[11] uptake of cadmium is calculated and disposition of the different species of cadmium described in quantitative terms. Basic knowledge on cadmium metabolism and species of cadmium present in the biological system has been brought together in a mechanistic model for cadmium[12,13]. Cadmium is initially taken up as cadmium-albumin and this form is taken up by the liver. The metal induces the synthesis of MT in the liver and constitutes part of the molecule. Cadmium is subsequently released in the form of MT to the blood stream and filtered through the glomerular membrane in the kidney. After that it is taken up by the brushborder of the renal tubule and MT is catabolized in the lysosomes of the cell. Released cadmium induces new synthesis of MT in the renal cells and is as well free to interact with sensitive sites. Membrane damage occurs in a similar way as assumed for free radicals and MT serves as a scavenger for cadmium and free radicals. In addition to 20 cysteine residues there is also one methionine present in the MT-molecule. The MT molecule consists of two metal clusters with three metals closer to the N-terminal and four at the C-terminal of the molecule. In previous studies[14] it was shown by autoradiography that Cd when given as chloride by intravenous injection mainly bound to albumin, was taken up by the liver while Cd as MT was distributed to the kidney. Observation time and route of exposure are important factors which may determine which effects that are observed as a result of exposure. Since renal damage and related effects such as uremia takes longer time to develop it is important to have observation times of several days for MT exposed animals[14].

Metallothionein has several functions in the biological system such as transport of metals eg Cd, Cu, Zn, detoxification of metals, protection from metal toxicity, storage of metals, metabolism of essential metals and functions related to the immune response. Cadmium may cause various adverse health effects depending on exposure route, duration and intensity[15]. IARC classified Cd as belonging to group I-a human carcinogen[16]. After exposure via food on life-time basis cadmium is expected to cause renal effects. The metabolic model for cadmium[17] takes account of exposure route as well as different species of Cd in blood i.e. CdAlb, CdMT and Cd in blood cells. From this model and data on proteinuria in population groups exposed to cadmium from Sweden and Japan the relationship between daily intake of cadmium and development of proteinuria after 50 years of exposure was estimated. It was shown by these calculations that an intake of 55 ug Cd/day can be expected to give rise to 1 % of the population with proteinuria. In the Cadmibel[18] study it was however shown that renal dysfunction develops at even lower concentration than previously accepted. This means that a higher percentage of the population than the one just mentioned might have renal effects due to cadmium exposure in the general environment at the intake level which was referred to. Vulnerable groups suffering from diabetes and osteoporosis likely are to be more sensitive to metal exposure than persons in general.

4 CRITICAL ORGAN AND CRITICAL EFFECTS

Metabolic models are specific for each metal. The metabolic models for inorganic and organic lead differ as a result of the differences in lead kinetics - metabolism of these forms of the metal. The critical organ for inorganic lead is the nervous system and bone marrow[8] while for the organic species it is the central nervous system. Critical effects are as well related to age. For children, adverse health effects on the central nervous system occur at blood concentration of 10 µg/100ml (0.48 µmol Pb/L)[19]. Adults develop hematological symptoms at 5 times higher concentration[8]. Aluminum is difficult to analyze in the low concentration found in biological system due to high surrounding concentration. The brain and bone are regarded as critical organ[20]. The various species of mercury as previously mentioned needs further explanation. The metabolic model for inorganic mercury deals with Hg^0 and Hg^{2+} in the model at the same time due to that mercury vapor is oxidized to mercuric ion by biochemical processes in the red blood cells. Mercury vapor is mobile and crosses blood-brain barrier and placenta. Inorganic mercury is present in the blood stream partly as MT and is taken up in the kidney. The distribution of methylmercury has been shown by autoradiograghic techniques[21] to be taken up in the central nervous system both in female and in fetuses. The biological half-time of various metal compounds varies greatly among metals.

For inorganic lead the biological half-time in blood is approximately 3 weeks, in tissue 35-40 days and the fraction retained in the bone 5-20 years[22]. Cadmium has, partly due to its binding to metallothionein, an extreme long biological half-time of 5-15 years in the liver, 30 years in muscle and in the kidney 10-30 years[23,24]. The biological half-time for cadmium is related to synthesis of MT. However MT synthesis seems to decrease with age. Methylmercury has a biological half time of 65-70 days and no compartment related difference is reported. Inorganic mercury has quite different biological half-times varying from 50 days to several years in different parts of the brain. In renal tissue values of 100 days have been reported[7]. In summary, partly because of metabolic differences the critical organ varies with species of the metal but also characteristics of the population such as age can be of importance.

5 PROTECTIVE MECHANISMS

By various mechanisms the cell may be protected from adverse effects. In persons with dental amalgam fillings, intraoral exposure to mercury as mercury vapor decreased after intake of certain foodstuff eg egg[10]. The mechanistic explanation is likely due to high content of SH-groups in the egg protein. Mercury was shown by gel chromatography of renal tissue to exist as either MT[25] or to exist in the same protein fraction as selenium in the cell[26]. Mercury-selenium complex and metallothionein bound mercury protects the cell from toxic effects. Metal binding proteins have been reported to serve as protecting agents. The localization in the cell is important. Lead may be bound to protein and is recognized histologically as inclusion bodies. MT in cytoplasm may serve as detoxificant. Any change of pH in the lysosomes might have dramatic influence on the toxicity of heavy metals. In addition of to previously mentioned protective mechanisms, interaction with other metals and trace elements eg Se, protein intake, iron status, genetic polymorphism, methylation and demethylation, influence as well the toxicity and intake of various forms of heavy metals. With new and increased knowledge of methods for metal analyses there are new possibilities for further studies and it is necessary to employ these methods in future research programs.

References:

1. M. Nordberg, *Environ. Health Perspect.*, 1984, **54**, 13.
2. M. Nordberg, I. Nuottaniemi, M. G. Cherian, G. F. Nordberg, T. Kjellström and J. S. Garvey, *Environ. Health Perspect.*, 1986, **65**, 57.
3. G. F. Nordberg and S. Skerfving, Biological monitoring, carcinogenicity and risk assessment of trace elements. *Scand. J. Work, Environ. and Health,* 1993, **19** Suppl 1, 1.
4. G. F. Nordberg, Factors influencing metabolism and toxicity of metals. *Environ. Health Perspect.*, 1978, **25**, 1.
5. G. F. Nordberg, R. A. Goyer and T. W. Clarkson, *Environ. Health Perspect.*,1985, **63**, 169.
6. WHO, Environmental Health Criteria, 1990, 101, Methylmercury, World Health Organization, Geneva
7. WHO, Environmental Health Criteria, 118, Inorganic mercury, 1991, World Health Organization, Geneva
8. L. Friberg, G. F. Nordberg and V. Vouk, "Handbook on The Toxicology of Metals". Elsevier, Amsterdam, 1986, Vol II, Chapter 1, 3, 14.
9. D. H. Petering and B. A. Fowler, *Environ. Health Perspect.*, 1986, **65**, 217.
10. A-M Aronsson, B. Lind, M. Nylander and M. Nordberg, *BioMetals*, 1989, **2** 25. 11. T. Kjellström and G. F. Nordberg, *Environ. Res.*, 1978, **16**, 248.
12. G. F. Nordberg, *Environ. Physiol. Biochem.,* 1972, **2**, 7
13. M. Nordberg, Environ. Res., 1978, **15**, 381
14. M. Nordberg and G. F. Nordberg, *Environ. Health Perspect.*, 1975, **12**, 103.

15. G.F. Nordberg and M. Nordberg, Biological monitoring of cadmium. "Biological Monitoring of Toxic Metals" T. W. Clarkson, L. Friberg, G. F. Nordberg and P. R. Sager, Plenum Publishing Co., New York. 1988 p. 151.

16. IARC, Monographs On The Evaluation Of Carcinogenic Risks To Humans, **58**, Beryllium, Cadmium, Mercury, and Exposures in the Glass Manufacturing Industry, 1993, IARC, Lyon, France

17. G. F. Nordberg and T. Kjellström, *Environ Health Perspect.*, 1979, **28**, 211.

18. A. Bernard. "Cadmium in the Human Environment:Toxicity and Carcinogenicity, G. F. Nordberg, L. Alessio, R. F. M. Herber, IARC Publications No 118, 1992, p.15.

19. CDC, Preventing lead poisoning in young children. A Statement by the Centers for Disease Control, 1991, Atlanta GA, USA

20. T. W. Clarkson, L. Friberg, G. F. Nordberg and P.R. Sager, "Biological Monitoring of Toxic Metals", Plenum Publishing Co., New York, 1988.

21. M. Berlin, L-G Jerksell and G. Nordberg, *Acta pharmacol. et toxicol.*, 1965, **23**, 312.

22. L. Gerhardsson, R. Attewell, D. R. Chettle, V. Englyst, N-G Lundström, G. F. Nordberg, H. Nyhlin, M. C Scott and A. C. Todd, *Archives Environ. Health*, 1995, **48**, 3 147.

23. G. F. Nordberg, T. Kjellström and M. Nordberg, "Cadmium and Health", I. L. Friberg, C-G Elinder, T. Kjellström and G. F. Nordberg, CRC Press Boca Raton, Fl, USA, 1985, p. 103.

24. WHO, Environmental Health Criteria, 134, Cadmium, 1992, World Health Organization, Geneva

25. M. Nordberg, B. Trojanowska and G. F. Nordberg, *Environ. Physiol. Biochem.*, 1974, **4** 149.

26. L. Björkman, B. Palm, M. Nylander and M. Nordberg, *Biol. Trace Element Res.*, 1994, **40**, 225.

CADMIUM BINDING COMPOUNDS IN WHEAT: OCCURRENCE AND PARTIAL CHARACTERIZATION

J. Brüggemann[1], N. Tümmers[1], H.P. Thier[2] and Th. Betsche[1]

1. Federal Centre for Cereal, Potato and Lipid Research Institute for Biochemistry and Analytics of Cereals, P.O.Box 1354, D-32703 Detmold, Federal Republic of Germany
2. University Münster, Institute for Food Chemistry, Piusallee 7, D-48147 Münster

1 INTRODUCTION

Since 1975 we have monitored the Cd concentrations in the German wheat crop and have analyzed more than 5 000 wheat samples. All samples contained Cd. From this and other findings [1] we conclude that Cd is a constituent of wheat which is taken up by the roots and is transported to the other parts of the plant, including the grains. It is not clear, however, whether this Cd occurs as a free ion or bound to other compounds. In this paper we present results showing the occurrence of Cd-binding compounds in wheat grains.

2 MATERIALS AND METHODS

Grains, roots or leaves from wheat were homogenised in buffers of high (1.2 mol/l NH_4-acetate pH 7.0) and low (0.05 mol/l Tris buffer pH 8.0) ionic strength and centrifuged (8340 g for 30 min, 4 °C). The pellet was resuspended and centrifuged again. This step was repeated twice and the supernatants were pooled. In some experiments reducing agent (up to 100 mmol/l dithiothreitol or mercaptoethanol or dithionite) was added to the extraction buffer. Analysis of Cd in the homogenate, supernatant and residue was carried out after wet digestion by electrothermal atomic absorption spectrometry with Zeeman correction (ZAAS).

The main insoluble components (starch and bran-gluten) were separated by density centrifugation and Cd content was determined. The pellet from the extraction of wheat grains with high ionic strength buffer was resuspended in the same buffer, layered on 10 ml of aqueous CsBr solution (1 g/ml), and centrifuged at 5 000 g for 30 min at 22 °C (Hettich Roto Silenta 5300 centrifuge; swing out rotor 5066, r = 160 mm).

To detect and purify Cd binding compounds, supernatants were subjected to dextran gel chromatography (Sephadex G-25 or G-100). The fractions were analyzed by ZAAS and elution was done with the same buffers used for extraction. The fractions from each Cd peak were pooled separately, concentrated by ultrafiltration using Amicon membranes. The 15 kD compound was dialysed against water, freeze dried and resuspended in weak buffer for isoelectric focusing using Bio-Lyte 3-10 (BIORAD).

3 RESULTS

3.1 Distribution of Cd

The distribution of Cd in fractions of soluble protein, gluten, bran and starch is shown in Fig. 1. About 75% of the total Cd was in the soluble fraction. Aqueous buffers with high ionic strength proved to be more effective in extracting soluble Cd binding compounds than buffers of low ionic strength [2]. Density centrifugation was used to separate gluten-bran from starch in order to localize the insoluble Cd (Fig. 1). The starch fraction contains little Cd and protein (less than 10 % of the Cd in the whole wheat grain). In the gluten-bran fraction, the Cd concentration is three times as high as in the whole wheat grain (on a wet weight basis). But the percentage of total Cd in the gluten protein fraction is only about 30 %. This means that most of the Cd is bound to the soluble proteins of the supernatant (Fig. 1).

3.2 Cd binding compounds that are soluble in buffers of low ionic strength

The Cd compounds were separated by dextran gel chromatography (Fig. 2-4). Using freshly harvested wheat grains, 70 % of the extractable Cd was bound to a 15 kD compound (Fig. 2). By contrast, additional Cd binding compounds with MW of 32, 65, and >100 kD were detected in wheat stored for about 7 years at 10°C (Fig. 3). Cd binding compounds of similar size have been found in Arabidopsis thaliana [4,5]. The portion of Cd in the > 100 kD fraction increased with storage age at the expense of the 15 kD fraction. In wheat stored for 9 years, the high MW fraction (> 100 kD) contained nearly 50 % of the soluble Cd. The concentration of glutathion decreased from an initial value of 43 nmol/g (± 1.4) to 6 nmol/g (± 0.8), while cystein remained nearly constant (see acknowledgements).

The extraction of wheat grains was also done using buffer with reducing agents. One sole compound of 15 kD was detected. This observation suggests that the higher MW compounds were cleaved by 10 or 100 mmol/l mercaptoethanol or dithionite. The 15 kD Cd binding compound from grains was not further cleaved to smaller fragments of the size typical for photochelatins in reducing conditions (< 5 kD). The isoelectric point of the purified 15 kD compound was pI = 4.5.

Cd binding compounds of similar MW's were isolated from leaves and roots of the wheat plant with the same buffer. However, some of the 15 kD compound from roots was cleaved by highly concentrated reducing agents to smaller compounds which could be phytochelatins.

3.3 Cd, Ni and Zn binding compounds that are soluble in buffers of high ionic strength

Cd binding compounds of low MW (< 1 kD) were extracted from wheat grains when 1.2 mol/l NH_4-acetate buffer, pH 7.0, was used. It is thus possible that these compounds are phytochelatines. Fig. 5 shows that Ni and Zn elute at the same position as Cd in the gel filtration using Sephadex G-25. Zn was also detected in a higher MW fraction (about 1 kD). The amount of Zn in the different fractions was variety-dependent and was correlated with the total Zn concentration.

Because the Zn was eluted from the column together with phytate, we conclude in accordance with Ellis and Morris [3], that Zn phytate occurred in the supernatant (Fig. 6). When the homogenate was stored 16 hours at room temperature, the Zn phytate peak disappeared and a Zn peak at a lower MW increased correspondingly. When Zn^{2+} was added to the wheat grain homogenate, the Zn phytate peak was increased. It is proposed from these findings that Zn binding compounds from the wheat grain are degraded and Zn phytate is formed when buffers of high ionic strength are used. However, no Cd or Ni was detected near the phytate peak suggesting that the Cd binding compounds are more stable than Zn binding compounds. The other compounds in the low MW metal binding fraction are believed to be phytochelatins. Once these compounds can be purified, their amino acid composition will be determined for unequivocal identification.

References

1. H. Lorenz, H.D. Ocker, J. Brüggemann, P. Weigert, and M. Sonneborn, Z. Lebensm Unters Forsch, 1986, **183**, 402.
2. H. Ditters, Untersuchungen zur Bindung von Cd, Ni und Zn in Getreidekörnern. Universität Münster, Dissertation, 1991.
3. R. Ellis, and E.R. Morris, Cereal Chem. 1981, **58**, 367.
4. R.Howden, P.B. Goldsbrough, C.R. Andersen, and C.S.Cobbett, Plant Physiol, 1995, **107**, 1059.
5. R. Howden, C.R. Andersen, P.B. Goldsbrough, and C.S. Cobbett, Plant Physiol, 1995, **107**, 1067.

Acknowledgements

Glutathion analysis was conducted by W. Grosch, H. Wieser, Deutsche Forschungsanstalt für Lebensmittelchemie, Garching. The help of these colleagues is gratefully acknowledged.

Legends

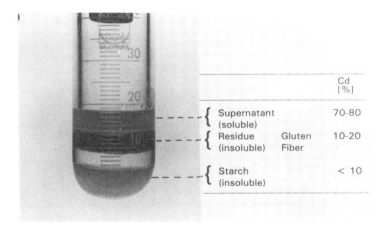

Figure 1 *Distribution of buffer-soluble Cd (supernatant) and buffer-insoluble Cd (fiber-gluten and starch). Values mean percentages of the total Cd applied to the gradient.*

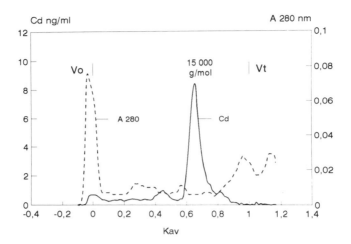

Figure 2 *Molecular weights (MW) of Cd binding compounds from wheat grain stored for 1 year (gelfiltration using Sephadex G-100 (5 x 100 cm) and 0.05 mol/l Tris buffer, pH 8.0 as elution buffer, Cd (—), A 280 nm (- -)).*

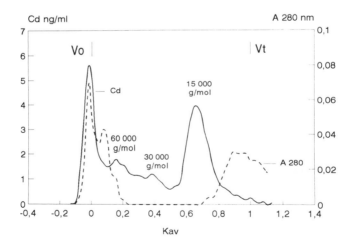

Figure 3 *MW of Cd binding compounds from wheat grain stored 7 years (see Fig. 2 for details).*

Contaminants in Food and Diets 421

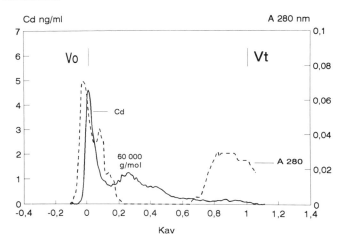

Figure 4 *MW of Cd binding compounds from wheat grain stored 9 years (see Fig. 2 for details).*

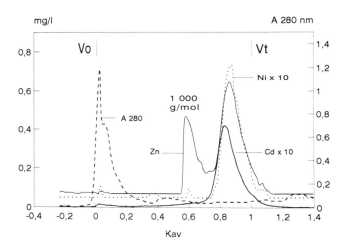

Figure 5 *MW of Cd (—), Ni (···), and Zn (—) binding compounds from wheat grain (Gelfiltration using Sephadex G-25 (5 x 100 cm) and 1.2 mol/l NH_4- acetate buffer, pH 7.0 as elution buffer, (- -) A 280 nm).*

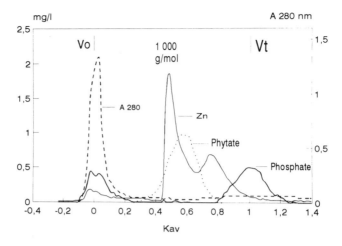

Figure 6 *MW of Zn (—) binding compounds, phosphorus (—) and phytate (···) from wheat grain (see Fig. 5 for details).*

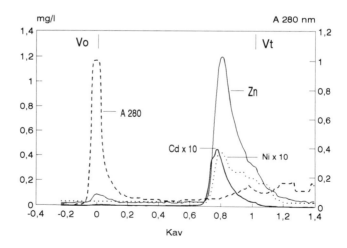

Figure 7 *MW of Cd (—), Ni (···), and Zn (—) binding compounds from wheat grain after incubation of the homogenate for 16 h at 20 °C (see Fig. 5 for details).*

LEAD, CADMIUM, MERCURY, COPPER AND ZINC CONTENT IN POLISH POWDERED MILK AND PRODUCTS FOR INFANTS AND CHILDREN

Maria Wojciechowska-Mazurek, Kazimierz Karlowski, Jorma Kumpulainen*, Krystyna Starska, Elzbieta Bruliñska-Ostrowska, Kazimiera Zwiek-Ludwicka,
National Institute of Hygiene, Department of Food Research, Chocimska 24, 00-791 Warsaw, Poland
*Laboratory of Food Chemistry, Agricultural Research Centre of Finland, SF-31600 Jokioinen, Finland

1 INTRODUCTION

Contamination of food with heavy metals poses a significant health problem. Lead, cadmium and mercury are considered to be the most dangerous contaminants of food both because of their toxic properties, abundance and the hazard they create for the population.

Food products for infants and small children should meet special requirements in respect to their health quality. Milk and other infant formulas are the main, and in practice the only source of food for infants and children; therefore, even the relatively low levels of noxious metals in those products may lead to significant intakes as compared with the Provisional Tolerable Weekly Intake (PTWI) of metals in the daily diet. Moreover, children are a particularly susceptible group of populationo to metals [6,9,15] because:
- absorption of metals from the alimentary tract for children is several-fold higher than for adults;
- metabolic processes are more rapid in young organisms; absorption of food and of contaminants is, therefore, higher per unit of body weight;
- small children do not have a fully developed circulatory-cerebral barrier, causing particular threat to the central nervous system;
- children do not have fully developed detoxification systems, e.g. the enzymatic or alimentary tract microflora.

Investigations conducted in many countries, including Poland, have shown that as compared with adults, children consume significantly higher per unit body weight quantities of metals in the daily diet and frequently (particularly in industrial areas) had lower safety margins between the reported intake and amounts tolerated by the organism (PTWI) [1,7,8,10,11-14,16,19,25,27].

Only few countries have legislation which provides requirements as to the maximum permissible levels of noxious metals in food products made for infants and small children and for the raw materials used for such production.

Ordinance of the Polish Minister of Health and Human Welfare which specifies a list of additives permitted in foods also gives a specification of the maximum levels of contaminants allowed in such products [26], including heavy metals such as Pb, Cd, Hg, Zn, Cu - see Table 1.

Table 1. *Permitted levels of Pb, Cd, Hg, Zn and Cu in Poland (mg/kg), (according to Ordinance of the Minister of Health and Human Welfare, March 31, 1993)*

	Pb	Cd	Hg	Zn	Cu
Milk and milk/cereal products	0.1	0.01	0.005	35	5
Cereal and cereal/fruit products	0.15	0.01	0.01	30	6
Vegetable/meat products	0.2	0.01	0.01	30	4
Juices	0.1	0.01	0.01	5	2
Pomaces	0.1	0.01	0.01	20	5

The need to have those products meet special requirements has been emphasized recently by the Codex Committee on Food Additives and Contaminants of FAO/WHO Codex Alimentarius Commission [3].

In 1992-1994 the National Institute of Hygiene, Department of Food Research, in cooperation with the Voivodship Sanitary and Epidemiological Stations (VSES) conducted systematic investigations on the content of lead, cadmium, mercury, copper and zinc in various groups of Polish products made for infants and children. At the same time, content of those metals in powdered milk used in infant and child formulas was investigated within the FAO European Research Network on Trace Elements. In total, about 2600 samples of those products were tested.

2 MATERIALS AND METHODS

Investigations were performed using the following products:
1) Whole powdered milk (class „Extra", according to Polish Standard PN-92/A-86024. Powdered Milk) for preparing infant formulas;
2) Infant and children products in the following groups:
    a) milk infant formulas:
        Bebiko „0"
        Bebiko „1"
    b) milk cereal and milk/cereal/fruit products, such as:
        Bebiko 2R
        Bebiko 2GR
        milk/rice porridge
        milk/rice/fruit porridge
    c) cereal and cereal/fruit products
    d) vegetable/meat products
    e) fruit and vegetable products:
        BoboFrut juices
        BoboFrut pomaces.

2.1. Collection and preparation of samples

2.1.1 Powdered milk. Whole powdered milk samples were taken in accordance with the guidelines of the Subnetwork on Trace Element Status in Food (FAO European Cooperative Research Network on Trace Elements).

The samples were taken by the VSES directly at 10 dairy plants throughout Poland in two periods: in September-October (green fodder) and February-March (dried fodder), in 1991-1994. Samples weighing 0.5 kg were taken at random out of current production every day during 5 successive days over a period of two weeks with one week break in-between. The collective sample was sent to the Department of Food Research, National Institute of Hygiene which prepared the final collective samples for analysis. Two parallel collective samples for each sampling period were sent to the Agricultural Research Centre of Finland which analyzed them for heavy metals content, within the FAO European Cooperative Research Network on Trace Elements program.

2.1.2 Products for infants and children. Products were sampled in 1992-1994 directly from four Polish manufacturing plants and from the commercial network in 23 voivodships.

2.2. Determination of heavy metals

2.2.1. Powdered milk. Investigations were performed at the Agricultural Research Centre of Finland, Laboratory of Food Chemistry, Finland, for representative samples of Polish powdered milk. Zn and Cu were determined by flame AAS after digesting with concentrated HNO_3. Samples for Pb and Cd analyses were digested by heating overnight in concentrated HNO_3 and determined by GFAAS using $(NH_4)H_2PO_4$ as the matrix modifier. Detailed data on the methods employed and their accuracy are available elsewhere[28].

2.2.2. Products for infants and children. Analyses were performed by laboratories in the VSES using flame AAS after dry ashing of samples at 400°C. Zn and Cu were measured directly in the mineralizate solution; Pb and Cd were analyzed after extraction of metal complexes with ammonium-1-pyrrolidine-dithiocarbamate (APDC) to methyl isobutyl ketone.

Hg was determined after wet mineralization of the sample in H_2SO_4 and HNO_3, adding V_2O_5 as the oxidation catalyst. The "cold vapour" flameless AAS method was employed for analytical measurements.

2.3. Quality control

The laboratories in the VSES which performed analyses of samples of products for infants and children participate in proficiency testing programs organized regularly since 1991 by the National Institute of Hygiene 2-3 times each year. Such programs are conducted in accordance with recommendations of the Joint UNEP/FAO/WHO Food Contamination Monitoring Program (GEMS/Food). These proficiency tests also included determination of Pb, Cd, Cu, Zn and Hg in powdered milk. Moreover, the National Institute of Hygiene provides detailed guidelines for sampling and analysis methodologies to participants of the investigations.

Analytical methods developed and published by the National Institute of Hygiene have been verified as to their accuracy and precision using certified reference materials and during interlaboratory testing. All samples were analyzed in duplicate.

3 RESULTS AND DISCUSSION

Analytical results are presented in Tables 2-7. Table 2 presents the results obtained at the Laboratory of Food Chemistry of the Agricultural Research Centre of Finland for metal contents in representative samples of Polish powdered milk. Lead and cadmium levels found were low, i.e. Pb 23.7-32.2 µg/kg, Cd <1-4.3 µg/kg, well below the levels considered in Poland as permissible. Content of copper was also low, at 0.11-0.26 mg/kg. However, the content of zinc was relatively high at 35.8-42.0 mg/kg. Usually, higher content of all analyzed metals was reported in samples taken during the spring season, i.e. when cows are fed with dry fodder.

The levels reported by the Polish laboratories for metal contents (Tables 3 to 7) for most of Polish products for infants and children remained below the maximum permissible levels [24]. Raw materials, particularly cereals, vegetables and fruits used in production are specially selected, are characterized by variable metals content, which depends principally on the local environmental pollution and soil conditions [12,17,18,22,23,26].

Table 2. *Content of Pb, Cd, Cu and Zn in Polish powdered milk in 1991-1994*

	*	Pb µg/kg	Cd µg/kg	Cu mg/kg	Zn mg/kg
Fall 91	I	29.8	2.6	-	35.8
	II	27.1	2.5	-	36.4
Spring 92	I	30.4	2.7	-	36.2
	II	26.5	2.5	-	36.7
Fall 92	I	28.8	2.7	0.18	37.2
	II	27.9	3.2	0.17	37.1
Spring 93	I	27.6	3.4	0.21	38.9
	II	32.2	4.3	0.26	38.0
Fall 93	I	26.6	<1	0.14	42.0
	II	23.7	<1	0.11	38.2
Spring 94	I	30.0	<1	0.26	40.0
	II	31.0	<1	0.18	40.4

*) I, II - parallel collective samples

Table 3. *Content of lead in Polish products for infants and children, mg/kg*

Product	Year	No. of samples	Median	Mean	90th percentile
Bebiko "0" and Bebiko "1"	1992	59	0.02	0.03	0.07
(milk infant formulas)	1993	45	0.01	0.03	0.09
	1994	26	0.01	0.01	0.04
Bebiko 2R and Bebiko2GR	1992	68	0.04	0.05	0.14
(milk/rice/buckwheat infant formulas)	1993	105	0.01	0.02	0.08
	1994	87	0.01	0.03	0.06
Milk/rice and milk/rice/fruit products	1992	173	0.05	0.06	0.15
	1993	174	0.03	0.04	0.10
	1994	142	0.03	0.04	0.08
Milk/wheat/fruit products	1992	104	0.03	0.04	0.12
	1993	79	0.02	0.03	0.10
	1994	14	<0.01	0.02	0.11
Cereal and cereal/fruit products	1992	78	0.04	0.07	0.17
	1993	53	<0.01	0.03	0.10
	1994	49	0.01	0.03	0.08
Vegetable/meat products	1992	38	<0.01	0.01	0.05
	1993	44	0.02	0.03	0.07
	1994	78	0.02	0.02	0.05
Juices	1992	421	0.015	0.03	0.10
	1993	362	0.01	0.02	0.06
	1994	290	<0.01	0.02	0.04
Pomaces	1992	21	0.02	0.06	0.14
	1993	21	0.05	0.05	0.10
	1994	17	0.03	0.04	0.06

Lowest lead and cadmium contents, well below maximum permissible values, were reported in fruit/vegetable juices and in milk products. The average content of lead in these groups of products did not exceed 0.03 mg/kg, and the median was 0.02 mg/kg; the respective values for cadmium are 4 and 2 µg/kg. Low lead contents were also often reported in vegetable/meat products. However, the 90th percentile content of Pb and Cd in cereal, cereal/fruit, milk/cereal/fruit products and pomaces slightly exceeded permitted levels. In some 15% of products in which cereals constitute a significant component the content of cadmium exceed 10 µg/kg, though in all years of investigations the median value was much lower. The highest average content of cadmium (13 µg/kg) was reported in the group of cereal and cereal/fruit products in 1992. Bearing in mind the recognized acccumulation of cadmium in cereals[2,20,21] and results of domestic testing of cereals, it is currently being considered in Poland to increase the permissible cadmium contents of such products to 20 µg/kg.

Results for lead (median of <0.01-0.05 mg/kg, average 0.01-0.07 mg/kg) allow to reduce the permitted content of this metal in all products for infants and children circulated on the Polish market to 0.1 mg/kg. The currently permitted higher content of lead in vegetable/meat products (0.2 mg/kg) was associated with the type packaging used, i.e. cans for which use of lead-containing solder was permitted. As these packagings are being presently withdrawn from use in Poland, in line with FAO/WHO recommendations[3], there is no reason and purpose in defining separate maximally permitted levels for canned food. Research conducted in the USA showed that the replacement of soldered cans gives several-fold reduction in lead content in canned food; use of soldered cans was finally abandoned in the USA in 1993. Similarly, lead content in canned evaporated milk in the period from 1982 to 1986 decreased from 0.11 mg/kg to under 0.01 mg/kg[5]. Similar results were also received in Canada [6].

Table 4. *Content of cadmium in Polish products for infants and children, µg/kg*

Product	Year	No. of samples	Median	Mean	90th percentile
Bebiko "0" and Bebiko "1"	1992	54	2	4	10
(milk infant formulas)	1993	45	<1	3	10
	1994	28	2	4	10
Bebiko 2R and Bebiko2GR	1992	75	2	4	11
(milk/rice/buckwheat infant formulas)	1993	105	3	5	10
	1994	87	3	4	11
Milk/rice and milk/rice/fruit products	1992	180	5	7	20
	1993	172	5	5	13
	1994	142	5	6	13
Milk/wheat/fruit products	1992	114	6	11	30
	1993	78	6	8	15
	1994	14	9	8	21
Cereal and cereal/fruit products	1992	74	6	13	42
	1993	49	1	3	10
	1994	49	7	8	24
Vegetable/meat products	1992	36	1	5	15
	1993	44	6	8	26
	1994	78	8	10	15
Juices	1992	411	1	4	11
	1993	363	1	3	10
	1994	290	<1	3	8
Pomaces	1992	20	10	12	35
	1993	21	7	6	17
	1994	17	6	8	31

Table 5. *Content of mercury in Polish products for infants and children, μg/kg*

Product	Year	No. of samples	Median	Mean	90th percentile
Bebiko "0" and Bebiko "1"	1992	20	2	3	8
(milk infant formulas)	1993	14	1	2	10
	1994	13	1	2	5
Bebiko 2R and Bebiko2GR	1992	27	1	2	6
(milk/rice/buckwheat infant formulas)	1993	48	1	2	6
	1994	27	3	3	6
Milk/rice and milk/rice/fruit produts	1992	67	1	2	5
	1993	61	1	2	9
	1994	62	2	3	7
Milk/wheat/fruit products	1992	26	5	4	9
	1993	25	<1	2	7
	1994	9	-	1	-
Cereal and cereal/fruit products	1992	10	2	2	8
	1993	12	<1	1	3
	1994	25	2	5	10
Vegetable/meat products	1992	18	<1	1	1
	1993	7	-	<1	-
	1994	3	-	1	-
Juices	1992	199	<1	1	4
	1993	175	<1	1	5
	1994	102	<1	1	3
Pomaces	1992	10	<1	1	2
	1993	3	-	<1	-
	1994	5	-	4	-

Content of mercury in the tested products were low, with mean values from <1 to 5 μg/kg and the 90th percentile values were from 1 to 10 μg/kg and does not pose a health hazard.

Reported copper and zinc values were well below maximum permitted levels[24]. The median content of copper is from 0.2 to 1.2 mg/kg and the 90th percentile value is from 0.3 to 3.2 mg/kg. The median content for zinc in milk, milk/cereal and milk/cereal/fruit is from 13.5 to 18.3 mg/kg and the 90th percentile value is from 17.5 to 23 mg/kg; for vegetable and fruit products these values were 0.7- 1.0 and 1.1-2.8 mg/kg, respectively.

Table 6. *Content of zinc in Polish products for infants and children, mg/kg*

Product	Year	No. of samples	Median	Mean	90th percentile
Bebiko "0" and Bebiko "1"	1992	60	15.8	14.6	21.6
(milk infant formulas)	1993	47	15.4	14.2	17.5
	1994	28	13.5	13.9	18.8
Bebiko 2R and Bebiko2GR	1992	80	18.3	17.6	23.0
(milk/rice/buckwheat infant formulas)	1993	105	17.4	17.4	21.4
	1994	87	18.0	16.8	23.0
Milk/rice and milk/rice/fruit products	1992	183	14.8	14.2	20.5
	1993	179	16.0	15.5	20.0
	1994	141	17.5	17.1	22.1
Milk/wheat/fruit products	1992	117	14.5	13.2	19.3
	1993	82	15.1	14.6	19.2
	1994	12	16.3	16.2	22.2
Cereal and cereal/fruit products	1992	78	9.4	9.3	16.6
	1993	54	9.5	9.5	14.3
	1994	49	10.7	11.4	20.2
Vegetable/meat products	1992	40	14.2	13.4	27.4
	1993	44	5.0	5.9	15.4
	1994	78	4.7	5.4	9.3
Juices	1992	430	0.7	0.7	1.1
	1993	379	0.7	0.8	1.2
	1994	291	0.7	0.8	1.2
Pomaces	1992	20	0.9	1.1	2.8
	1993	21	0.8	1.3	2.5
	1994	17	1.0	1.1	2.4

Table 7. *Content of copper in Polish products for infants and children, mg/kg*

Product	Year	No. of samples	Median	Mean	90th percentile
Bebiko "0" and Bebiko "1"	1992	60	0.30	0.29	0.44
(milk infant formulas)	1993	47	0.21	0.21	0.31
	1994	28	0.27	0.26	0.40
Bebiko 2R and Bebiko2GR	1992	79	0.70	0.70	1.00
(milk/rice/buckwheat infant formulas)	1993	105	0.60	0.65	1.00
	1994	87	0.64	0.62	0.92
Milk/rice and milk/rice/fruit products	1992	183	0.82	0.81	1.22
	1993	179	0.90	1.03	2.17
	1994	142	1.02	1.17	1.98
Milk/wheat/fruit products	1992	117	0.72	0.70	1.00
	1993	82	0.74	0.83	1.69
	1994	14	0.74	0.82	2.55
Cereal and cereal/fruit products	1992	77	1.08	1.23	2.50
	1993	54	1.02	1.20	2.70
	1994	49	1.21	1.37	3.25
Vegetable/meat products	1992	40	0.47	0.66	1.10
	1993	43	0.38	0.34	0.58
	1994	78	0.43	0.48	0.64
Juices	1992	430	0.20	0.23	0.37
	1993	380	0.22	0.24	0.38
	1994	291	0.23	0.26	0.49
Pomaces	1992	21	0.25	0.37	1.18
	1993	21	0.29	0.31	0.53
	1994	17	0.33	0.35	0.50

Majority of Polish products for infants and children had relatively low metals content, within the required Polish limits; therefore, even if contamination of water used for dilution is included, the total intake of Pb, Cd and Hg should remain considerably lower than those tolerated by organism (PTWI).

Acknowledgements

The authors wish to express their gratitude to all participating Voivodship Sanitary-Epidemiological Stations.

References

1. Berglund M., Vahter M., Slorach S., Chem. Speciat. Bioavailab., 1991, **3**, 69.
2. Brüggemann J., Ocker H. D., Trends in Trace Elements, Bulletin of the FAO Research Network on Trace Elements, 1989, **1**, 29.
3. CAC FAO/WHO, CX/FAC 95/18, 1994.
4. CAC FAO/WHO, CX/FAC 95/16, 1995, Add. 1.
5. CAC FAO/WHO, CX/FAC 95/18, 1995, Add.2.
6. Dabeka R.W., McKenzie A.D., J. Assoc. Off. Anal. Chem., 1987, **70**, 754.
7. Dabeka R.W., McKenzie A.D., Food Add. Contam. 1988, **5**, 333.
8. Dabeka R.W., McKenzie A.D., J. Assoc. Off. Anal. Chem. International 1992, **75**, 386.
9. Dudek B., Medycyna Pracy, 1993, **44**, 101 (Suppl. 1).
10. Gunderson E.L., J. Assoc. Off. Anal. Chem. 1988, **71**, 1200.
11. Joint UNEP/FAO/WHO Food Contamination Monitoring and Assesment Programme (GEMS/Food), Assessment of Dietary Intake of Chemical Contaminants, 1992.
12. Karlowski K., Wojciechowska-Mazurek M., Chem. Speciat. Bioavailab. 1991, **3**, 21.
13. Kumpulainen J. in Monitoring Dietary Intakes, Ian Macdonald ed., Springer Verlag Berlin-Heidelberg, 1991, 61.
14. Slorach S., Jorhem L., Becker W., Chem. Speciat. Bioavailab. 1991, **3**, 13.
15. Smith M.A., Grant L.D., Sors A.I., ed. Lead Exposure and Child Development, Kluwer Academic Publishers, Dordrecht/Boston/London, 1989.
16. Szymczak J., Regulska B., Roczn. PZH 1987, **38**, 230.
17. Tahvonen R., Kumpulainen J., Fresenius' Z. Anal. Chem. 1991, **340**, 242.
18. Tahvonen R., Kumpulainen J., Food Add. Contam., 1993, **10**, 245.
19. UNEP/GEMS Environment Library No 5, The Contamination of Food, 1992.
20. Wiersma D., van Goor B. J., van der Veen N. G., J. Agric. Food Chem., 1986, **34**, 1067.
21. Wolnik K. A., Fricke F. L., Capar S. G., Braude G. L., Meyer M. W., Satzger R. D., Bonnin E., J. Agric, Food Chem., 1983, **31**, 1240.
22. Wojciechowska-Mazurek M., Zawadzka T., Kar3owski K., Starska K., Æwiek-Ludwicka K., Brulińska-Ostrowska E., Roczn. PZH, 1995, **44**, 223.
23. Zalewski W., Oprz1dek K., Syrocka R., Lipińska J., Jaroszyńska J., Roczn. PZH 1994, **45**, 19.
24. Ordinance of the Ministry of Health and Human Welfare on specification of permitted additives and technical impurities in food products and condiments, 31.03.1993, Monitor Polski Nr 22, poz. 233, 1993.
25. Zawadzka T. Brulińska-Ostrowska E., Æwiek K., Roczn. PZH 1987, **38**, 113.
26. Zawadzka T., Mazur H., Wojciechowska-Mazurek M., Starska K., Brulińska-Ostrowska E., Æwiek K., Umińska R., Bichniewicz A., Roczn. PZH 1990, **41**, 111 i 132.
27. Zawadzka T., Wojciechowska-Mazurek M., Brulińska E., Roczn. PZH 1986, **37**, 473.
28. Becker, W. and Kumpulainen, J. Br. J. Nutr. 1991, **66**, 151

LEVELS AND TRENDS OF PCBS, ORGANOCHLORINE PESTICIDE RESIDUES AND CARCINOGENIC OR MUTAGENIC PAH COMPOUNDS IN FINNISH AND IMPORTED FOODS AND DIETS

Veli Hietaniemi

Agricultural Research Centre of Finland, Food Research Institute / Laboratory of Food Chemistry, FIN-31600 Jokioinen, Finland

1 INTRODUCTION

Although the use of most organochlorine pesticide compounds (OCPs) and PCB has either been reduced or banned in recent years [1], residues of these lipophilic persistent compounds can still be found in fat-containing foods, such as fish, edible fats and oils, meats and dairy products [2,3,4,5,6,7]. Due to the lipophilic nature of most neutral organochlorine compounds, they accumulate in adipose tissue, the liver, kidneys, the central nervous system and breast milk [4,8].

Polycyclic aromatic hydrocarbons (PAHs) are generated by the incomplete combustion of organic material [9]. Fossil fuels, traffic and industry are the principal sources of PAHs. Besides the above-mentioned environmental pollutants PAHs are also formed during certain processes of the food industry and particularly by home preparation of foods by smoking and grilling.

The aim of the present study was to determine the contents of OCPs, PCBs and PAHs in representative samples of Finnish milk and cereals, in domestic and imported cheeses, breakfast cereals, egg and meat products, fish and vegetables.

2 MATERIALS AND METHODS

2.1 Standards and reagents

2.1.1 PCBs and OCPs. Solid organochlorine pesticide standards of hexachlorobenzene (HCB), hexachlorocyclohexanes α-, β- and γ-HCH (lindane), p,p'-DDT, o,p'-DDT, p,p'-DDE, o,p'-DDE, p,p'-DDD and o,p'-DDD, heptachlor, aldrin, dieldrin, endrin, oxychlordane, β-heptachlor epoxide (β-Hepo), cis- and trans-chlordane, transnonachlor, α- and β-endosulfan and endosulfan-sulphate were purchased from Dr. Ehrenstorfer (Augsburg, Germany) and Promochem (Ulricehamn, Sweden).

Solid PCB congener standards, numbered according to the IUPAC numbers 8, 15, 20, 28, 35, 52, 53 (int. std.), 77, 81, 101, 105, 114, 118, 126, 128, 138, 149, 153, 156, 169, 170 and 180, were supplied from Dr. Ehrenstorfer (Augsburg, Germany) and Promochem (Ulricehamn, Sweden). When calibrating the GC-MS or GC-MSD for the determination of OCPs and PCBs, standard solutions were prepared at five different concentrations ranging from 0.01 to 5 µg/ml. Mixtures were stored at - 20 °C and used at room temperature.

2.1.2 PAHs. Phenanthrene (Phe), anthracene (Ant), fluoranthene (Flu), pyrene (Pyr), benz(a)anthracene (BaA), triphenylene (Tri), chrysene (Chr), benzo(a)pyrene (BaP), benzo(e)pyrene (BeP), benzo(ghi)perylene (BghiP), benzo(b)fluoranthene (BbF), benzo(j)fluoranthene (BjF), benzo(k)fluoranthene (BkF), dibenz(a,h)anthracene (BahA), indeno(1,2,3-cd)pyrene (IcdP) and b,b'-binapthyl (int. std.) were supplied from Dr. Ehrenstorfer (Augsburg, Germany). When calibrating the GC-MSD for the determination of PAHs, standard solutions were prepared at six different concentrations: 0.1...10 µg/ml. Mixtures were stored at -20 °C and used at room temperature.

2.2 Apparatus and analytical conditions

2.2.1 PCBs and OCPs. A Hewlett Packard Model 5890 gas chromatograph (GC) equipped with a Model 5971 mass selective detector (MSD) or a Model 5989 MS Engine mass spectrometer (MS) were employed for the determinations. Operating conditions were as follows: for MS 25 m x 0.32 mm i.d. silica capillary column with film thickness of 0.25 µm (PAS-1701) or for MSD 30 m x 0.25 mm i.d. silica capillary column with film thickness of 0.25 µm (HP-5); helium carrier gas 0.5 - 1.0 ml/min; injector and detector temperature 250 °C; temperature programme: 70 °C (1.0 min), 70-160 °C (18 °C/min), 160 °C (0.5 min), 160-210 °C (30 °C/min), 210 °C (1.5 min), 210-240 °C (4 °C/min), 240 °C (7 min), ionizing voltage 70 eV and injection (splitless) volume 1-2 µl.

Ions monitored for the detection of OCPs were: 284/286 (HCB); 183/181 (HCHs); 235/237 (p,p'-DDT, o,p'-DDT, p,p'-DDD and o,p'-DDD); 246/318 (p,p'- and o,p'-DDE); 272/274 (heptachlor); 263/261 (aldrin, dieldrin and endrin); 387/389 (oxychlordane); 353/355 (β-Hepo); 375/373 (cis- and trans-chlordane); 409/407 (transnonachlor); 241/339 (α- and β-endosulfan) and 387/385 (endosulfan-sulphate).

Ions monitored for the detection of PCB congeners were: 222/224 (PCBs 8 and 15; Cl_2); 256/258 (PCBs 20, 28 and 35; Cl_3); 292/290 (PCBs 52, int. std. 53, 77 and 81; Cl_4); 326/324 (PCBs 101, 105, 114, 118 and 126; Cl_5); 360/362 (PCBs 128, 138, 149, 153, 156 and 169; Cl_6) and 394/396 (PCBs 170 and 180; Cl_7).

2.2.2 PAHs. A Hewlett Packard Model 5890 gas chromatograph equipped with a Model 5971 mass selective detector was employed for the analysis. Operating conditions were as follows: 30 m x 0.25 mm i.d. silica capillary column with film thickness of 0.25 µm (HP-5); helium carrier gas 0.5 ml/min; injector and detector temperatures 250 °C and 260 °C, respectively; temperature programme: 70 °C (0.5 min), 70-200 °C (15 °C/min), 200 °C (2 min), 200-250 °C (15 °C/min), 250 °C (10 min), 250-260 °C (10 °C/min); ionizing voltage 70 eV and injection (splitless) volume 1 or 2 µl.

Following ions were monitored for the detection of polycyclic aromatic hydrocarbons: 177.9/151.9 (Phe and Ant); 202.0/100.7 (Flu); 201.9/101.0 (Pyr); 228.0/113.0 (BaA); 228.0/113.1 (Tri); 227.9/113.0 (Chr); 252.0/125.8 (BbF); 252.0/125.2 (BjF); 252.0/126.0 (BkF); 252.0/125.0 (BeP); 252.0/126.0 (BaP); 278/279 (BahA); 276.0/137.7 (IcdP and BghiP) and 254.0/126.0 (int. std. β,β'-binapthyl).

2.3 Samples

Regionally representative Finnish milk, cheese and meat samples were obtained from the principal dairies and slaughter-houses, respectively. Correspondingly, domestic rye bread and rye flour samples were collected from nationally or regionally important bakeries. Imported cheese, yolk powder and beef tenderloin samples were collected from the National Veterinary and Food Research Institute.

In addition, domestic and imported breakfast cereals, vegetables, fish and fish products were collected from the major Finnish wholesalers. Altogether 1300 samples pooled from 10 000 individual samples of milk and milk products, cereals and breakfast cereals, egg and meat products, fish and vegetables were analysed.

2.4 Extraction and clean-up

OCPs and PCB congeners were isolated and cleaned-up by solvent extraction and concentrated sulphuric acid treatment or by gel permeation chromatography [2,6]. PAHs were isolated and cleaned-up by saponification, various liquid-liquid extraction systems and silica gel column clean-up [10]. The purified sample extracts were then analysed by GC-MSD or GC-MS.

3 RESULTS AND DISCUSSION

3.1 PCBs and OCPs

The results showed that concentrations of PCBs and OCPs in domestic milk and milk products, edible oil and margarine samples were negligible [2,11]. Concentrations of OCPs, such as p,p'-DDE, p,p'-DDD and PCBs were higher in imported cheese samples than in domestic cheese samples [2]. Contents of PCBs in domestic egg samples were slight, but α-HCH concentrations in some samples were relatively high. PCB and OCP concentrations in meat and meat product samples [11] were low on average, except in some imported beef tenderloin samples which had quite high concentrations of p,p'-DDE (160 µg/kg fat). However, Σ PCB congeners in imported and domestic beef tenderloin samples were slight and at the same concentration level (21-28 µg/kg fat).

Differences in PCB and OCP concentrations in various fish samples were relatively large [11]. Higher PCB and OCP contents were encountered in Baltic herring (Σ PCB 109 µg/kg and Σ DDT 97 f.w. on average) and Baltic salmon (Σ PCB 199 µg/kg and Σ DDT 147 f.w. on average) samples. On the other hand, the lowest contents of Σ PCBs (0.2-18 µg/kg f.w.) and OCPs (Σ DDT 0.2-6.3 µg/kg f.w.) were found in imported tuna fish, sole, mackerel, flounder and in Finnish whitefish, perch and vendace samples. Altogether, PCB and OCP concentrations in fish and in all samples analysed were well below the present tolerance limits established by FAO/WHO Codex Alimentarius [12] or by the Finnish Ministry of Trade and Industry. For example, Finland has established official tolerance limits (mg/kg f.w.) for PCB congeners in fish and fish products as follows: PCB 28 0.6, PCB 52 0.1, PCBs 101, 118, 138, 153 and 180 0.2 and other individual congeners 0.6. The tolerance limit of 2.0 mg/kg f. w. for Σ PCB has also been established. Average intake of total PCBs investigated was determined to be approximately 1.6 µg/d per capita [11]. This is a lower intake or at the same level as those reported earlier from other countries [13,14,15]. Fish and meat products were the principal sources of the PCBs studied.

3.2 PAHs

Regional differences in PAH concentrations in Finnish milk samples were encountered [16]. Highest PAH contents were found in samples from industrialized areas where large-scale manufacturing industries employ fossil fuels. On the other hand, lowest PAH contents were found in milk samples from Lapland and Ahvenanmaa. PAH concentrations in domestic cheese samples were slight (1-4 µg/kg) and regional differences were not encountered. However, substantial differences in PAH concentrations (2-126 µg/kg) were found in various imported cheese samples, depending on the country and type of cheese. Contents of PAH in imported cheese samples were higher on average than in domestic cheese samples. Relatively large differences, depending on the country and various lots of yolk powder, were also detected in PAH concentrations (4-189 µg/kg) in imported yolk powder samples. Contents of domestic bulk liquid egg samples [16] were at the same level (38 µg/kg) as on average in imported yolk powder samples.

Average Σ PAH concentrations in meat and meat product samples were slight (7-70 µg/kg) with a few exceptions. Σ PAHs in imported and domestic beef tenderloin samples were low and at the same concentration level (13-15 µg/kg). However, the results showed that attention should also be paid to the study of PAHs in the raw materials of meat and meat products and not focus only on PAHs in grilled and smoked products. In addition, PAHs in edible oil and margarine samples were low. Σ PAHs in imported edible oil samples (26-134 µg/kg) were higher than in domestic edible oil samples (10-24 µg/kg). Average PAH contents were low in the vegetable group. Highest Σ PAH concentrations were found in imported butterhead lettuce and paprika samples (75-90 µg/kg).

Lowest Σ PAH concentrations in all analysed food samples were detected in fish, such as various imported canned fish, flounder, sole, whitefish, vendace and in domestic perch, pike and rainbow trout samples. On the other hand, highest Σ PAH concentrations in fish samples were encountered in imported sardine samples (22 µg/kg).

An interesting result was that carcinogenic compounds such as benz(a)antrachene, chrysene, benzofluoranthenes, benzo(a)pyrene and indeno(1,2,3-cd)pyrene were found in traditionally dried rye bread and corresponding flour samples, and in the meat group, edible oil and margarine samples. However, most of the compounds detected were not carcinogenic, such as phenanthrene, anthracene, fluoranthene and pyrene. According to the previous studies and official tolerance limits established by many countries, only smoked and grilled products together with benzo(a)pyrene (BaP) have been considered important. Finland has issued guidelines for BaP with a tolerance limit of 1 µg/kg in smoke aroma preparations. Conversely, in Austria, PAH compounds in smoke aroma preparations are prohibited and the tolerance limit for BaP in smoked foods is 1 µg/kg. In addition, Germany has established tolerance limits of 1 µg/kg for BaP in meat and cheese. The average intake of the PAHs investigated was determined to be approximately 18 µg/d per capita [17]. Cereals, meat, fruits and vegetables were the most important sources of the PAHs investigated. This seemed to be in good agreement with those reported earlier [18,19,20].

References

1. E. C. Voldner and Y.-F. Li *The Science of the Total Environment* **160/161** (1995) 201-210.

2. V. Hietaniemi and J. Kumpulainen *Food Additives and Contaminants* **11** (1994) 685-694.

3. F. Goni, E. Serrano, J. M. Ibarluzea, M. Dorronsoro and L. G. Galdeano *Food Additives and Contaminants* **11** (1994) 387-395.

4. R. Vaz *Organochlorine contaminants in Swedish foods of animal origin and in human milk 1973-1992. Ph.D. Dissertation*, Uppsala University, 1993.

5. P. Fürst, C. Fürst and K. Wilmers *Chemosphere* **25** (1992) 1039-1048.

6. R. Moilanen, J. Kumpulainen and H. Pyysalo *Annales Agriculturae Fenniae* **25** (1986) 177-184.

7. J. Falandysz, S. Tanabe and R. Tatsukawa *The Science of the Total Environment* **145** (1994) 207-212.

8. L. G. M. Th. Tuinstra, M. Huisman and E.R. Boersma *Chemosphere* **29** (1994) 2267-2277.

9. K. D. Bartle, M. L. Lee and S. A. Wise, *Chemical Society Reviews* **10** (1981) 113.

10. G. Grimmer and H. Böhnke, *Journal of the Association of Official Analytical Chemists* **58** (1975) 725.

11. A. Mustaniemi, V. Hietaniemi, A. Hallikainen and J. Kumpulainen *Intake of PCB-compounds via food. Publications of the National Food Administration*, Helsinki 1995.

12. FAO/WHO *Codex Maximum Limits for Pesticide Residues, Codex Alimentarius Volume XIII*, 1986

13. G. Moy, F. Käferstein, Y. M. Kim, Y. Moterjemi and F. Quevedo *Archiv für Lebensmittelhygiene*, **44** (1993) 22-56.

14. UNEP/FAO/WHO *Assessment of dietary intake of chemical contaminants. Joint UNEP/FAO/WHO Food Contamination Monitoring and Assessment Programme*, 1992.

15. H. G. Gorchev and C. F. Jelinek *Bulletin of the World Health Organization* **63** (1985) 945-962.

16. V. Hietaniemi *PAH-compounds . Publications of the National Food Administration*, Helsinki 1993.

17. V. Hietaniemi and E.-L. Kupila *Analytical Sciences* **7** (1991) 979-982.

18. H. A. M. G. Vaessen, A. A. Jekel and A. A. M. M. Wilbers *Toxicological and Environmental Chemistry* **16** (1988) 281.

19. J. Tuominen, *Determination of polycyclic aromatic hydrocarbons by gas chromatography/mass spectrometry and method development in supercritical fluid chromatography, Technical Research Centre of Finland, Publications 60*, Espoo, Finland 1990.

20. B. Larsson *Polycyclic aromatic hydrocarbons in Swedish foods. Ph.D. Dissertation*, Uppsala University, 1986.

LEVELS OF ORGANOCHLORINE AND ORGANOPHOSPHORUS PESTICIDE RESIDUES IN GREEK HONEY

D. Tsipi[1], A. Hiskia[1] and M. Triantafyllou[2]

1. General Chemical State Laboratory, 16 An. Tsoha, 11521, Athens, Greece
2. Agricultural University of Athens, 75 Iera str. 11855, Athens, Greece

1 INTRODUCTION

The monitoring of trace levels of pesticide residues in various food commodities is considerably important for the human health protection. In the case of honey the contamination may be caused by the use of pesticides directly to the beehive against the Varroa mite diseases[1] or indirectly following the use of pesticides in agriculture. The presence of organochlorine pesticides in honey has also been reported in several papers[2].

The aim of the present study was a) the development of a simple and rapid clean up method for the isolation of organochlorine pesticide residues from the honey and b) the determination of 17 organochlorine and 10 organophosphorus pesticide residues in honey samples produced by different Greek companies.

A clean up step by solid phase extraction (C18) followed by gas capillary chromatography with ECD detection was performed for the determination of organochlorine pesticide residues in honey samples.

The mean recovery was estimated at three concentration levels.

The determination of organophosphorus pesticide residues carried out with a multiresidue method[3] followed by gas capillary chromatography with two different special detectors FPD and NPD.

The mean recovery was estimated at two concentration levels.

The results of this study provide information for estimating the residue levels of 27 pesticides in 42 honey samples.

2 MATERIALS AND METHODS

2.1 Analysis of organochlorine pesticide residues in honey samples

2.1.1. Apparatus. a) Gas chromatograph: Varian Model 3400 equipped with ECD, split/splitless injection port and autosampler Model 8200 cχ, with program for the evaluation of GC runs (DAPA). b) GC column: 30m x 0.32mm id DB-1 fused silica capillary column (J&W Scientific Inc.), film thickness 0.25μm. The chromatographic conditions are listed in Table 1.

Table 1 *Chromatographic conditions for column DB-1*

Conditions	DB-1
TEMPERATURE PROGRAM	
Initial temp. (°C)\ Time (min)	80\1
Programming rate (°C/min)	8
1st step temp (°C)\ Time (min)	218\18
Programming rate (°C/min)	4
2nd step temp (°C)\ Time (min)	250\10
Inlet temperature (°C)	250
Detector temperature (°C)	300
FLOW RATES	
Carrier gas (ml/min)	1,5
Make up gas (ml/min)	20
SPLITLESS INJECTION	
Valve time on (sec)	42"

2.1.2 Materials. During clean up step, hexane, methanol, free of pesticides were used. For the recovery experiments and for the standarization of the detector standard solutions of 17 organochlorinated pesticides were prepared.

2.1.3 Method of analysis. The analysis of organochlorine pesticides in honey samples was performed by gas chromatography with ECD using solid phase extraction for isolation.

The clean up procedure was as follow: 10 gr of honey sample was dissolved in 50 ml of methanol and the mixture was stirred for an hour. Then 25 ml of the above solution was diluted in 2L distilled water, in pH 2. The mixture was passed through the C18 column (Waters Ass., part no. 51910) pre-treated with 10ml methanol followed by 5 ml of water. The C18 column was fitted to a glass column (25cm x 1cm) which was connected with a 2L flask reservoir. After the sample volume had passed through the column, the cartridge was dried for 30min under stream of nitrogen and the organochlorine pesticides were eluted with 10 ml of n-hexane. The extract was concentrated to 5 ml and transferred to an autosampler vial for GC-ECD analysis.

2.2 Analysis of organophosphorus pesticide residues in honey samples

2.2.1 Apparatus. a) Gas chromatograph: (1) Carlo Erba Model Mega 2 equipped with FPD, split/splitless injection port and autosampler Model A200S, with program for the evaluation of GC runs (Chrom-Card). (2) Hewlett-Packard Model 5890 II, equipped with NPD, split/splitless injection port and autosampler Model 7673, with program for the evaluation of GC runs (HPCHEM).

b) GC column: (1) 30m x 0.32mm id DB-1 fused silica capillary column (J&W Scientific Inc.), film thickness 3μm. (2) 50m x 0.32mm id CP-Sil 13 CB fused silica capillary column (Chrompack) fused silica capillary column (Chrompack), film thickness 0.4μm. The chromatographic conditions for the two columns are listed in Table 2.

Table 2 *Chromatographic conditions for two different columns.*

Conditions	DB-1	CP-SIL 13 CB
TEMPERATURE PROGRAM Initial temp. (ºC)\ Time (min) Programming rate (ºC/min) 1st step temp (ºC)\ Time (min) Programming rate (ºC/min) 2nd step temp (ºC)\ Time (min)	80\1 12 270\43	80\1 12 270\30
Inlet temperature (ºC)	250	250
Detector temperature (ºC)	250	290
FLOW RATES Carrier gas (ml/min) Make up gas (ml/min)	1,2 30	0,9 (He) 25 (He)
SPLITLESS INJECTION Valve time on (sec)	60"	60"

2.2.2 Materials. During clean up hexane, dichloromethane, water and acetone free of pesticides were used. For recovery experiments and for the standarization of the detector standard solutions of 10 organophosphorus pesticides were prepared.

2.2.3 Method of analysis. The analysis of organoposphorus pesticides in honey samples was performed by gas chromatography with FPD and NPD.

The clean up procedure was as follow: Weigh 50 g of honey sample into a high speed blender jar and add 50 ml water, 30 g NaCl and 200 ml acetone. Blend at high speed for 1-2 min. Add 150 ml dichloromethane and blend for 1-2 min again. Tranfer the organic phase into a 400ml beaker and let it settle down. Tranfer the upper layer through a funnel with glasswool and sodium sulfate into a 400 ml measuring cylinder. Determine the volume. Take an aliquot 175 ml and reduce the volume to 3-5 ml. Add 5ml acetone and reconcentrate.

3 RESULTS AND DISCUSSION

3.1. Determination of organochlorine pesticides in honey.

3.1.1 GC separation. Retention times (RT) of 17 organochlorine pesticides were determined individually on DB-1 and are given in Table 3.

The detection limits were determined in real samples, from the minimum detectable amount of pesticides (S\N: 3/1) and are presented in Table 3.

Table 3 *Retention times (RTs) and detection limits of 17 organochlorine pesticides on DB-1 column and ECD.*

N	Organochlorine Pesticides	Detection limits (µg/L)	RT (min)
1	A-HCH	≤0.080	12.67
2	LINDANE	≤0.002	13.57
3	D-HCH	≤ 0.160	14.22
4	HEPTACHLOR	≤ 0.008	15.54
5	ALDRIN	≤ 0.160	16.43
6	HEPTACHLOR EPOXIDE	≤ 0.016	17.44
7	A-ENDOSULFAN	≤ 0.080	18.42
8	4,4 DDE	≤ 0.160	19.20
9	DIELDRIN	≤ 0.160	19.25
10	ENDRIN	≤ 0.160	19.88
11	B-ENDOSULFAN	≤ 0.160	20.18
12	4,4 DDD	≤ 0.160	20.70
13	ENDRIN ALDEHYDE	≤ 0.160	20.87
14	ENDOSULFAN SULFATE	≤ 0.160	21.88
15	4,4DDT	≤ 0.032	22.27
16	METHOXYCHLOR	≤ 0.080	24.01
17	ENDRIN KETONE	≤0.160	25.61

3.1.2. Recoveries. The mean recoveries and the standard deviations of 17 organochlorine pesticides, at three fortification levels, are given in Table 4.

Table 4 Mean recoveries of 17 organochlorine pesticides and relative standard deviations at 3 different fortification levels in honey (n=3).

N	fortif. level 20 µg / kg	fortif.level 10 µg / kg	fortif. level 4 µg / kg
1	66.4 ±5.7	70.7 ± 4.5	79.8 ± 12.1
2	65.1 ± 8.3	93.0 ± 8.1	85.7 ± 9.5
3	120.0 ± 5.6	84,9 ± 4.2	107.2 ± 12.7
4	65.5 ± 3.3	84.9 ± 9.7	76.9 ± 3.3
5	60.7 ± 5.1	87.1 ± 7.3	72.4 ± 2.4
6	75.0 ± 6.7	82.2 ± 3.6	81.1 ± 3.7
7	71.2 ± 3.6	72.6 ± 1.1	55.6 ± 5.8
8	79.9 ± 5.7	76.6 ± 6.7	78.8 ± 2.5
9	83.2 ± 5.1	102.8 ± 4.3	80.0 ± 6.3
10	107.9 ± 2.7	93.8 ± 3.4	87.5 ± 4.7
11	71.9 ± 4.5	73.6 ± 6.5	87.3 ± 7.8
12	115.7 ± 6.7	100.8 ± 6.9	77.3 ± 7.3
13	80.2 ± 3.1	104.3 ± 1.2	48.0 ±7.9
14	73.7 ± 2.1	91.7 ± 3.5	72.6 ± 4.1
15	79.0 ± 3.5	101.7 ± 1.8	124.8 ±2.2
16	81.3 ± 2.7	72.9 ± 2.7	53.1 ± 3.1
17	74.2 ± 1.5	82.7 ± 4.1	78.5±4.3

The recoveries of all organochlorine pesticides at 20 µg/kg fortification level were between 65.5-120%. The recoveries of the same pesticides were better at 10 µg /kg fortification level, ranging from 70.7 to 104.3%.The recoveries at the lower fortification level were between 53.1-124.8%. The average recovery was 82.2 ±15.9.

3.2 Determination of organophosphorus pesticides in honey.

3.2.1 GC separation. Retention times (RT) of 10 organophosphorus pesticides were determined on two different columns and two different detectors and are given in Table 5.

Table 5 *Retention times, RTs, of 10 organophosphorus pesticides on two different columns and different detectors.*

N	Organophosphorus Pesticides	RT (min) DB-1 FPD	RT (min) CP-SIL13 CB NPD
1	Dimethoate	31.20	21.490
2	Diazinon	34.08	21.877
3	Malaoxon	37.87	25.924
4	Parathion Me	38.83	26.359
5	Malathion	41.70	28.423
6	Fenthion	42.95	29.425
7	Parathion Et	43.50	29.334
8	Methidathion	48.78	34.878
9	Carbophenothion	61.08	42.664
10	Coumaphos	—	60.565

3.2.2 Recoveries. Recoveries of 10 organophosphorus pesticides and the relative standard deviations in honey samples, at two fortification levels, are given in Table 6.

Table 6 *Mean recoveries of 10 organophosphorus pesticides and relative standard deviations at two different fortification levels in honey.*

N	Organophosphorus Pesticides	Fortif. level 0.1 mg/Kg	Fortif. level 0.01 mg/Kg
1	Dimethoate	121.6 ± 8.7	114.4 ± 8.1
2	Diazinon	49.2 ± 3.0	51.6 ± 1.3
3	Malaoxon	116.5 ± 9.3	117.3 ± 11.9
4	Parathion Me	91.3 ± 9.9	99.34 ± 7.7
5	Malathion	81.1 ± 5.3	74.6 ± 4.3
6	Fenthion	81.2 ± 5.7	100.3 ± 2.6
7	Parathion Et	91.3 ± 9.9	67.1 ± 11.1
8	Methidathion	69.6 ± 2.2	74.1 ± 9.6
9	Carbophenothion	67.4 ± 2.6	64.2 ± 3.0
10	Coumaphos	72.9 ± 6.9	81.3 ± 7.9

The recoveries of organophosphorus pesticides were very efficient, except diazinon (49.2-51.6%). The average recovery was 84.3% ±21.0.

3.2.3 Results. The levels of 17 organochlorine pesticide residues in honey samples are presented in Table 7.

Table 7 *Levels of 17 organochlorine pesticides residues in 42 honey samples. Percentage % of samples with positive value and samples with value near max value. Max value, min value.*

N	% samples with positive value	max value (µg/Kg)	min value (µg/Kg)	% samples with value near max value
1	59.5	1.20	0.18	11.9
2	78.6	1.20	0.03	11.9
3	11.9	0.71	0.22	2.3
4	73.8	2.24	0.07	19.0
5	2.3	0.09	0.09	2.3
6	21.4	2.38	0.05	2.3
7	47.6	1.52	0.16	11.9
8	28.5	1.10	0.07	11.9
9	40.5	1.90	0.26	21.4
10	33.3	1.47	0.09	19.0
11	2.5	0.12	0.12	2.3
12	21.4	1.29	0.10	2.3
13	11.9	1.34	0.49	2.3
14	2.3	1.16	0.16	2.3
15	19.0	4.75	0.20	2.3
16	2.3	0.08	0.08	2.3
17	0.0	0.00	0.00	0.0

The concentation leves of 17 organochlorine compounds were found to be between 4.75-0.03 ppb.

The levels of 10 organophosphorus pesticides which were determined in 42 honey samples produced during 1992/93, are presented in Table 8. The concentrations are given in mg/Kg.

Table 8 *Levels of organophosphorus pesticide residues in 42 honey samples.*
Percentage % of samples with positive value and samples with value near max value. Max value ,min value.

N	% samples with positive value	max value (mg/Kg)	min value (mg/Kg)	% samples with value near max value
1	4.8	0.25	0.01	2.4
2	0	0	0	0
3	21.4	2.10	0.02	2.4
4	0	0	0	0
5	4.8	0.04	0.04	4.8
6	0	0	0	0
7	0	0	0	0
8	0	0	0	0
9	0	0	0	0
10	0	0	0	0

3.3 Conclusions

The proposed GC-ECD method (C18 Sep-Pak) is suitable for the reliable determination of organochlorine pesticide residues in honey without a further clean up step.

With that method, the 17 organochlorine pesticides which were studied, where determined down to 0.2µg/Kg in honey samples.

The presence of malathion (+malaoxon) residues in some honey samples might be attributed to the use of this substance in the beehive against the bees diseases.

The presence of dimethoate residues (4.8%), in honey samples may be caused by spraying from the air, due to the agricultural practice for olive tree protection.

References

1. A. T. Thrasyvoulou, N. Pappas,1988, J. of Agric. Res. **27(1)**, 55.
2. A. M. Fernandez Muino, J. Simal Lozano 1991, Analyst **116**, 269.
3. H. Steinwandter, 1985, Fresenius Z. Anal. Chem. **322**, 752.

Index of Authors

Adlercreutz H., 349
Akimoto K., 241
Alfthan G., 161
Arend A., 60
Aro A., 168
Atalay M., 36
Benitez L., 393
Bianchini F., 73
Birk.R., 83
Björkhem I., 11
Brüggemann J., 417
Chopra M., 150
Dragsted L.O., 365
Dubner D., 46
Dubrovshchic O.I., 130
Dujic I.S., 199
Favier A., 113
Freitas J.P., 27
Gaspar J., 290
Gebicki J.M., 87
Giray B., 208
Grønbæk M., 260
Hietaniemi V., 432
Hägg M., 213, 225
Häkkinen S.H., 298
Ito Y., 373
Kamal-Eldin A., 230
Kansanen L.A., 386
Kazdova L., 31
Korpela H., 110
Kvíèala J., 177
Lacsamana M., 55
Larson D., 22
Luoma P.V., 123
MacPherson A., 172
Maiani G., 264
Mangoni di S.Stefano C., 41
Miller N.J., 69, 256
Mira L., 281
Mälkki Y., 311
Nordberg G.F., 405
Nordberg M., 412
Nyyssönen K., 18
Osato J.A., 273
Persson-Moschos M., 195

Pietta P.G., 249
Plaami S.P., 328, 338
Priemé H., 78
Prosky L., 303
Rabe E., 322
Rambeck W.A., 156
Rauma A-L., 145, 389
Romera J.M., 221
Salonen J.T., 3
Sandström B., 317, 334
Sasaki R., 52
Scherat T.G., 286
Sharoni Y. , 378
Stahl W., 95, 102
Thompson L.U., 356
Tsipi D., 437
Törrönen A.R., 135
Viegas-Grespo A.M., 188
Vrana A., 141
Weigert P., 401
Wojciechowska-Mazurek M., 423
Yamashita K., 236
Ylä-Herttuala S., 7

Subject Index

Alcohol intake, 261
Alfatocopherol (see Vitamin E)
Allicin, 288
Anthocyanins, 256
Antioxidant activity
   and comparison of TAA methods, 70, 71
   and ferrylmyoglobin/ABTS assay, 69
Antioxidant vitamins
   and HPLC, 221
   and infant formulas, 221
   and intake, 106
   and serum concentrations, 147
   and supplementation, 113
   and surgical stress, 130
Antioxidants, 221, 230, 237, 241
   and antagonistic effects, 117
   and mortality from ischaemic heart disease, 123
   and prevention of cancer, 52
   and prevention of cardiovascular diseases, 4, 7
   and prooxidant effects, 117
   and risk, 116
   and synergistic effects, 118
Apigenin, 251
Ascorbic acid, 393
Ascorbic acid (see Vitamin C)
Atherogenesis (see Cardiovascular disease)
ß-carotene
   and antioxidant activity, 96
   and antioxidative effect, 117
   and canning, 213
   and cooking, 214
   and daily intake, 252
   and HPLC, 136, 146, 151, 251
   and packaging, 216
   and respiratory burst of human phagocytes, 393
   and response to chronic supplementation, 135
   and serum concentrations, 137
   and storage, 218
ß-glucan, 328
ß-tocopherol (see Vitamin E)
Black currant extract, 386
Cadmium binding compounds in wheat, 418, 422
Caffeic acid, 264, 299
Caffein, 264
Calcium panthothenate, 130
Canthaxanthin, 96

Cardiovascular disease
   and dietary fiber, 319
   and fish oils, 60
   and LDL oxidation, 7, 18, 41
   and natural antioxidants, 4
   and oxidative stress, 3, 18, 22, 31, 40
   and oxysterols, 11
   and vitamin E, 35
   development, 7, 41
Carotenoid esters, 99
Carotenoids
   and actions, 95
   and antioxidants, 96
   and associations, 95
   and bioavailability, 95
   and biokinetics, 97
   and biological activity, 95
   and canning, 213
   and freezing, 214
   and functions, 95
   and gap junctional communications, 96
   and intake, 102, 152
   and plasma concentrations, 104, 152
Cathechins, 252, 256, 264, 299
Chlorophyllin
   and respiratory burst of human phagocytes, 393
   antimutagenic effect, 393
Cholesterol
   and enzymatic method, 146
   and mortality from ischaemic heart disease, 126
   and plasma concentrations, 153
Cinnamic acid, 256
Coumaric acid, 299
Coumarins
   7-hydroxylation of, 389
   and P-450 isoenzyme CYP2A6, 389
Cyanidin, 256
Deltatocopherol (see Vitamin E)
Diabetes mellitus
   and antioxidant therapy, 31
   and exercise, 22
   and oxidative stress, 22, 27, 31
Dietary fiber
   and associated components, 334
   and cancer, 319
   and coronary heart disease, 319

Subject Index

and lipid metabolism, 317
and mineral bioavailability, 319, 334
and obesity, 319
and phytic acid, 317, 328, 338
contents, 327, 328, 338
enrichment of baked goods, 327
gastrointestinal effects, 318
health implications, 317, 328, 338
intakes, 311, 327
labelling, 303, 327
methods, 303, 327
physico-chemical characteristics, 317
recommendations, 311, 317, 322
sources, 311
Dietary intakes of heavy metals
and carcinogenic effects, 409, 410
and health effects, 405
and risk assessment, 405, 407
and risk assessment - deterministic effects, 407, 408
Dietary n-3 polyunsaturated fatty acids, 60
Ellagic acid, 299
Epicatechin gallate, 264
Epicathechin, 256, 264, 299
Epigallocatechingallate, 264
Epigallocathechin, 264
Ferulic acid, 299
Flavan-3-ols, 256
Flavanones, 256
Flavonoids
and antioxidants, 249, 256, 273
and berries, 298
and cardiovascular heart diseases, 250
and CE, 251
and free radicals, 249, 273
and HPLC, 251, 298
and inhibition of AHH activity, 386
and inhibition of carcinogen activating enzyme CYP1A1, 386
and intake, 252
and lipid peroxidation, 250, 274
and oxidative stress, 249
and phenoxyl radicals, 249
and prooxidant activities, 290
Flavonol aglycones, 300
Free fatty acids, 143
Galangin, 290
and chromosomal aberration, 293
Gallic acid, 299
Gammatocopherol (see Vitamin E)
Garlic extract
and gluthation, 287
and oxidative stress, 286
and thiobarbituric acid reactive substances (MDA), 287

GEMS/FOOD-EURO
and a WHO/FAO/UNEP approach towards better quality foods in Europe, 401
and analytical quality assurance, 402, 403
and evaluation of exposure, 403
and programme components, 402
Glucose, 143
Glutathione
and exercise, 36
and oxidative stress, 36
and vitamin E, 37
Glutathione peroxidase, 147
Gluthation, 287
Heavy metals, copper and zinc , 425, 430
Hesperetin, 256
Hydroxycinnamic acids, 300
Indole- glucosinolates, 365
Insulin resistance syndrome
and antioxidant therapy, 31
and cardiovascular disease, 31
Iodometric peroxide assay
and advantages and disadvantages, 87, 88
and fraction of peroxide found, 91
Iron
bioavailability, 334
influence of dietary fiber to bioavailability, 334
influence of phytic acid to bioavailability, 334
Isoflavonoids, 349
and colon cancer, 349
anticancer effects of , 349
epidemiology, 349
metabolism, 349
occurrence, 349
Isothiocyanates, 365
Kaempferol, 256, 290, 299
and chromosomal aberration, 292
and DNA damage, 292
LDL, 3, 7, 18, 41
LDX oxidation, 150
Lead, 208
Lignans
and colon cancer, 349, 356
and colonic becteria, 356
and mammary tumorigenesis, 356
anticancer effects of , 349
biological effects, 356
enterodiol, 349, 356
enterolactone, 349, 356
epidemilogy, 349
flaxseeds, 356
mammalian, 349, 356
matairesinol, 349, 356
mechanism of action, 349, 356
metabolism, 349
occurrence, 349, 356
plant , 349, 356

secoisolariciresinol, 349, 356
  urinary, 356
Lipid peroxidation
  and cardivascular disease, 3, 7, 45
  and diabetes, 27
  and irradiation, 46
  and lipid peroxide and superoxide dismutase levels in serum and cancer mortality in Japan, 52
  and myoglobin, 282
  and rat liver microsomes, 275
Living food diet
  and antioxidant status, 145
  and coumarin 7-hydroxylation, 389
Lutein, 151
Lycopene, 96
  and the growth of DMBA induced rat mammarytumors in vivo, 378
  and the growth of MCF-7 mammary cancer cells in vitro, 378
  anticarcinogenic properties, 378
  antioxidant activity, 378
  effect on expression of a gene encoding connexin 43, 378
  intervention in autocrine growth factor action, 378
  mechanism of anticancer action, 378
Malvidin, 256
Mineral bioavailabilit, 334
Myoinositol phosphates
  and dietary fiber, 328, 338
  contents, 328, 338
  intakes, 338
Myricetin, 290, 299
  and chromosomal aberration, 292
  and DNA damage, 292
N-3 fatty acids
  and glucoregulation, 141
  and lipid transport, 141
Organic contaminats, 434, 435
Oxidative DNA damage and repair
  and DNA damage and repair, 73
  and HPLC prepurification and GC-MS analysis, 73, 75
  and quantitation of 5-HMUra and 5-HMdUrd, 75, 77
Oxidative DNA injury
  and ELISA vs. HPLC, 81
  and measurement of oxidative DNA injury in humans, 78, 79
Oxidative stress
  and administration of farmorubicin and alpha-tocopherol, 85, 86
  and atherosclerosis, 18
  and cardiovascular disease, 3, 7, 31
  and indicators of oxidative stress in blood, 83, 84
  in diabetes mellitus, 22
  induced by exercise, 22, 36
Pesticide residues, 443
Phenolic acids
  and berries, 298
  and HPLC, 298
Phenols, 258
Physical exercise
  and diabetes mellitus, 36
  and oxidative stress, 36
Phytic acid
  and dietary fiber, 317, 328, 338
  and mineral bioavailability, 317, 328, 334, 338
  and zinc bioavailability, 334
  contents, 328, 338
  intakes, 338
Phytoestrogens
  and breast cancer, 349
  and prostate cancer, 349
  biological effects, 356
Plant compounds
  and induction of detoxifying enzymes, 365
  and induction of antioxidative enzymes, 365
  and induction of cytochrome P-450 monooxygenases, 365
  and induction of phase II enzymes, 365
  and modulation of activities of xenobiotic metabolizing enzymes, 365
Polyphenols, 258
  and green tea, 264
  and HPLC, 264
Protein peroxidation, 55
Pyridoxine (see Vitamin B6)
Quercetin, 251, 256, 299
  and chromosomal aberration, 292
Resistant starch, 312
Retinol (see vitamin A)
Riboflavin (see Vitamin B2)
Rutin, 257, 273
  and chemiluminescence assay, 274
  and free radical overproduction, 276
  and lipid peroxidation, 276
  and nonheme iron, 276
Selenium
  and biomarker, 161
  and cardiovascular risk, 188
  and cholesterol, 190
  and free radicals, 117
  and glutathione peroxidase, 169
  and lead, 208
  and mortality from ischaemic heart disease, 126
  and selenoproteins, 195
  in food, 172, 199
  in hair, 177, 201
  in plasma, 161, 174, 195, 205

in serum, 169, 177, 189, 208
in urine, 177
intake, 161, 172, 177, 195, 202
status in populations, 177, 188, 208
supplementation, 168, 172, 195
Serum carotenoid levels, 373
Serum zeaxanthin /lutein levels, 373
Sesame seed, 230, 236
Sesamin, 230, 236, 241
   and carcinogenesis, 233
   and HPLC, 242
   and lipid peroxidation, 242
   and liver microsomes, 241
   and mass spectrum, 244
   and rat liver, 231
   and rat lungs, 231
   and rat plasma, 231
   and tocopherols, 233
Sesamolin, 230, 236
Silibinin
   and antioxidative activity, 281
   and DNA damage, 281
   and lipid peroxidation, 281
Silibinin dihemisuccinate, 281
Strawberry extract, 386
Superoxide dismutas, 147
SUVIMAX study, 113
Taxifolin, 256
Thiamine (see Vitamin B1)
Thiobarbituric acid reactive substances (MDA), 287
Tocopherols (see Vitamin E)
Tocotrienols (see Vitamin E)
Toxicity and intakes of heavy metals
   and critical organ and critical effects, 414
   and intake of metals, 413, 414
   and metals in food, 412, 413
   and protective mechanisms, 415
Trace elements, 113
Transition metals, 4
Triglycerides, 142
Trolox Equivalent Antioxidant Capacity, 257
Vegan diet
   and coumarin 7-hydroxylation, 389
   and coumarins, 389
Vegetarian diet, 145
Vitamin A
   and daily intake, 252
   and HPLC, 222, 251
   and infant formulas, 223
   and mortality from ischaemic heart disease, 126
   and vitamin retention, 219
Vitamin B1 / thiamine
   and cooking, 214
   and freezing, 214
   and storage, 218
Vitamin B2 / riboflavin
   and freezing, 214
   and storage, 218
Vitamin B6 / pyridoxine
   and cooking, 214
   and freezing, 214
   and storage, 218
Vitamin C
   and uptake of mercury vapor, 156
   and canning, 213
   and cooking, 214
   and coronary heart disease, 4
   and daily intake, 252
   and freezing, 213
   and HPLC, 146, 222
   and infant formulas, 223
   and packaging, 215
   and storage, 218
Vitamin E
   and antioxidant activity, 110
   and breakfast cereals, 227
   and cardiovascular disease, 4
   and cooking, 215
   and epidemiological studies, 111
   and fish, 228
   and glucoregulation, 141
   and HPLC, 142, 146, 222, 225, 231, 251
   and infant formulas, 223
   and insulin resistance syndrome, 31
   and intake, 110, 252
   and LDL oxidation, 110
   and lipid transport, 141
   and mortality from ischaemic heart disease, 126
   and müslies, 227
   and oxidative stress, 37
   and rat liver, 231
   and rat lungs, 231
   and rat plasma, 231
   and requirement, 111
   and sesame seed, 236
   and supplementation, 142
   and Vitamin K deficiency, 111
Vitamins
   and canning, 213
   and cooking, 214
   and freezing, 213
   and packaging, 215
   and processing, 213
   and storage, 218
Wine
   and cardiovascular disease, 260
   and cerebrovascular disease, 260
   and consumption, 261
Zinc
   bioavailability, 334
   influence of dietary fiber to bioavailability, 334
   influence of phytic acid to bioavailability, 334